Progress in Optical Science and Photonics

Volume 37

Indexed by Scopus

The purpose of the series Progress in Optical Science and Photonics is to provide a forum to disseminate the latest research findings in various areas of Optics and its applications. The intended audience are physicists, electrical and electronic engineers, applied mathematicians, biomedical engineers, and advanced graduate students.

Carlos Marques · Raja V. L. N. Sridhar ·
Santosh Kumar
Editors

Advanced Optical Sensors for Aerospace Applications

Springer

Editors
Carlos Marques
Department of Physics
CICECO—Aveiro Institute of Materials
University of Aveiro
Aveiro, Portugal

Department of Physics
VSB—Technical University of Ostrava
Ostrava, Czech Republic

Raja V. L. N. Sridhar
Photonics Systems Area
Laboratory for Electro-Optics Systems
(LEOS), U. R. Rao Satellite Centre (URSC)
Indian Space Research Organization (ISRO)
Bengaluru, Karnataka, India

Santosh Kumar
Department of Electronics
and Communication Engineering
Centre of Excellence for Nanotechnology
Koneru Lakshmaiah Education Foundation
Guntur, Andhra Pradesh, India

ISSN 2363-5096 ISSN 2363-510X (electronic)
Progress in Optical Science and Photonics
ISBN 978-981-95-1625-4 ISBN 978-981-95-1626-1 (eBook)
https://doi.org/10.1007/978-981-95-1626-1

This Springer imprint is published by the registered company Springer Nature Singapore Pte Ltd.
The registered company address is: 152 Beach Road, #21-01/04 Gateway East, Singapore 189721, Singapore

Preface

In the modern era of aerospace advancement, the fusion of optical sensing technologies with space and aviation applications is creating unprecedented possibilities in exploration, safety, and sustainability. From guiding satellites through the deep reaches of space to monitoring the physiological well-being of pilots in flight, optical sensors are at the heart of numerous breakthroughs that are shaping the future of aerospace systems. This book, *Advanced Optical Sensors for Aerospace Applications*, provides a cohesive and detailed overview of cutting-edge research, engineering practices, and practical applications involving optical sensors across multiple aerospace domains. As our ambitions extend beyond Earth's orbit and into the vastness of deep space, the demand for reliable, autonomous, and intelligent sensor systems grows exponentially. The chapters in this volume collectively present a multidisciplinary perspective, covering innovations in navigation, remote sensing, material analysis, sustainability, and human-centered monitoring. Aimed at researchers, engineers, technologists, and advanced students, this book brings together the latest findings and trends from both academic and industrial viewpoints. It outlines how optical sensing technologies are supporting safer, more efficient, and more autonomous aerospace operations. Whether in orbit, in the atmosphere, or in the cockpit, optical sensors are proving to be indispensable tools for the next generation of aerospace missions.

Organization of This Book

The book begins with Chapter "Introduction to Optical Sensors in Aerospace Engineering", which introduces the fundamentals of optical sensors and their broad relevance in aerospace systems. It lays the groundwork by exploring the core principles, materials, and designs that underpin modern optical sensing technologies. Chapter "Basic Concepts of Light and Optics for Aerospace Sector" describes the basic concepts of light and optics for aerospace sector, focusing on their fundamental principles and materials. It covers a wide range of devices, including photodetectors, Fiber Bragg Gratings, photonic crystal fibers (including plasmon resonance in fibers), diffraction gratings, MEMS, piezo-optical sensors, interferometers, and spectrometers, highlighting their applications in areas discussed in the next chapters.

Following that, Chapter "Optical Sensors for Structural Health Monitoring" explores the role of optical sensors in structural health monitoring (SHM) for aerospace applications, highlighting their capability to measure strain, temperature, vibration, and other key parameters in real time. Special attention is given to acoustic emission techniques and their integration into aircraft structures to detect damage, enhance safety, and support the reliable use of advanced materials like composites. Chapter "FBG Sensor Applications in Aerospace Engineering" dives into the implementation of distributed sensing systems, highlighting how fiber-based and photonic sensors can provide real-time structural health monitoring of large aerospace systems.

Chapter "Optical Sensors for Aircraft Structural Health Monitoring" provides an overview of distributed and quasi-distributed optical fiber sensing technologies for aircraft structural health monitoring. Emphasis is placed on FBG-based sensing for discrete high-resolution measurements, OFDR for continuous monitoring with sub-millimeter resolution, and emerging high-scattering optical fibers that enhance backscattering for advanced, low-cost distributed sensing solutions.

Next, Chapter "Digital Image Correlation Techniques for Structural Analysis in Aerospace" discusses the role of hyperspectral and multispectral imaging technologies for environmental observation, material identification, and mission support, enabling more precise aerial and satellite-based Earth observation. Chapter "Advancements in Optical Sensors for Avionics Systems" discusses advancements in optical sensors are revolutionizing avionics systems by offering high-precision, lightweight, and EMI-immune solutions ideal for the demanding conditions of aircraft environments. These innovations are enabling the transition to "fly-by-light" architectures. The journey continues in Chapter "Optical Sensors for Stellar-Based Satellite Positioning: Precision Navigation Through Single-Star Tracking", which explores optical sensors for stellar-based satellite positioning. This chapter delves into the principles of single-star tracking and how celestial references can provide autonomous navigation for deep-space missions where GPS is no longer viable.

Chapter "Laser Spectroscopic Techniques for Aerospace Remote Sensing" expands on remote material analysis using laser spectroscopic techniques such as LIBS, LIF, and Raman spectroscopy. These complementary methods offer powerful insights into the elemental and molecular composition of aerospace materials and planetary surfaces. Moving from deep space back to Earth, Chapter "Remote Sensing Technologies in Aerospace: Applications and Opportunities" presents a detailed discussion on satellite-based remote sensing technologies for monitoring marine and coastal ecosystems. Applications include pollution tracking, climate modeling, and biodiversity assessment using instruments such as MODIS, MSI, SAR, and OLCI. Shifting the focus to orbital sustainability, Chapter "Optical Sensor Innovations for Space Sustainability: Managing Space Debris in Aerospace Applications" explores innovations in optical sensors for managing space debris. It discusses how high-resolution imaging and spectral analysis tools are being used to track and mitigate space debris to ensure safe and sustainable operations in crowded orbits. In Chapter "Towards Psychophysiological Monitoring of Aircraft Pilot Crew", the discussion moves to psychophysiological monitoring of aircraft pilot crews. This chapter introduces wearable and non-invasive sensor technologies used to assess pilot fatigue, stress, and cognitive load during flight, contributing to safer and more responsive cockpit environments. Chapter "Emerging Trends and Future Directions in Aerospace Optical Sensing" looks ahead at emerging trends and future directions in aerospace optical sensing. This chapter outlines advances in Fiber Bragg Grating sensors, miniaturized devices for IoT and biomedical use, and the potential for in-space manufacturing of high-purity optical fibers aboard platforms like the International Space Station. These innovations point toward a future where optical sensors play a central role not only in spacecraft systems but also in broader aerospace manufacturing and diagnostics. Finally, Chapter "Optical Metamaterials and Artificial Intelligence in Aerospace Sensing" highlights the transformative role of optical metamaterials, thermal and infrared sensors, and artificial intelligence in advancing aerospace sensing technologies.

Together, these fourteen chapters offer a panoramic view of how optical sensors are revolutionizing aerospace systems—from Earth's atmosphere to interstellar space, from mechanical health monitoring to human physiological sensing, and from sustainability initiatives to next-generation scientific exploration.

Aveiro, Portugal — Carlos Marques
Bengaluru, India — Raja V. L. N. Sridhar
Guntur, India — Santosh Kumar

Acknowledgments

The authors and editors of *Advanced Optical Sensors for Aerospace Applications* would like to extend their heartfelt gratitude to all those who played a significant role in the successful completion of this book. This work represents the culmination of the efforts, support, and encouragement of many individuals and institutions, to whom we are deeply indebted. First and foremost, we express our sincere thanks to **Springer Nature Press** for their trust in this project and for providing a global platform to share this important work with the scientific community. We are especially grateful to the editorial team for their professionalism, guidance, and commitment throughout the publication process. Their attention to detail and constructive feedback have helped ensure the highest quality of this volume. We would also like to thank our contributing authors and colleagues whose expertise, insights, and collaborative spirit significantly enriched the content of this book. Their valuable contributions and stimulating discussions were instrumental in shaping the scope and depth of each chapter.

Carlos Marques extends his sincere gratitude to the University of Aveiro, CICECO—Aveiro Institute of Materials (Portugal), and VSB—Technical University of Ostrava (Czechia) for their support and encouragement in fostering academic collaboration. Their contribution was instrumental in shaping a cohesive and insightful overview of advanced optical sensors, tailored to the multifaceted challenges of aerospace research and technology. Carlos Marques would like to acknowledge his institute and his family members for their support and encouragement. The research was co-funded by the financial support of the European Union under the REFRESH—Research Excellence For REgion Sustainability and High-tech Industries project number CZ.10.03.01/00/22003/0000048 via the Operational Programme Just Transition. This work was developed within the scope of the projects CICECO-Aveiro Institute of Materials, UID/50011/2025, UIDB/50011/2020 (DOI 10.54499/UIDB/50011/2020), UIDP/50011/2020 (DOI 10.54499/UIDP/50011/2020) and LA/P/0006/2020 (DOI 10.54499/LA/P/0006/2020) financed by national funds through 30 the FCT/MCTES (PIDDAC).

Raja V. L. N. Sridhar would like to express his gratitude to Dr. K. V. Sriram, director of the Laboratory for Electro-Optics Systems (LEOS), and to M. Sankaran, director of the U. R. Rao Satellite Centre, ISRO, for their encouragement and support in collaborating with academia experts in bringing out a cohesive and comprehensive overview into the intricate world of advanced optical sensors catering to the diverse domains of aerospace research and technology. Mr. Sridhar Raja would like to extend his affable gratitude to his wife, K. V. Jyothi, and his son, Gnana Mayuk, and his daughter, Jayalekhya, for their understandable support and cooperation, as well as to his beloved parents.

Santosh Kumar would like to express his sincere gratitude to Shri Koneru Satyanarayana Garu, honorable president of the Koneru Lakshmaiah Education Foundation (deemed to be University), Vaddeswaram, Andhra Pradesh, India, for their continuous support, research infrastructure, and encouragement throughout the preparation of this book. Dr. Kumar would also like to express his heartfelt gratitude to his wife, Dr. Ragini Singh, and son, Ayaansh Singh, for their unwavering support, encouragement, and cooperation, as well as to his loving parents.

This book is a reflection of the shared dedication, hard work, and passion of everyone involved. We hope that it will serve as a valuable resource for researchers, engineers, and students and inspire further advancements in the field of optical sensing for aerospace applications.

Carlos Marques
Raja V. L. N. Sridhar
Santosh Kumar

About This Book

This book presents a comprehensive exploration of the evolving role of optical sensing technologies within modern aerospace systems. As aerospace missions grow increasingly complex—ranging from deep space exploration to environmental monitoring and human factors analysis—optical sensors have become vital for ensuring precision, autonomy, safety, and sustainability. This interdisciplinary volume brings together cutting-edge research and case studies from leading experts across academia, research institutions, and space agencies. It addresses the full spectrum of optical sensor technologies and their integration into aerospace applications, spanning from stellar-based satellite navigation to remote sensing and space debris management. Through a blend of theoretical foundations, engineering principles, and practical insights, the book examines how these sensor systems are designed, deployed, and optimized for the extreme conditions of aerospace environments. The volume also investigates the future trajectory of optical sensing, highlighting the growing role of AI, machine learning, and space-based manufacturing in improving sensor intelligence, adaptability, and autonomy.

Key topics include:

- Optical sensors for stellar-based satellite positioning and autonomous navigation in deep space
- Laser spectroscopic techniques (LIBS, LIF, and Raman) for material characterization in aerospace.
- Remote sensing technologies for monitoring coastal and marine ecosystems.
- Advanced optical systems for space debris detection, tracking, and sustainability.
- Psychophysiological monitoring of pilot crew using non-invasive optical sensors.
- Fiber Bragg Grating (FBG) sensors for structural health monitoring of aircraft and spacecraft.
- Space-based optical fiber production and its implications for sensor technologies.
- Miniaturization and ruggedization of optical components for extreme environments.
- AI-enhanced sensor data processing for predictive analytics and autonomous control.

Designed for aerospace engineers, optical scientists, researchers, graduate students, and industry professionals, *Advanced Optical Sensors for Aerospace Applications* serves as a definitive reference for anyone engaged in the design, development, and application of optical sensing systems in aerospace science and engineering.

Contents

Editors and Contributors

About the Editors

Carlos Marques is Expertise on optical fiber sensors and optical devices for optical communications and sensing. He received the Ph.D. degree in Physics Engineering from the University of Aveiro (UA), Portugal, in 2013. He was a Marie Curie (MC) fellow in the Aston Institute of Photonics Technologies, Aston University, UK until March 2016. After that, he was a research scientist (Assistant and Principal Researcher (CEEC)) in the Physics Department and I3N, University of Aveiro, Aveiro, Portugal. Currently, he is a professor in Physics Department, University of Aveiro, Portugal, and full member of CICECO—Aveiro Institute of Materials, University of Aveiro, Portugal. Also, he is a researcher of REFRESH project at VSB Technical University of Ostrava, Department of Physics. His work is focused in sensors and devices. His main interests include innovative optical devices and structures for sensing, integrated optics, low-cost interrogation systems, eHealth, aviation, data acquisition, predictive analytics, and optical transmission data. He authored or co-authored 5 books, 7 book chapters, 250+ journal peer-review papers, more than 100 conference papers with emphasis on optical communications, data processing and sensing. According to Google Scholar, journal/conference articles have received 14,000+ citations (h-index: 76). He was an organization member of IMOC2019 and AOP2024 conferences, in Aveiro. He will be the chair of EWOFS2027 in Aveiro. He was awarded with several awards as researcher, reviewer, and editor. He is an editor-in-chief of *IEEE Sensors Reviews.* Also, he was/is Associate Editor of prestigious scientific journals: *Opto-Electronic Advances Journal*, *OPTICA Optics Express*, *IEEE Sensors Journal* (2018–2024), *Optical Fiber Technology* (Elsevier) (2017–2024), and he also acts as referee of several high-level international books/journals (Wiley, Nature, OPTICA (former OSA), CRC, IEEE, Elsevier), receiving awards by review/editorial quality. He is a regular expert evaluator of international research projects of different countries such as Romania (UEFISCDI agency), Belgium (F.R.S.-FNRS), UK (via Multiply project—Marie Curie actions), Portugal (ANI), Serbia (Fond za nauku) and Poland (NCN agency), including European Commission. He is a regular TPC member of some international and reputable conferences of optics and photonics field, and he is regular invited to be session chairman in some international conferences. Last ten years, he participated/leaded in several research projects (+25) including national projects from FCT agency, European projects including Marie Curie actions (Individual Fellowship POSSIBLE project ID: 628604 (with Airbus and Meggit from UK), ITN Tripod ID: 608382, POSPORI project no 844618), FET (InstaBat ID: 955930), and other international from Brazil (CAPES, FAPES, CNPq, etc. (see project lists and the active ones in projects section). Currently, he is the coordinator professor of Technologies for Aerospace Energy Systems and also Space Optical Systems from Aerospace Engineering degree as well the coordinator professor from Physics Department of Introduction to Aerospace Engineering. He was a coordinator professor of Optical Technologies, Lasers and Photonics, from Physics Engineering doctoral programme, Physics Engineering Integrated Master, respectively, and was co-coordinator professor of Photonics from Physics Engineering Master. He is a lecturer in Mechanics and Oscillations and also in Topics in Doctoral Programme of Physics Engineering. He is a coordinator professor from Physics Department of Aerospace Engineering. He was/is supervisor of more than 70 students from research fellows (pos-doc and junior investigators), Ph.D. students, master students, and B.Sc. students.

Raja V. L. N. Sridhar is a senior scientist at the U. R. Rao Satellite Centre (URSC) of the Indian Space Research Organization (ISRO), India. He has been serving the Laboratory for Electro-Optics Systems (LEOS), one of the premier ISRO units known for the development of precise and accurate navigation sensors, high-quality 1.2 m class optics, optical assemblies, detectors, and scientific instruments for terrestrial and interplanetary missions. He received the M.Tech. degree in the Laser Technology Program from the IIT Kanpur, India, in 2004, and the M.Sc. degree in Physics from ANU, Andhra Pradesh, India. He worked as teaching faculty in the physics department for postgraduate studies at TJPS College, Guntur, Andhra Pradesh, for two years before joining IIT Kanpur. After his M.Tech. degree, he was a research fellow at the IIT Kanpur in the Department of Physics until January, 2006 and has been working at the ISRO since then. He is currently the Research Scholar in the Faculty of Engineering and Technology at M. S. Ramaiah University of Applied Sciences (MSRUAS), Bengaluru, India. His main interests include atomic and molecular physics, laser spectroscopy (LIBS, Raman, and LIF), laser-based instrumentation for space research, and data analytics. He is the principal investigator for the Laser-Induced Breakdown Spectroscopy (LIBS) instrument aboard the Chandrayaan-3 Pragyaan rover and the system engineer for the scientific instruments, namely, LAP (Lyman Alpha Photometer) aboard the MOM1 (India's maiden Mars Orbiter Mission), EELAA (Earth-Exospheric Lyman Alpha Analyzer) aboard the INS-1B, and the team member of the 1st Indian Laser Altimeter (LLRI) flown in the Chandrayaan-1 mission. His contributions further include the realization of spin-off technology developments, viz., photon-counting lidar, miniature Raman spectroscope, space-qualified solid-state Q-switched Nd:YAG laser, miniaturized Er:Glass laser, dual-wavelength lidar sensor, etc. He is currently leading the science team for the Raman Spectrometer that is going to be part of the near-future ISRO-JAXA lunar mission and the LIBS-Raman spectrometer for the future Indian Mars landing mission. Sridhar Raja was/is supervisor for more than 25 students of various institutes/universities, catering to their master's and bachelor's program projects and internships. He has authored/co-authored over 50 articles in prestigious journals and conferences, i.e., *OPTICA Applied Optics*, *Optics Letters*, *Applied Physics A* (Springer), *IEEE Sensors*, *Journal of Physics D* (IOP Science), *Advances in Space Research* (Elsevier), *IETE Journal of Research* (Taylor & Francis), *Current Science*, *Journal of Spacecraft Technology*, and conferences like SPIE, OSI, IAC, LPSC, AOGS, COSPAR, COLA, ICONTOP, IEEE CONECCT, IEEE TENCON, PLANEX, NLS, etc. He is a Life Member of the Indian Laser Association (ILA), Astronautical Society of India (ASI), and the ISSE (Indian Society of Systems for Science and Engineering). He is the recipient of the ISRO Young Scientist Merit Award and a three-time ISRO Team Excellence Awardee.

Dr. Santosh Kumar is a Professor in the Department of Electronics and Communication Engineering at Koneru Lakshmaiah Education Foundation (KLEF), India. He also serves as the Lab Head of the Centre of Excellence for Nanotechnology (CoE-NANO), Group Head of the Nanotechnology Research Group (NTRG), and Advisor—R&D at KLEF. He earned his Ph.D. from IIT (ISM) Dhanbad in 2014 and was previously with Liaocheng University, China (2018–2023). Recognized among the world's top 2% scientists by Stanford University for four consecutive years, Dr. Kumar has supervised 20 M.Tech. dissertations and nine Ph.D. candidates. He has published over 550 research articles in prestigious SCI journals, including 130 IEEE journal papers, with more than 11,000 citations and an h-index of 65. His 100 papers are recognized under various Sustainable Development Goals (SDGs) according to Scopus. His research has been featured in leading journals such as *Biosensors and Bioelectronics*, *Journal of Lightwave Technology*, *Optics Express*, *Optics Letters*, *Applied Physics Letters*, *ACS Applied Nano Materials* and *Biosensors*. His expertise spans fiber optic sensors, photonics and plasmonic devices, nano and biophotonics, waveguides, interferometers, and the Internet of Things. He has authored seven scholarly books with CRC Press (Taylor & Francis), IEEE-Wiley, and Springer Nature. Dr. Kumar was recently granted a patent for his groundbreaking optical fiber sensing technology and has published 15 patents on biosensing applications. A highly regarded expert, he has reviewed over 2700 manuscripts for top-tier SCI journals published by IEEE, Elsevier, Springer, OPTICA, SPIE, and *Nature*. His collaborative research extends across institutions in India, China, Portugal, Iraq, Kazakhstan, Brazil, France, and Italy.

Dr. Kumar is a Fellow of SPIE and IETE, a Life Fellow of the Optical Society of India (OSI), and a Senior Member of IEEE, OPTICA, and SPIE. He is an OPTICA Traveling Lecturer and serves as the Chair of the Optica Optical Biosensor Technical Group, USA. He is also an Associate Editor for *Scientific Reports (Nature)*, *IEEE Internet of Things Journal*, and *Biomedical Optics Express*. His recent accolades include the Best Performing Associate Editor Award from *IEEE Sensors Journal, USA*, the Best IEEE JLT Paper Award from USA, and the Foreign Distinguished Professor Award from China.

Contributors

V. B. Ankith Department of AIML, PESITM Shivamogga, Virupina Koppa, Karnataka, India

Leonardo Azevedo CERENA/DER, Instituto Superior Técnico (IST), Universidade de Lisboa, Lisbon, Portugal

V. R. Balaji Centre for Healthcare Advancement, Innovative and Research, Vellore Institute of Technology, Chennai, Chennai, India

Chiara Bellezza Prinsi Department of Electronics and Telecommunications, Politecnico Di Torino, Torino, Italy

Aurora Bellone Department of Electronics and Telecommunications, Politecnico Di Torino, Torino, Italy

Gaurav Kumar Bharti School of Electrical, Electronics & Communication Engineering, Galgotias University, Noida, Uttar Pradesh, India

Vaishnavi P. Bhat Department of ISE, PESITM, Shivamogga, Karnataka, India

Beatriz Biguino Faculdade de Ciências, MARE—Marine and Environmental Sciences Centre/ARNET—Aquatic Research Network, Universidade de Lisboa, Lisboa, Portugal

Wilfried Blanc Université Côte d'Azur, CNRS, INPHYNI, Nice, France

Ana C. Brito Faculdade de Ciências, MARE—Marine and Environmental Sciences Centre/ARNET—Aquatic Research Network, Universidade de Lisboa, Lisboa, Portugal;
Departamento de Biologia, Faculdade de Ciências, Universidade de Lisboa, Lisboa, Portugal

Ana Sofia Antunes Calado Instituto de Telecomunicações e Instituto Superior Técnico, Universidade de Lisboa, Lisboa, Portugal

Vedika Chavan Department of Atomic and Molecular Physics, Manipal Academy of Higher Education, Manipal, Karnataka, India

Hugo Humberto Plácido da Silva Instituto de Telecomunicações e Instituto Superior Técnico, Universidade de Lisboa, Lisboa, Portugal

Paulo Sérgio de Brito André Instituto de Telecomunicações e Instituto Superior Técnico, Universidade de Lisboa, Lisboa, Portugal

João M. Dias CESAM—Centre for Environmental and Marine Studies, Physics Department, University of Aveiro, Campus de Santiago, Aveiro, Portugal

Miguel Fernandes Lisbon School of Engineering (ISEL/IPL), Rua Conselheiro Emídio Navarro, Lisboa, Portugal;
CTS—Centre of Technology and Systems and Associated Lab of Intelligent Systems (LASI), Caparica, Portugal

Tiago Filipe Rodrigues Fernandes Instituto de Telecomunicações e Instituto Superior Técnico, Universidade de Lisboa, Lisboa, Portugal

Afonso Ferreira Faculdade de Ciências, MARE—Marine and Environmental Sciences Centre/ARNET—Aquatic Research Network, Universidade de Lisboa, Lisboa, Portugal

T. R. Harsha Department of AIML, PESITM Shivamogga, Virupina Koppa, Karnataka, India

Rajini V. Honnungar Department of Electronics and Communication Engineering, RNS Institute of Technology, VTU, Bengaluru, India

M. A. Ibrar Jahan Department of Electronics and Communication Engineering, RNS Institute of Technology, VTU, Bengaluru, India

Manish Kalra Department of Electronics and Communication Engineering, GLA University, Mathura, Uttar Pradesh, India

Farhan Khan Department of AIML, PESITM Shivamogga, Virupina Koppa, Karnataka, India

Arnaldo Leal-Junior Mechanical Engineering Department, Federal University of Espirito Santo, Brazil and Department of Cybernetics and Biomedical Engineering, VSB–Technical University of Ostrava, Ostrava, Czech Republic

Carina L. Lopes CESAM—Centre for Environmental and Marine Studies, Physics Department, University of Aveiro, Campus de Santiago, Aveiro, Portugal

Paulo Lourenço Lisbon School of Engineering (ISEL/IPL), Rua Conselheiro Emídio Navarro, Lisboa, Portugal;
CTS—Centre of Technology and Systems and Associated Lab of Intelligent Systems (LASI), Caparica, Portugal

Carlos Marques Department of Physics, CICECO—Aveiro Institute of Materials, University of Aveiro, Aveiro, Portugal;
Department of Physics, VSB—Technical University of Ostrava, Ostrava, Czech Republic

Anna Mauro Department of Electronics and Telecommunications, Politecnico Di Torino, Torino, Italy

Renato Mendes +ATLANTIC CoLAB, Museu das Comunicações, Lisbon, Portugal;
Underwater Systems and Technology Laboratory (LSTS), Associate Laboratory for Energy, Transports and Aerospace (LAETA), Faculdade de Engenharia, Universidade do Porto (FEUP), Porto, Portugal

Jan Nedoma Department of Telecommunications, Faculty of Electrical Engineering and Computer Science, VSB—Technical University of Ostrava, Ostrava, Czech Republic

Sandro C. Oliveira Physics Department, CICECO—Aveiro Institute of Materials, University of Aveiro, Aveiro, Portugal

Massimo Olivero Department of Electronics and Telecommunications, Politecnico Di Torino, Torino, Italy

Rajath R. Pai Department of AIML, PESITM Shivamogga, Virupina Koppa, Karnataka, India

Fufei Pang Key Laboratory of Specialty Fiber Optics and Optical Access Networks, School of Communication and Information Engineering, Shanghai University, Shanghai, China

Usha Raghunathagouda Patil Department of EEE, PESITM, Sivamogga, Karnataka, India

Guido Perrone Department of Electronics and Telecommunications, Politecnico Di Torino, Torino, Italy

K. Prabhanjana Department of AIML, PESITM Shivamogga, Virupina Koppa, Karnataka, India

Américo S. Ribeiro CESAM—Centre for Environmental and Marine Studies, Physics Department, University of Aveiro, Campus de Santiago, Aveiro, Portugal

S. J. Rudresha Department of EEE, PESITM, Sivamogga, Karnataka, India

Than Singh Saini Department of Physics, National Institute of Technology Kurukshetra, Kurukshetra, Haryana, India

Magda C. Sousa CESAM—Centre for Environmental and Marine Studies, Physics Department, University of Aveiro, Campus de Santiago, Aveiro, Portugal

Srujana Department of Electronics and Communication Engineering, IEEE Member, Bengaluru, India

Łukasz Święch Department of Aerospace Engineering, Faculty of Mechanical Engineering and Aeronautics, Rzeszow University of Technology, Rzeszów, Poland

Likewin Thomas Department of AIML, PESITM Shivamogga, Virupina Koppa, Karnataka, India

V. K. Unnikrishnan Department of Atomic and Molecular Physics, Manipal Academy of Higher Education, Manipal, Karnataka, India

Alberto Vallan Department of Electronics and Telecommunications, Politecnico Di Torino, Torino, Italy

S. K. Vinay Department of ISE, PESITM, Sivamogga, Karnataka, India

Heming Wei Key Laboratory of Specialty Fiber Optics and Optical Access Networks, School of Communication and Information Engineering, Shanghai University, Shanghai, China

Archana Yadav Department of Electronics and Communication Engineering, Integral University, Lucknow, India

Dengwei Zhang State Key Laboratory of Extreme Photonics and Instrumentation, Zhejiang University, Hangzhou, China

Liang Zhang Key Laboratory of Specialty Fiber Optics and Optical Access Networks, School of Communication and Information Engineering, Shanghai University, Shanghai, China

Shutong Zhangyang Key Laboratory of Specialty Fiber Optics and Optical Access Networks, School of Communication and Information Engineering, Shanghai University, Shanghai, China

Shijie Zheng MOE Key Laboratory of High-Speed Railway Engineering, School of Civil Engineering, Southwest Jiaotong University, Chengdu, China

Mengshi Zhu Key Laboratory of Specialty Fiber Optics and Optical Access Networks, School of Communication and Information Engineering, Shanghai University, Shanghai, China

Abbreviations

AI	Artificial Intelligence
CMOS	Complementary Metal-Oxide Semiconductor
ESA	European Space Agency
ETM+	Enhanced Thematic Mapper Plus
FBG	Fiber Bragg Grating
GPS	Global Positioning System
IoT	Internet of Things
ISRO	Indian Space Research Organisation
ISS	International Space Station
LEO	Low Earth Orbit
LIBS	Laser-Induced Breakdown Spectroscopy
LIF	Laser-Induced Fluorescence
ML	Machine Learning
MODIS	Moderate Resolution Imaging Spectroradiometer
MSI	Multi-Spectral Instrument
NASA	National Aeronautics and Space Administration
OLCI	Ocean and Land Colour Instrument
PMT	Photomultiplier Tube
SAR	Synthetic Aperture Radar
SNR	Signal-to-Noise Ratio
TM	Thematic Mapper
UAV	Unmanned Aerial Vehicle
UV	Ultraviolet

Introduction to Optical Sensors in Aerospace Engineering

Sandro C. Oliveira, Jan Nedoma, and Carlos Marques

Abstract The aerospace industry demands an incredible level of precision to ensure the risk of disaster is minimized. Multiple sensing systems are needed to provide an accurate overall assessment and facilitate decision-making. Thus, this introductory chapter aims to provide an overview of optical sensors and their crucial role in modern aerospace applications. It begins by introducing some of the key specific requirements and challenges in aerospace environments (namely extreme temperatures and radiation exposure), and provides some insights on the advantages of optical over traditional electronic-based sensing. This is followed by a brief outline of some fundamental areas where optical sensors can be applied to aerospace applications, namely in navigation, structural health monitoring, environmental sensing, among others. This chapter introduces the scope of this book, while paving the way for a more comprehensive and focused discussion in subsequent chapters.

Keywords Aircraft · Fiber optics · Optics · Spacecraft · Sensors

S. C. Oliveira (✉)
Physics Department, CICECO—Aveiro Institute of Materials, University of Aveiro, Aveiro, Portugal
e-mail: sandrorcruz@ua.pt

J. Nedoma
Department of Telecommunications, Faculty of Electrical Engineering and Computer Science, VSB—Technical University of Ostrava, Ostrava, Czech Republic

C. Marques
Department of Physics, CICECO—Aveiro Institute of Materials, University of Aveiro, Aveiro, Portugal

Department of Physics, VSB—Technical University of Ostrava, Ostrava, Czech Republic

C. Marques et al. (eds.), *Advanced Optical Sensors for Aerospace Applications*, Progress in Optical Science and Photonics 37,
https://doi.org/10.1007/978-981-95-1626-1_1

1 Introduction

The aerospace industry, which encompasses commercial and military aircraft, space launch and in-space systems, missiles, satellites, and general aviation, is driven by the dual objectives of developing cleaner transportation methods and reducing operating costs [1]. Economic pressures, arising from rising fuel prices combined with stringent government-mandated environmental regulations, have spurred researchers to explore innovative approaches for creating lighter and more efficient aircraft. In this regard, it is well recognized that the utilization of lightweight alloys significantly diminishes fuel consumption [2].

Advances in aerospace systems are closely intertwined with progress in materials and processing technologies. Modern components are anticipated to be more durable, efficient, and to deliver enhanced performance characteristics. For instance, gas turbine engines for aircraft are now being designed to operate at higher pressures and temperatures, thereby generating greater thrust. Comparable considerations extend to spacecraft, where in-space propulsion and power systems critically depend on breakthroughs in material science [3].

Although aluminum has traditionally served as the primary material for aerospace applications, there is now a substantial demand for new materials that combine high strength, low weight, non-corrosiveness, recyclability, and resistance to ultraviolet radiation and impacts [4]. In recent years, there has been a notable increase in the use of composite laminates. Natural composites, such as wood and fabrics, were employed as early as the Wright Brothers' Flyer 1, and over time, more robust materials have been developed. The discovery of carbon fiber at the Royal Aircraft Establishment in Farnborough, UK, in 1964, marked a significant milestone that paved the way for its major contribution to aircraft structures [5]. Today, carbon fiber reinforced polymers (CFRP) and glass fiber reinforced polymers (GFRP) are celebrated for their exceptional thermal–mechanical properties, which include high specific strength and stiffness, while also being lightweight, corrosion-resistant, and effective thermal insulators [6]. These outstanding properties have facilitated the application of such materials in engine casings, wing sections, tail planes, fuselages, and control surfaces. In fact, composite materials are estimated to constitute around 50% of the weight of modern aircraft, such as the Boeing 787 and Airbus 350, where key structural components like main frames, fuselage/wing skins, and associated stringers are manufactured from CFRP [7, 8].

This shift towards composite materials has concurrently generated a need for lightweight sensors and cabling that are inherently immune to electromagnetic interference. Within the aerospace sector, a multitude of sensing parameters—including strain, temperature, pressure, acceleration, and vibration—must be meticulously evaluated at various stages throughout a component's lifecycle, ranging from fabrication to testing and during onboard missions [9]. The selection of a sensor is heavily dependent on the specific measurand and the intended application. Both aircraft and spacecraft must be meticulously designed and tailored to meet the precise requirements of their missions, from material selection and design through to the final manufacturing process. It is, therefore, essential to closely monitor these systems to ensure successful operation, with sensor specifications ultimately determined by factors such as mission altitude, operational lifetime, sensor placement within the craft, and the particular measurement parameters required [10]. Moreover, it is important to note that optical sensors, especially optical fiber sensors, offer distinct advantages over traditional electronic sensors. These optical systems are particularly attractive due to their lightweight and miniaturized design, inherent immunity to electromagnetic interference, and their capacity to measure multiple parameters simultaneously, all of which are highly desirable features in harsh operational environments [11].

Space remains one of the most challenging frontiers, characterized by an environment that is exceptionally unforgiving. Nevertheless, following the decline in orbital launches after the space race, there has been an increase in launches, especially on the commercial side (Fig. 1) and it is anticipated that the coming decades will witness significant advancements, particularly in the development of automated, smart, and reusable spacecraft that simultaneously ensure optimal health and safety conditions for astronauts [12, 13]. Achieving these ambitious goals will require the deployment of robust sensing networks capable of withstanding the harsh conditions inherent to space—including vacuum, extreme temperatures, and high levels of radiation—as well as meeting the stringent size constraints of space travel.

This chapter aims to introduce the topic of optical sensing for aerospace applications, highlighting the harsh conditions that these technologies need to withstand, while also providing some introductory context to specific fields of application that will be further explored in the following chapters.

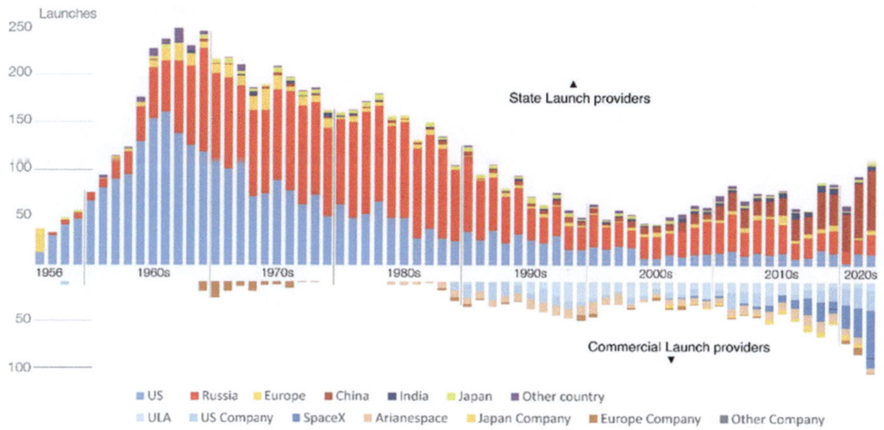

Fig. 1 Orbital launches from 1956 to 2022 from the main operators [12]

2 Aspects of the Space Environment

The harsh conditions that air and spacecraft are subject to heavily condition the types of materials that may be used for aerospace applications. This section intends to provide some insight into the specific challenges that need to be overcome while developing systems that need to withstand this harsh environment while maintaining unaltered performance and, in the case of sensors, provide accurate monitorization.

Several advanced materials have been developed for a plethora of applications ranging from transportation on manned spacecrafts to communications in satellites. These materials are subject to many environmental threats that degrade them overtime, including vacuum, solar ultraviolet (UV) radiation, charged particle (ionizing) radiation, plasma, atomic oxygen, temperature extremes, thermal cycling and impacts from micrometeoroids and orbital debris. The effects these threats have on the materials vary greatly according to the materials themselves, their thickness, and stress levels. Also to be considered is the specific mission environment, including the altitude at which the materials are used and for how long they remain there [14]. Figure 2 evidences the changes that can happen during a space mission, highlighting the differences between pre- and post-flight pictures of the same component.

2.1 *Vacuum*

The vacuum level depends on the missions' specifications, mainly the spacecraft altitude. A lot of remote sensing missions lie between 200 and 1000 km in altitude—the so-called Low Earth Orbit (LEO). Here, vacuum is typically 10^{-9}–10^{-11} Torr outside the spacecraft and 10^{-6}–10^{-7} Torr inside spacecraft [15]. The hard vacuum of

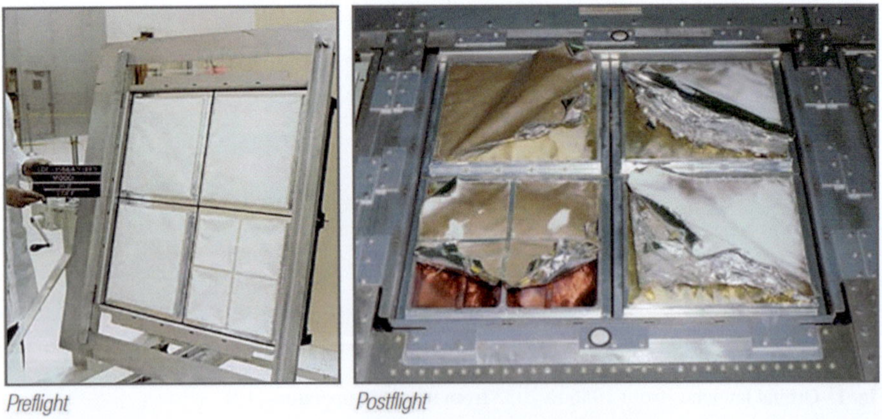

Fig. 2 Preflight and postflight long duration exposure facility M0001 heavy ions in space experiment, indicating atomic oxygen erosion and ultraviolet degradation [14]

Fig. 3 Photograph of contamination on a spacecraft as a result of outgassing on aluminum panel [16]

space can result in outgassing, which is the release of volatiles from materials. The outgassed molecules can then deposit on surrounding surfaces (Fig. 3) [16]. This molecular contamination can end up affecting the performance of many systems on the spacecraft by causing loss of dimensional stability and some other detrimental effects on material properties [17]. Dealing with outgassing should be done in advance, by choosing appropriate materials, and ensuring that they are screened and/or treated accordingly [18].

2.2 Atomic Oxygen

Atomic oxygen (AO) is predominantly formed in the LEO through the process of photo-dissociation of diatomic oxygen (O_2) induced by short-wavelength ultraviolet (UV) solar radiation, (wavelengths below 243 nm). This phenomenon is of considerable scientific and practical importance, as atomic oxygen represents the most significant component of the space environment at the altitude of the ISS in terms of its contribution to material degradation [19].

The issue of atomic oxygen shielding has emerged as a critical area of interest in the study of the low Earth orbit environment, primarily due to the highly reactive and aggressive nature of this atomic species. Atomic oxygen exhibits a pronounced tendency to interact with and degrade organic materials commonly present on the external surfaces of spacecraft. This interaction often manifests as physical damage, including erosion and corrosion, which can severely compromise the integrity of exposed materials [20]. Notably, atomic oxygen is known to oxidize a wide range of metals, with silver, copper, and osmium being particularly susceptible. Furthermore, atomic oxygen reacts vigorously with materials containing carbon, nitrogen, sulfur, or hydrogen bonds, leading to the erosion of many polymeric materials. Polymers that incorporate fluorine, such as Teflon®, demonstrate a synergistic reaction with atomic oxygen, wherein their reactivity increases progressively with prolonged exposure to UV radiation. Even materials that are ostensibly protected by atomic oxygen-resistant coatings are not immune to degradation, as atomic oxygen can induce undercutting erosion at sites where defects or imperfections in the protective coating are present [21].

In the context of advanced materials, carbon-based composites, such as carbon/carbon (C/C) and carbon/silicon carbide (C/SiC), have garnered significant attention within the aerospace industry due to their exceptional thermomechanical properties. These materials are capable of withstanding the extreme thermal and mechanical stresses encountered during atmospheric re-entry, making them highly suitable for applications such as thermal protection systems (TPS), nozzle throats, and ablation-resistant structures. Their minimal thermal expansion and remarkable chemical stability at elevated temperatures further enhance their suitability for reusable space components. However, for long-duration missions, the prolonged exposure of these materials to the highly oxidative environment of LEO presents a substantial challenge. The interactions between atomic oxygen and these advanced materials can lead to significant degradation over time, potentially compromising the performance, structural integrity, and functionality of spacecraft systems and subsystems. Consequently, addressing these interactions is essential to ensure the reliability and longevity of materials and components utilized in space missions.

2.3 Radiation

The space environment subjects spacecraft and their components to a spectrum of radiative hazards that can profoundly affect their performance and longevity. In particular, ultraviolet (UV) and ionizing radiation represent two critical sources of degradation. Statistically, the anomalies caused by radiation account for approximately 40% of the total problems induced by the space environment [22].

UV Radiation

While the Earth's atmosphere effectively filters out the majority of the sun's harmful radiation, materials operating in the space environment, which lacks such protective shielding, are directly exposed to the damaging effects of solar photons. Among these, ultraviolet (UV) radiation poses a particularly significant threat, especially to polymeric materials, where it can induce either cross-linking (resulting in hardening) or chain scission (leading to weakening). Additionally, UV radiation under high vacuum conditions can generate oxygen vacancies in oxide materials, often causing noticeable color changes. Given that many spacecraft components are fabricated from polymers, which are inherently vulnerable to UV-induced photodegradation [23], prolonged exposure to such radiation can result in chain scission, reduced molecular weight, embrittlement, and a subsequent decline in mechanical strength. This degradation not only compromises structural elements but also adversely impacts the functionality of coatings and adhesives critical to spacecraft construction.

The interaction of UV photons with material surfaces frequently initiates photochemical reactions, such as oxidation or other chemical alterations, which can form surface films. While these films may occasionally provide a protective barrier, they more commonly accelerate material aging by altering surface properties and potentially catalyzing further degradation processes [24]. Although protective coatings are routinely applied to mitigate environmental hazards, UV radiation can progressively degrade these layers over time. The deterioration of such coatings may expose underlying materials to additional damage from other space environment factors, including atomic oxygen and thermal cycling. Furthermore, sensitive electronic components are especially susceptible to UV-induced degradation, as UV exposure can create charge traps within insulating materials, impairing their performance and reliability.

Particulate or Ionizing Radiation

In the context of space exploration, three primary sources of naturally occurring charged particle radiation can be identified: galactic cosmic rays (GCRs), which are high-energy particles originating from beyond the solar system; particles trapped within Earth's magnetic field, commonly referred to as the Van Allen Belts; and solar energetic particle (SEP) events, such as solar flares (Fig. 4) [25]. When these primary radiation sources interact with a spacecraft's hull, they generate secondary particles, including protons, neutrons, and heavy ions, which often possess even higher energies [26]. Among these sources, GCRs and SEPs are generally regarded as the most significant contributors to ionizing radiation, particularly at altitudes at or above those of commercial aviation (approximately 8 km). GCRs, which originate from outside the solar system, are predominantly composed of energetic protons (p+), with smaller proportions of alpha particles (α) and heavier ions. SEPs, in contrast, originate from the Sun and share a similar composition to GCRs, being primarily protons but with a reduced fraction of heavier ions. Collectively, GCRs and SEPs are often categorized under the broader term cosmic rays [24].

The effects of particulate radiation on materials vary depending on the specific polymer. Exposure can lead to cross-linking or chain scission, processes analogous to UV-induced damage, ultimately resulting in polymer embrittlement. In avionics systems, the impact of radiation is more pronounced, manifesting as single-event upsets, bit errors, and latch-ups. To mitigate these risks, strategies such as the use of radiation-hardened ("rad-hard") avionics, the incorporation of shielding around sensitive electronics, and the implementation of error-correction circuitry or triple-module redundancy—where two correct process results override a corrupted one—are commonly employed [27].

On Earth, the average annual radiation dose from natural terrestrial sources and medical procedures is typically less than 0.005 Sievert (Sv). For comparison, a standard chest radiograph delivers approximately 0.02 μSv, while exposure during

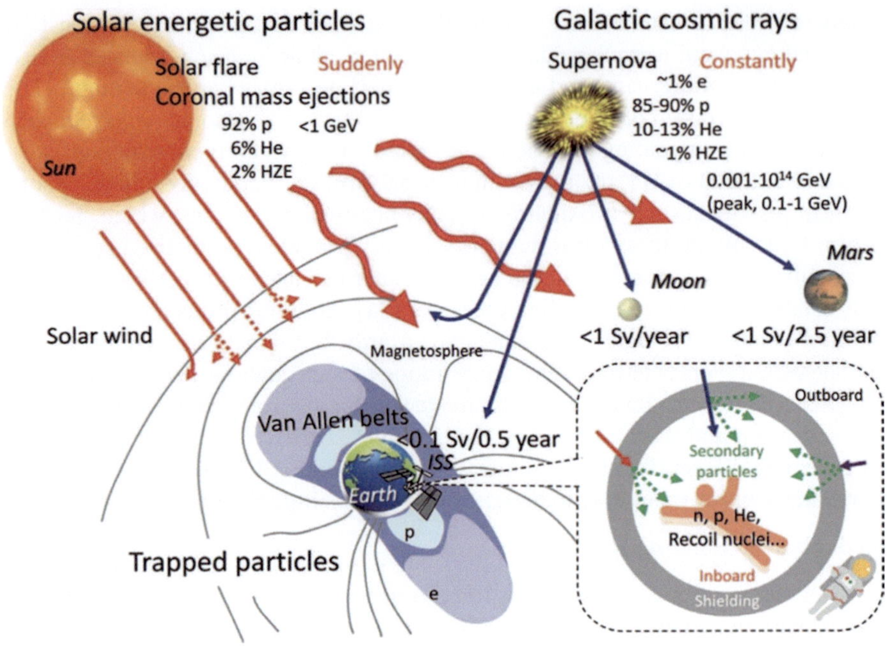

Fig. 4 Different types of ionizing radiation in the space environment [27]

commercial air travel ranges from 0.3 to 5.7 μSv per hour. At orbital altitudes comparable to that of the International Space Station, astronauts are exposed to an annual dose-equivalent of about 0.3 Sv [28].

For astronauts conducting missions exclusively in low Earth orbit, the cumulative career radiation dose is unlikely to exceed 1 Sv. However, as mission durations increase and destinations extend farther from Earth, radiation exposure is expected to rise significantly. For instance, astronauts embarking on future exploratory missions to Mars are projected to receive an estimated 1 Sv of radiation during a round trip. Over the course of these 2- to 3-year missions, there is a high probability of encountering at least one solar flare, which could elevate their exposure to as much as 5 Sv [26].

2.4 *Temperature Extremes and Thermal Cycling*

Among all the phenomena which are hostile to the spacecraft, the temperature-induced anomalies account for 11% [29]. The Sun and, to a lesser extent, the Earth are the most prominent sources of heat to in-orbit spacecraft. This heating is due to direct solar radiation, Earth-emitted radiation and Earth-reflected radiation (albedo). Solar radiation is the heat fluxes coming from the sun, while albedo radiation is the fraction of solar radiation reflected from the Earth back into space and Earth IR is the radiation emitted from the earth at low energy to space [30].

The temperature of the spacecraft's component varies dramatically taking into account if the surface is turned towards the sun (very high temperatures), or facing away from it, thus being in the shadow (very low temperatures). As a consequence, the in-orbit spacecraft experiences large temperature differences and gradients, leading to thermal deformation and vibration, and bringing about anomalies like signal errors therewith. To avoid this, precautionary arrangements need to be made, namely the adoption of appropriate materials and hardware capable of enduring such harsh temperature extremes. the degree to which a material experiences thermal cycling temperature extremes depends on the following: its thermo-optical properties (solar absorptance and thermal emittance), its view of the sun, its view of Earth, its view of other surfaces of the spacecraft, durations of time in sunlight and in shadow and its thermal mass and the influence of equipment or components that produce heat [31, 32].

These cyclic temperature variations often oscillate between − 120 °C and + 120 °C, but high solar absorptance with low infrared emittance will contribute to greater temperature swings. Considering an object in low orbit of the earth (at similar altitudes to the ISS, for example), which travels at around Mach 25, approximately 8 km per second. This is the case of the ISS for example, which orbits the Earth approximately once every 92 min. While the crew controls the temperature in the crew compartment of the space shuttle and modules maintaining an average temperature between 21 and 23 °C, much greater extremes occur outside the spacecraft. When on the sun-lit side of the earth, the temperature on the spacecraft or space station reaches over 100 °C. Forty-five minutes later, during a night pass through earth's dark shadow, temperatures plunge to below − 100 °C [33]. It is subject to sixteen of these thermal cycles a day, as it moves in and out of the sunlight during its orbit, which can pose significant damage to the materials, if not adequate.

Although protective coatings may be applied, these are at risk of degrading if there is a mismatch in the CTE between the coating and the substrate. Such harsh conditions may result in cracking, peeling, spalling or formation of pinholes in the coatings, which in turn expose the underlying materials to every other damaging agent present in the space environment, such as radiation or atomic oxygen (Fig. 5).

2.5 *Micrometeoroid/Orbital Debris Impact*

Space debris originates from human space activities, whereas micrometeoroids have their origins in asteroids and comets. The impacts caused by both of these entities have the potential to inflict significant damage [22]. In fact, space debris constitutes approximately 95% of all human-made objects that are currently orbiting in outer space [34]. This category includes various non-functional entities such as fragmented components and other elements that remain in orbit around the Earth or eventually re-enter the Earth's atmosphere [35].

Fig. 5 Postflight photographs displaying thin film samples from an experiment on the ISS with catastrophic failure of both coatings related to thermal cycling and atomic oxygen erosion [14]

In LEO, the typical velocities of space debris range from about 7–10 km/s [36]. When impacts occur at these extreme speeds, they are referred to as "hypervelocity impacts" (HVIs). During such impacts, the speed is so high that the material strength of the colliding objects becomes negligible compared to their inertial forces [37]. Consequently, even relatively small particles, with sizes spanning from approximately 10 μm to a few millimeters, can generate sufficiently high impact energies to damage orbiting spacecraft. In contrast, larger objects exceeding 1–10 cm in size have the potential to cause catastrophic destruction or severe damage to their targets [38].

The impact of high-velocity space debris and micrometeoroids on spacecraft structures poses a significant risk to both the structural integrity and operational performance of the systems. This is particularly concerning as successive impact events can progressively weaken spacecraft materials, increasing their vulnerability over time [39].

The presence of space debris also fluctuates in response to variations in the solar cycle. As solar activity intensifies, the Earth's atmosphere becomes more heated, which in turn increases atmospheric drag on orbiting space debris, altering their trajectories and eventual fate. To mitigate the risk posed by macroscopic particles, it is crucial to develop sophisticated expert systems dedicated to forecasting and protection. These systems should be based on a combination of observational data, computational modeling, and advanced simulations. While large space debris objects are actively tracked, allowing for the execution of avoidance maneuvers such as those performed by the ISS, there remains no effective method to prevent collisions with smaller debris. Most of the impacts recorded on returned spacecraft experiments have involved small debris, typically resulting in craters with diameters of 0.5 mm or less.

2.6 Biological Effects

On top of the damage to the materials used in aerospace missions, it is equally important to consider the damage done to the crew that operates these missions. So far, more than 500 astronauts have been sent into space to conduct space exploration missions, who are inevitably faced with a harsh space environment when conducting research activities [40].

After more than 50 years of space exploration, a large number of studies have proved that the space environment has adverse effects on organisms. Astronauts carrying out space exploration missions may present changes in gene regulation and genomic integrity, as well as molecular and cellular biological changes such as mitochondrial dysfunction. These changes will increase health risks, including muscle atrophy and bone loss, carcinogenesis, immune disorders, changes in vascular function, and cognitive impairment [41]

The orbit of a spacecraft in low orbit around the earth is about 90 min, which means 16 day/night cycles in 24 h. Such a light–dark cycle pattern is inconsistent with the rhythm of human life on the earth. This seriously affects the biological clock of astronauts, resulting in a decline in sleep quality, appetite, learning and memory capacity [42, 43]. It is essential to provide astronauts with sensing tools to monitor important health markers to ensure that they get the proper treatment they need, but also to further expand on the knowledge of the effects of a trip to space on the human body.

3 Advantages of Optical Sensing

Optical methods are being increasingly used in multiple fields, including sensing applications, due to the continuous advancement of innovative, low-cost equipment, namely light sources and detectors. Coupled with other scientific advances, these technologies enable the development of high-performance optical sensors characterized by low detection limits, high specificity, and sensitivity. There are multiple principles that can be exploited, tailored to specific needs, and this versatility supports diverse applications in pharmaceutical research, medical diagnostics, environmental monitoring, agriculture, industry, food safety, and particularly the aerospace industry [44].

When compared with traditional electronic sensors, optical sensors, particularly those based on fiber-optics have several advantages in aerospace applications. One key advantage is the ability to use these sensors in harsh environments, such as space, because of the fiber's immunity to electromagnetic interference, ensuring accuracy and reliability in aerospace environments [45, 46]. Space radiation is approximately 2500 times more intense than ground radiation, which impacts electronic system reliability [47]. Extreme temperatures and radiation exposure also degrade solder joints, increasing the risk of electronic failure in spacecraft [48]. Their lightweight and compact design makes them ideal where weight and space are critical considerations. The compact structure enables seamless integration into space-constrained systems without compromising performance [49].

Fiber Bragg Grating (FBG) sensors, in particular, are well-suited for harsh environments requiring temperature or strain monitoring. FBG sensors exhibit high sensitivity and precision in monitoring temperature, strain, and vibration, which are essential for assessing structural safety. Their precise measurements contribute to improved aircraft maintenance and safety protocols [50, 51]. Their thin and lightweight nature (hundreds of μm) allows seamless integration into various materials without structural alteration. Additionally, FBG sensors effectively measure temperatures near 0 °C, facilitating ice detection [43]. These attributes make fiber-optic sensors particularly advantageous in aerospace applications, where traditional sensors may be less effective [52].

While traditional sensors remain preferred in some cases due to established reliability and lower initial costs, optical sensors may also offer long-term benefits such as reduced maintenance and superior performance in extreme conditions, making them increasingly viable for aerospace applications.

4 Examples of Optical Sensor Applications in Aerospace

Optical systems have the potential to continue to revolutionize the aerospace industry, offering significant advantages in areas such as communication, environmental sensing, and structural and human health. Multiple different technologies can be employed for this purpose and there is a lot of research actively being undertaken to further develop such techniques. This section aims to introduce some of the most important areas of aerospace engineering that benefit from the use of optical sensors.

4.1 Navigation

Robust identification of a vehicle's position is essential for ensuring effective navigation. In this context, optical remote sensing cameras play an indispensable role by providing critical data for space altitude measurements and offering visual insights that underpin scientific research. Notably, recent technological advances have led to considerable improvements in the performance of these cameras, with significant enhancements in temporal, spatial, and spectral resolution. Furthermore, in an effort to reduce overall mass—and consequently lower fuel consumption costs—multispectral integrated optical systems have emerged as a prevailing trend in deep space exploration. This branch of research encompasses space activities aimed at investigating celestial bodies both within and beyond our solar system. In light of these developments, this paper describes the design and construction of a common-aperture multispectral imaging system capable of simultaneously capturing data across the ultraviolet, visible, mid-wave, and long-wave infrared spectral bands [53].

Gyroscopes, on the other hand, are fundamental for maintaining spacecraft attitude and orbital control, and they are indispensable for a myriad of missions. Historically, traditional mechanical gyroscopes have been largely replaced by fiber optic gyroscopes (FOGs) due to their enhanced mechanical robustness, reduced weight, lower power consumption, shorter reaction times, and superior accuracy [54–56]. These advanced devices are based on the principle of the Sagnac interferometer—a configuration named after the French physicist Georges Sagnac—which consists of a circular optical path along which two interfering beams propagate in opposite directions [57]. In the absence of rotational motion, the resulting interference fringes remain stationary; however, when the system rotates, the fringe pattern shifts in direct proportion to the angular velocity, thereby enabling precise measurement of the spacecraft's rotational dynamics.

4.2 *Environmental Monitoring*

Temperature

Aerospace vehicles, particularly aircraft, are exposed to a wide range of extreme temperatures, spanning from as low as − 150 °C to as high as 2000 °C. Such extreme thermal conditions pose significant challenges for the use of conventional unpackaged commercial optical fibers, which are commonly employed for temperature sensing but are generally unable to withstand such intense temperature variations. Nevertheless, these fibers have demonstrated potential in certain extreme applications, such as monitoring heat pipes [58], indicating their adaptability under specific conditions. Despite this, the limitations of existing fiber technologies necessitate the development of innovative approaches to enable the effective use of optical fibers as reliable sensors in these demanding environments.

Among the various optical fiber sensing technologies, Fiber Bragg Gratings (FBGs) have emerged as particularly promising for temperature sensing applications. Extensive research has been conducted to evaluate their stability, accuracy, and sensitivity under diverse operating conditions [59]. For instance, researchers have demonstrated the capability of FBGs to achieve a linear response within a temperature range of 100–1000 °C, with a consistent sensitivity of approximately 15.7 pm/°C for both radiation-resistant FBGs (RFBGs) and conventional FBGs [60]. These findings underscore the potential of FBGs as robust and precise temperature sensors, even in the face of extreme thermal environments encountered in aerospace applications. However, further advancements and refinements are required to fully optimize their performance and ensure their reliability across the entire spectrum of operational temperatures experienced by aerospace vehicles.

Radiation

Dosimetry, the measurement of ionizing radiation dose absorbed by matter or tissue, is of special interest for space applications, where different types of particles, broad energy spectra and solar activity need to be accounted for. Optical fibers have been used for these applications, such as a phosphorus-doped, single-mode optical fiber dosimeter, with an embedded interrogator operating at 1610 nm [61]. While under γ- and X-rays, the proposed system evidenced a linear response between 10^{-3} and 10^2 Gy(SiO_2), with a performance comparable to other standard-sized systems.

Optical fiber-based dosimeters have already been deployed on the ISS, under the scope of the LUMINA project (Fig. 6). P-doped fibers were used for this purpose, and before deployment, a theoretical model was developed and experimental tests were carried out at CERN, which allowed for the calibration curve of the flight model to be obtained [62].

Gas and Particulate Matter

Although the majority of materials used in the interior of manned spacecraft are designed to be non-flammable, they can still produce smoke when subjected to overheating. Fires in spacecraft pose a particularly catastrophic risk due to the severely limited options available for extinguishing them in such confined and isolated environments. Consequently, smoke detectors must be capable of identifying smoke generated through processes such as oxidative pyrolysis, including smoldering, which is a common precursor to open flames. Spacecraft fires differ significantly from those encountered on Earth, as highlighted by studies conducted on the ISS [63]. This underscores the need for smoke detectors to be specifically optimized for the unique conditions of space environments. A robust and reliable system comprising

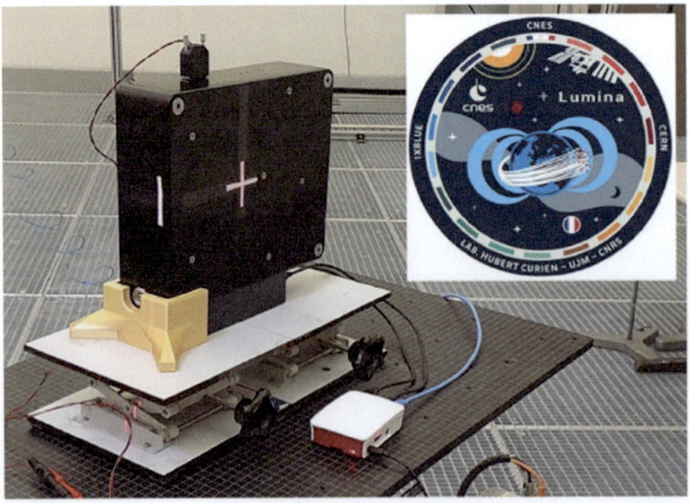

Fig. 6 LUMINA dosimeter during the experimental tests at CERN [62]

small, low-power sensors capable of measuring various key compounds associated with combustion could enable the early detection and precise localization of fires. Such a system would facilitate the timely deployment of fire suppressants, thereby mitigating potential disasters.

Absorption spectroscopy represents a promising approach for detecting and quantifying the concentrations of gases relevant to combustion and pyrolysis, as many of these gases exhibit distinct absorption features in the mid-infrared range. For instance, Terracciano et al. developed and tested a CO_2 sensor based on a 4.2 μm light-emitting diode [64]. By employing amplitude modulation, the sensor was able to determine species concentration, while frequency modulation provided insights into ambient temperature. The authors further suggested that this principle could be extended to detect other hazardous gases, demonstrating the versatility of the technique.

Another example of absorption spectroscopy in action is the Combustion Product Monitor, which utilizes a tunable laser. This six-channel system was developed as part of the Spacecraft Fire Safety Demonstration (Saffire) Project, aimed at measuring gas concentrations during low-gravity, reduced-pressure oxidative pyrolysis of materials relevant to spacecraft environments. The system was tested aboard a transfer vehicle returning from low Earth orbit and demonstrated its capability to measure gas-phase concentrations of critical compounds, including CO, HCl, HCN, HF, CO_2, and O_2, within specified thresholds. Importantly, the system performed effectively across a pressure range of 0.5–1 atm [65], highlighting its suitability for use in the variable conditions encountered in space missions. These advancements underscore the potential of absorption spectroscopy as a critical tool for enhancing fire safety in spacecraft, ensuring the protection of both crew and equipment.

4.3 Location

Low Earth orbits are experiencing increasing congestion, driven by the rapid growth in the launch of both commercial and institutional satellites. The number of annual launches has surged from approximately 100 in 2010 to more than ten times that figure in 2020. This dramatic rise can be attributed to several factors, including the availability of more cost-effective components and the growing trend toward the miniaturization of satellites. In addition to the escalating volume of active satellites, the density of space debris in LEO is also on the rise [66]. As a result, the efficient management of space traffic has become a critical issue, necessitating innovative solutions to ensure the sustainability of orbital environments. A fundamental step in addressing this challenge is the accurate identification of space objects, including both operational satellites and debris. The ability to identify satellites could enable institutions or commercial service providers to detect and mitigate potential collision threats, thereby enhancing the safety and reliability of space operations.

Currently, the identification of space objects relies primarily on RADAR and optical observation technologies. However, these methods have limitations, as a significant number of objects in LEO remain unidentifiable using existing techniques. To address this gap, alternative strategies have been proposed, including equipping satellites with specialized hardware to facilitate identification. One such approach involves the use of an extremely low-resource optical identifier, which incorporates a "license plate" system consisting of an array of laser diodes mounted on the satellite [67]. When the optical signal emitted by these diodes is received and decoded on Earth, each satellite can be uniquely identified. This method represents a promising advancement in space traffic management, offering a reliable and resource-efficient means of tracking and identifying satellites in increasingly crowded orbital environments.

4.4 Structural Health Monitoring

The initial approach to assessing structural health in aircraft was based primarily on log hours and landing cycles, with aircraft being retired once they reached a predetermined threshold. However, as scientific understanding advanced, a clear link was established between load cycles, fatigue, stress, and structural damage. This led to the realization that flight hours alone are not the sole determinant of structural integrity [68]. Over time, these insights have evolved into the concept of Structural Health Monitoring (SHM), which focuses on the continuous assessment of structural conditions to enhance reliability, safety, and ultimately reduce maintenance costs [12, 69]. All engineering infrastructures are subject to aging, and damage inevitably occurs as a result of the loads they endure. Consequently, regular inspection and maintenance are critical to predicting and extending their operational lifespan. For aircraft, which operate under particularly harsh conditions, periodic and scheduled inspections are essential to ensure safe and efficient operation. While the cost of personnel required to perform these inspections is significant, the economic impact of aircraft downtime during these time-intensive procedures is even greater [70].

In parallel with the remarkable growth in air and space travel in recent years, there has been a corresponding increase in the need to ensure that these modes of transportation are affordable, safe, and environmentally sustainable [39]. In this context, the use of composite materials has become indispensable for designing high-strength, lightweight aircraft structures. These materials contribute significantly to reducing fuel consumption and emissions without compromising performance. However, composite materials are inherently more complex than traditional metallic structures, raising concerns about their durability and safety. This is where SHM plays a crucial role, enabling the continuous monitoring of components to prevent catastrophic failures.

A variety of sensors have been explored for SHM applications, including strain gauges, accelerometers, ultrasonic sensors, and passive acoustic sensors. However, many of these sensors face limitations, such as high costs, difficulties in deployment over large structures, or an inability to withstand harsh environments and operate reliably over extended periods. Among the potential solutions, fiber optic

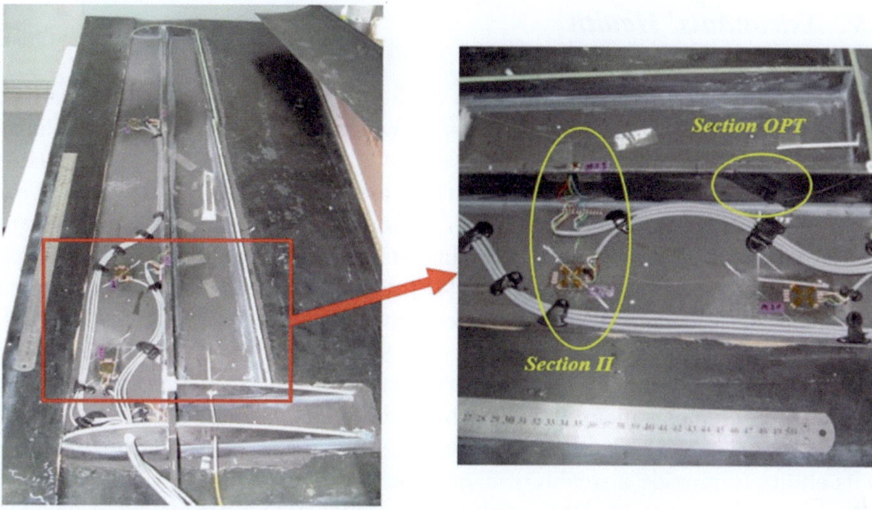

Fig. 7 FBG strain gauges integrated into a wing structure [75, 76]

sensors [71], particularly those based on Fiber Bragg Gratings (FBGs) [72], have emerged as a highly promising technology (Fig. 7). FBGs offer several advantages, including higher sensitivity, immunity to electromagnetic interference, and durability. Additionally, their multiplexing capability significantly reduces the complex wiring required by traditional electrical strain gauges and accelerometers. Beyond FBG-based systems, fiber optic sensing techniques utilizing Rayleigh or Brillouin scattering in standard telecom fibers have opened the door to distributed SHM systems. In these systems, the entire fiber acts as a sensor, capable of detecting strain changes with resolutions in the centimeter or even sub-centimeter range [73].

The implementation of SHM systems is critical to ensuring that aerospace structures can perform effectively and endure the extreme conditions they encounter throughout their operational lives. Aerospace applications typically demand monitoring systems that are lightweight, compact, low power, and immune to electro magnetic interference [39]. However, a significant challenge lies in the fact that current manufacturing processes for air and spacecraft do not routinely incorporate embedded smart sensors. Retrofitting SHM systems into existing structures can be complex, though surface-mounted sensors may offer a viable solution in certain cases [74]. As the aerospace industry continues to evolve, the integration of advanced SHM technologies will be essential to maintaining safety, efficiency, and sustainability in air and space travel.

4.5 Astronauts' Health

Space represents an exceptionally harsh environment that exposes astronauts to a range of unique health challenges, many of which are not encountered on Earth. Among the most prominent physiological effects are body fluid redistribution, muscle atrophy, and loss of bone density, all of which are primarily attributed to the microgravity conditions experienced in space [77]. Additionally, the increased exposure to cosmic radiation poses significant risks, including a higher likelihood of developing neurodegenerative diseases and cancer [78]. Given the limited access to medical equipment and specialized healthcare personnel in space, the prospect of longer-duration missions underscores the urgent need for advanced health monitoring tools. Such tools must be simple to use, compact, and autonomous, enabling astronauts to perform measurements independently while minimizing the need for external intervention. In this context, breath analysis has emerged as a highly promising field due to its ability to provide a wealth of diagnostic information through a non-invasive and user-friendly approach. Recognizing its potential, NASA has already developed a breath analysis system tailored for space applications. Known as the Portable Unit for Metabolic Analysis (PUMA), this wearable device includes a breathing mask and is capable of quantifying key metabolic parameters such as fatty acid oxidation rates, carbohydrate oxidation rates, oxygen (O_2) consumption, and carbon dioxide (CO_2) production [79]. The CO_2 measurements, in particular, are achieved using an optical sensor based on Non-Dispersive Infrared (NDIR) spectroscopy. Beyond this, researchers such as Silva et al. have highlighted the potential of optical-based sensors to detect other critical substances, including acetone [80], ammonia [81], and methane [82], all of which could be adapted for use in aerospace applications.

In-flight clinical decision-making can be significantly enhanced by the availability of immediate and comprehensive diagnostic information, which can be obtained from small biological samples such as blood or other bodily fluids. These samples can be collected easily and frequently, enabling the measurement of a wide range of biomarkers, including blood counts, hormones, enzymes, nucleic acids, proteins, and other critical diagnostic parameters. Furthermore, cytometry offers the advantage of high levels of assay multiplexing, allowing for the simultaneous measurement of multiple diagnostic and biological markers. This capability increases the throughput and informational content of each sample analysis, making it a valuable tool for space missions. Recently, Rea et al. reported the successful demonstration of a miniature cytometry-based analyzer (rHEALTH ONE) aboard the ISS (Fig. 8) [83]. This system was specifically customized to meet the stringent size, weight, and power constraints of spaceflight, incorporating alignment-free optics to ensure its survival during the rigors of a rocket launch. The analyzer utilizes two lasers (405 and 532 nm) to analyze each sample, with the resulting photon counts collected by five detectors at 10-microsecond intervals. The data is then processed using the rHEALTH Capture software program, marking a significant advancement in on-orbit biomedical analysis capabilities.

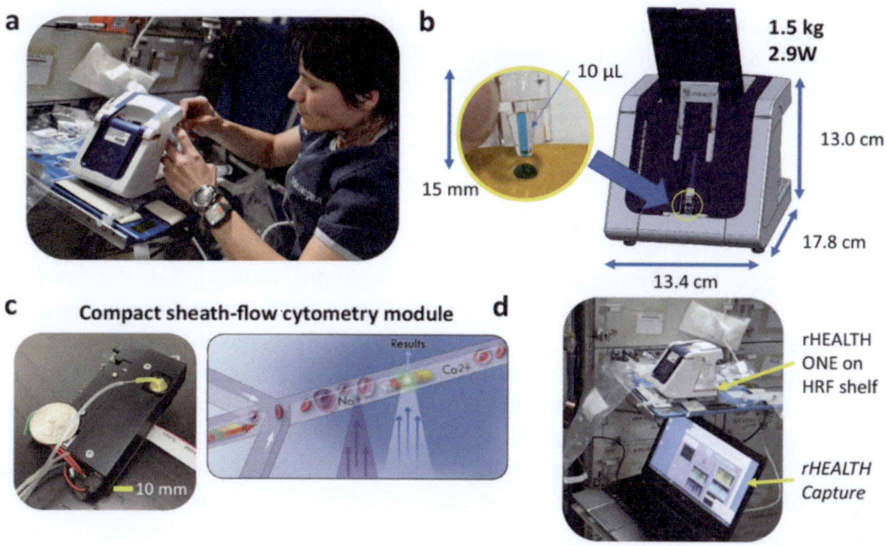

Fig. 8 Overview of the rHEALTH ONE experiment on the ISS. **a** The rHEALTH ONE during device setup at the ISS. **b** Detail of the sample positioning on the device. **c** Sheath-flow cytometry module, with diagram of particle analysis through the two lasers (405 and a 532 nm). **d** On-orbit test runs [83]

Another critical health issue associated with the microgravity environment is Spaceflight-Associated Neuro-Ocular Syndrome (SANS), a condition characterized by unique neuro-ophthalmic changes observed in astronauts during and after long-duration spaceflights. The clinical manifestations of SANS include optic disc edema (ODE), chorioretinal folds, hyperopic refractive shifts, posterior globe flattening, and thickening of the retinal and retinal nerve fiber layers [84]. To better understand this condition, a study has been conducted to examine changes in the optic nerve head and surrounding tissues using preflight and postflight Optical Coherence Tomography (OCT) scans from ISS astronauts [85]. This research has provided valuable insights into the structural changes associated with SANS, contributing to the development of potential countermeasures.

Notably, OCT technology has already been deployed in space since NASA installed the Heidelberg Spectralis OCT system on the ISS [86]. Astronauts undergo baseline OCT examinations before their missions, with follow-up scans conducted during and after their time in space to monitor ocular changes over the course of long-duration missions. More recently, an upgraded version of the system, the Heidelberg Spectralis "OCT2", was activated on the ISS [87]. This advanced system offers improved sensitivity, higher resolution, and enhanced depth imaging, in addition to being more user-friendly. These technological advancements provide a foundation for a deeper understanding of SANS, which NASA has identified as one of the highest-priority risks to astronaut health during space missions. By leveraging these tools, researchers and medical professionals can work toward developing effective strategies to mitigate the adverse effects of long-duration spaceflight on human health.

5 Final Remarks

Given the expected increase in aerospace activity in the coming years, the development and integration of new and improved technologies is essential to continue to make these explorations easier and more effective. The inherent advantages of using optical technologies—such as the resilience to harsh environments and the ability to construct lightweight systems, while maintaining a high level or precision—position them as transformative solutions to modernize the aerospace industry.

This chapter has established a comprehensive foundation for understanding the critical role of optical sensors in aerospace applications. The harsh conditions of the space environment were examined in detail, with emphasis on the extreme temperature fluctuations, intense radiation levels, and plasma interactions that pose significant challenges to sensor durability and functionality. The discussion underscored the importance of robust sensor design and innovative materials in overcoming these adversities, thereby ensuring that optical sensors can perform optimally even under the most demanding conditions.

A brief overview of the advantages of using optical sensor technology has been provided, followed by a series of application examples—from navigation sensors to advanced health monitoring—served to illustrate the practical implementations of optical sensors in aerospace platforms. These examples not only demonstrated the transformative potential of optical sensors but also provided a contextual framework for the technical and design challenges addressed in subsequent chapters.

In summary, the topics covered in this introductory chapter lay the groundwork for a deeper exploration of optical sensor technologies in aerospace settings. The forthcoming chapters will delve into the intricate technical details, address specific design considerations, and explore emerging trends that are driving advancements in this field. The insights presented here are intended to serve as a valuable reference, setting the stage for a detailed examination of the innovations that are shaping the future of aerospace sensor technology.

References

1. M.G.R. Sause, E. Jasiūnienė (eds.), *Structural Health Monitoring Damage Detection Systems for Aerospace* (2021). https://doi.org/10.1007/978-3-030-72192-3
2. A.T. Oyewo, O.O. Oluwole, O.O. Ajide, T.E. Omoniyi, M. Hussain, A summary of current advancements in hybrid composites based on aluminium matrix in aerospace applications. Hybrid Adv. **5**, 100117 (2024). https://doi.org/10.1016/J.HYBADV.2023.100117

3. B. Blakey-Milner et al., Metal additive manufacturing in aerospace: a review. Mater. Des. **209**, 110008 (2021). https://doi.org/10.1016/J.MATDES.2021.110008
4. R.K. Agarwal, Grand challenges in aerospace engineering. Front. Aerosp. Eng. **3**, 1383934 (2024). https://doi.org/10.3389/FPACE.2024.1383934
5. C. Soutis, P. Irving, Polymer composites in the aerospace industry, in *Polymer Composites in the Aerospace Industry*. (2014), pp. 1–520. https://doi.org/10.1016/C2013-0-16303-9
6. F. Ciampa, P. Mahmoodi, F. Pinto, M. Meo, Recent advances in active infrared thermography for non-destructive testing of aerospace components. Sensors **18** (2):609 (2018). https://doi.org/10.3390/S18020609
7. S. Rana, R. Fangueiro, Advanced composites in aerospace engineering, in *Advanced Composite Materials for Aerospace Engineering*. (2016), pp. 1–15. https://doi.org/10.1016/B978-0-08-100037-3.00001-8
8. Y. Li, Y. Xiao, L. Yu, K. Ji, D. Li, A review on the tooling technologies for composites manufacturing of aerospace structures: materials, structures and processes. Compos. Part A Appl. Sci. Manuf. **154**, 106762 (2022). https://doi.org/10.1016/J.COMPOSITESA.2021.106762
9. G. Hegde, S. Asokan, G. Hegde, Fiber Bragg grating sensors for aerospace applications: a review. ISSS J. Micro Smart Syst. **11**(1):257–275 (2022). https://doi.org/10.1007/S41683-022-00101-Z
10. I. Mckenzie, S. Ibrahim, E. Haddad, S. Abad, A. Hurni, L.K. Cheng, Fiber optic sensing in spacecraft engineering: an historical perspective from the European space agency. Front Phys. **9**(719441) (2021). https://doi.org/10.3389/fphy.2021.719441
11. A. Sposito, R.D. Pechstedt, Optical sensors for aerospace applications: Brake temperature sensors and fuel pump pressure sensors for aircraft, in *2016 IEEE Metrology for Aerospace (MetroAeroSpace)*. (2016), pp. 97–101. https://doi.org/10.1109/MetroAeroSpace.2016.7573193
12. A. Rovera, A. Tancau, N. Boetti, M. Dalla Vedova, P. Maggiore, D. Janner, Fiber optic sensors for harsh and high radiation environments in aerospace applications. Sensors **23**, 2512 (2023). https://doi.org/10.3390/s23052512
13. J.C. Ince et al., Overview of emerging hybrid and composite materials for space applications, **6**, 130 (2023). https://doi.org/10.1007/s42114-023-00678-5
14. N. International Space Station Research Integration Office, A Researcher's Guide to: Space Environmental Effects
15. A. Anwar et al., Vacuum effect on spacecraft structure materials. Int. Conf. Aerosp. Sci. Aviat. Technol. **16**, no. AEROSPACE SCIENCES & AVIATION TECHNOLOGY, ASAT-16—May 26–28, 2015, 1–10 (2015). https://doi.org/10.21608/ASAT.2015.22908
16. L.A. Rogers et al., Outgassing environment of spacecraft: an overview. IOP Conf. Ser. Mater. Sci. Eng. **611**(1), 012071 (2019). https://doi.org/10.1088/1757-899X/611/1/012071
17. E. Grossman, I. Gouzman, Space environment effects on polymers in low earth orbit. Nucl. Instrum. Methods Phys. Res. B **208**(1–4), 48–57 (2003). https://doi.org/10.1016/S0168-583X(03)00640-2
18. K. You, C. Zhou, L. Ding, Y. Wang, Construction robotics in extreme environments: from earth to space. Engineering (2025). https://doi.org/10.1016/J.ENG.2024.11.037
19. Z. Hooshangi, S.A. Hossein Feghhi, R. Saeedzadeh, The effects of low earth orbit atomic oxygen on the properties of Polytetrafluoroethylene. *Acta Astronaut* **119**, 233–240 (2016). https://doi.org/10.1016/J.ACTAASTRO.2015.11.031
20. M. Tagawa, K. Yokota, Atomic oxygen-induced polymer degradation phenomena in simulated LEO space environments: how do polymers react in a complicated space environment? Acta Astronaut. **62**(2–3), 203–211 (2008). https://doi.org/10.1016/J.ACTAASTRO.2006.12.043
21. S.W. Samwel, Low earth orbital atomic oxygen erosion effect on spacecraft materials. Space Res. J. **7**(1), 1–13 (2014). https://doi.org/10.3923/SRJ.2014.1.13
22. Y. Lu, Q. Shao, H. Yue, F. Yang, A review of the space environment effects on spacecraft in different orbits. IEEE Access **7**, 93473–93488 (2019). https://doi.org/10.1109/ACCESS.2019.2927811

23. Y. Zheng et al., Overview, progress and next steps for our understanding of the near-earth space radiation and plasma environment: science and applications. Adv. Space Res. (2024). https://doi.org/10.1016/J.ASR.2024.05.017
24. W.K. Tobiska, M.M. Meier, D. Matthiae, K. Copeland, Characterizing the variation in atmospheric radiation at aviation altitudes in *Extreme Events in Geospace: Origins, Predictability, and Consequences.* (2018), pp. 453–471. https://doi.org/10.1016/B978-0-12-812700-1.00018-2
25. L.C. Simonsen, T.C. Slaba, P. Guida, A. Rusek, NASA's first ground-based Galactic Cosmic Ray Simulator: enabling a new era in space radiobiology research. PLoS Biol. **18**(5), e3000669 (2020). https://doi.org/10.1371/JOURNAL.PBIO.3000669
26. R. Thirsk, A. Kuipers, C. Mukai, D. Williams, The space-flight environment: the international space station and beyond. CMAJ. Can. Med. Assoc. J. **180**(12), 1216–1220 (2009). https://doi.org/10.1503/cmaj.081125
27. J. Chancellor et al., Everything you wanted to know about space radiation but were afraid to ask. J. Environ. Sci. Health C **39**(2), 113–128 (2021). https://doi.org/10.1080/26896583.2021.1897273
28. L.W. Townsend, R.J.M. Fry, Radiation protection guidance for activities in low-earth orbit. Adv. Space Res. **30**(4), 957–963 (2002). https://doi.org/10.1016/S0273-1177(02)00160-6
29. X.Y. Ji, J. Wang, Statistical analysis of spacecraft failure in full-life based on STED, in *2019 IEEE 10th International Conference on Mechanical and Aerospace Engineering, ICMAE 2019.* (2019), pp. 89–96. https://doi.org/10.1109/ICMAE.2019.8880888
30. R. Almehisni, F. Al Naimat, Heat transfer influence of solar panel on spacecraft, in *2018 Advances in Science and Engineering Technology International Conferences, ASET 2018.* (2018), pp. 1–6. https://doi.org/10.1109/ICASET.2018.8376815
31. P. Blumenfeld, A. Boca, P. Sharps, D. Russell, H. Ogawa, H. Toyota, Thermal design for high-temperature high-irradiance testing of space solar panel components, in *Conference Record of the IEEE Photovoltaic Specialists Conference.* (2013), pp. 647–651. https://doi.org/10.1109/PVSC.2013.6744234
32. R. Darbali-Zamora, N. Cobo-Yepes, E.I. Ortiz-Rivera, Design considerations based on the effects of varying temperature conditions on the efficiency of size constrained electronic power supplies for Cubesat applications, in *Proceedings of the 17th InterSociety Conference on Thermal and Thermomechanical Phenomena in Electronic Systems, ITherm 2018.* (2018), pp. 778–787. https://doi.org/10.1109/ITHERM.2018.8419570
33. D.J. Newman, Life in extreme environments: how will humans perform on Mars? Gravit. Space Biol. **13**(2), 35–48 (2000). Accessed 15 Feb 2025. [Online]. Available: https://go.gale.com/ps/i.do?p=AONE&sw=w&issn=1089988X&v=2.1&it=r&id=GALE%7CA176373045&sid=googleScholar&linkaccess=fulltext
34. S. Tonetti, S. Cornara, M. Faenza, O. Verberne, T. Langener, G. Vicario de Miguel, Active debris removal and space debris mitigation using hybrid propulsion solutions. Astrophys. Space Sci. Proc. **52**, 163–180 (2018). https://doi.org/10.1007/978-3-319-69956-1_10
35. J.N. Pelton, *Space Debris and Other Threats from Outer Space* (2013). https://doi.org/10.1007/978-1-4614-6714-4
36. J. Mespoulet, P.L. Héreil, H. Abdulhamid, P. Deconinck, C. Puillet, Experimental study of hypervelocity impacts on space shields above 8 km/s. Procedia Eng **204**, 508–515 (2017). https://doi.org/10.1016/J.PROENG.2017.09.748
37. M. Liu, Z. Su, Q. Zhang, R. Long, Modeling hypervelocity-impact-induced shock waves for characterizing orbital debris-produced damage. J. Appl. Mech. Trans. ASME **83**(8) (2016). https://doi.org/10.1115/1.4033679
38. M.E. De Simone, C. Andreades, M. Meo, F. Ciampa, Smart composite detector of orbital debris and micrometeoroids particles. Mater. Today Proc. **34**, 202–209 (2021). https://doi.org/10.1016/J.MATPR.2020.02.796
39. R. Di Sante, Fibre optic sensors for structural health monitoring of aircraft composite structures: recent advances and applications. Sensors **15**(8), 18666–18713 (2015). https://doi.org/10.3390/S150818666

40. Y. Xu, W. Pei, W. Hu, A current overview of the biological effects of combined space environmental factors in mammals. Front. Cell. Dev. Biol. **10**, 861006 (2022). https://doi.org/10.3389/FCELL.2022.861006/BIBTEX
41. F.E. Garrett-Bakelman et al., The NASA twins study: a multidimensional analysis of a year-long human spaceflight. Science **364**(6436) (2019). https://doi.org/10.1126/SCIENCE.AAU8650/SUPPL_FILE/AAU8650_TABLE_S9.XLSX
42. J. Taillard, P. Sagaspe, P. Philip, S. Bioulac, Sleep timing, chronotype and social jetlag: impact on cognitive abilities and psychiatric disorders. Biochem. Pharmacol. **191**, 114438 (2021). https://doi.org/10.1016/J.BCP.2021.114438
43. C. von Gall, The effects of light and the circadian system on rhythmic brain function. Int. J. Mol. Sci. **23**(5), 2778 (2022). https://doi.org/10.3390/IJMS23052778
44. M. Lepore, I. Delfino, Optical sensors technology and applications. Sensors **22**(20), 7905. https://doi.org/10.3390/S22207905
45. S. Liu, S. Xu, K. Zhang, G. Qin, Power electronics in extreme environments. **12781**, 1278102 (2023). https://doi.org/10.1117/12.2686704
46. A. Hassan, Y. Savaria, M. Sawan, Electronics and packaging intended for emerging harsh environment applications: a review. IEEE Trans. Very Large Scale Integr. VLSI Syst. **26**(10), 2085–2098 (2018). https://doi.org/10.1109/TVLSI.2018.2834499
47. D. Kobayashi, K. Hirose, "How Harsh is Space?-Equations that connect space and ground VLSI, in *Digest of Technical Papers—Symposium on VLSI Technology*, vol. 2023 (2023). https://doi.org/10.23919/VLSITECHNOLOGYANDCIR57934.2023.10185386
48. Q. Guan et al., Research progress on the solder joint reliability of electronics using in deep space exploration. Chin. J. Mech. Eng. (English Edition) **36**(1), 1–13 (2023). https://doi.org/10.1186/S10033-023-00834-4
49. A. Aimasso, C.G. Ferro, Proposal of a standard method to define a best practice for bonding fbg sensors for aerospace use, in *2023 IEEE 10th International Workshop on Metrology for AeroSpace (MetroAeroSpace).* (2023), pp. 498–502. https://doi.org/10.1109/MetroAeroSpace57412.2023.10189989
50. A. Aimasso, M.D.L.D. Vedova, P. Maggiore, Innovative sensor networks for massive distributed thermal measurements in space applications under different environmental testing conditions, in *2022 IEEE 9th International Workshop on Metrology for AeroSpace (MetroAeroSpace).* (2022), pp. 503–508. https://doi.org/10.1109/MetroAeroSpace54187.2022.9856275
51. A. Aimasso, M.D.L. Dalla Vedova, P. Maggiore, Analysis of FBG sensors performances when integrated using different methods for health and structural monitoring in aerospace applications, in *2022 6th International Conference on System Reliability and Safety (ICSRS).* (2022), pp. 138–144. https://doi.org/10.1109/ICSRS56243.2022.10067392
52. A. Cutolo et al., Innovative photonic sensors for safety and security, Part II: aerospace and submarine applications. Sensors **23**(5), 2417 (2023). https://doi.org/10.3390/S23052417
53. J. Cao et al., Optical design and fabrication of a common-aperture multispectral imaging system for integrated deep space navigation and detection. Opt. Lasers Eng. **167**, 107619 (2023). https://doi.org/10.1016/J.OPTLASENG.2023.107619
54. H. Xu, L. Wang, Y. Zu, W. Gou, Y. Hu, Application and development of fiber optic gyroscope inertial navigation system in underground space. Sensors **23**(12), 5627 (2023). https://doi.org/10.3390/S23125627
55. J. Nayak, Fiber-optic gyroscopes: from design to production [Invited]. Appl. Opt. **50**(25):E152–E161 (2011). https://doi.org/10.1364/AO.50.00E152
56. I. Mckenzie, N. Karafolas, Fiber optic sensing in space structures: the experience of the European Space Agency. **5855**, 262–269 (2005). https://doi.org/10.1117/12.623988
57. J. Spytek, L. Ambrozinski, I. Pelivanov, Non-contact detection of ultrasound with light—review of recent progress. Photoacoustics **29**, 100440 (2023). https://doi.org/10.1016/J.PACS.2022.100440
58. S. Kabashima, T. Ozaki, N. Takeda, S. Kabashima, T. Ozaki, N. Takeda, Structural health monitoring using FBG sensor in space environment. SPIE **4332**, 78–87 (2001). https://doi.org/10.1117/12.429644

59. A. Aimasso, M.D.L. Dalla Vedova, P. Maggiore, G. Quattrocchi, Study of FBG-based optical sensors for thermal measurements in aerospace applications, *J. Phys. Conf. Ser.* **2293**(1), 012006 (2022). https://doi.org/10.1088/1742-6596/2293/1/012006
60. Q. Tian et al., Optical fiber sensor with double tubes for accurate strain and temperature measurement under high temperature up to 1000 °C. IEEE Sens. J. **22**(12), 11710–11716 (2022). https://doi.org/10.1109/JSEN.2022.3167950
61. A. Meyer et al., Toward an embedded and distributed optical fiber-based dosimeter for space applications. IEEE Trans. Nucl. Sci. **70**(4), 583–589 (2023). https://doi.org/10.1109/TNS.2022.3226194
62. D. Di Francesca et al., Low radiation dose calibration and theoretical model of an optical fiber dosimeter for the International Space Station. Appl. Opt. **62**(16), E43–E50 (2023). https://doi.org/10.1364/AO.483560
63. M.E. Meyer et al., Evaluation of spacecraft smoke detector performance in the low-gravity environment. Fire Saf. J. **98**, 74–81 (2018). https://doi.org/10.1016/j.firesaf.2018.04.004
64. A.C. Terracciano et al., Hazardous gas detection sensor using broadband light-emitting diode-based absorption spectroscopy for space applications. New Space **6**(1), 28–36 (2018). https://doi.org/10.1089/space.2017.0044
65. M. Fradet, R.M. Bendig, R.M. Briggs, The Combustion Product Monitor instrument for the spacecraft fire safety demonstration project, in *49th International Conference on Environmental Systems (ICES)*, Boston, Massachusetts: Pasadena, CA: Jet Propulsion Laboratory, National Aeronautics and Space Administration, Apr. 2019. Accessed 27 Sept 27 2024. [Online]. Available: http://dataverse.jpl.nasa.gov/dataset.xhtml?persistentId=hdl:2014/51087
66. ESA—ESA Space Environment Report 2024. Accessed 12 Sept 2024. [Online]. Available: https://www.esa.int/Space_Safety/Space_Debris/ESA_Space_Environment_Report_2024
67. D.M. Palmer, R.M. Holmes, Extremely low resource optical identifier: a license plate for your satellite. J. Spacecr Rockets **55**(4), 1014–1023 (2018). https://doi.org/10.2514/1.A34106
68. S. Zafar Stellantis, N. shafi Khan, A. Ali Khan, N. Khan, History, current status and challenges to structural health monitoring system in aviation field. J. Space Technol. **20**(1) (2014). [Online]. Available: https://www.researchgate.net/publication/267182948
69. C.R. Farrar, K. Worden, An introduction to structural health monitoring. Philos. Trans. R. Soc. A Math. Phys. Eng. Sci. **365**(1851), 303–315 (2007). https://doi.org/10.1098/RSTA.2006.1928
70. I. García, J. Zubia, G. Durana, G. Aldabaldetreku, M.A. Illarramendi, J. Villatoro, Optical fiber sensors for aircraft structural health monitoring. Sensors **15**(7), 15494–15519 (2015). https://doi.org/10.3390/S150715494
71. K.O. Hill, G. Meltz, Fiber Bragg grating technology fundamentals and overview. J. Lightwave Technol. **15**(8), 1263–1276 (1997). https://doi.org/10.1109/50.618320
72. B. Culshaw, A. Kersey, Fiber-optic sensing: aA historical perspective. J. Lightwave Technol. **26**(9), 1064–1078 (2008). https://doi.org/10.1109/JLT.0082.921915
73. X. Bao, L. Chen, Recent progress in distributed fiber optic sensors. Sensors **12**(7), 8601–8639 (2012). https://doi.org/10.3390/S120708601
74. H. Rocha, C. Semprimoschnig, J.P. Nunes, Sensors for process and structural health monitoring of aerospace composites: a review. Eng. Struct. **237**, 112231 (2021). https://doi.org/10.1016/J.ENGSTRUCT.2021.112231
75. Ł Święch, Calibration of a load measurement system for an unmanned aircraft composite wing based on fibre Bragg gratings and electrical strain gauges. Aerospace **7**(3), 27 (2020). https://doi.org/10.3390/aerospace7030027
76. Y. Pei, T. Liao, Y. Pei, J. Xu, H. Lin, T. Ning, FBG strain sensor applied in harsh environment of aerospace, in *2018 IEEE 3rd Optoelectronics Global Conference (OGC)*. (2018), pp. 81–84. https://doi.org/10.1109/OGC.2018.8529994
77. A. Roda et al., Advanced biosensors for monitoring astronauts' health during long-duration space missions. Biosens. Bioelectron. **111**, 18–26 (2018). https://doi.org/10.1016/J.BIOS.2018.03.062
78. T. Elias Abi-Ramia Silva, F. Burisch, A.T. Güntner, Gas sensing for space: health and environmental monitoring. TrAC Trends Anal. Chem. **177**, 117790 (2024). https://doi.org/10.1016/J.TRAC.2024.117790

79. D.L. Dietrich et al., Portable unit for metabolic analysis, US-Patent-10,076,268, Sept 18, 2018. Accessed 27 Sept 2024. [Online]. Available: https://ntrs.nasa.gov/citations/20180006275
80. F. Bovey et al., Breath acetone as a marker of energy balance: an exploratory study in healthy humans. Nutrition Diabetes **8**(1), 1–4 (2018). https://doi.org/10.1038/s41387-018-0058-5
81. G. Neri, A. Lacquaniti, G. Rizzo, N. Donato, M. Latino, M. Buemi, Real-time monitoring of breath ammonia during haemodialysis: use of ion mobility spectrometry (IMS) and cavity ring-down spectroscopy (CRDS) techniques. Nephrol. Dial. Transplant. **27**(7), 2945–2952 (2012). https://doi.org/10.1093/NDT/GFR738
82. Y. Lin, D. Manalili, A. Khodabakhsh, S.M. Cristescu, Real-time measurement of CH_4 in human breath using a compact CH_4/CO_2 sensor. Sensors **24**(4) (2024). https://doi.org/10.3390/S24041077
83. D.J. Rea et al., Single drop cytometry onboard the International Space Station. Nat. Commun. **15**(1), 1–15 (2024). https://doi.org/10.1038/s41467-024-46483-6
84. J. Ong et al., Spaceflight associated neuro-ocular syndrome: proposed pathogenesis, terrestrial analogues, and emerging countermeasures. Br. J. Ophthalmol. **107**(7), 895–900 (2023). https://doi.org/10.1136/BJO-2022-322892
85. N. Patel, A. Pass, S. Mason, C.R. Gibson, C. Otto, Optical coherence tomography analysis of the optic nerve head and surrounding structures in long-duration international space station astronauts. JAMA Ophthalmol. **136**(2), 193–200 (2018). https://doi.org/10.1001/JAMAOPHTHALMOL.2017.6226
86. A.G. Lee et al., Spaceflight associated neuro-ocular syndrome (SANS) and the neuro-ophthalmologic effects of microgravity: a review and an update. npj Microgravity **6**(1), 1–10 (2020). https://doi.org/10.1038/s41526-020-0097-9
87. B. Soares et al., Imaging in spaceflight associated neuro-ocular syndrome (SANS): current technology and future directions in modalities. Life Sci. Space Res. (Amst) **42**, 40–46 (2024). https://doi.org/10.1016/J.LSSR.2024.04.004

Basic Concepts of Light and Optics for Aerospace Sector

Paulo Lourenço and Miguel Fernandes

Abstract Aerospace and aeronautical systems are becoming ever more complex due to an increasing demand for the integration of multipurpose functionality and the availability of advanced technological solutions. These systems are required to perform several concurrent tasks such as operational parameters monitoring (temperature, pressure, strain, vibration, acceleration, altitude, etc.) and structural integrity diagnosis and prognosis, to name a few. All these functionalities require dedicated sensing hardware with the corresponding increase in complexity, cost, weight and failure rate probability. The progress made in optics and photonics in the last decades has provided a diverse range of sensing solutions, which present a reliable alternative to other sensing principles due to their reduced dimensions and cost, low power, flexibility, electrochemical inertness, immunity to electromagnetic interference, high sensitivity, and multipoint and remote sensing capabilities. This way, optical sensors are starting to play an important role in spacecraft and earth orbiting manmade objects, either in local or remote sensing applications. These sensors are typically involved with positional control of the spacecraft and detection of remote objects or environmental conditions locally or in the distance. These sensors can be viewed as active, or passive depending on their ability to illuminate the object of interest or just receive its emitted or scattered light respectively. In either situation the detector is usually composed of one or an array of optical detectors depending on the type of measurement we want to achieve either a single signal or a spectrum. In this chapter, we will be addressing advanced optical sensors being implemented in the aerospace industry with emphasis on the fundamental principles they rely on and the used materials. To this end, we will be describing the working principles of a wide range of highly sensitive advanced photodetectors, fibre Bragg gratings, photonic crystal fibres, diffraction gratings, optical microelectromechanical structures, piezo-optical sensors and transducers, interferometers and spectrometers. All these devices have found application in the aerospace industry and are being used to perform a

P. Lourenço (✉) · M. Fernandes
Lisbon School of Engineering (ISEL/IPL), Rua Conselheiro Emídio Navarro, Lisboa, Portugal
e-mail: plourenco@deetc.isel.ipl.pt

CTS—Centre of Technology and Systems and Associated Lab of Intelligent Systems (LASI), Caparica, Portugal

C. Marques et al. (eds.), *Advanced Optical Sensors for Aerospace Applications*,
Progress in Optical Science and Photonics 37,
https://doi.org/10.1007/978-981-95-1626-1_2

wide range of functionalities, e. g. implemented in applications such as structural health monitoring, gyroscopes, inertial navigation systems, etc. Our approach will remain as broad and reader friendly as possible, leaving out mathematical abstractions whenever possible, and often relying on simulations and graphical interpretations to demonstrate the involved phenomena. This way, we intend to establish an easy to follow reading experience and, hopefully, a seamless connection to subsequent chapters of the book.

Keywords Photon sensing · Single photon detectors · Photon amplifiers · Optical sensing · Diffraction gratings · Optical interference

1 Introduction

Advanced optical sensors are now a technological requirement in the modern aerospace industry. This is mainly due to their ability to operate in extremely harsh environments, their unprecedent precision and their immunity to Electromagnetic (EM) interference. These sensing devices are often used in critical functions such as structural health monitoring, environmental analysis, remote detection, navigation, and more. Optical sensors, in contrast to conventional ones, exploit light-matter interaction to quantify chemical, biological and physical properties with high resolution and sensitivity [1]. Modern sensing systems in the aerospace industry have had significant improvements in autonomous operations, flight safety and mission efficiency, by integrating advanced optical sensors in their production lines.

Advanced optical sensors base their operation on several photonic and optical phenomena which are underlined by fundamental principles, enabling highly accurate and real-time measurements. Highly significant key principles include:

- Optical fibers as sensors—uses the propagation of EM radiation through optical fibers and related phenomena (Bragg reflection, evanescent waves, diffraction…) to quantify variations in temperature, strain and pressure, with evident advantages in durability and weight reduction for aerospace applications [2];
- Spectroscopy—enables the determination of gas species concentration, chemical composition and temperature distribution through the analysis of the interaction of EM radiation with materials [3];
- Interferometry—the exploitation of the interference phenomenon allows efficient precision metrology and aerospace structural health monitoring, by enabling the measurement of minute deviations in strain, displacement and refractive index [4];
- Photonic crystal and nanophotonic sensors—nanostructured materials enhance the interaction between light and matter, hence enabling highly sensitive detection capabilities for mechanical and chemical properties [5];

- Laser Doppler velocimetry—uses lasers and Doppler shift principles to quantify fluids flow velocity and turbulence, essential for aerodynamics and atmospheric research;
- Plasmonic and quantum sensing—introduces plasmonic resonances and quantum measuring techniques in aerospace applications such as navigation, communication and high resolution imaging [6].

Optical sensors are being used with extended capabilities, due to the advancements in miniaturization, photonic signal processing and miniaturization, in both manned and unmanned aerospace systems. Moreover, these sensors present excellent characteristics for missions in outer space and next generation monitoring systems, since they are able to endure extreme temperatures, dynamic environment conditions and exposure to radiation [7].

To address the fundamental principles of advanced optical sensors our approach will remain as broad and reader friendly as possible, leaving out mathematical abstractions whenever possible, and often relying on simulations and graphical interpretations to demonstrate the involved phenomena. This way, we intend to establish an easy to follow reading experience and, hopefully, a seamless connection to subsequent chapters of the book.

The remainder of this chapter consists of two main sections, Photon Detectors and Optical Sensing Structures, each divided in several subsections and where are approached the fundamental operating principles of each considered exemplary device:

1. Photon Detectors:—Several types of photon detectors will be approached.

    a. Photo multiplier tube—provides fast response times and high sensitivity by using the photoelectric effect and dynodes to amplify weak optical signals;
    b. Single-photon avalanche diode—device able to detect single photons with high timing resolution through a semiconductor based detector operating in Geiger mode;
    c. Electron-multiplying charge-coupled device—a charge-coupled device with on-chip amplification, ideal for imaging applications in low light conditions;
    d. Superconducting nanowire single-photon detector—provides ultra-fast response and near unity detection efficiency;
    e. Micro-channel plate—photo-electrons amplification capability provided by a dense array of microscopic channels and used in time-resolved spectroscopy, and high-speed imaging;
    f. Transition edge sensor—capable of measuring individual photons energy through variation of electrical resistance at superconducting transition edge (cryogenic device);
    g. Quanta image sensor—image sensor capable of detect and counting single photon with high temporal and spatial resolution;

2. Optical Sensing Structures:—exploit light/matter interaction to provide precise measurement of environmental and material properties.

a. Optical fibers—used in communications and sensing, for they are able to guide optical waves with insignificant losses over large distances;
b. Fiber Bragg gratings—mainly used in temperature and strain sensing and consisting of periodic variations of the refractive index of optical fibers to make them behave as wavelength selective reflectors;
c. Diffraction gratings—used in spectroscopy and base their operation on interference and the dispersion of light into its constituent wavelengths;
d. Optical interference—used in interferometry by exploiting the phase difference between light waves to enhance detection sensitivity;
e. Piezoelectric effect—often integrated in extremely precise optical systems to assure reliability, for this effect converts an electric signal in minute displacement and vice-versa;
f. Surface plasmon resonance—used to detect minute variations of the refractive index in the environment adjacent to the transducer through charge density oscillations at the metal/dielectric separation interface.

2 Photon Detectors

2.1 Photo Multiplier Tube (PMT)

Photomultiplier tubes (PMTs) have been the workhorse of highly sensitive optical detection since their invention in the 1930s, derived from Albert Einstein explanation of the photoelectric effect, early studies of secondary emission by Austin and Starke [8] and early developments in photoelectric and vacuum tube technologies in the early twentieth century. The first commercially available photomultiplier was the type 931 developed by Zworykin and Rajchman of RCA. Several developments followed with a wide range of applications [9, 10].

In its simplest form, a photomultiplier tube is composed of a photocathode where incident photons having sufficient energy will eject electrons by the photoelectric effect. An electron multiplication stage, generally composed of a series of dynodes, biased at sequentially varying voltages, amplifying the electron signal via secondary emission. And finally, an anode collecting the charge generated by the dynodes and producing a measurable electrical signal proportional to the incoming light. A schematic representation of a typical photomultiplier tube is given in Fig. 1. The device is usually encapsulated in a evacuated glass tube with an optical window with different materials depending on the wavelength of the incoming radiation. Other geometries were developed and commercialized by Hamamatsu, one of the most historic and important companies in the field [11]. The most commonly used window materials are lime glasses and borosilicate glasses with a cut-off wavelength of ~ 300 nm for visible applications or UV-transparent borosilicate glasses, fused silica or sapphire (ultraviolet grade) with cut-off wavelengths lower than 190 nm, for UV applications.

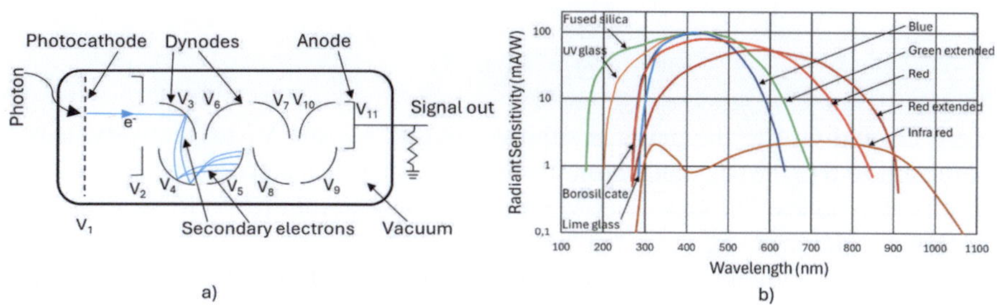

Fig. 1 **a** Schematic representation of a photomultiplier tube. **b** Cathode radiant sensitivity of photomultiplier tubes with typical window materials

The electrically active elements inside the device are biased at different potentials starting from the lower one, V_1, applied to the photocathode grid, up to the higher, V_{11}, applied to the anode. Different biasing schemes can be applied using positive or negative high voltages applied to the anode or to the cathode, respectively, being the intermediate voltages generated by a series of different resistors creating a multiple output voltage divider. The grounded anode type of biasing, presented in the schematic, is the most common one as the measurement of the output current is simpler, since no high voltage is present in this node. The voltage difference between adjacent dynodes can be constant or varied, increasing from cathode to anode, having influence on device gain and linearity.

The overall device conversion characteristic is named anode sensitivity, expressed in A/W or A/lm, is the product of cathode sensitivity, collection efficiency and gain and is dependent on the supply voltage and cathode and anode materials. Cathode radiant sensitivity, usually expressed in mA/W, is the photocathode current produced in response to an incident light power at a specific wavelength. The quantum efficiency of the device is related to the cathode radiant sensitivity by:

$$QE(\%) = \frac{S_k \times 1240}{\lambda} \times 100\%$$

where S_k is the radiant sensitivity in mA/W at the given wavelength and λ is the wavelength in nm of the incident light. The spectral sensitivity characteristic is presented in Fig. 2. The lower limit of the spectral response is defined by the cut-of wavelength of the window material presented before, while the upper limit depends on the photocathode material. PMT photocathodes are usually made of alkali antimonides like SbKCs, SbNa2K, GaAs or RbCsSb. The response of the multialkali cathodes can be tailored to extend to the green or red regions. Nevertheless, the spectral range of radiation detected by PMTs can be expanded using a scintillator screen between the radiation source and the input window. The gain or current amplification, G, of a PMT is the ratio of the anode current to the photocathode current. The gain shows a power dependence on the supply voltage is given by:

$$\frac{G_2}{G_1} = \left(\frac{V_2}{V_1}\right)^{\alpha N}$$

where G_2 and G_1 are the gains at supply voltages V_2 and V_1 respectively, α is a coefficient (0.6 to 0.8) dependent on the dynode material and geometry, and N is the number of dynodes. The gain of a PMT typically ranges from 10^4 to 10^8 with applied voltages ranging from 500 to 2000 V.

The extreme miniaturization of the PMT was achieved using the MEMS technology giving origin to the micro photomultiplier (microPMT or μPMT). The microPMTs are aimed at a wide application areas like high performance medical portable devices for Point-of-Care testing, flow cytometry, Raman spectroscopy and semiconductor wafer inspection [12].

Particle physics is the largest field of application of PMTs [13] with some single experiments involving detectors using thousands of PMTs. One of the best examples is the Super-Kamiokande where 11,146 PMTs (50 cm) and 1885 PMTs (20 cm) were

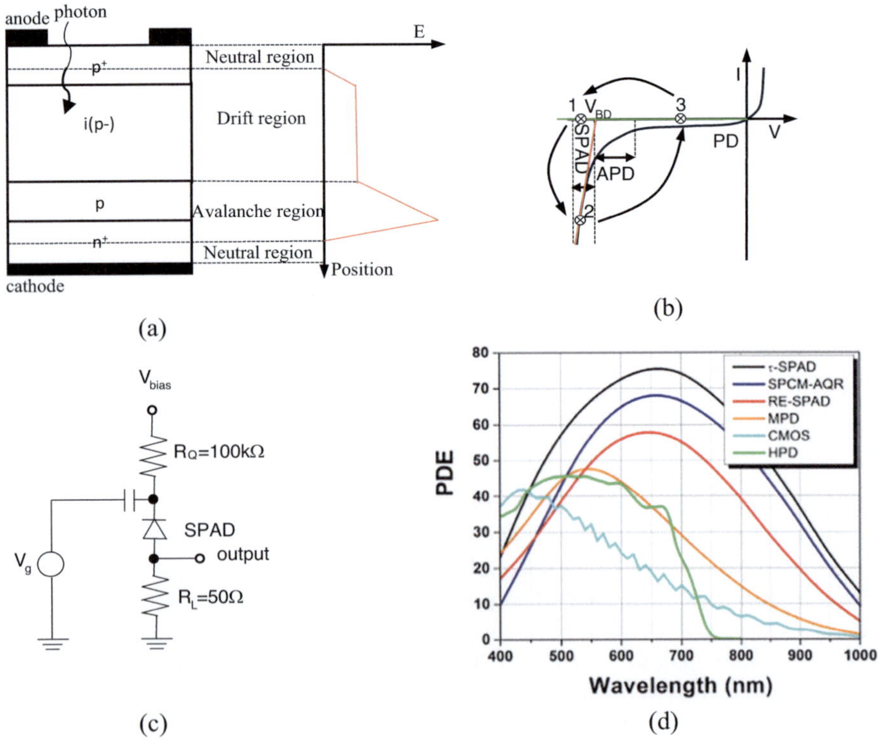

Fig. 2 **a** Schematic representation of a SPAD. **b** I–V characteristic of a SPAD showing the different working regions and operating points in a detection event. **c** Passive quenching circuit example. Example PDEs from different commercial devices fabricated with different technologies [19] (printed with author's permission)

installed in a water pool in the world's largest Cherenkov neutrino detector. Another area is medical imaging like Positron Emission Tomography (PET) [14, 15].

2.2 Single-Photon Avalanche Diode (SPAD)

The single-photon avalanche diode (SPAD) sometimes also called also called Geiger-mode avalanche photodiode (GM-APD) is a solid-state device that with its high gain is a direct concurrent of PMTs in photon detection. The device belongs to the same family as the simple p–n junction diode and shares behaviors of other elements of the family like photodiodes and avalanche photodiodes (APDs). The device has several advantages with respect to PMTs, like, smaller footprint, lower operating voltage and implementation compatible with standard complementary-metal-oxide semiconductor (CMOS) technology [16]. The SPAD fabrication in commercial CMOS processes enabled the development of imaging arrays [17] and a broad range of applications integrating in the same chip the photon detection and conversion and the signal processing. A recurrent approach consists in the combination of SPAD arrays with FPGA-based host boards, relying on the natively digital SPAD data output, this way real-time algorithms like autocorrelation and lifetime calculations, can be implemented near the sensor.

Figure 2a presents one possible structure of an SPAD while its typical I–V characteristic SPAD is sketched in Fig. 2b. As seen in the sketch, after equilibrium the electric field varies across the device, on top and bottom we have two undepleted regions where the field is null extending over some part of the heavily doped p and n layers. A low field drift region follows extending over a thicker low doping layer. This is the absorption layer where electron hole pairs generated by the incident photons drift towards the contacts. A thicker layer increases the probability of the photons being absorbed. The carriers that drift towards the cathode then find a high electric field region where they are accelerated creating other electron hole pairs in an avalanche effect. The operating process of an SPAD can be explained by the I-V characteristics in Fig. 2b. Initially a reverse bias higher than the breakdown voltage (Vbd) is applied (point 1). This biases the device in the Geiger or SPAD region where no free carriers are present in the depletion region and therefore no current is flowing. In this condition, the absorption of a single photon with the creation of an electron hole pair can initiate the impact-ionization process in the high electrical field region, giving rise to a high self-sustaining avalanche current (point 2). At this point, the detector is in an avalanche mode and any photons absorbed have little influence on the current, meaning that the device gets completely blind while this current is sustained. To interrupt the avalanche current, the biasing of the device must be decreased, bringing it back to the normal photodiode region (point 3), this can be done by an external quenching circuit. The quenching can be performed by a single high-value resistor (see Fig. 2c) or a more complex circuit including transistors, usually referred to as active quenching circuit (AQC) [18]. To make it ready for the next optical event the SPAD must be reset to its initial bias condition (point 1) with a reverse bias exceeding the breakdown voltage.

One important parameter of the SPAD is the Photon Detection Efficiency (PDE), that is simply the probability of a photon impinging on the active area of the detector give rise to an avalanche current that can be detected by the electronics. This topic was covered by research works in the 1970s [20] extending to the actuality with the presentation of extensive general models [21]. For an optical event to give rise to a current detected by the electronics in a SPAD several conditions have to be met: first, a photon must reach the active area of the device and be absorbed; then one of the photogenerated carriers must reach the high-field region and trigger the avalanche event. This way, the PDE can be expressed as the product of four different terms:

$$PDE = \int_{0}^{l} (1 - R)\alpha e^{-\alpha x} P_{coll}(x) P_{trig}(x) dx$$

where R is the reflection coefficient at the surface, α is the absorption coefficient of the light in the material, $P_{coll}(x)$ and $P_{trig}(x)$ are the collection and triggering probabilities, respectively, and x is the absorption depth. Some example PDE curves from different manufacturers can be found in Fig. 2d τ-SPAD (PicoQuantGmbH), SPCM-AQR (Excelitas Technologies), RE-SPAD (MPD). CMOS (MPD); HPD (Hamamatsu). More information about this subject can be found in the literature [22].

2.3 *Electron-Multiplying Charge-Coupled Devices (EM-CCD)*

Charge-Coupled Devices (CCDs) were invented in 1969 by Bell Labs, with the aim to replace the bulky and power eager vacuum tube imager. The main idea behind the CCD was the shift of charge across a semiconductor structure using a series of adjacent Metal Oxide Semiconductor (MOS) gates (see Fig. 3a). The minority carriers in the semiconductor (e.g., electrons), electrostatically trapped by a positive voltage applied to a gate, can be dragged along in the semiconductor by pulsing each gate in sequence. The electron signal charge can either be electrically injected, resulting in delay line or shift register, or locally created by light in the case of the CCD image sensor. The pixels in the CCD can be distributed over a single line or in a 2D array, when light is focused on the surface of the device, the locally produced charge is transferred across the structure in a line-column fashion controlled by a sequence of clock signals, exiting by a single charge amplifier (see Fig. 3b). This serialized voltage information of the light intensity impinging on each pixel can then be rearranged to reconstruct the original image. The spectral sensitivity of any pixel can be modified by the interposition of filters, image intensifiers or the use of special

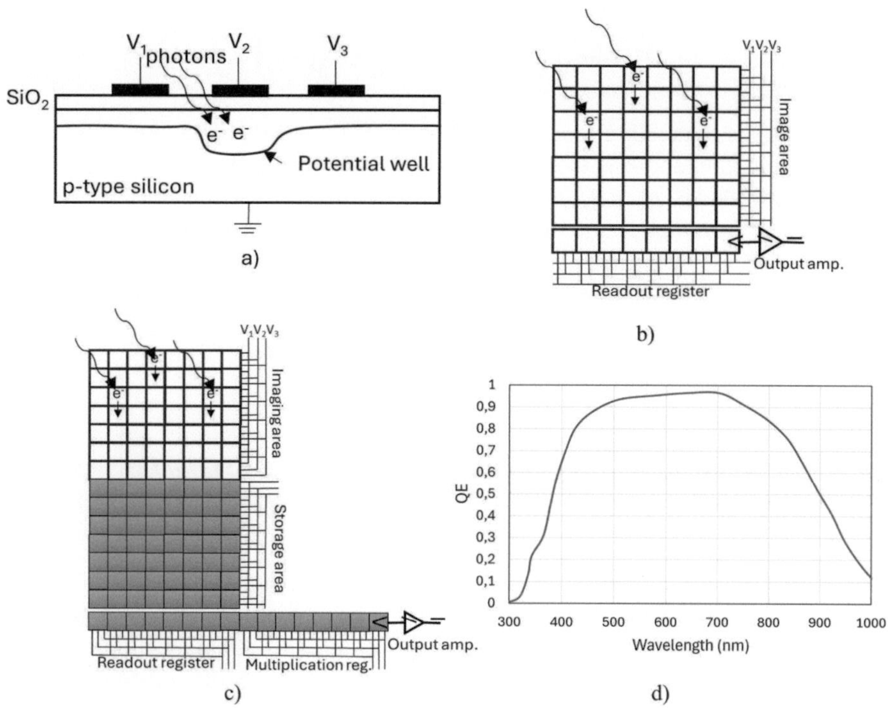

Fig. 3 **a** Detail of the charge shifting technique in a CCD, **b** architecture of a 2D CCD array with its main components, **c** architecture of the EM-CCD and **d** quantum efficiency of back illuminated EM-CCDs

photosensor structures. The quantum efficiency of CCDs is very high in the visible, extending to the UV and IR spectrum as can be seen in (see Fig. 3d).

The EM-CCD inherits the photon conversion and shifting technique from the CCD, adding a specialized electron multiplication register before the charge to voltage amplifier, enabling the detection of single-photon events without the need for external image intensifiers (see Fig. 3c). The use of CCDs to detect very low light intensities like a single photon is undermined by the fact that the noise contribution of the whole photodetection circuit is much larger than the one electron produced through the interaction of one photon with the semiconductor material. In order to increase the number of electrons produced by one photoelectron a multiplication effect can be obtained using avalanche breakdown on the semiconductor, if we have a large enough field the electrons will acquire enough energy to release secondary electrons. As one knows this effect is not easy to control, and the avalanche gain may vary, introducing a kind of multiplication noise. To mitigate this noise one can use several multiplication stages with a lower multiplication factor on each one, this is the main idea behind the EM-CCD [23].

The noise referred to the image area of the EM-CCD is given by:

$$\sigma_{eff} = \sqrt{\left(F^2(S + S_{dark}) + \frac{\sigma^2_{readout}}{M^2}\right)}$$

where $\sigma_{readout}$ is the readout noise, S is the signal, S_{dark} is the dark signal. M is the multiplication gain and F is the noise factor. As one can see in the equation, if the multiplication gain is set to a sufficiently high value the readout noise can be virtually eliminated, even at high frame rates where the readout noise increases. The application of EM-CCDs range from surveillance, where at low light levels they still give superior performance than GaAs, because of the higher inherent quantum efficiency of silicon, LIDAR [24], adaptive optics for astronomy and space applications [25] and even direct or indirect particle and synchrotron radiation detection.

2.4 Superconducting Nano-Wire Single-Photon Detector (SNSPD)

The SNSPD evolved starting from the first demonstration of single-photon detection, at a wavelength of 0.81 μm, with current-biased superconducting microbridges by G. Goltsman in 2001 [26]. This work follows other well established and already achieving high performances superconducting devices like the superconducting quantum interference devices (SQUIDs) [27] and transition edge sensors (TESs) [28]. The emerging interest in quantum communications pushed the development of the SNSPD with the first commercial system for quantum key distribution (QKD) made available in 2006 [29]. The continuous development of the technology and materials led to the possibility of integration with photonic waveguides presented in 2011 [30] enabling high on-chip detection efficiency and contributing to the development of the so called Photonic Integrated Circuit (PIC). Current state-of-the-art SNSPDs combine a very high (near-unity) detection efficiency over a large spectral range, low dark counts, short dead times, and picosecond time resolution [31].

The SNSPD consists of a thin (5–10 nm thickness) and narrow (50–100 nm width) superconducting nanowire patterned from a thin layer of material. A meander-like structure (see inset in Fig. 4c) with a length of the order of hundreds of micrometers is nanofabricated to create a square or circular pixel with high detection efficiency. The wire is then cooled to a temperature less than $0.5T_c$ and biased (see Fig. 4b)) close to the device's critical current, and low noise amplifiers and counting electronics are used to detect single-photon events outputting the correspondent voltage pulses.

The critical current is defined as the maximum current at which the nanowire can sustain superconductivity. A photon incident on the nanowire may be absorbed and break hundreds of Cooper-pairs because its energy is usually two to three orders magnitude higher than the binding energy of a Cooper-pair and reduces the local critical current below that of the bias current. This results in the formation of a localized non-superconducting region, or hotspot, (Fig. 4a-ii) which repels the supercurrent into the superconducting path around the hotspot. The increase in the current

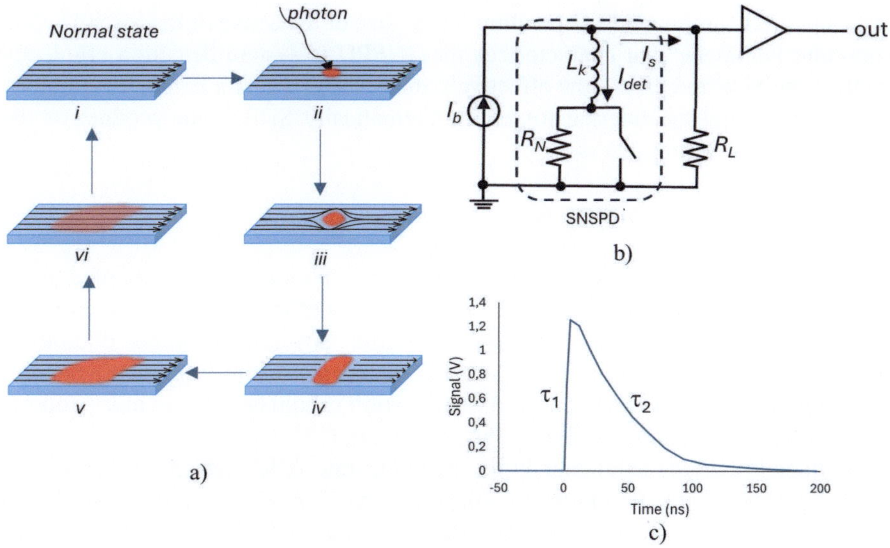

Fig. 4 **a** Schematic of the detection cycle; **b** a simplified circuit model of an SNSPD and readout circuit; and **c** the output signal of an SNSPD upon a detection event

on the edges of the hotspot may exceed the critical current creating an enlargement of the hotspot until it eventually reaches the edges, creating a resistive section across the nanowire (Fig. 4a-iii, iv). The resistance of this section (several hundred ohms) will then enlarge in the direction of the nanowire because of local heating caused by the Joule effect (Fig. 4a-v). This resistance becomes typically larger than the input impedance of the readout amplifier or load resistor, forcing most of the bias current to be shunted to the amplifier resulting in a peak of the output voltage (Fig. 4c). With the reduction of the current in the nanowire and thermal relaxation via the substrate, the resistive part of the nanowire will cool (Fig. 4a-vi); and, when original conditions are met again, will disappear (Fig. 4a-i). The SNSPD returns to the original state and can detect the next incoming photon.

Figure 4b presents the typical model of the SNSPD, the parallel of a switch and a resistor (R_N) in series with an inductor. The switch, when it opens, simulates the detection event triggered by a photon. R_N simulates the resistance of the resistive section formed in the nanowire. The series inductance represents the combined effect of a geometrical and kinetic inductance (L_k). With the latter generally being much higher for typical geometries thus representing the SNSPD inductance. The readout circuit may take the form of a simple load resistor that converts the current into a voltage which is then amplified. A current source biases the nanowire.

The typical shape of the pulse in the output of the amplifier is presented in Fig. 4c, the signal has a sharp rising edge with a time constant $\tau_1 = L_k/(R_L + R_N)$ followed by a slow decay with a time constant $\tau_2 = L_k/R_L$, being R_L the impedance of the readout circuit typically 50 Ω. τ_1 is in the order of a 1 ns and τ_2 from several nanoseconds to a few hundred nanoseconds, depending on the size of the active of the SNSPD. One important parameter that characterizes the SNSPD is system detection efficiency (SDE). The SDE indicates how effectively the SNSPD detector can detect photons while considering the internal losses. Mathematically SDE is the product of the efficiencies of several processes:

$$SDE = \eta_{coupling} \cdot \eta_{absorption} \cdot \eta_{intrinsic}$$

where $\eta_{coupling}$ represents the fraction of the incident photons that are coupled to the active area of the detector, $\eta_{absorption}$ is the absorption efficiency of the detector area coupled photons, and $\eta_{intrinsic}$ is the triggering efficiency of the absorbed photons to generate a detectable electric signal. As the SDE depends on the different efficiencies to be close to the ideal value of 1, every calculation should be close to unity; reports in the literature present an SDE of 98% at 1550 nm [32].

Another important parameter is the counting rate (CR), which describes how fast the detector can react to the incoming photons. The maximum counting rate is physically limited to the order of 10 GHz by the thermal relaxation time between the nanowire and the substrate after the absorption of a photon, which are usually some tens of ps. The limiting factor of the CR is the large kinetic inductance of the SNSPD, which slows the process of the nanowire current's recovery to the initial value after hotspot generation. The CR is then given by reciprocal of the total time delay:

$$CR = \frac{1}{\tau_1 + \tau_2} \approx \frac{R_L}{L_k}$$

Nevertheless, after an event the current recovery on SNSPD is a gradual process, and the SDE also recovers gradually. Thus, the SDE does not completely recover when the next photon arrives at a repetition rate of CR., which makes the number of counts usually smaller than the product CR x SDE. But this also means that the SNSPD may count the incoming photons with a higher repetition rate than CR if we accept a lower efficiency. One way of increasing the CR is by reducing L_k, which can be done simply by a single nanowire into multiple nanowires. A few different structures, with arrays (Wollman et al. 2019) and other similar structures were presented, attaining a CR of 1.5 GHz with an SDE of $\sim$ 12% at photon flux of 1.26×10^{10} photons/s [33].

Since the invention of SNSPD in 2001, it has seen applications in a broad range of fields, starting from the first application in the diagnostic of VLSI CMOS circuits in 2013 [34]. Then followed several applications in different fields such as Quantum Infrastructure, biological fluorescence deep-space communication [35], and light detection and ranging (LIDAR), including satellite laser ranging. The applications in QI are expected to increase because of the actual momentum gathered by quantum communication, quantum computation and the characteristics of the SNSPD.

2.5 *Micro-Channel Plate (MCP)*

The MCP in its simplest form is a thin disk consisting of millions of micro glass tubes (channels) (6–25 μm) fused in parallel with each other. Each channel acts as an independent electron multiplier. The MCP offers much faster time response than the other discrete dynodes. It also features good immunity from magnetic fields and two-dimensional detection ability when using an array of anodes.

Detectors with MCPs are widely used in applications where photon/particle counting with high spatial (~ 10 μm) and temporal resolution (~ 10 ps) with low readout noise and high dynamic range is required. The incoming photon/particle is first converted into an electron by a photocathode. The number of ejected electrons is then amplified by secondary emission inside each channel up to ~ 10^8 electrons by a single MCP or a stack of several MCPs, depending upon the voltage across the tube and its length/diameter ratio. The typical biasing voltage is usually ~ 1 kV per microchannel plate. (See Fig. 5). The technique used for the conversion into an initial electron depends on the type of radiation/particles we want to detect. Photocathodes are typically used for the detection of soft X-rays, UV and visible light photons, while alpha particles, ions and electrons directly produce secondary electrons upon interaction with the input surface of the MCP, usually coated with a thin nichrome electrode. Bulk glass can even be doped with heavy elements like Gd in order to produce secondary electrons from neutron impacts. When the number of electrons of a pulse in a channel reaches a certain level the channel saturates, which limits the gain, but it reduces gain fluctuations appreciably which can be seen as an advantage. This effect is useful when detecting very low current pulses in photon counting. When compared to discrete dynode PMTs, MCPs have several advantages like:

- Faster response. As the distance the electrons travel inside the channel is short (1 to 2 mm) and the high electric field is high (1–20 kV/cm), the transit time is extremely short. The transit-time fluctuations are also small due to the aspect ratio of the channel. The response pulse duration is shorter (FWHM ≈ 0.2 ns) and the time resolution δt ($<$ 100 ps) is better;
- Insensitivity to magnetic fields. Because of the short travel, the high electric field, and the constraint imposed by the microchannels, the electron paths are marginally affected by magnetic fields up to 0.1 T;

Fig. 5 Typical structure of a MCP

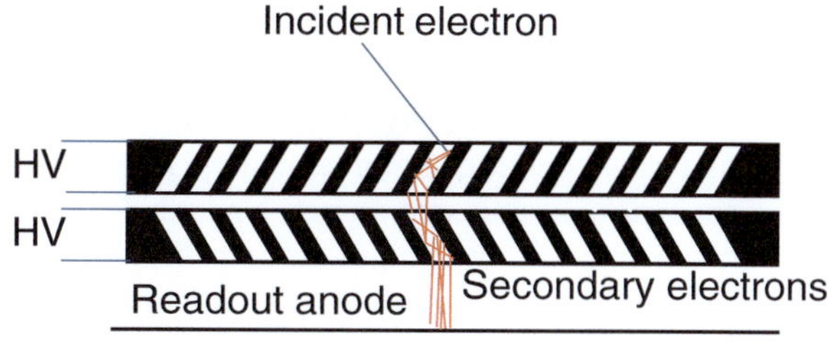

- Spatial resolution. The honeycomb structure of a microchannel plate makes it possible to convert an optical image on the cathode into a corresponding electron image on the anode by simple proximity focusing. This is the principle used in the image intensifier tubes by replacing the anode with a phosphorous screen. Spatial information can be recovered by using a pixeled anode.

Nevertheless, microchannel-plate photomultipliers also have some limitations that can make them inadequate for some applications. The most important are:

- Limitations on count rate and linearity in the pulse mode, due to the limitation on the average current at the output to a few $\mu A/cm^2$ and the long recovery time of the channel.
- Limitations of pulse charge capacity per channel (5×10^{-14} to 5×10^{-13} C) due to the small size of the channel. This limits the maximum gain to about 10^6 and also the current that can be delivered per pulse, which causes a lengthening in the shape of the pulse.
- Limitations of useful life. The large total surface of a microchannel plate makes it difficult to outgas the channels completely, even at high temperature. When operating the internal electron bombardment then generates ions that bombard the cathode and quickly reduce its sensitivity. This results in a useful life much smaller than that of a conventional photomultiplier.

MCPs are an essential block of the Medipix/Timepix chips [36] with possible applications in areas like dosimetry, education, microscopy, particle tracking, photon detection, medical imaging, etc. A long time application field is the image intensifier equipment [37], and a variety of photon counting applications that requires high timing and spatial resolution, such as Ion time-of-flight (TOF), electron spectroscopies, analytical and space instruments, and MCP-based photomultipliers such as the Large-Area Picosecond Photodetector [38].

2.6 Transition Edge Sensor (TES)

The Transition Edge Sensor (TES) is a highly sensitive cryogenic sensor capable of detecting very small amounts of energy. It is simply a resistive film that has superconducting properties at low temperature operated close to its transition temperature Tc. In this situation a very small change in temperature causes a large change in resistance which is externally sensed by a SQUID sensor coupled to a coil in series with the TES. The first experiments about the measurement potential of the superconductor transition were performed by D. H. Andrews in the 1940s with the demonstration of the first transition-edge bolometer simply composed by a current biased tantalum wire sensitive to an infrared signal. Nevertheless, the difficulties in stabilizing the temperature in a narrow margin around the superconducting transition region and the difficulty of signal readout from a low impedance source, kept the device form further developments for about 50 years. Another problem was the self-heating of

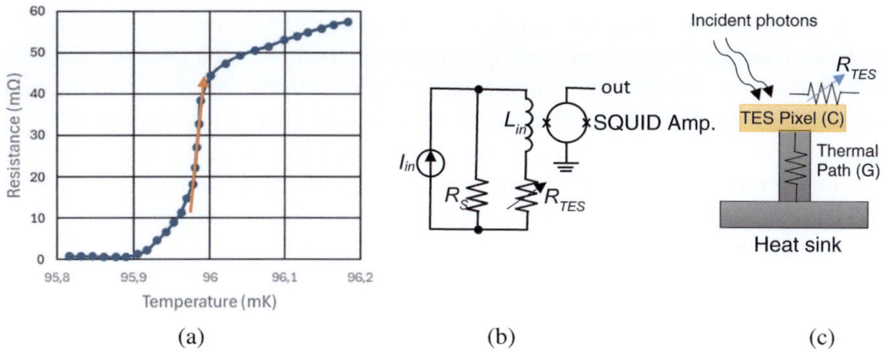

Fig. 6 **a** The transition of a superconducting film (Mo and Cu bilayer), **b** readout circuit For the TES device, **c** sketch of a typical TES pixel

the device due to the Joule effect that could easily lead the TES in a resistive state due to thermal runaway. With the solution of the thermal runaway problem in 1995 by K. D. Irwin by using voltage bias in the TES [39] and using Superconducting Quantum Interference Devices (SQUID) for signal detection, the TES detectors finally gained widespread interest [40].

As stated before, the detection principle of the TES is quite simple, observing Fig. 6a, representing the variation of the resistance with temperature for a bilayer of molybdenum and copper in the vicinity of the transition temperature, $T_c = 96$ mK, a very small variation in temperature can change its resistance from 0 to it normal value $R_n \approx 58$ mΩ. Using this steep change in the resistance to the temperature, a very sensitive thermometer can be created.

The typical readout circuit is depicted in Fig. 6b. A current source I_{in} biased the parallel of the TES (R_{TES}) with a shunt resistor R_S with a resistance much smaller than the TES. As most of the current now flows through the shunt resistor, the TES is voltage biased and the thermal runaway is avoided, as is easily demonstrated. The Joule power dissipation in the TES is: $P_J = V^2/R(T)$. When the signal power increases, its temperature also rises up to the R_n value which makes the Joule power decrease, lowering the temperature. This negative feedback loop stabilizes the temperature. The coil L_{in} is part of the TES current measurement setup being magnetically coupled to the SQUID amplifier. The electrical time constant of the system is set by the shunt resistance value in conjunction with the device and bias circuit inductance. Figure 6c shows the typical TES pixel structure. The sensor contains three basic elements: an absorber with a defined heat capacity (C), a thermometer to measure the temperature variation (TES), and a thermal link to a heat sink to dissipate the absorbed energy (G). The absorber can be simply a metal film like tungsten for operation in the visible, NIR and UV regimes, or multilayer optical cavity that works as a trap for photons achieving a high detection efficiency (~95%). In order to increase the sensitivity, the absorber should have low heat capacity with respect to the TES, this way a given absorbed energy will produce a larger change in the TES resistance. As for the thermometer, for maximal sensitivity, the TES should have low heat capacity and a narrow transition. A Tungsten thin film is a popular choice, but bilayer and multilayer structures of different materials can be used to adjust the desired T_c value. The thermal conductance (G) between the absorber and the heat sink also plays an important role in the device characteristics. A low thermal conductance is necessary to ensure that incident energy is not directly dissipated to the heat sink, on the other hand after the energy has been absorbed it is necessary to cool the TES back to below the transition temperature, so a higher thermal conductance will speed up the recovery.

The figure of merit for TES and other microcalorimeter resistive thermometers is called α, given by:

$$\alpha = \frac{T}{R}\frac{dR}{dT}$$

where T is the temperature and R is the resistance of the TES. The energy resolution of a TES can be approximated by:

$$\Delta E \approx 2.35\sqrt{4kT^2\frac{C}{\alpha}}$$

where k is Boltzmann's constant, T is the TES temperature, C is the microcalorimeter's heat capacity, α was defined previously. So, for better energy resolution, the devices should be operated at the lower temperature possible, and have a low heat capacity and a high α.

The main application field of TES bolometers is in the astronomy either in ground, airborne or space platforms [41, 42]. Extensive applications are found also in nuclear and particle physics [43, 44]. Quantum optics and quantum communication are emerging areas where TES bolometers also attracted the attention of researchers [45, 46]. TESs can be developed to work as excellent Soft X-ray detectors for many applications at beam-line facilities [47]. Properties like near unity quantum efficiency, a very good energy resolution, in the order of a few eV at 1–10 keV and wide bandwidth makes this sensors interesting for many applications such as X-ray Emission and Absorption Spectroscopy (XES and XAS) and Resonant Inelastic X-ray Scattering (RIXS) [48].

2.7 *Quanta Image Sensors (QIS)*

Quantum image sensors leverage the principles of quantum mechanics to achieve unprecedented levels of sensitivity and precision in imaging applications. QISs have been implemented using different electron devices, including impact ionization-gain devices, like SPADs, and low-capacitance, high conversion-gain devices, such as modified CMOS image sensors (CIS) with very low noise readout signal chains. The first references appear in 2004 as a forward look at where image sensors could evolve

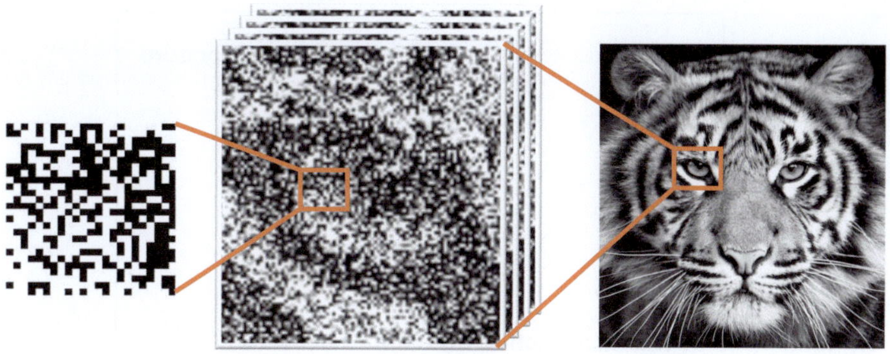

Fig. 7 Simulation of **a** raw jot data; **b** same at lower magnification; **c** after processing cubicles to form grey scale image

in the next decades with the possibility of surpassing the sub-diffraction limit [49]. The QIS concept consists of an array of specialized pixels referred to as jots that are binary in nature, indicating the arrival of a single photoelectron, or not. Originally it was envisioned to consist of millions of small-pitch, low Full Well Capacity (FWC), ideally one photoelectron, jots readout at high frame rates, and thus very high bit rates. Although each pixel is binary, a sequence of the one-bit frames contains the information needed to reconstruct the image see Fig. 7.

Following the concept, image pixels can be created from the sum of a small x–y-t "cubicle" of bits from the jot data cube (see Fig. 7b). The dimensions of the cubicle will determine the spatial and temporal resolution of the output image, being all jot values within the cubicle weighted equally. The maximum signal-to-noise ratio (SNR) of the image pixel (assuming that all data was taken using the same illumination and readout conditions) is determined by the volume of the cubicle. For example, a cubicle of size $16 \times 16 \times 16$ of single-bit QIS jots summed together would have a maximum value of 4096 being the maximum SNR $\sqrt{4096} = 64$.

The image formation on the QIS has a statistical nature depending on the arrival of photons and the photoelectrons they produce that are well described by Poisson arrival statistics. If we define the quanta exposure H as the average number of photoelectrons collected by a jot over some integration period, the bit density D of jots (fraction per bit ensemble that have logic value "1") gives rise to an S-shaped curve if D is plotted as a function of log(H), reflecting a relationship similar to the one found by Hurter and Driffield in 1890, given by:

$$D = 1 - e^{-H}$$

This characteristic QIS-response curve is shown in Fig. 8. This behavior is also similar to the behavior of an ensemble of avalanche detectors in a Si photomultiplier and also observed in SPADs and in subsequent analysis of binary sensors [50].

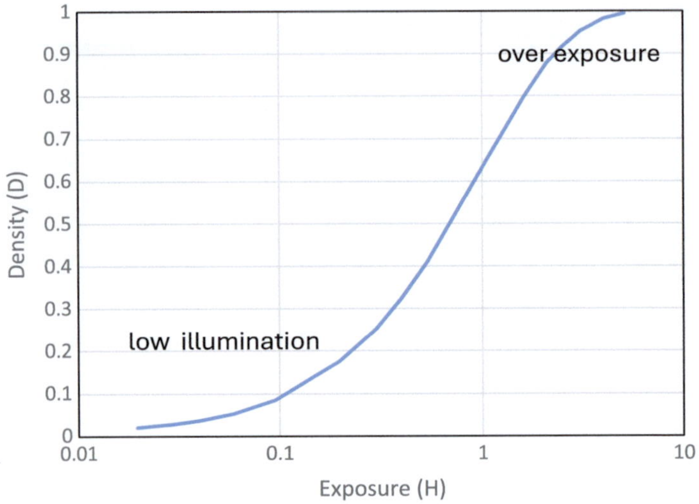

Fig. 8 Single-bit QIS bit density as a function of quanta exposure, in D-log(H) format

The exposure-referred (SNR_H) is more meaningful for non-linear devices than for the readout of a single-bit QIS and is given by:

$$SNR_H = \sqrt{M}\frac{H}{\sqrt{e^H - 1}}$$

where M is the number of jots in the ensemble or cubicle used for the read signal sum. For normal readout, the SNR_H reaches a maximum value at $H \cong 1.6$ and the maximum value of SNR_H is approximately $0.8\sqrt{M}$. For multi-bit QIS, maximum SNR_H grows as $\sim \sqrt{FWC}$. The QIS being able to detect and count individual photons with extreme precision and high-speed operation with many applications envisaged in areas requiring accurate photon counting.

3 Optical Sensing Structures

3.1 Optical Fibers

The refractive index is an intrinsic property of materials and characterizes the propagation speed of EM radiation when traversing them. It also impacts both the speed and direction of the transmitted beam, as it propagates through different media, and is of fundamental importance in optics, physics and engineering. The refractive index is mathematically defined as:

$$n = \frac{c}{v}$$

where n is the refractive index, c is the propagation speed of EM radiation in vacuum and v is the propagation speed of EM radiation (phase velocity) within a given material. Furthermore, the propagation of EM radiation through media of different refractive indexes experiences refraction and this phenomenon is governed by Snell's law:

$$n_1 \sin \theta_1 = n_2 \sin \theta_2$$

In previous equation, n_1 and n_2 are the refractive indexes of the two different media, and θ_1 and θ_2 are, respectively, the incident and refracted angles w.r.t. the normal at the point of incidence. Radiation coming from n_1 towards n_2 at an angle, will tend to move away from the normal as $n_2 > n_1$, and closer as $n_1 > n_2$.

Moreover, the refractive index is wavelength dependent, for it is not constant throughout the EM spectrum. This phenomenon is designated dispersion and may be confirmed when white light goes through a prism or a fine grating, and we observe the separation into its constituent colors on the transmitted radiation. Also, despite being able to propagate over large distances, EM radiation does not travel indefinitely. There are losses or absorption (mainly in metals) associated with the propagation, which occur due to the complex nature of the refractive index. As such, a more comprehensive and realistic description of the refractive index will be:

$$n = n_r + ik$$

where n_r is the real part (associated with refraction) and ik is the imaginary part (associated with loss/absorption) of the complex refractive index.

All dielectric waveguides are built in a way that a material, with a higher refractive index than the surrounding medium, allows EM wave propagation based on the total internal reflection phenomena. In this case, the central medium acts as a light "trap" within which optical rays remain confined by multiple internal reflections at the boundaries; a propagating optical wave travels through the waveguide. This behavior, together with the medium's low loss characteristics, are then exploited to make light conduits, enabling light transportation from one location to another (sometimes hundreds or even thousands of kilometers apart from each other) with low propagation losses.

Optical fibers are intrinsically flexible waveguides made from glass or plastic and able to propagate EM energy from one point to another. This technology has revolutionized data communication over long distances and is the cornerstone of contemporaneous Information and Communications Technologies (ICT). These waveguides are essentially formed by two components:

- Core—the central part of the waveguide with a higher refractive index than the outer material and where most of the propagating radiation is confined due to successive bouncing off on the interface between media;
- Cladding—the surrounding material of the embedded core which refractive index is lower than the core's material, hence favoring propagation of the EM radiation through total internal reflection.

There are several different types of optical fibers, but the ones often used in most applications may be classified in two different types:—Step index and graded index optical fibers. The former may be subsequently divided into single or multimode optical fibers and the latter are usually relevant in multimode applications. Step index optical fibers are the commonly used waveguiding technology, where the single mode type has a thinner core diameter (~ 8 to 10 μm) [51], hence able to only propagate the first mode, and finding widespread use in long distance (several tens of km) sensing and communications. Step index multimode optical fibers, with a wider core diameter (~ 50 to 62.5 μm) [52], are able to propagate simultaneously multiple modes of EM radiation, though for shorter distances (units of km) when compared to single mode optical fibers. Graded index optical fibers are essentially multimode and are characterized by having the core's refractive index engineered in a way to ensure that all different propagating modes reach the end point simultaneously. Figure 9 presents a diagram of the three different optical fibers, with details describing the propagation of EM radiation through each of them.

3.2 Fiber Bragg Gratings

Fiber Bragg Grating (FBG) consists of a section of an optical fiber where the core's refractive index has been periodically modulated. These periodic variations of the refractive index in the core of the optical fiber are highly selective and are able to reflect some wavelengths of the incoming signal, while transmitting the rest of the signal's bandwidth components. This highly selective wavelength filtering finds use in telecommunications, laser and sensing applications.

These FBGs base their operation on Bragg's law, which states that EM radiation of a particular wavelength is reflected due to constructive interference and the other wavelength components are transmitted through. The reflected wavelength is designated the Bragg wavelength and is given by [53]:

$$\lambda_B = 2n_{eff}\Lambda$$

where λ_B is the Bragg wavelength, n_{eff} is the modal index of the propagating mode and Λ is the periodic modulation of the refractive index or grating period. Figure 10 shows a diagram that represents a step index optical fiber with an embedded FBG in its core and operating as a single wavelength filter ($\lambda 3$).

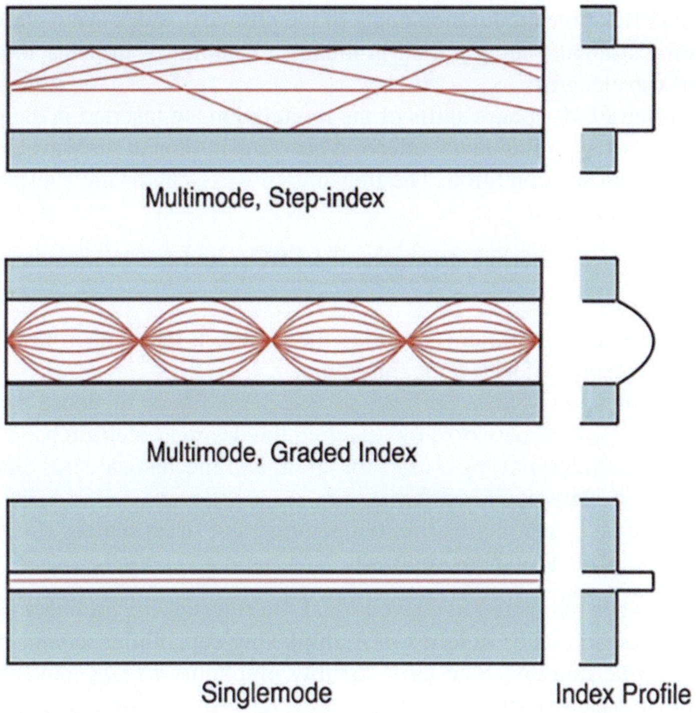

Fig. 9 Diagram of the optical fibers; top: multimode step index; middle: multimode graded index; bottom: single mode step index (printed with author's permission from https://www.thefoa.org/tech/ref/basic/fiber.html)

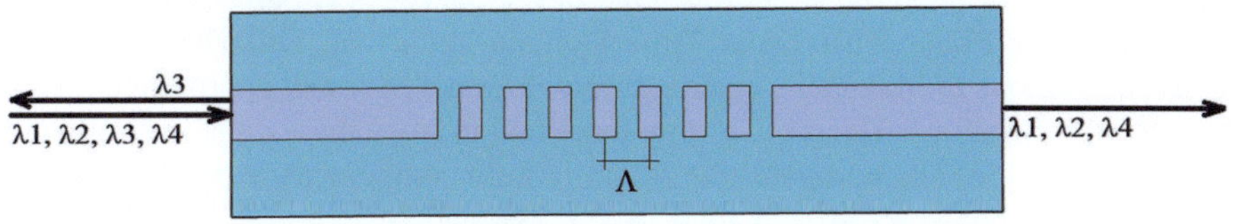

Fig. 10 Representation of an optical fiber with the core's refractive index modulated by a rectangular grating and operating as a wavelength filter

There are different types of FBGs and they are classified in terms of the characteristics of their gratings:

- Uniform FBG—consists of a grating with a uniform modulation and reflecting a particular wavelength;
- Tilted FBG—the modulation of the grating planes are tilted, which facilitates coupling between the core propagating and cladding modes;
- Chirped FBG—the grating modulation of the core does not follow a constant spacing in length, originating a wider bandwidth reflected spectrum;

- Apodised FBG—the modulation depth of the refractive index is not constant over the grating's length, being varied gradually to minimize the side lobes of the reflected wavelengths;
- Phase-shifted FBG—phase shifts of the modulation are inserted in the grating at given locations, resulting in no reflection but transmission of the wavelengths that satisfy the imposed condition. The transmitted wavelengths are subtracted from the reflected spectrum.

These wavelength filtering structures, the FBGs, find use in many applications related to:

- Lasers and optical amplifiers—used as the feedback mechanism in fiber based lasers and in erbium-doped fiber amplifiers;
- Telecommunications—they perform as excellent filters in dense wavelength division multiplexing networks by selecting the desired spectrum band;
- Structural health monitoring—used for strain and mechanical stress monitoring in buildings, bridges and aircrafts;
- Optical sensing—used in pressure, temperature and strain sensing, for their high sensitivity to the external environment.

Moreover, when compared to conventional filtering/sensing architectures, FBGs show superior performance in terms of multiplexing capabilities, compactness and weight, high precision and long-term stability, immunity to EM interference and endurance to harsh environments. Nevertheless, fabrication cost and complexity are still high and present some challenges, namely in unusual grating design.

To illustrate the operation of an FBG and its filtering capabilities, as close as possible to physical reality, a passive photonic simulation platform has been used. RSoft [54] is a software platform that implements several packages for the analysis and simulation of photonic structures. One of the packages is FullWAVE [55] that implements the Finite Differences Time Domain algorithm. FullWAVE provides rigorous solutions to Maxwell's equations and does not have any approximations or theoretical restrictions. The most common method to solve these equations is based on Yee's mesh [56] and computes the $\vec{E}$ and $\vec{H}$ field components at points on a grid and where grid points are spaced apart by Δx, Δy, and Δz.

This approach is a direct solution method for Maxwell's time-dependent curl equations. It is based on the volumetric sampling over a period of time, of the unknown electric and magnetic near-fields distribution in the structure and the surrounding environment of the simulated workspace. Space sampling is at subwavelength resolution, and it is the user's responsibility to set this parameter, in order to properly sample the highest near-field spatial frequencies that are thought to be relevant to the physics of the problem. Typically, the sampling should be set at 10 or 20 samples per the lowest wavelength to ensure the numerical stability of the algorithm.

Resuming, the finite differences time domain method is a marching-in-time procedure that simulates the continuous electromagnetic waves in a finite spatial region, through sampled numerical data from the wave and being iterated in a computer data space. Time-stepping continues as the numerical data evolves in the space grid and

Table 1 Characteristics configuration for simulations

Operating wavelength	1310 nm
Input signal bandwidth	43 nm
Core refractive index	1.45213
Core diameter	8.5 μm
Mode field diameter	9.2 μm
Cladding refractive index	1.44692
Grating periods	500

causally links the physics of the simulated region. By introducing no restrictions or approximations, this method tends to be an intensive time and computer resources consumer as the dimensions of the simulated workspace increase (namely, in 3D projects) (Taflove and Hagness 1995).

Our intent is to demonstrate the filtering wavelength capabilities of the FBGs and how this may be translated to a quantification of mechanical stress, strain and temperature. The usual length of the grating that modulates the refractive index in FBGs varies from 1 to 10 mm [57], however we considered a shorter length in our simulations due to the highly demanding computational resources of the finite differences time domain algorithm. For the simulation we selected the optical characteristics of an often used single mode optical fiber (SMF-28), operating within the zero absorption O band, namely the 1310 nm wavelength. Table 1 describes the characteristics configuration considered for the finite differences time domain simulations:

The simulation considers a Gaussian input field with ~ 43 nm bandwidth that is launched within the core of an SMF-28 optical fiber and propagates along its length. The optical fiber has three sections where the core's refractive index is modulated with a sinusoidal variation amplitude reaching the core and cladding's refractive indexes. Figure 11 depicts the first 5 μm of the periodic modulation of the refractive index embedded in one of the three sections.

Next, we simulated the configuration described in Table 1 when considering an FBG structure formed by three grating sections, each separated by 10 μm of unmodulated fiber core from the following one. Each of these grating sections has been calculated to filter off a thin wavelength band from the incoming spectrum and they are all 500 periods long. Real FBGs are much longer than 500 periods (usually ~ 10,000 periods at this operating wavelength), only the finite differences time domain algorithm is highly demanding with computer resources and if we were to use 10,000 periods in each of the three grating sections, simulation time would become prohibitively long. Anyhow, using 500 periods for each grating section is enough to clearly demonstrate the intended functionality, as we may undoubtedly observe in Fig. 12 and where we readily identify three peaks with approximately 3 nm bandwidth, each reflected by the corresponding grating section.

Figure 12 demonstrates outstanding wavelength filtering precision, with excellent selectivity and suitable to be used in dense wavelength division multiplexing networks. In sensing applications, previous characteristics are of utmost importance,

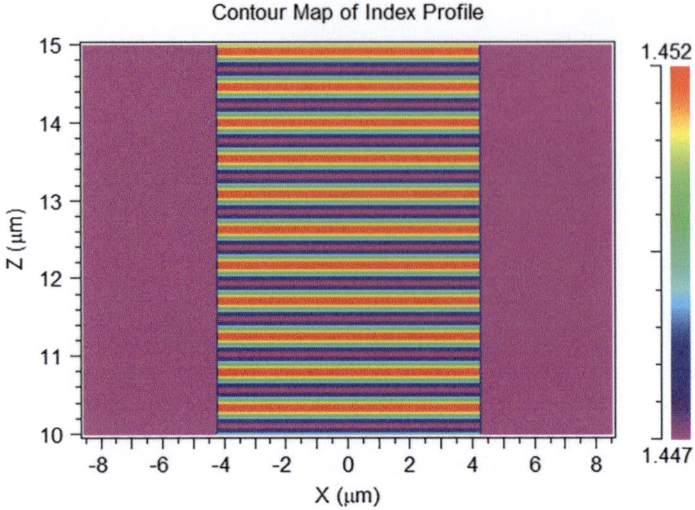

Fig. 11 Representation of the first 5 μm of the grating with a periodic variation of the refractive index due to the embedded sinusoidal modulation

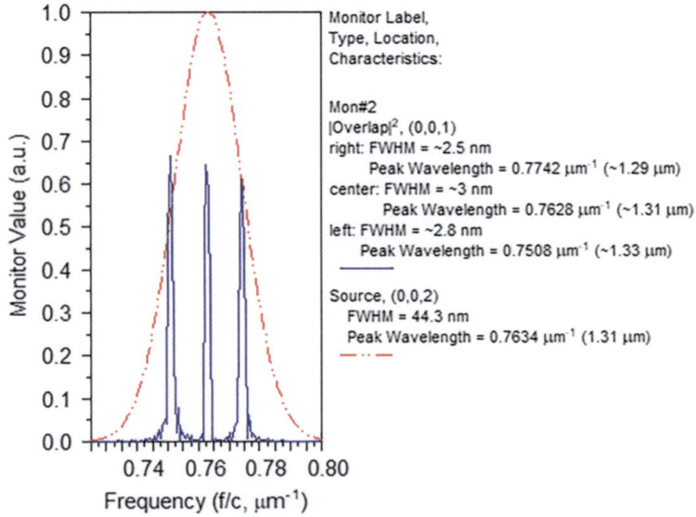

Fig. 12 Results obtained by the simulation of an FBG with 3 grating sections, each 500 periods long

but high sensitivity is definitely a paramount requirement. To this end, by assuming a perturbation of the system (e. g. mechanical stress, strain or temperature variation) corresponding to an elongation of ~ 5 nm in a given grating period, due to mechanical stress or strain (similar effect may also be achieved by change in the

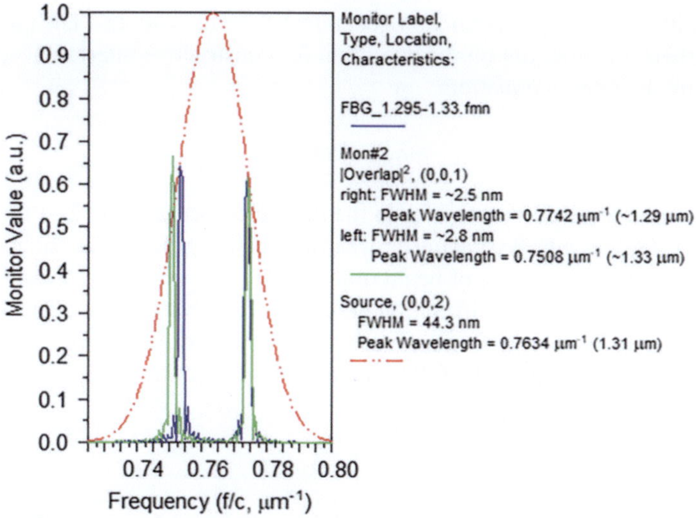

Fig. 13 Results obtained of propagation superimposed over both, a 5 nm elongated grating and the unperturbed structure

refractive index due to temperature variation—see Bragg's wavelength equation), we conducted further simulations in these conditions to evaluate the response of the structure. Figure 13 depicts the results obtained for the simulated conditions, where we have superimposed the results of propagation through an FBG with two grating sections of 500 periods each and where the green line represents the unperturbed structure and the blue line the 5 nm elongated grating. As one may observe, these results present a highly sensitive response to minute variations of the propagation conditions, which are contemporaneously exploited by advanced optical sensors.

3.3 Diffraction Gratings

An optical component which presents a regular structure and splits incoming EM radiation into its different constituent wavelengths, has been designated a diffraction grating. These diffracted wavelengths form a series of beams which directions depend on the spacing of the grating's pattern and the wavelength of incident energy. This diffraction property finds applications in optical communications, lasers and spectroscopy.

The operating principle of diffraction gratings is supported by the physical principle of constructive and destructive interference of EM waves. When a beam of EM energy strikes a structure formed by closely spaced grooves, each pair groove/trench behave as an additional source of waves (the Huygens-Fresnel principle [53]) which interfere with the incoming beam and generate diffraction patterns. Diffraction gratings base their functionality on the condition for constructive interference, which is governed by the grating equation:

$$m\lambda = d \sin \theta_m$$

where m is an integer corresponding to the diffracted beam order, λ is the incident wavelength, d represents the spacing between alternating grooves or trenches (period) and θ_m is the diffraction angle of beam order m. Another important parameter when designing high performance diffraction gratings is Resolving Power. The Resolving Power (R) of a diffraction grating represents its ability to differentiate closely spaced wavelengths and is given by the following equation:

$$R = \frac{\lambda}{\Delta\lambda}$$

where R is the Resolving Power, λ is the wavelength being diffracted and $\Delta\lambda$ represents the minimum wavelength range that can be resolved. Moreover, the Resolving Power varies linearly with the diffraction order and the number of illuminated grooves or trenches of the grating:

$$R = mN$$

with m and N being, respectively the diffracted order and the number of grooves/trenches illuminated. For instance, a given diffraction grating with 1000 groves or trenches ($N = 1000$) and operating at the 650 nm wavelength in the third diffraction order ($m = 3$), would be able to differentiate spectral lines 0.217 nm apart:

- $R = 3000$
- $\Delta\lambda = \frac{650}{3000} = 0.217\,\text{nm}$

Diffraction gratings are, essentially, classified into two main types:—Transmission and reflection gratings. In the former, EM radiation goes through small and closely spaced transparent apertures and deviates from the direction of the incoming beam due to diffraction. For the latter, incoming radiation is reflected from a closely spaced structured surface at an angle that differs from specular reflection.

To demonstrate the physical phenomenon underlying diffraction gratings functionality we selected two examples. The first one consists of using these structures to eliminate reflection from a particular surface, at a given operating wavelength. The second example considers a checkerboard structure to divide the incoming radiation into a series of well-defined and spatially separated beams of diffracted energy.

For the anti-reflection surface we selected a substrate with an imposed diffraction grating, calculated for the intended functionality. The distribution of the refractive index distribution and the propagation of the electric field (y-component) in the structure are depicted in Fig. 14, where it is assumed an incident plane wave normal to an infinitely wide substrate and diffraction grating:

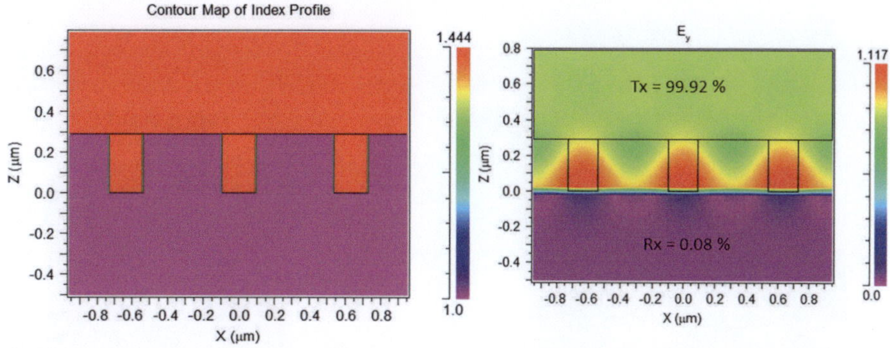

Fig. 14 Antireflection structure; left: Refractive index distribution; right: Electric field propagation in the structure

This structure consists of an antireflection surface using its zero order diffraction to reduce the reflection at the optical interfaces defined between air and fused silica. The substrate material is fused silica and the surface has an inscribed grating with a period of ~ 0.634 μm. The grating depth is ~ 0.29 μm with a filling factor of 0.3 and the operating wavelength is 1310 nm.

As can be observed, this structure is able to minimize reflection from an incident plane wave, at normal incidence and operating wavelength. The first diffraction order of the incoming wave is almost completely transmitted (99.92%) and reflection is almost inexistent (0.08%).

Regarding the second example related to diffraction, we considered a 3D checkerboard pattern to divide the incoming beam of radiation into a series of diffracted beams, which may be used in structured light research [58]. Figure 15 depicts the implemented structure, which is considered to be repeating itself infinitely along the X and Y cartesian coordinates, and the gaussian beam that is launched from the substrate upwards (directed at Z+). The refractive index of the checkerboard structure was set to 1.5, the surrounding environment to air, the working wavelength of the input field to 500 nm (visible range of the EM spectrum) and the $1/e$ width to ~ 10 μm. The structure is a raised square groove grating over a flat substrate of the same material, as can be easily perceived given the image perspective.

The simulation was conducted considering above mentioned parameters (modeling real life physical conditions) and the far-field components of the input field, after propagation through the 3D checkerboard structure, were obtained and analyzed. The results obtained characterize the far-field and are presented in Fig. 16, where we are able to identify both the transmitted and reflected diffraction orders.

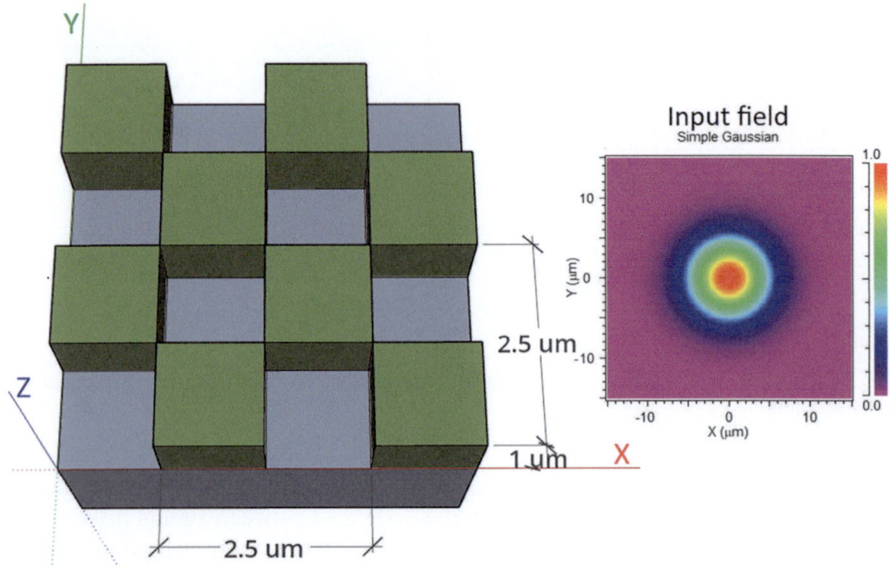

Fig. 15 3D checkerboard structure (left) and the gaussian input field (right)

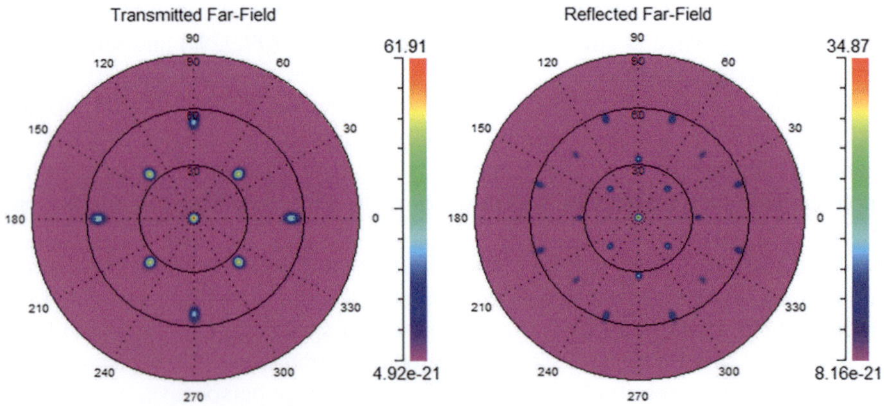

Fig. 16 Far-field transmitted (left) and reflected (right) diffracted orders after propagation through a 3D checkerboard structure

3.4 Optical Interference

When two or more optical waves are at the same location in space and time they superimpose and form a resultant wave, which might have lower, equal or greater amplitude than any of the constituent waves. This is a widely observed phenomenon in many areas of scientific research, covering acoustics, optics and quantum mechanics. The understanding of interference is of crucial importance for the advancement of several scientific and technological areas, namely telecommunications, fundamental physics, imaging and sensing [53].

Interference may be, essentially, grouped into two main classes:

- Destructive interference—this phenomenon takes place when out of phase waves interfere with each other, partial or totally cancelling the resultant wave;
- Constructive interference—when waves superimpose in phase, each component wave contributes to the resultant wave with the sum of individual amplitudes.

The mathematical representation of the resultant amplitude of two waves of the same amplitude and propagation constant, sharing the same space and time location, may be expressed by the following equation:

$$A_r = 2A\cos\left(\frac{\Delta\phi}{2}\right)$$

where A is the individual waves' amplitude and $\Delta\phi$ is the phase difference between waves.

Interference may be found in a wide range of applications in science and technology. Devices such as interferometers (Michelson and Mach-Zender) which are able to measure minute spatial displacements or variations of the refractive index [59], thin film interference resulting in anti-reflective coatings [60], noise cancellation in acoustics to reduce unwanted sound [61] and quantum interference with the demonstration of the wave-particle duality in light and matter [62] through the double-slit experiment.

Interference has widespread applications across several disciplines, continuing to drive scientific and technological innovation and reinforcing its significance in theoretical and applied physics, with paramount importance in areas related to precision measurement, communications and sensing.

To demonstrate the phenomenon of interference we selected modelling one of the most known experiments, the Mach–Zehnder interferometer. This experiment consists of verifying the waves' superposition principle by separating an incoming beam of EM radiation into two paths, propagate them independently by a given distance, join the paths again at a subsequent location in space and observe the resultant wave generated. This may be easily modelled by simulating a single mode propagating waveguide that opens up through s-bend waveguides into two independent paths, propagate the two waves by a given distance and rejoin them, again with s-bend waveguides, to monitor the resultant wave's amplitude at the end of a few millimeters waveguide. Figure 17 shows a diagram of the optical setup of a Mach–Zehnder interferometer on the left and the implemented model in our simulation platform on the right. The configuration settings in this simulation considered a working wavelength of 1300 nm, the optical circuit elements and the surrounding environment are assumed to have refractive indexes of 3.3847 and 3.378, respectively.

The optical circuit consisted of an input waveguide and all s-bends 2 mm long, independent paths 4 mm long and the output waveguide 10 mm long. Two independent paths were configured in our simulation platform to emulate as much as possible

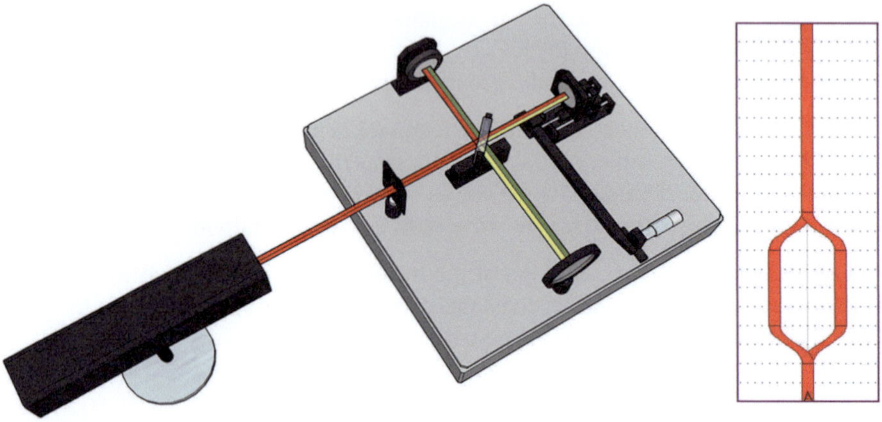

Fig. 17 Mach–Zehnder interferometer; left:—diagram of the optical setup; right:—representation of the simulated optical model

physical reality. One of the paths contains the input waveguide, the left s-bends and the output waveguide. The other path contains the input waveguide, the right s-bends and the output waveguide. Figure 18 presents the results obtained by propagating the fundamental mode of the waveguide throughout the structure.

It is clearly evident the cancellation of the output power in Fig. 18 on the left, for one sees no power present in the output waveguide and which is also confirmed by the graph that monitors power amplitude in both the independent paths. Moreover, on the right of Fig. 18, the graph depicts the dependence of the Mach–Zehnder instrument with the length of independent light paths.

3.5 *Piezoelectric Effect*

There are certain materials which are able to generate an electric charge when subjected to mechanical stress. This phenomenon has been designated the piezoelectric effect. The effect has been discovered by the Curie brothers in 1880 and, not only do these materials generate an electric charge when stressed, but the opposite is also true:—By applying an electric field to these materials, they deform. This has been used in many energy harvesting, actuating and sensing systems.

The piezoelectric effect is due to the asymmetry in the crystal structure of the material. Under normal conditions, i.e. no electric field or mechanical force applied, electric charges are distributed in a way that the mean electric potential is zero and crystal structures are symmetric within the material. When stress is applied, displacement takes place and symmetry is broken, hence electric potential is generated [63].

The direct piezoelectric effect describes the generated electric displacement due to strain and may be explained by the following equation:

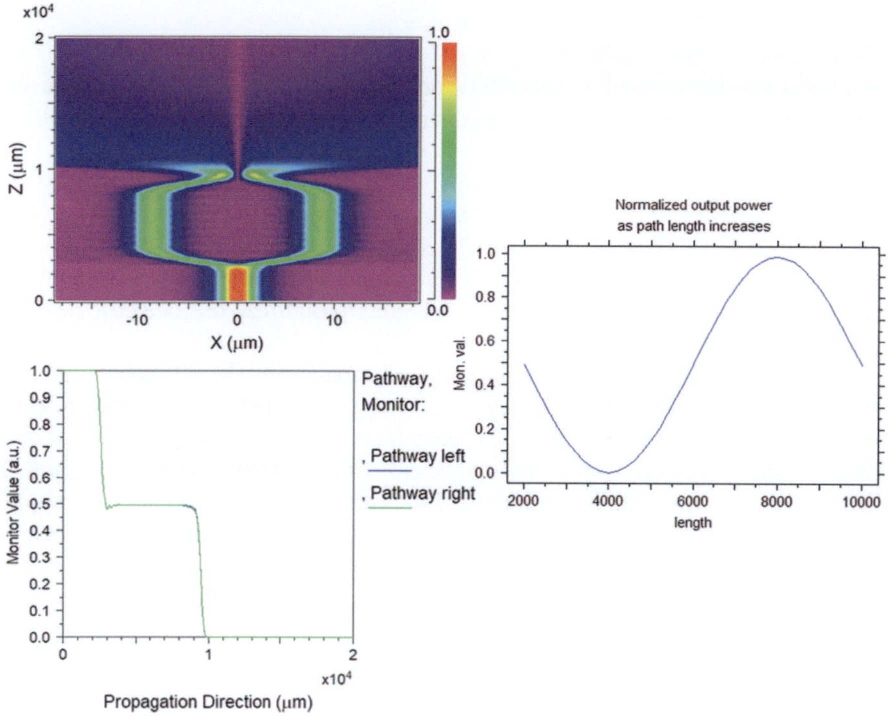

Fig. 18 Mach–Zehnder operation; left top: propagation throughout the structure; left bottom: graph showing monitored power along each path; right: output power as the length of the independent arms is varied from 2 to 10 mm

$$D = dT + \varepsilon E$$

where D is the electric displacement field (c/m^2), d is the piezoelectric strain coefficient (C/N), T is the applied mechanical stress (N/m^2), ε is the dielectric permittivity of the material (F/m) and E is the applied electric field (V/m).

The inverse piezoelectric effect describes the induced mechanical strain in response to an applied electric field and may be described by the equation:

$$S = sT + dE$$

where S is mechanical strain and s is the material's strain coefficient. Previous equations together with dielectric properties and elasticity are the foundation for modelling piezoelectric materials [64]. The common approach is to use the finite elements analysis to simulate the interplay between electric and mechanical domains. The modelling consists of using constitutive equations from continuum mechanics and thermodynamics, namely the Navier equation:

$$\nabla \cdot \sigma + f = \rho \ddot{u}$$

where σ is the stress tensor, f is the force per unit volume, ρ is the material density and u is the field displacement. Charge distribution is governed by the electrostatic equation:

$$\nabla \cdot D = \rho_f$$

where ρ_f is the free charge density. Previous coupled equations are solved by numerical methods, such as the finite elements method, to optimize piezoelectric devices in several applications [65].

Widely used piezoelectric materials include quartz (SiO_2), lead zirconate titanate (PZT), barium titanate ($BaTiO_3$) and polyvinylidene fluoride (PVDF). Recent research in piezoelectric materials has been directed at lead-free alternatives and sodium bismuth titanate (NBT) and potassium sodium niobate (KNN) are now a sustainable substitute [66], nanogenerators for energy harvesting which have applications in self-powered sensors and biomedical implants [67] and 2D piezoelectric materials, such as molybdenum disulfide (MoS_2), which are exploring flexible electronic and miniaturized devices [68].

3.6 Surface Plasmon Resonance

Surface plasmon resonance is an optical phenomenon that takes place at the interface separating a dielectric and a metal (usually a glass prism with its base coated with a thin film of gold or silver) when a beam of electromagnetic radiation interacts with the free electrons of the metal. This interaction excites the free electrons of the metal and generates surface plasmons. Surface plasmons are characterized by the oscillation of the conduction electrons in a metallic surface (propagating wave), which induces the fluctuation of the electric charges at the surface, resulting in the EM field oscillation and propagation [69]. On resonance, these electron charge density waves (surface plasmon polaritons), result from the energy absorption at the resonant wavelength subtracted from the spectrum of the incident EM field. Hence, the intensity of reflected/transmitted spectrum decreases at the resonant wavelength. By observing this minimum in the spectral response of the system, one is able to correlate both the measurand and its concentration.

This phenomenon is a quantum optical effect, which results from the collective oscillation of free electrons at the separation interface of metal/dielectric structures. These electronic oscillations are excited by an EM field striking the metal/dielectric interface at a specific angle [70]. One of the most efficient optical setups coupling an EM wave into surface plasmons is the Kretschmann configuration. Numerical simulations may be used to model this phenomenon and opens up performance optimization of sensing capabilities, together with plasmonic behavior insights.

The surface plasmon resonance phenomenon takes only place when the propagation constant of the incident photon matches the plasmon propagation constant. This condition is analytically described by the following expression:

$$k_{sp} = k_0 \sqrt{\frac{\varepsilon_m \varepsilon_d}{\varepsilon_m + \varepsilon_d}}$$

where k_{sp} is the plasmon propagation constant, $k_0 = {}^{2\pi}/_{\lambda_0}$ is the propagation constant of the incident wave (λ_0 is the wavelength of the incident wave in vacuum), ε_m is the permittivity of the metal and ε_d is the permittivity of the dielectric [59]. The angle of resonance is obtained by calculating the dispersion relation and applying it into an arcsine function. The Kretschmann configuration optical setup consists of a prism of optical glass with a thin layer of gold or silver coated over the prism's base. Incident EM field at an angle greater than the critical angle, reflects at the prism's base by total internal reflection and couples with the surface plasmons at the metal/glass interface, resulting in a decrease in reflected intensity at resonance

Sensors based on the surface plasmon resonance phenomenon are overly sensitive to variations in the refractive index of the surrounding environment and they have been thoroughly investigated by the research community. This resulted in commercial applications contemporaneously in use for chemical and biomolecular detection because of their label free features, high sensitivity and real-time monitoring capabilities.

Contemporaneous surface plasmon resonance based sensors rely on noble metals such as gold or silver. The often used metal for the transducer in these devices is gold, for it has better chemical stability and exhibits a higher shift of the resonance peak wavelength to variations of the sensed analyte. Although, its resonance curve possesses a wide spectral width which worsens its performance, namely when operating on lower energies. Sensors utilizing silver as the active metal benefit from a narrower spectral width but oxidation occurs due to silver's poor chemical stability. Moreover, both gold and silver cannot be deposited directly over a substrate, and a transitional material is required [71], which increases the complexity of the manufacturing process.

To demonstrate the physical phenomenon, we have considered modelling the Kretschmann optical configuration [72] using the finite differences time domain algorithm, for this computational approach uses no approximations to solve Maxwell's equations in space and time, and provides an intensity distribution profile of the reflected EM field which can be hardly distinguished from the results obtained by real life physical instruments. The simulated workspace consisted of a hemispherical dielectric object (emulating the optical glass prism) with a metallic segment (with gold's refractive index at 650 nm) placed next to the object's base and of the same length. The hemispherical object radius = 4 μm in length and the metallic segment is 8 μm long and 30 nm wide. An incident gaussian field at the operating wavelength of 650 nm is placed at a fixed location and directed at the curved surface of the dielectric hemisphere. The simulation consisted of rotating the set formed by the hemispherical

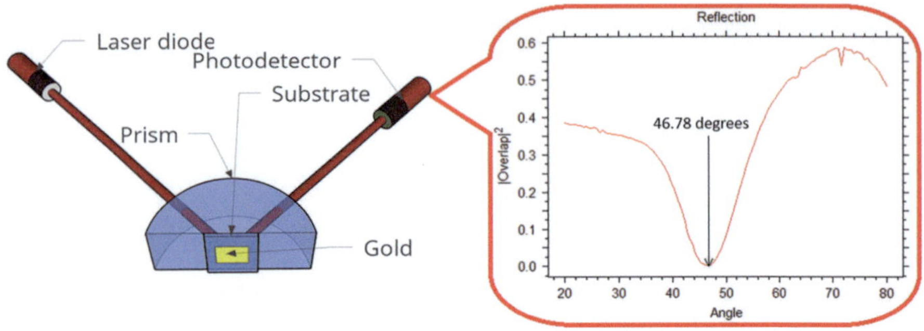

Fig. 19 Diagram of the Kretschmann optical configuration and the simulation result in the graph on the right

and the metallic segment as if it were only one object and by monitoring the reflected intensity. Figure 19 shows a diagram of the Kretschmann configuration setup.

As may be observed in Fig. 19, by monitoring the reflected intensity of the laser beam while rotating the set prism/glass substrate/gold layer we get a graph as the one depicted on the right of the above figure. This simulation considered a fixed location laser beam directed at the prism's curved surface. The angle of incidence is varied by rotating the set prism/glass substrate/gold layer from 20° to 80°. Previous angles are measured with respect to the horizontal axis and not to the norma. As such, following analysis regarding the interpretation of the graph in Fig. 19 considers the start and finishing angles as 80° and 20°, respectively.

At start and as the incidence angle is iterated towards the end, we observe an initial increase of the reflected intensity until a maximum is reached (~ 72°). This maximum corresponds to the critical angle and marks the total internal reflection condition at the prism's base. From this point onwards, no light is transmitted through the prism's base and the incident beam is totally reflected. However, as the incident angle continues to be iterated, one notices a decrease in the reflected intensity. The reflected intensity keeps decreasing until it reaches a minimum (~0) before it starts increasing again. This minimum in intensity is achieved at a precise angle, the resonance angle, and since there is no incident beam transmitted through the prism's base, almost all incoming energy is absorbed by the electrons at the gold's surface:—This phenomenon has been designated Surface Plasmon Resonance. Moreover, if the environment next to the outside gold surface is slightly altered, the resonance angle changes accordingly, which is extremely useful for sensing purposes.

Acknowledgements This research was supported by Portuguese national funds provided by CTS and LASI: Portuguese FCT program, Center of Technology and Systems (CTS) UIDB/00066/2020/ UIDP/00066/2020, and by IPL project IPL/IDI&CA2024/REFINDEX_ISEL.

References

1. C. Pendão, I. Silva, Optical fiber sensors and sensing networks: overview of the main principles and applications. Sensors **22**, 7554 (2022). https://doi.org/10.3390/s22197554
2. B. Lee, Review of the present status of optical fiber sensors. Opt. Fiber Technol. **9**, 57–79 (2003). https://doi.org/10.1016/S1068-5200(02)00527-8
3. F.F. Martín, M.V. Llopis, J.C.C. Rodríguez, A.L. Martínez, A.S. Cabezuelo, M.T. Fernández-Arguelles, J.M. Costa-Fernández, Optoelectronic instrumentation and measurement strategies for optical chemical (bio)sensing. Appl. Sci. **11**, 7849 (2021). https://doi.org/10.3390/app11177849
4. P. Hariharan, *Basics of Interferometry*. (Elsevier, 2007). https://doi.org/10.1016/B978-0-12-373589-8.X5000-7
5. J.D. Joannopoulos, S.G. Johnson, J.N. Winn, R.D. Meade, *Photonic Crystals: Molding the Flow of Light*. (Princeton University Press, 2011). https://doi.org/10.2307/j.ctvcm4gz9
6. V. Giovannetti, S. Lloyd, L. Maccone, Advances in quantum metrology. Nat. Photonics **5**, 222–229 (2011). https://doi.org/10.1038/nphoton.2011.35
7. M.A. Butt, X. Mateos, R. Piramidowicz, Photonics sensors: a perspective on current advancements, emerging challenges, and potential solutions (Invited). Phys. Lett. A **516**, 129633 (2024). https://doi.org/10.1016/j.physleta.2024.129633
8. L. Austin, H. Starke, Ueber die Reflexion der Kathodenstrahlen und eine damit verbundene neue Erscheinung secundärer emission. Ann. Phys. **314**, 271–292 (1902). https://doi.org/10.1002/andp.19023141003
9. F.-H. Marshall, J.W. Coltman, A.I. Bennett, The photo-multiplier radiation detector. Rev. Sci. Instrum. **19**, 744–770 (1948). https://doi.org/10.1063/1.1741156
10. H.O. Anger, Scintillation camera. Rev. Sci. Instrum. **29**, 27–33 (1958). https://doi.org/10.1063/1.1715998
11. Photomultiplier Tubes: Basics and Applications, 4th ed. (2017). https://www.hamamatsu.com/content/dam/hamamatsu-photonics/sites/documents/99_SALES_LIBRARY/etd/PMT_handbook_v4E.pdf. Accessed 14 Feb 2025
12. T. Szczesniak, M. Grodzicka, M. Moszynski, M. Szawlowski, J. Baszak, MicroPMT–A new photodetector for gamma spectrometry and fast timing? IEEE Trans. Nucl. Sci. **61**, 2687–2693 (2014). https://doi.org/10.1109/TNS.2014.2357772
13. M.A. Unzueta, W. Mixter, Z. Croft, J. Joseph, B. Ludewigt, A. Persaud, Position sensitive alpha detector for an associated particle imaging system. (2019), p. 050005. https://doi.org/10.1063/1.5127697
14. G.B. Ko, J.S. Lee, Performance characterization of high quantum efficiency metal package photomultiplier tubes for time-of-flight and high-resolution PET applications. Med. Phys. **42**, 510–520 (2015). https://doi.org/10.1118/1.4903897
15. W. Jiang, Y. Chalich, M.J. Deen, Sensors for positron emission tomography applications. Sensors **19**, 5019 (2019). https://doi.org/10.3390/s19225019
16. A. Rochas, M. Gani, B. Furrer, P.A. Besse, R.S. Popovic, G. Ribordy, N. Gisin, Single photon detector fabricated in a complementary metal–oxide–semiconductor high-voltage technology. Rev. Sci. Instrum. **74**, 3263–3270 (2003). https://doi.org/10.1063/1.1584083
17. A. Rochas, M. Gosch, A. Serov, P.A. Besse, R.S. Popovic, T. Lasser, R. Rigler, First fully integrated 2-D array of single-photon detectors in standard CMOS technology. IEEE Photonics Technol. Lett. **15**, 963–965 (2003). https://doi.org/10.1109/LPT.2003.813387
18. S. Cova, M. Ghioni, A. Lacaita, C. Samori, F. Zappa, Avalanche photodiodes and quenching circuits for single-photon detection. Appl. Opt. **35**, 1956 (1996). https://doi.org/10.1364/AO.35.001956
19. X. Michalet, A. Ingargiola, R.A. Colyer, G. Scalia, S. Weiss, P. Maccagnani, A. Gulinatti, I. Rech, M. Ghioni, Silicon photon-counting avalanche diodes for single-molecule fluorescence spectroscopy. IEEE J. Sel. Top. Quantum Electron. **20**, 248–267 (2014). https://doi.org/10.1109/JSTQE.2014.2341568

20. W.G. Oldham, R.R. Samuelson, P. Antognetti, Triggering phenomena in avalanche diodes. IEEE Trans. Electron Devices **19**, 1056–1060 (1972). https://doi.org/10.1109/T-ED.1972.17544
21. L. Pancheri, D. Stoppa, G.-F. Dalla Betta, Characterization and modeling of breakdown probability in sub-micrometer CMOS SPADs. IEEE J. Sel. Top. Quantum Electron. **20**, 328–335 (2014). https://doi.org/10.1109/JSTQE.2014.2327791
22. A. Gulinatti, I. Rech, M. Assanelli, M. Ghioni, S. Cova, A physically based model for evaluating the photon detection efficiency and the temporal response of SPAD detectors. J. Mod. Opt. **58**, 210–224 (2011). https://doi.org/10.1080/09500340.2010.536590
23. J. Hynecek, Impactron-a new solid state image intensifier. IEEE Trans. Electron Devices **48**, 2238–2241 (2001). https://doi.org/10.1109/16.954460
24. J.-L. Bezy, W. Leibrandt, A. Heliere, P. Silvestrin, C.-C. Lin, P. Ingmann, T. Kimura, H. Kumagai, The ESA Earth Explorer EarthCARE mission, in ed. by J.J. Butler. (2005), p. 58820F. https://doi.org/10.1117/12.619438
25. S.M. Tulloch, Application of L3 technology to wavefront sensing. (2004), p. 1167. https://doi.org/10.1117/12.550349
26. G.N. Gol'tsman, O. Okunev, G. Chulkova, A. Lipatov, A. Semenov, K. Smirnov, B. Voronov, A. Dzardanov, C. Williams, R. Sobolewski, Picosecond superconducting single-photon optical detector. Appl. Phys. Lett. **79**, 705–707 (2001). https://doi.org/10.1063/1.1388868
27. J. Clarke, A.I. Braginski, eds., *The SQUID Handbook*. (Wiley, 2004). https://doi.org/10.1002/3527603646
28. A.J. Miller, S.W. Nam, J.M. Martinis, A.V. Sergienko, Demonstration of a low-noise near-infrared photon counter with multiphoton discrimination. Appl. Phys. Lett. **83**, 791–793 (2003). https://doi.org/10.1063/1.1596723
29. R.H. Hadfield, J.L. Habif, J. Schlafer, R.E. Schwall, S.W. Nam, Quantum key distribution at 1550 nm with twin superconducting single-photon detectors. Appl. Phys. Lett. **89** (2006). https://doi.org/10.1063/1.2405870
30. J.P. Sprengers, A. Gaggero, D. Sahin, S. Jahanmirinejad, G. Frucci, F. Mattioli, R. Leoni, J. Beetz, M. Lermer, M. Kamp, S. Höfling, R. Sanjines, A. Fiore, Waveguide superconducting single-photon detectors for integrated quantum photonic circuits. Appl. Phys. Lett. **99** (2011). https://doi.org/10.1063/1.3657518
31. C.M. Natarajan, M.G. Tanner, R.H. Hadfield, Superconducting nanowire single-photon detectors: physics and applications. Supercond. Sci. Technol. **25**, 063001 (2012). https://doi.org/10.1088/0953-2048/25/6/063001
32. D.V. Reddy, A.E. Lita, S.W. Nam, R.P. Mirin, V.B. Verma, Achieving 98% system efficiency at 1550 nm in superconducting nanowire single photon detectors, in *Rochester Conference Coherence Quantum Opticcs*. (OSA, Washington, D.C., 2019), p. W2B.2. https://doi.org/10.1364/CQO.2019.W2B.2
33. W. Zhang, J. Huang, C. Zhang, L. You, C. Lv, L. Zhang, H. Li, Z. Wang, X. Xie, A 16-Pixel interleaved superconducting nanowire single-photon detector array with a maximum count rate exceeding 1.5 GHz. IEEE Trans. Appl. Superconductivity **29**, 1–4 (2019). https://doi.org/10.1109/TASC.2019.2895621
34. A. Korneev, A. Lipatov, O. Okunev, G. Chulkova, K. Smirnov, G. Gol'tsman, J. Zhang, W. Slysz, A. Verevkin, R. Sobolewski, GHz counting rate NbN single-photon detector for IR diagnostics of VLSI CMOS circuits. Microelectr. Eng. **69**, 274–278 (2003). https://doi.org/10.1016/S0167-9317(03)00309-5
35. M.E. Grein, A.J. Kerman, E.A. Dauler, M.M. Willis, B. Romkey, R.J. Molnar, B.S. Robinson, D.V. Murphy, D.M. Boroson, An optical receiver for the Lunar Laser Communication Demonstration based on photon-counting superconducting nanowires, in M.A. Itzler, J.C. Campbell ed. by (2015), p. 949208. https://doi.org/10.1117/12.2179781
36. A.S. Tremsin, J.V. Vallerga, O.H.W. Siegmund, J. Woods, L.E. De Long, J.T. Hastings, R.J. Koch, S.A. Morley, Y.-D. Chuang, S. Roy, Photon-counting MCP/Timepix detectors for soft X-ray imaging and spectroscopic applications. J. Synchrotron Radiat. **28**, 1069–1080 (2021). https://doi.org/10.1107/S1600577521003908

37. C.B. Johnson, C.E. Catchpole, C.C. Matle, Microchannel plate inverter image intensifiers. IEEE Trans. Electron Devices **18**, 1113–1116 (1971). https://doi.org/10.1109/T-ED.1971.17339
38. M.J. Minot, B.W. Adams, M. Aviles, J.L. Bond, T. Cremer, M.R. Foley, A. Lyashenko, M.A. Popecki, M.E. Stochaj, W.A. Worstell, M.J. Wetstein, J.W. Elam, A.U. Mane, O.H.W. Siegmund, C. Ertley, H.J. Frisch, A. Elagin, E. Angelico, E. Spieglan, Large Area Picosecond Photodetector (LAPPD—Pilot production and development status. Nucl. Instrum. Methods Phys. Res., Sect. A **936**, 527–531 (2019). https://doi.org/10.1016/j.nima.2018.11.137
39. K.D. Irwin, An application of electrothermal feedback for high resolution cryogenic particle detection. Appl. Phys. Lett. **66**, 1998–2000 (1995). https://doi.org/10.1063/1.113674
40. K.D. Irwin, G.C. Hilton, Transition-edge sensors. (2005), pp. 63–150. https://doi.org/10.1007/10933596_3
41. D. Barret, T. Lam Trong, J.-W. den Herder, L. Piro, X. Barcons, J. Huovelin, R. Kelley, J. M. Mas-Hesse, K. Mitsuda, S. Paltani, G. Rauw, A. Rożanska, J. Wilms, M. Barbera, E. Bozzo, M. T. Ceballos, I. Charles, A. Decourchelle, R. den Hartog, J.-M. Duval, F. Fiore, F. Gatti, A. Goldwurm, B. Jackson, P. Jonker, C. Kilbourne, C. Macculi, M. Mendez, S. Molendi, P. Orleanski, F. Pajot, E. Pointecouteau, F. Porter, G.W. Pratt, D. Prêle, L. Ravera, E. Renotte, J. Schaye, K. Shinozaki, L. Valenziano, J. Vink, N. Webb, N. Yamasaki, F. Delcelier-Douchin, M. Le Du, J.-M. Mesnager, A. Pradines, G. Branduardi-Raymont, M. Dadina, A. Finoguenov, Y. Fukazawa, A. Janiuk, J. Miller, Y. Nazé, F. Nicastro, S. Sciortino, J.M. Torrejon, H. Geoffray, I. Hernandez, L. Luno, P. Peille, J. André, C. Daniel, C. Etcheverry, E. Gloaguen, J. Hassin, G. Hervet, I. Maussang, J. Moueza, A. Paillet, B. Vella, G. Campos Garrido, J.-C. Damery, C. Panem, J. Panh, S. Bandler, J.-M. Biffi, K. Boyce, A. Clénet, M. DiPirro, P. Jamotton, S. Lotti, D. Schwander, S. Smith, B.-J. van Leeuwen, H. van Weers, T. Brand, B. Cobo, T. Dauser, J. de Plaa, E. Cucchetti, The Athena X-ray Integral Field Unit (X-IFU), in J.-W.A. den Herder, T. Takahashi, M. Bautz, eds. ed. by (2016), p. 99052F. https://doi.org/10.1117/12.2232432
42. E. Shirokoff, B.A. Benson, L.E. Bleem, C.L. Chang, Hsiao-Mei Cho, A.T. Crites, M.A. Dobbs, W.L. Holzapfel, T. Lanting, A.T. Lee, M. Lueker, J. Mehl, T. Plagge, H.G. Spieler, J.D. Vieira, The south pole telescope SZ-receiver detectors. IEEE Trans. Appl. Superconductivity **19**, 517–519 (2009). https://doi.org/10.1109/TASC.2009.2018229
43. M. Borghesi, B. Alpert, M. Balata, D. Becker, D. Bennet, E. Celasco, N. Cerboni, M. De Gerone, R. Dressler, M. Faverzani, M. Fedkevych, E. Ferri, J. Fowler, G. Gallucci, J. Gard, F. Gatti, A. Giachero, G. Hilton, U. Koster, D. Labranca, M. Lusignoli, J. Mates, E. Maugeri, S. Nisi, A. Nucciotti, L. Origo, G. Pessina, S. Ragazzi, C. Reintsema, D. Schmidt, D. Schumann, D. Swetz, J. Ullom, L. Vale, An updated overview of the HOLMES status. Nucl. Instrum. Methods Phys. Res. Sect. A **1051**, 168205 (2023). https://doi.org/10.1016/j.nima.2023.168205
44. G. Agnolet, W. Baker, D. Barker, R. Beck, T.J. Carroll, J. Cesar, P. Cushman, J.B. Dent, S. De Rijck, B. Dutta, W. Flanagan, M. Fritts, Y. Gao, H.R. Harris, C.C. Hays, V. Iyer, A. Jastram, F. Kadribasic, A. Kennedy, A. Kubik, K. Lang, R. Mahapatra, V. Mandic, C. Marianno, R.D. Martin, N. Mast, S. McDeavitt, N. Mirabolfathi, B. Mohanty, K. Nakajima, J. Newhouse, J.L. Newstead, I. Ogawa, D. Phan, M. Proga, A. Rajput, A. Roberts, G. Rogachev, R. Salazar, J. Sander, K. Senapati, M. Shimada, B. Soubasis, L. Strigari, Y. Tamagawa, W. Teizer, J.I.C. Vermaak, A.N. Villano, J. Walker, B. Webb, Z. Wetzel, S.A. Yadavalli, Background studies for the MINER coherent neutrino scattering reactor experiment. Nucl. Instrum. Methods Phys. Res. Sect. A **853**, 53–60 (2017). https://doi.org/10.1016/j.nima.2017.02.024
45. C.J. Chunnilall, I. Pietro Degiovanni, S. Kück, I. Müller, A.G. Sinclair, metrology of single-photon sources and detectors: a review. Opt. Eng. **53**, 081910 (2014). https://doi.org/10.1117/1.OE.53.8.081910
46. T. Gerrits, A. Lita, B. Calkins, S.W. Nam, Superconducting transition edge sensors for quantum optics. (2016), pp. 31–60. https://doi.org/10.1007/978-3-319-24091-6_2
47. S.-J. Lee, C.J. Titus, R. Alonso Mori, M.L. Baker, D.A. Bennett, H.-M. Cho, W.B. Doriese, J.W. Fowler, K.J. Gaffney, A. Gallo, J.D. Gard, G.C. Hilton, H. Jang, Y. Il Joe, C.J. Kenney, J. Knight, T. Kroll, J.-S. Lee, D. Li, D. Lu, R. Marks, M.P. Minitti, K.M. Morgan, H.Ogasawara, G.C. O'Neil, C.D. Reintsema, D.R. Schmidt, D. Sokaras, J.N. Ullom, T.-C. Weng, C. Williams,

B.A. Young, D.S. Swetz, K.D. Irwin, D. Nordlund, Soft X-ray spectroscopy with transition-edge sensors at Stanford Synchrotron Radiation Lightsource beamline 10–1. Rev. Sci. Instr. **90** (2019). https://doi.org/10.1063/1.5119155
48. J. Uhlig, W.B. Doriese, J.W. Fowler, D.S. Swetz, C. Jaye, D.A. Fischer, C.D. Reintsema, D.A. Bennett, L.R. Vale, U. Mandal, G.C. O'Neil, L. Miaja-Avila, Y.I. Joe, A. El Nahhas, W. Fullagar, F. Parnefjord Gustafsson, V. Sundström, D. Kurunthu, G.C. Hilton, D.R. Schmidt, J.N. Ullom, High-resolution X-ray emission spectroscopy with transition-edge sensors: present performance and future potential. J. Synchrotron Radiation **22**, 766–775 (2015). https://doi.org/10.1107/S1600577515004312
49. E. Fossum, J. Ma, S. Masoodian, L. Anzagira, R. Zizza, The quanta image sensor: every photon counts. Sensors **16**, 1260 (2016). https://doi.org/10.3390/s16081260
50. G. Bondarenko, B. Dolgoshein, V. Golovin, A. Ilyin, R. Klanner, E. Popova, Limited Geiger-mode silicon photodiode with very high gain. Nucl. Phys. B Proc.Suppl. **61**, 347–352 (1998). https://doi.org/10.1016/S0920-5632(97)00585-9
51. ITU-T G.651.1. https://www.cepetel.org.ar/wp-content/uploads/2021/07/T-REC-G.651.1-201811-IPDF-E.pdf. Accessed 7 Feb 2025 (2018)
52. ITU-T G.652. https://img.antpedia.com/standard/pdf/1/2501/ITU-TG.652 (08_2024).pdf. Accessed 7 Feb 2025 (2024)
53. B.E.A. Saleh, M.C. Teich, *Fundamentals of Photonics*. (Wiley, New York, USA, 1991). https://doi.org/10.1002/0471213748
54. RSoft, Synopsys RSoft Solutions. https://www.synopsys.com/optical-solutions/rsoft.html. Accessed 30 July 2024 (n.d.)
55. FullWAVE, FullWAVE—Synopsys. https://www.synopsys.com/photonic-solutions/rsoft-photonic-device-tools/passive-device-fullwave.html. Accessed 29 Dec 2019 (n.d.)
56. K. Yee, Numerical solution of initial boundary value problems involving Maxwell's equations in isotropic media. IEEE Trans. Antennas Propag. **30**, 141–147 (1966). https://doi.org/10.1007/978-3-319-61185-3_17
57. T.K. Panda, A.B. Mukesh kumar Behera, S. Polei, Performance comparison of Fiber Bragg Grating for different Grating length, Apodization function and Chirp function, in *2020 International Conference Computer Science Engineering and Applications*. (IEEE, 2020), pp. 1–5. https://doi.org/10.1109/ICCSEA49143.2020.9132943
58. J. Xiong, K. Yin, K. Li, S.-T. Wu, Holographic optical elements for augmented reality: principles, present status, and future perspectives. Adv. Photonics Res. **2**, 1–15 (2021). https://doi.org/10.1002/adpr.202000049
59. M. Born, E. Wolf, A.B. Bhatia, P.C. Clemmow, D. Gabor, A.R. Stokes, A.M. Taylor, P.A. Wayman, W.L. Wilcock, *Principles of Optics*. (Cambridge University Press, 1999). https://doi.org/10.1017/CBO9781139644181
60. E. Hecht, *Optics 5th Edition* (2016).
61. L.L. Beranek, I.L. Ver, Noise and vibration control engineering. IEEE Power Eng. Rev. **12** (2006). https://doi.org/10.1109/MPER.1992.161424
62. R.P. Feynman, *Feynman Lectures on Physics*. (1977), pp. xix–xx
63. H. JAFFE, Piezoelectric ceramics. J. Am. Ceramic Soc. **41**, 494–498 (1958). https://doi.org/10.1111/j.1151-2916.1958.tb12903.x
64. F.I. Fedorov, *Theory of Elastic Waves in Crystals*. (Springer US, Boston, MA, 1968). https://doi.org/10.1007/978-1-4757-1275-9
65. T. Li, P.S. Lee, Piezoelectric energy harvesting technology: from materials, structures, to applications. Small Struct. **3** (2022). https://doi.org/10.1002/sstr.202100128
66. K. Uchino, *Advanced piezoelectric materials*. (Woodhead Publishing Limited, 2010). https://doi.org/10.1533/9781845699758
67. E. Delgado-Alvarado, J. Martínez-Castillo, L. Zamora-Peredo, J.A. Gonzalez-Calderon, R. López-Esparza, M.W. Ashraf, S. Tayyaba, A.L. Herrera-May, Triboelectric and piezoelectric nanogenerators for self-powered healthcare monitoring devices: operating principles, challenges, and perspectives. Nanomaterials **12**, 4403 (2022). https://doi.org/10.3390/nano12244403

68. W. Wu, L. Wang, Y. Li, F. Zhang, L. Lin, S. Niu, D. Chenet, X. Zhang, Y. Hao, T.F. Heinz, J. Hone, Z.L. Wang, Piezoelectricity of single-atomic-layer MoS2 for energy conversion and piezotronics. Nature **514**, 470–474 (2014). https://doi.org/10.1038/nature13792
69. S. Hayashi, T. Okamoto, Plasmonics: visit the past to know the future. J. Phys. D: Appl. Phys. **45** (2012). https://doi.org/10.1088/0022-3727/45/43/433001
70. J. Homola, S.S. Yee, G. Gauglitz, Surface plasmon resonance sensors: review. Sens. Actuators, B Chem. **54**, 3–15 (1999). https://doi.org/10.1016/S0925-4005(98)00321-9
71. D. Yuan, Y. Dong, Y. Liu, T. Li, Design of a high-performance micro integrated surface plasmon resonance sensor based on silicon-on-insulator rib waveguide array. Sensors (Switzerland) **15**, 17313–17328 (2015). https://doi.org/10.3390/s150717313
72. E. Kretschmann, H. Raether, Notizen: radiative decay of non radiative surface plasmons excited by light. Zeitschrift für Naturforschung A **23**, 2135–2136 (1968). https://doi.org/10.1515/zna-1968-1247

Optical Sensors for Structural Health Monitoring

Heming Wei, Shutong Zhangyang, Shijie Zheng, Mengshi Zhu, Liang Zhang, Fufei Pang, and Dengwei Zhang

Abstract In the realm of structural health monitoring (SHM) for aerospace applications, optical sensors have gained significant prominence due to their ability to deliver real-time measurements of critical parameters, including strain, deformation, temperature, vibration, and pressure. These measurements are essential for evaluating the structural integrity of aerospace components. Numerous optical sensing technologies, such as optical interferometers, are specifically developed for SHM purposes. Among these, acoustic emission (AE) stands out as a vital technique extensively utilized in SHM to identify damage and estimate the remaining service life of aerospace structures. AE arises from the rapid release of energy within a material, generating transient elastic waves, also referred to as stress waves. These emissions, distinguished by unique amplitude and frequency characteristics, are associated with failure mechanisms such as crack growth and deformation, making AE a powerful tool for damage assessment and localization. In this chapter, optical sensors for SHM are introduced, which can be integrated into modern aircraft design enhances safety, reliability, and efficiency, addressing critical challenges associated with aging aircraft and emerging materials like composites.

H. Wei (✉) · S. Zhangyang · M. Zhu · L. Zhang · F. Pang
Key Laboratory of Specialty Fiber Optics and Optical Access Networks, School of Communication and Information Engineering, Shanghai University, Shanghai, China
e-mail: hmwei@shu.edu.cn

S. Zheng
MOE Key Laboratory of High-Speed Railway Engineering, School of Civil Engineering, Southwest Jiaotong University, Chengdu, China

D. Zhang
State Key Laboratory of Extreme Photonics and Instrumentation, Zhejiang University, Hangzhou, China

C. Marques et al. (eds.), *Advanced Optical Sensors for Aerospace Applications*, Progress in Optical Science and Photonics 37,
https://doi.org/10.1007/978-981-95-1626-1_3

1 Introduction to SHM in Aerospace

SHM is a proactive approach to assess the integrity of aerospace structures (e.g., aircraft, spacecraft) in real-time using sensor networks and data analytics [1]. It aims to detect damage early, enhance safety, and optimize maintenance by transitioning from scheduled inspections to condition-based monitoring. Ensuring safety and reliability in aerospace systems is paramount, as system failures can result in severe outcomes. The increasing use of advanced materials, such as carbon-fiber-reinforced polymers (CFRPs), coupled with the aging of aircraft fleets, underscores the need for real-time, high-precision monitoring of structural conditions [2–4]. Although the aerospace industry has always prioritized safety, historical incidents underscore the limitations of traditional inspection methods. Many pivotal incidents exposed the vulnerabilities of periodic manual inspections and galvanized the industry to seek more robust monitoring solutions.

For decades, Non-Destructive Inspection (NDI) methods, including ultrasonic testing, X-ray imaging, and eddy current analysis, have been the primary approaches for evaluating structural integrity [5–7]. While effective, these methods required aircraft to be grounded, involved labor-intensive processes, and often failed to detect hidden or evolving damage. For instance, corrosion in concealed joints or micro-cracks in composite materials could remain undetected until reaching critical stages [8]. Among these methods, Ultrasonic sensing methods are one of the most powerful techniques for SHM as ultrasound can propagate through materials to detect internal flaws—such as cracks, delamination, or corrosion—with high sensitivity and spatial resolution, even in complex geometries or layered composite structures [9–11]. By analyzing changes in wave velocity, amplitude, or reflection patterns, these methods identify subsurface defects that are invisible to visual inspections or traditional non-destructive testing (NDT). In aerospace applications, ultrasonic methods are particularly effective for monitoring bonded repairs, detecting fatigue cracks in aluminum airframes, and verifying the integrity of additive-manufactured components [12]. Their ability to operate in real time—combined with emerging AI-driven signal processing—makes ultrasonic SHM indispensable for ensuring the safety of aging aircraft fleets and next-generation CFRP-dominated structures, despite challenges such as signal attenuation in heterogeneous materials or the need for precise calibration. However, one of the key components is sensors, which are based on traditional piezoelectric transducers. The problem is that they face limitations in harsh aerospace environments, including susceptibility to electromagnetic interference (EMI), limited multiplexing capability, and poor durability under extreme temperatures or mechanical loads. Furthermore, the fabrication of piezoelectric-based technologies presents considerable challenges, particularly when dealing with ultrasonic frequencies in the range of tens of MHz, especially if the requirement is for an array of sensors.

In comparison, optical sensors are increasingly being used in SHM for aerospace applications, where they provide critical solutions into the integrity and performance of various aerospace structures. Optical sensors, especially optical fiber sensors can monitor various physical parameters, which can help engineers detect early signs of failure or fatigue before they lead to catastrophic damage. In this chapter, we will introduce the fundamentals, systems and applications of optical sensors for SHM.

2 Fundamentals of Optical Sensing Technologies

2.1 Fiber Bragg Grating Sensors

Fiber Bragg Grating Sensors are often named short-period fiber gratings because of their short-period structure. These gratings exhibit a narrow and precise reflectance spectral bandwidth, as well as high reflectance properties that can be flexibly adapted to the writing conditions. In particular, the main function of FBG is to reflect light field modes propagating in the same direction, as shown in Fig. 1. Based on this unique mechanism, they are clearly classified as an important type of reflective bandpass filter.

The operational principle of fiber Bragg gratings (FBGs) relies on ultraviolet-induced photoinscription technology, which establishes a spatially modulated refractive index profile within the optical fiber's core region. This periodic modulation generates wavelength-selective reflection characteristics, enabling the optical waveguide to exhibit specific spectral filtering properties. The photonic structure demonstrates selective retroreflection at the phase-matched wavelength (λ_B), while permitting transmission of other spectral components with minimal attenuation. Environmental perturbations, including thermal variations, mechanical stresses, and hydrostatic pressures, induce modifications in two fundamental grating parameters: the spatial period (Λ) of the refractive index modulation and the effective modal index (n_{eff}). These parametric alterations manifest as measurable shifts in the characteristic Bragg resonance condition, expressed as $\lambda_B = 2_{neff}\Lambda$ (as presented in Chap. 2).

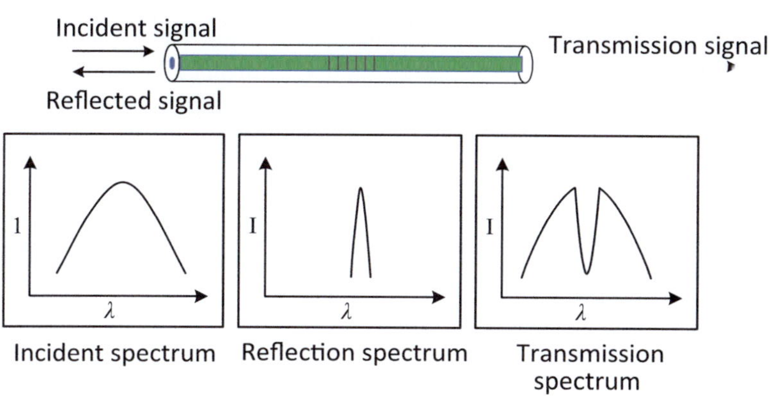

Fig. 1 FBG transmission schematic diagram

Through precise monitoring of spectral displacement ($\Delta\lambda_B$) in the reflected optical signal, quantitative assessment of external stimuli can be achieved with high sensitivity [13]. This equation establishes the quantitative relationship between two key structural parameters: n_{eff} (core's effective refractive index) and Λ (periodic modulation spacing). Notably, environmental variables including mechanical deformation and thermal fluctuations [14, 15] induce measurable alterations in these intrinsic parameters, thereby modifying the spectral characteristics of the reflected signal. By deeper derivation of Bragg law, we can conclude that the amount of drift in the centre wavelength of the fibre Bragg grating due to strain and temperature variations will follow a specific relational equation:

$$\Delta\lambda_B = 2\left(\Delta\frac{\partial n_{\text{eff}}}{\partial l} + n_{\text{eff}}\frac{\partial\Lambda}{\partial l}\right)\Delta l + 2\left(\Lambda\frac{\partial n_{\text{eff}}}{\partial T} + n_{\text{eff}}\frac{\partial\Lambda}{\partial T}\right)\Delta T. \tag{1}$$

Equation (1) mathematically characterizes the dual modulation mechanism induced by mechanical deformation in optical fibers. The initial component quantifies the strain-induced modifications in waveguide properties, encompassing both the photoelastic variation of refractive index and the geometric alteration of grating periodicity [13]. Subsequent formulations in Eqs. (2) and (3) provide explicit expressions for the strain sensitivity coefficient and the effective photoelastic constant (P_e), respectively. These equations incorporate fundamental material parameters, including the strain-optic tensor elements p_{11} and p_{12}, along with the Poisson ratio (ν) that characterizes the mechanical response of the fiber medium:

$$\Delta\lambda_B = \lambda_B(1 - P_e)\varepsilon_z, \tag{2}$$

$$P_e = \frac{n_{\text{eff}}}{2}\left[p_{12} - \nu(p_{11} + p_{12})\right]. \tag{3}$$

For the common SiGe fiber material, the value of p_{11} is 0.113, p_{12} is 0.252, Poisson's ratio v is 0.16, and the effective refractive index n_{eff} reaches 1.482. Based on these specific parameter values as well as Eq. (3), we can accurately compute that, for a fiber Bragg grating with a central wavelength set to 1550 nm, whenever it is subjected to a micro-strain force, the center wavelength will be displaced by 1.2 pm.

In Eq. (1), the second term describes how temperature affects the grating wavelength shift, which can be described by the following equation:

$$\Delta\lambda_B = \lambda_B[\alpha_\Lambda + \alpha_n]\Delta T \tag{4}$$

where α_Λ represents the thermal expansion coefficient of the optical fibre [13], which is approximately equal to 0.55×10^{-6} for silicon materials. Based on these specific parameters, we can conclude that for a FBG with a central wavelength set to 1550 nm, the central wavelength will produce a displacement of approximately 13 pm for every 1 degree change in ambient temperature. This characteristic makes the FBGs show a wide range of application prospects in both the two major fields of optical fiber communication and sensing technology [16–18].

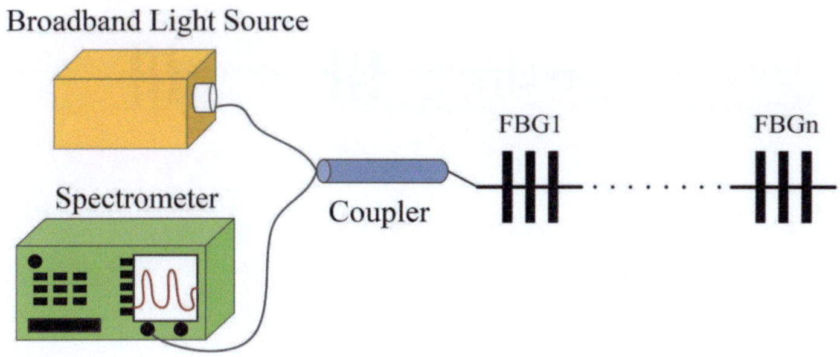

Fig. 2 Spectrometer detection method

As for signal demodulation, the demodulation system can be divided as the following methods:

Spectrometer detection method

The detection method of the spectrometer involves connecting the optical signal reflected back from the sensing field into the spectrometer to detect and record the wavelength drift. The measured physical quantity is then demodulated and determined based on the scale factor of the grating sensor to the measured physical quantity. The principle is shown in Fig. 2. Spectrometer detection is commonly used in FBG systems to measure the offset of the center wavelength of a fiber grating to monitor changes in strain, temperature or other physical quantities of the grating. The spectrometer detection method has merits of simple structure and wide demodulation range, however, the measuring frequency is limited.

Edge filtering method

Optical edge filters are characterized by their monotonic transmission profiles with well-defined transition regions, typically spanning spectral ranges between 10 and 100 nm. As illustrated in Fig. 3, the demodulation mechanism of this detection scheme exploits the linear response region of the filter's transmission curve. The operational principle relies on the spectral intensity conversion effect: when the reflected optical signal from a fiber Bragg grating (FBG) is spectrally aligned with the filter's linear transition zone, spectral shifts in the Bragg wavelength ($\Delta\lambda_B$) induce proportional variations in transmitted optical power. This wavelength-to-intensity conversion technique enables real-time monitoring of FBG spectral characteristics with microsecond temporal resolution, making it particularly suitable for dynamic strain analysis and multiplexed distributed sensing applications.

Matching grating detection method

The core principle of the matched grating detection method is to add a reference grating to an existing sensing system whose parameters (e.g. grating period, depth of refractive index modulation, etc.) are identical to those of the grating in the sensing system. When the sensing grating is affected by external physical factors

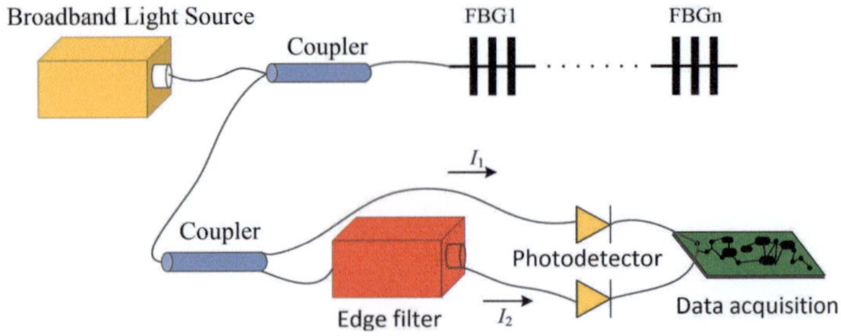

Fig. 3 Edge filtering method

(e.g. temperature and strain), the wavelength of its reflection or transmission spectrum will change. Figures 4 and 5 show the specific schematic diagrams of the reflective structure matching grating method and the transmissive structure matching grating method, respectively. The reference grating is fixed to a piezoelectric transducer (PZT) and the wavelength of the reference grating can be controlled by changing the voltage across the PZT. However, the matched grating detection method has some drawbacks such as limited sensitivity and limited Dynamic Response.

Scanning laser detection method

When a FBG is subjected to external factors such as temperature and strain, the wavelength of the reflected or transmitted light changes. Scanning lasers have the ability to output continuous laser light at wavelengths that cover the entire range of grating excursions band. As shown in Figs. 6 and 7, the basic principle of the scanning semiconductor laser detection method and the working principle of the scanning semiconductor laser detection method with standard equipment are illustrated. This method has high precision and high sensitivity.

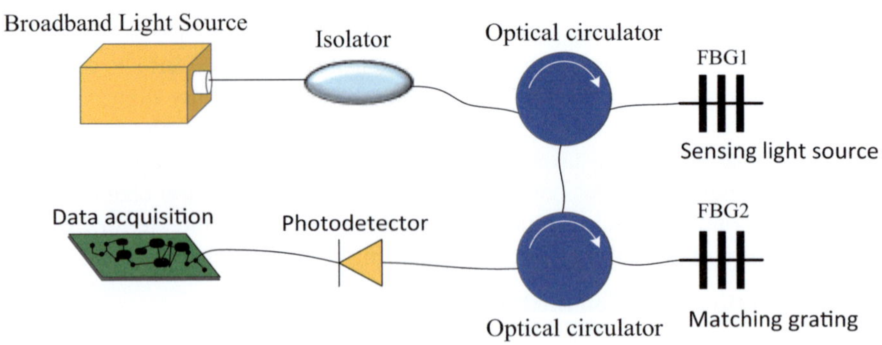

Fig. 4 Reflective structure matching grating method

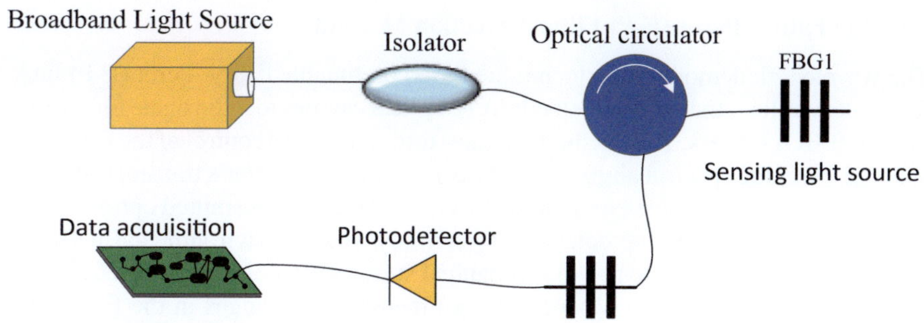

Fig. 5 Transmissive structure matching grating method

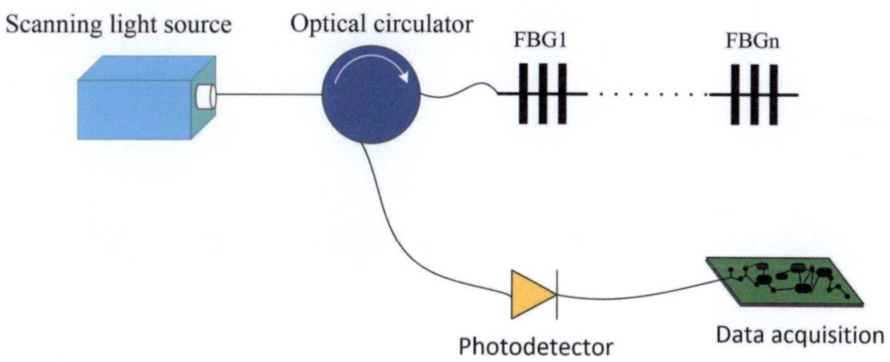

Fig. 6 Principle of the scanning semiconductor laser detection method

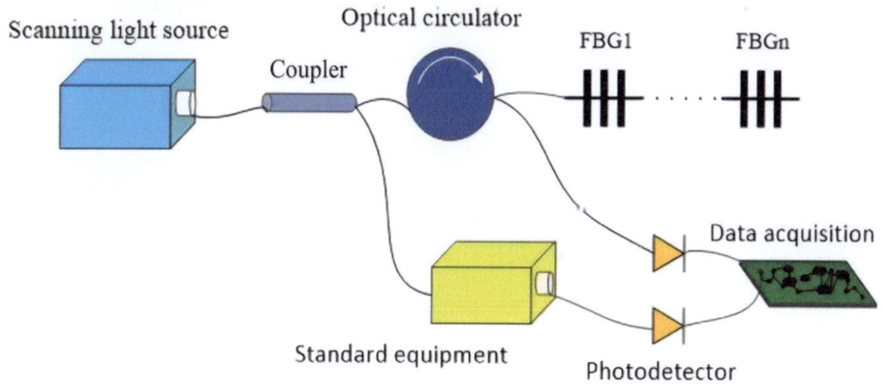

Fig. 7 Principle of the scanning semiconductor laser detection method (with standards)

Tunable Fabry–Perot (F-P) Filter Detection Method

The wavelength demodulation technique based on tunable Fabry–Perot (F-P) interferometers utilizes the spectral selectivity of optical cavities to interrogate fiber Bragg grating (FBG) sensors. This method operates through precise control of the F-P cavity spacing, enabling spectral alignment between the interferometer's transmission peak and the characteristic wavelength of the FBG's reflected or transmitted optical signal. As illustrated in Fig. 8, the detection system responds to environmental perturbations (e.g., thermal variations and mechanical stresses) by converting spectral shifts in the FBG's characteristic wavelength into measurable changes in the F-P filter's transmission spectrum. Through continuous monitoring of these spectral variations, quantitative evaluation of external physical parameters acting on the FBG can be achieved with high resolution.

These methods are analysed and compared in five aspects: accuracy and resolution, system complexity and cost, dynamic response speed, environmental adaptability and application areas. Table 1 shows the comparison table.

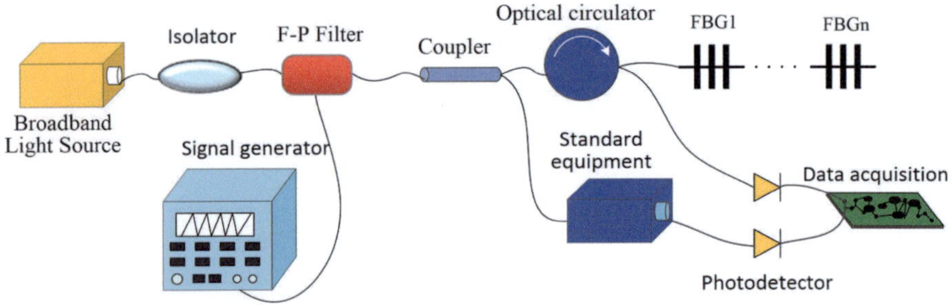

Fig. 8 Principle of tunable F-P filter detection method

Table 1 Comparison of fibre grating demodulation methods

Demodulation method	Accuracy and resolution	Areas of application	Environmental adaptation	Dynamic response speed application areas	System complexity and cost
Spectral detection	Low	Wide	Sensitive	Slowly	High
Edge filtering detection method	Low	Narrow	Stronger	Faster	Low
Matching spectrum detection method	Lower	Narrow	Sensitive	Slowly	Low
Tunable filter detection method	High	Wide	Stronger	Faster	High
Scanning laser detection method	High	Wide	Stronger	Fast	Middle

2.2 Interferometric Optical Sensors

Interferometric optical fiber sensing is a technology that uses the principle of light wave interference to detect the changes of external physical quantities. By detecting variations in the phase of light waves propagating through optical fibers, the parameters such as temperature, strain, pressure and vibration can be accurately measured.

Interference principle: two or more beams of coherent light propagate in the fiber and are superimposed to form interference fringes. The change of external physical quantity causes the change of fiber length and refractive index, resulting in optical path difference and phase difference, thus changing the interference light intensity.

Relationship between phase change and physical quantity: the relationship between phase difference $\Delta\phi$ and optical path difference ΔL is:

$$\Delta\Phi = \frac{2\pi}{\lambda}\Delta L \tag{5}$$

Within the theoretical framework, the parameter λ corresponds to the optical wavelength propagating through the waveguide medium. The total optical path variation (ΔL) is mathematically expressed as the superposition of two distinct contributions: the product of refractive index (n) and physical length change ($\Delta L_{physics}$), combined with the interaction between fiber length (L) and refractive index variation (Δn). In this formulation, n represents the core's effective refractive index, while L characterizes the geometric length of the optical waveguide.

Interference light intensity expression:

$$I = I_1 + I_2 + 2\sqrt{I_1 I_2}\cos(\Delta\Phi) \tag{6}$$

The phase information is extracted by detecting the change of light intensity or using phase demodulation technology (such as heterodyne detection and PGC algorithm).

2.3 Distributed Optical Fiber Sensing

Distributed fiber optic sensors utilize optical fibers as continuous sensing media, detecting changes in the external environment by analyzing variations in the characteristics of scattered light within the fiber. However, most of the distributed optical fiber sensing techniques are focused on temperature, strain and vibration measurement of aerospace structures as these methods shows merits of long-distance sensing but the resolution and dynamic response are limited.

3 Optical Acoustic Emission (AE) Sensors for Damage Detection

Aim to solve the evaluation problem, acoustic Emission (AE) is typically used for characterizing the material performance, which is a passive nondestructive testing (NDT) technique that captures transient elastic waves generated by sudden energy release within a material under stress. These waves arise from structural failure mechanisms such as microcrack formation, delamination in composites, or fatigue-induced dislocation movements. AE signals carry unique frequency signatures (typically 20 kHz–1 MHz) that correlate with specific damage modes, enabling early detection of defects before catastrophic failure.

In aerospace applications, AE monitoring is critical for detecting hidden flaws in complex geometries like wing spars, fuselage panels, and composite joints. Traditional piezoelectric AE sensors, while effective, face limitations in harsh environments (e.g., electromagnetic interference, high temperatures). Optical AE sensors overcome these challenges by leveraging light-matter interactions to detect strain waves with unparalleled sensitivity and bandwidth.

3.1 Adaptive Fiber Laser-Based FBG AE Sensors

The experimental configurations incorporate fiber Bragg gratings (FBGs) exhibiting identical optical parameters, including resonance wavelength (λ_B), full-width at half-maximum (FWHM), physical length, and reflectivity (RFBG). To ensure comparable system performance, the signal-to-noise ratio (SNR) of the amplified spontaneous emission from the erbium-doped fiber amplifier (EDFA) based system is precisely controlled through optimization of the 980 nm pump source, achieving equivalent noise characteristics to those observed in the reflective semiconductor optical amplifier (RSOA) implementation.

EDFA-FBG Fibre Laser Model

The laser configuration employs a closed-loop resonator architecture incorporating an erbium-doped gain medium, where a fiber Bragg grating (FBG) serves as the wavelength-selective element within the optical cavity. This design utilizes an erbium-doped fiber amplifier (EDFA) as the primary gain source, establishing a unidirectional lasing path through the ring-shaped waveguide structure. The pump light is transmitted clockwise, and the ASE (amplified spontaneous emission) light is transmitted counterclockwise and reflected by the FBG, forming a circular amplification. An iterative algorithm is used to solve the steady-state photon number and power distribution in combination with boundary conditions [19–21]. Figure 9a illustrates the schematic configuration of the erbium-doped fiber amplifier (EDFA) based laser system incorporating a fiber Bragg grating (FBG) as the wavelength-selective component. The diagram presents the complete optical setup, demonstrating the

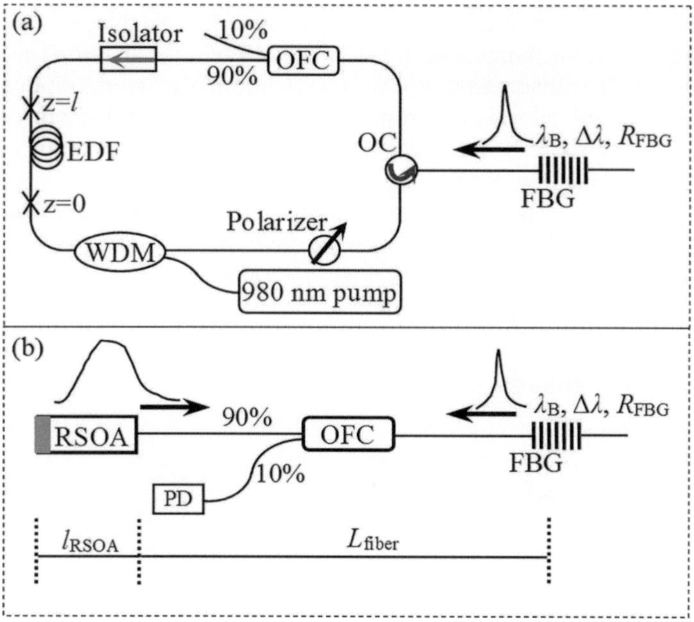

Fig. 9 Structural diagrams of **a** EDFA and **b** RSOA. Key components include: EDF (erbium-doped fiber), OFC (optical fiber coupler), OC (optical circulator), WDM (wavelength division multiplexing), and PD (photodetector) [32]

integration of the gain medium with the spectral filtering element within the laser cavity.

1. **Rate Equations**

Steady state conditions for the upper energy level particle number density $N_2(z,t)$ ($dN_2(z,t) / dt = 0$):

$$N_2 = \frac{W_p + W_{se}}{W_p+W_{se}+W_{sa}+\frac{1}{\tau_{21}}}, \tag{7}$$

where the spontaneous radiation lifetime τ_{21} of erbium is about 10 ms; rates of pumping, stimulated emission, and absorption:

$$W_p = \frac{\sigma_a(v_p)\eta_p P_p}{A_{\text{eff}} h v_p}, \tag{8}$$

$$W_{se} = \frac{\sigma_e(v) P_{\text{ASE}}}{A_{\text{eff}} h v}, \tag{9}$$

$$W_{sa} = \frac{\sigma_a(v) P_{\text{ASE}}}{A_{\text{eff}} h v}. \tag{10}$$

The mathematical formulation incorporates several key parameters: P_P represents the optical power of the pump source, while P_{ASE} denotes the amplified spontaneous emission power. The effective mode area (A_{eff}) characterizes the erbium-doped fiber's core dimensions, with σa and σe corresponding to the absorption and emission cross-sections of the gain medium, respectively. The parameter η_p quantifies the pump power confinement within the core region.

Under specific operational conditions where unidirectional pump propagation (clockwise direction) is established and the fiber Bragg grating reflects spontaneous emission, two boundary conditions are satisfied: the counter-propagating pump power P-p(0) becomes negligible, and the emission cross-section at the pump frequency $\sigma_e(v_p)$ approaches zero.

2. **Boundary Conditions**

Initial pumping power:

$$P_p^+(0) = P_0 \tag{11}$$

ASE power satisfied:

$$P_{ASE}^-(L) = R_{FBG}(v)\beta_{ring}P_{ASE}^+(L) \tag{12}$$

where RFBG is the FBG reflectivity and β_{ring} is the ring cavity loss. (13)

4 Reflective Semiconductor Optical Amplifier—Fiber Bragg Grating (RSOA-FBG) Linear Resonator Configuration

This laser architecture employs a straight-cavity design incorporating a reflective semiconductor gain element and a wavelength-selective fiber Bragg grating. The operational principle relies on electrical current injection to activate the RSOA's gain medium, while the FBG establishes optical feedback at a predetermined wavelength through selective reflection. The complete schematic representation of this linear resonator system is presented in Fig. 9b.

1. **Carrier density equation** [22, 23]:

$$\frac{dN}{dt} = \frac{I_{RSOA}}{eV} - R_{spon} - R_{stim}, \tag{14}$$

The mathematical formulation incorporates several fundamental parameters: N represents the concentration of charge carriers within the gain medium, while IRSOA denotes the applied injection current. The elementary charge constant (e) characterizes the quantum of electrical charge, and V corresponds to the spatial extent of the amplifying region. Additionally, R_{spon} quantifies the spontaneous radiative recombination rate, and R_{stim} describes the probability of stimulated emission events occurring within the semiconductor structure.

Rate of spontaneous radiation:

$$R_{spon} = AN + BN^2 + CN^3 \tag{15}$$

The recombination mechanism in semiconductor materials is characterized by three fundamental coefficients: A represents the rate of non-radiative carrier recombination through defect states, B corresponds to the bimolecular radiative recombination coefficient, and C quantifies the three-carrier Auger recombination probability. These parameters collectively describe the dominant carrier loss mechanisms in the active region of semiconductor devices.

Excited emission rate:

$$R_{\text{stim}} = \frac{\Gamma g}{d_w hc}(|E^+|^2 + |E^-|^2) \tag{16}$$

The mathematical formulation incorporates several physical parameters: Γ denotes the optical confinement factor within the gain medium, d_w represents the cross-sectional area of the amplifying region, and h corresponds to Planck's fundamental quantum constant. The parameter c characterizes the propagation velocity of electromagnetic waves in free space, while E ± describes the counter-propagating optical signals, including both the backward-traveling laser emission and the amplified spontaneous radiation within the reflective semiconductor optical amplifier (RSOA). The variable g quantifies the optical gain coefficient of the semiconductor material.

2. **Light field equations**

Forward and reverse light field transport equations:

$$\frac{\partial E^{\pm}}{\partial z} = \left[\frac{\Gamma g(1-i\alpha_{\text{LE}})}{2} - \frac{\alpha_{\text{int}}}{2}\right]E^{\pm} + \sqrt{R_{\text{sp}}}, \tag{17}$$

The model incorporates three critical parameters: α_{LE} represents the Henry factor characterizing spectral linewidth broadening, α_{int} quantifies the intrinsic optical losses within the gain medium due to scattering phenomena, and R_{sp} describes the spontaneous emission power spectral density exhibiting stochastic fluctuations with Gaussian noise characteristics [25].

Material gain g expression (InGaAsP material) [23]:

$$g = \frac{\pi e^2}{n_g \epsilon_0 c m_e m_{hh}} \left(\frac{m_e m_{hh}}{m_e + m_{hh}}\right)^{\frac{3}{2}} \sqrt{\frac{2}{\hbar^3}}(f_c - f_v), \tag{18}$$

The theoretical framework incorporates several fundamental parameters: m_e and m_{hh} represent the effective masses of charge carriers in the conduction and valence

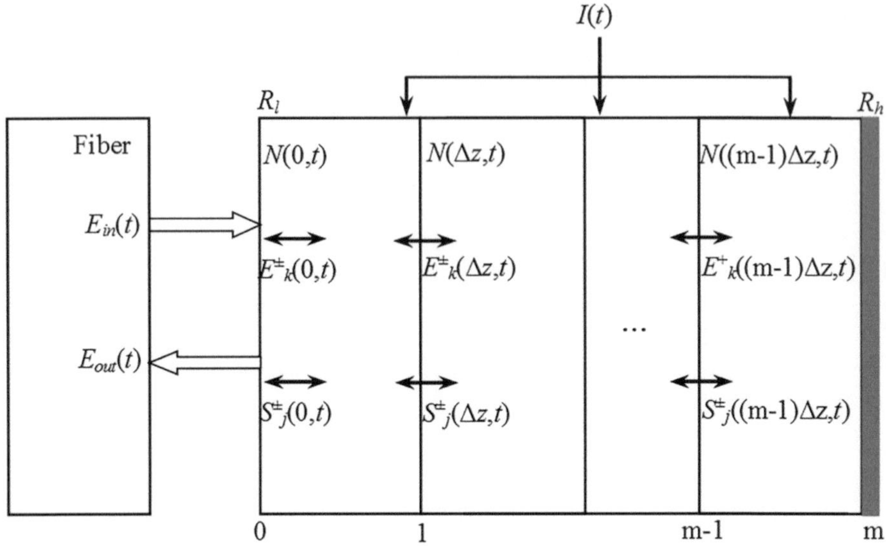

Fig. 10 The RSOA model for simulation [32]

bands, respectively, while n_g denotes the modal refractive index within the amplifying region of the reflective semiconductor optical amplifier (RSOA). Furthermore, $f_c(\lambda)$ and $f_v(\lambda)$ characterize the carrier population statistics through Fermi–Dirac distributions for both energy bands [24]. As illustrated in Fig. 10, the numerical simulation employs a discretization approach, dividing the RSOA into m-1 spatial segments. This computational model solves the coupled optical field and carrier density equations through iterative calculations, incorporating boundary conditions that account for input facet reflectivity and coupler transmission characteristics ($r_{coupler}$).

3. **Boundary conditions**

FBG reflection condition:

$$E^{+}(0) = R_{\text{coupler}} E^{-}(0) \tag{19}$$

$$E^{-}(\text{L}) = R_{\text{FBG}} E^{+}(\text{L}) \tag{20}$$

The system parameters include R_{FBG}, which characterizes the spectral reflectivity profile of the fiber Bragg grating, and rcoupler, representing the power coupling coefficient between the optical coupler and the reflective semiconductor optical amplifier (RSOA). These parameters play crucial roles in determining the optical feedback characteristics and power distribution within the laser cavity.

4.1 Static Performance Analysis

The investigation focuses on the spectral characteristics of amplified spontaneous emission (ASE) and signal outputs from both laser configurations under static conditions, where the fiber Bragg grating remains free from mechanical strain. In the numerical simulations, the reflective semiconductor optical amplifier (RSOA) operates with a direct current bias of 150 mA, and the optical cavity experiences a total loss of 5 dB. To establish comparable signal quality, the erbium-doped fiber amplifier (EDFA) based ring laser is optically pumped to achieve equivalent signal-to-noise ratio (SNR) performance relative to the RSOA-FBG linear cavity system.

The optical power emission can be mathematically expressed as:

$$I = \sum_{\lambda} \left| E_{PD}^{i}(t) \right|^2 \tag{21}$$

The temporal evolution of laser output characteristics, as depicted in Fig. 11a and b, demonstrates that both configurations reach stable operational conditions within millisecond timescales following pump or bias current initiation. Spectral characteristics are evaluated through numerical simulations incorporating 100 round-trip iterations, where each computational step corresponds to a complete photon circulation cycle within the resonator structure. Analysis reveals that the initial broadband amplified spontaneous emission undergoes significant spectral narrowing, ultimately achieving a linewidth of 0.25 nm that matches the reflection bandwidth of the fiber Bragg grating.

4.2 Dynamic Performance Characteristics

The transient responses of these laser systems exhibit significant variations owing to fundamentally different gain recovery mechanisms. Numerical simulations incorporate a dynamically strained fiber Bragg grating (central wavelength: 1550 nm, spectral bandwidth: 0.2 nm, reflectivity: 70%) experiencing periodic wavelength modulation at 10 kHz with peak-to-peak deviations of 10 pm (equivalent to ~ 8 με dynamic strain). Both systems maintain stable operation for minor wavelength perturbations below 5 pm, as evidenced in Fig. 12c and d. However, the EDFA-based system shows pronounced instability under larger dynamic strains (Fig. 12a), primarily due to its relatively slow gain recovery dynamics (millisecond timescale) that cannot effectively compensate for substantial spectral displacements. Conversely, the RSOA configuration, with its ultrafast gain recovery (sub-nanosecond response), demonstrates superior stability even under intense dynamic perturbations (Fig. 12b). The observed ~ 1 dB periodic power fluctuations originate from inherent gain spectrum non-uniformities in both systems. Notably, while the RSOA system exhibits enhanced dynamic stability, its lower overall gain results in reduced optical output power compared to the EDFA implementation.

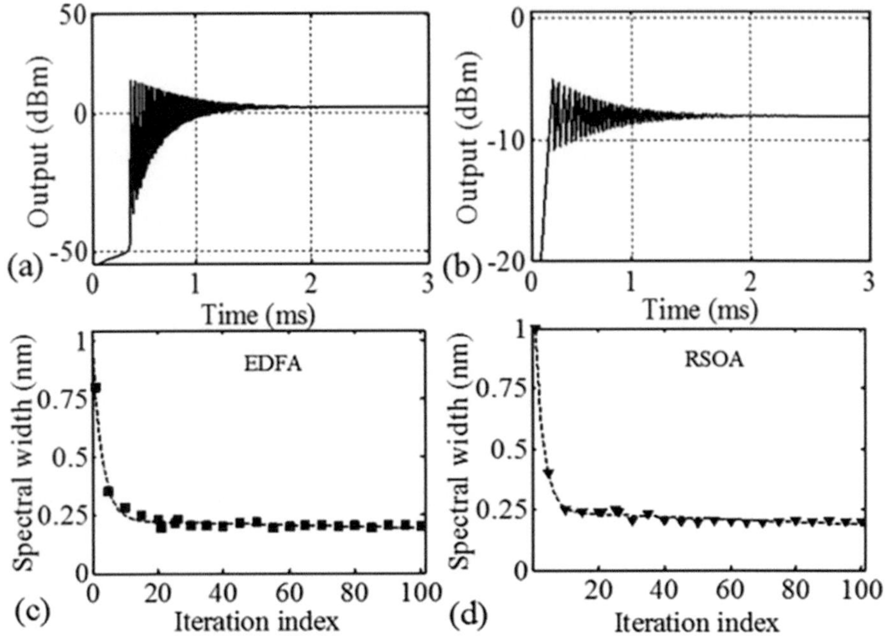

Fig. 11 Temporal output characteristics of the two laser systems after initiation: **a** EDFA-FBG, **b** RSOA-FBG, **c** amplified spontaneous emission (ASE) linewidth for EDFA-FBG, and **d** ASE linewidth for RSOA-FBG [32]

Experimental verification was conducted to corroborate the simulation results by analyzing the laser outputs under dynamic FBG perturbations. The fiber Bragg grating sensor, characterized by a 1544 nm center wavelength, 0.2 nm spectral bandwidth, and 70% reflectivity, was subjected to sinusoidal wavelength modulation at 10 kHz with peak deviations of 10 pm. As demonstrated in Fig. 12e and f, the EDFA configuration displays pronounced instability during large-amplitude perturbations, manifested as anomalous power spikes in the output signal. In contrast, the RSOA implementation maintains stable operation under identical dynamic conditions. For reduced modulation amplitudes, both systems exhibit consistent output characteristics with minimal power variations, in agreement with numerical simulation predictions.

4.3 FBG Phase Shift Demodulation

In the context of phase-shift demodulation for fiber Bragg grating sensors, these optical devices exhibit sensitivity to environmental factors including mechanical deformation and thermal variations. Such external stimuli induce measurable

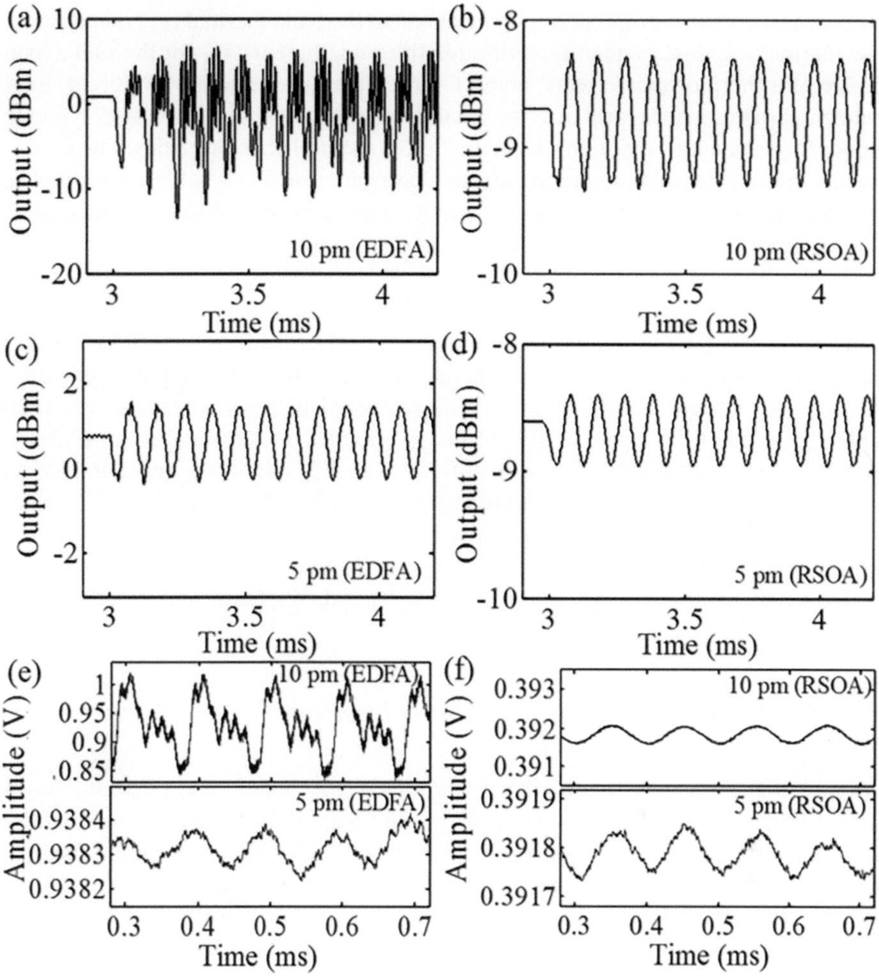

Fig. 12 Simulation and experimental results of EDFA and RSOA FBG lasers under a 10 kHz signal. The simulated dynamic spectral shifts are approximately 10 pm **a**, **b** and 5 pm **c**, **d**, while **e** and **f** depict the experimental outcomes [32]

displacements in the characteristic reflection spectrum. The relationship between the Bragg wavelength shift ($\Delta\lambda_B$) and external perturbations is [26]:

$$\frac{\Delta\lambda_B}{\lambda_B} = (1 - p)\varepsilon + (\alpha_\Lambda + \alpha_n)\Delta T \tag{22}$$

Within the theoretical framework, λ_B corresponds to the characteristic resonance wavelength of the fiber Bragg grating sensor, with $\Delta\lambda_B$ representing its spectral displacement caused by thermomechanical perturbations. The parameter p denotes the strain-optic coefficient, having a nominal value of 0.22, while ε characterizes the composite strain comprising both mechanical (ε_M) and thermal (ε_T) components. The thermal response is quantified through three coefficients: α_Λ for thermal expansion, α_n for thermo-optic effects, and ΔT for temperature variation in Kelvin. Environmental disturbances, primarily thermal fluctuations and ambient noise, typically induce low-frequency modulations in the FBG's spectral characteristics. The conversion from wavelength modulation $\delta\lambda(t)$ to phase variation $\delta(t)$ is achieved through an asymmetric Michelson interferometric configuration, as detailed in reference [27]:

$$\delta(t) = \frac{2\pi d \Delta\lambda_B}{\lambda_B^2} \tag{23}$$

The parameter d in the equation denotes the optical path difference (OPD) between the interferometer arms. From the mathematical relationship, it is apparent that operating within the quadrature region yields enhanced phase modulation sensitivity with increasing OPD, theoretically producing stronger interference fringe visibility. However, the spectral characteristics of the RSOA-FBG light source, characterized by finite spectral bandwidth, lead to reduced coherence length at larger OPD values. This fundamental limitation necessitates a comprehensive evaluation of demodulation system performance, as presented in subsequent analysis, to identify optimal configuration parameters that balance sensitivity and coherence requirements.

4.4 Demodulation Sensitivity Considerations

Optimal signal detection requires operation at the quadrature position, where optical intensity varies linearly with phase modulation. This operational condition is achieved through a closed-loop control system, employing proportional-integral-derivative (PID) feedback to precisely adjust the piezoelectric transducer position, thereby controlling the optical path difference (δd). When stabilized at the quadrature point, the photodetector response exclusively reflects the high-frequency spectral components of the fiber Bragg grating's modulated signal:

$$\Delta I(t) = 2\sqrt{I_R}\sqrt{I_S}|\gamma_{12}(d)|\sin[\delta_h(t)] \tag{24}$$

Typically, when the OPD (optical path difference) is zero, the value of γ_{12} equals 1. However, as the OPD increases, the complex degree of coherence undergoes a significant reduction.

The relationship between signal amplitude and optical path difference (OPD) is illustrated in Fig. 13 for fiber Bragg grating sensors exhibiting different spectral characteristics. Three distinct spectral displacement cases are examined: 0.1 pm (Fig. 13a), 10 pm (Fig. 13b), and 100 pm (Fig. 13c). At zero OPD, the system exhibits no phase modulation, resulting in the absence of detectable wavelength-demodulated signals. With increasing OPD, the signal amplitude initially reaches a peak value before experiencing gradual attenuation. Sensors with reduced spectral bandwidth demonstrate enhanced performance, characterized by greater optimal OPD values and higher demodulation signal amplitudes. However, this improved sensitivity comes at the cost of requiring longer sensor lengths, which inherently limits the high-frequency detection capability for dynamic strain measurements, particularly when monitoring stress wave propagation [28]. Figure 13d demonstrates this trade-off through comparative analysis of demodulator outputs for a 0.4 nm bandwidth FBG sensor under varying spectral displacements (0.1, 10, and 100 pm).

To quantitatively evaluate system performance, a sensitivity parameter is defined as:

$$S = \frac{\partial \Delta I(t)}{\partial \Delta \lambda_B} \tag{25}$$

The sensitivity varies across different OPDs, with the condition for maximum sensitivity being $\partial S/\partial d = 0$. When dealing with small strains (where $\Delta\lambda_B$ is on the order of pico-meters), the optimal OPD is determined as follows:

$$\Delta\lambda^2 d^2 = \frac{2\ln 2}{\pi^2}\lambda_B^4 \tag{26}$$

The mathematical formulation demonstrates a proportional dependence between the wavelength shift-path difference product ($\Delta\lambda \cdot d$) and the square of the Bragg

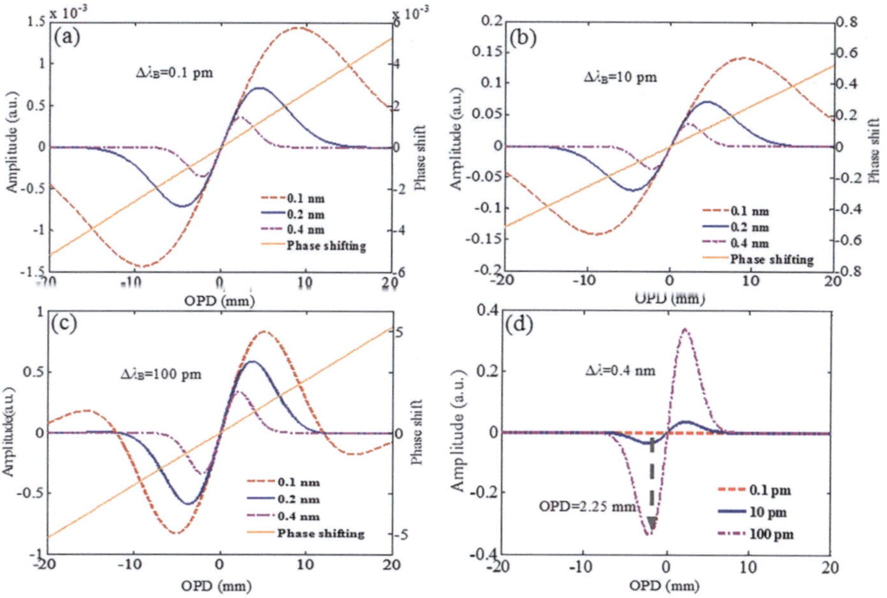

Fig. 13 Demodulator output for FBGs with spectral widths of 0.1 pm **a**, 10 pm **b**, and 100 pm **c**, each subjected to spectral shifts. Additionally, **d** shows the demodulator output versus OPD for an FBG with a 0.4 nm spectral width under spectral shifts of 0.1 pm, 10 pm, and 100 pm [33]

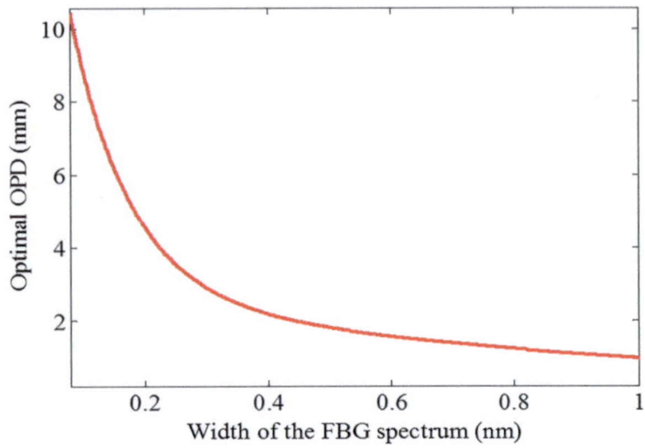

Fig. 14 Optimal optical path difference (OPD) as a function of the spectral width of the FBG sensor (λ_B = 1550 nm) for unbalanced interferometric techniques [33]

wavelength (λ^2_B). Figure 14 illustrates the inverse proportionality between optimal optical path difference and spectral bandwidth. For instance, a fiber Bragg grating sensor with 1550 nm center wavelength and 0.2 nm reflection bandwidth achieves maximum sensitivity at an OPD of 4.5 mm when subjected to minor mechanical deformations. The upper limit of detectable dynamic strain is fundamentally determined by the phase modulation constraint, where the induced phase variation must remain below π/2 radians to maintain measurement validity:

$$\Delta\lambda_B < \frac{\lambda_B^2}{4d} \tag{27}$$

For a fiber Bragg grating sensor operating at 1550 nm center wavelength with 0.2 nm spectral bandwidth, this physical constraint limits the maximum detectable wavelength shift ($\Delta\lambda_B$) to below 133.47 pm, corresponding to a mechanical strain of approximately 111.23 microstrains (με).

4.5 *Wavelength Demodulation*

Figure 15 illustrates the integration of an adaptive two-wave mixing (TWM) demodulator with the reflective semiconductor optical amplifier (RSOA) based light source. This configuration utilizes the dynamic holographic properties of photorefractive materials, which inherently suppress low-frequency noise through adaptive filtering [27]. The optical setup begins with laser amplification followed by beam splitting through a 1 × 2 coupler configured for 95/5 power distribution. The weaker beam component (5%), serving as the signal beam, and the stronger component (95%), functioning as the pump beam, propagate through asymmetric optical paths before reaching the iron-doped indium phosphide (InP:Fe) photorefractive crystal. The TWM phenomenon emerges when the pump beam undergoes diffraction from

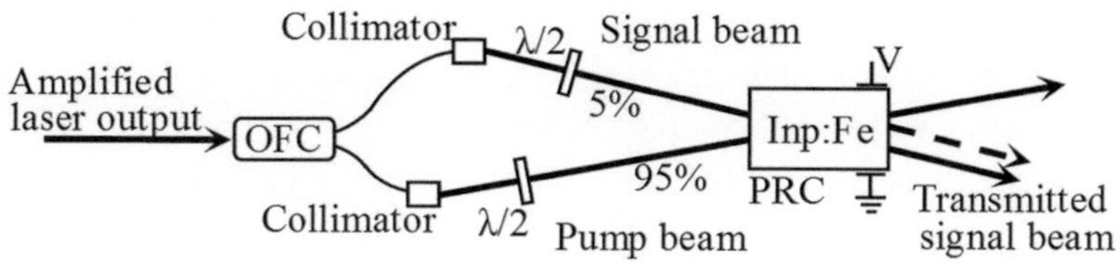

Fig. 15 Experimental setup wavelength demodulator [32]

the induced photorefractive grating, resulting in coherent interference with the transmitted signal beam. The intensity of this interference pattern can be mathematically described by the following expression [30]:

$$E_d(t) = E_0\{[\exp(\gamma L) - 1] + \exp[i\delta(t)]\} \tag{28}$$

Within the theoretical framework, E_0 corresponds to the signal beam's complex field amplitude, characterizing both magnitude and phase information. The parameter γ quantifies the energy transfer efficiency through the photorefractive medium, representing the diffraction grating's coupling strength. The interaction length within the photorefractive crystal is denoted by L, while δ(t) describes the time-dependent phase modulation imposed on the signal beam during propagation.

Within the TWM spectral demodulator, a spectral shift generated by the FBG sensor in the pump and signal beams traversing different paths leads to a phase shift in the signal. This phase shift, denoted as δ(t), is expressed by the following equation [27]:

$$\delta(t) = \frac{2\pi d}{\lambda_b^2}\Delta\lambda_B \tag{29}$$

Here, d represents the OPD. The mathematical formulation presented in Eq. (29) indicates a proportional relationship between optical path difference (OPD) and induced phase modulation. However, this relationship is fundamentally limited by the spectral characteristics of the laser source, where increasing OPD values result in interference pattern degradation due to finite spectral bandwidth (Δλ). Reference [29] provides the theoretical correlation between the FBG sensor's optimal OPD and its spectral bandwidth, expressed as:

$$d = \frac{\sqrt{2\ln 2}}{\pi}\frac{\lambda_B^2}{\Delta\lambda} \tag{30}$$

For dynamic strain measurement applications, a fiber Bragg grating sensor (λ_B = 1552 nm, Δλ = 0.2 nm) was mounted on a piezoelectric transducer (PZT) based stretching apparatus. TWM being inherently immune to quasi-static variations while sensitive to dynamic perturbations, was tested under 5 kHz oscillatory strain conditions with 2 με amplitude. Initial characterization involved evaluating wavelength-demodulated signals across different optical path differences (OPDs) using the RSOA laser configuration. Figure 16 demonstrates the experimental protocol: a pulsed DC

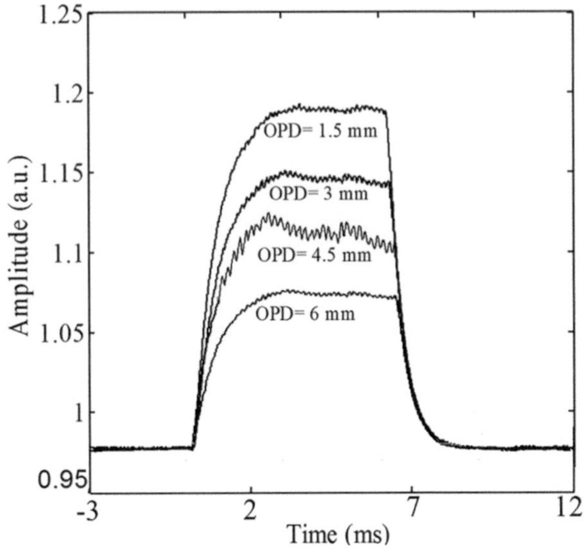

Fig. 16 Wavelength demodulated signal values [32]

field (0–7 ms) initialized the photorefractive grating formation, followed by continuous dynamic strain application. With OPD variation from 1.5 mm to 6 mm, the system exhibited decreasing TWM gain (attributed to coherence reduction) and increasing phase modulation (resulting from extended optical path difference). The experimentally determined optimal OPD of 4.5 mm agrees well with theoretical predictions from Eq. (30).

The dynamic performance characteristics of both laser configurations were systematically evaluated under varying strain conditions. Sinusoidal mechanical perturbations at 10 kHz frequency were applied with amplitudes of 2, 5, and 15$\mu\varepsilon$. Figure 17 presents the TWM demodulation results. Comparative analysis reveals superior signal amplitude in the EDFA's system compared to its RSOA counterpart. While both systems demonstrate proportional response to increasing strain amplitudes, the EDFA implementation exhibits instability at approximately 15$\mu\varepsilon$ strain levels, a phenomenon absent in the RSOA configuration.

The erbium-doped fiber amplifier (EDFA) system faces limitations in multiplexing multiple fiber Bragg gratings (FBGs) due to homogeneous broadening effects inherent to erbium ions [31]. In contrast, the reflective semiconductor optical amplifier (RSOA) based configuration supports multi-wavelength operation through its inhomogeneous broadening characteristics. This capability was experimentally demonstrated using a photorefractive two-wave mixing (TWM) demodulation scheme for multiplexed FBG sensing, as illustrated in Fig. 18. The experimental setup incorporated four elements ($\Delta\lambda = 0.2$ nm) with center wavelengths at 1544 nm, 1548 nm, 1552 nm, and 1564 nm, spaced approximately 0.5 m apart along a single optical fiber. The resulting output spectrum (Fig. 19) displays four distinct lasing peaks with signal-to-noise ratios (SNR) surpassing 40 dB. Dynamic strain measurements were conducted using the TWM interferometric demodulator,

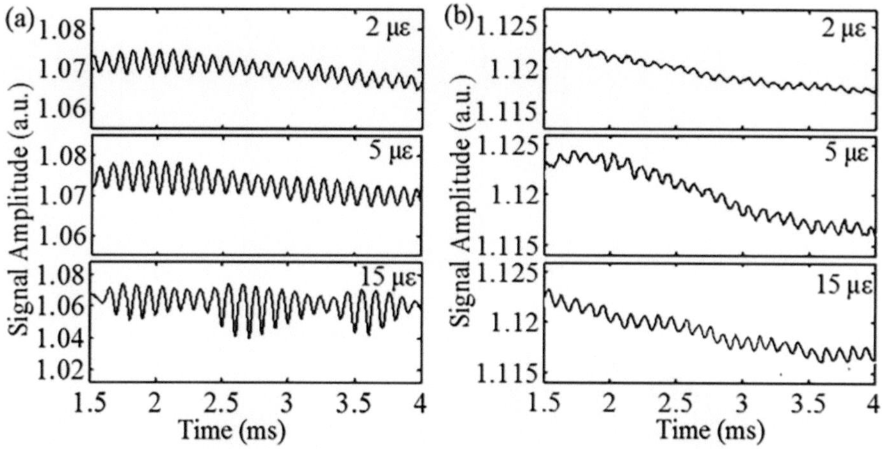

Fig. 17 Experimental demodulation results based on EDFA **a** and RSOA **b**, with a dynamic strain of 10 kHz and amplitudes of 2 με, 5 με, and 15 με [32]

which has established superior performance in wavelength-division multiplexing applications [27]. The free-space TWM output was fiber-coupled and subsequently separated using band-drop filters (BDF) with 4 nm bandwidths centered at 1544 nm, 1548 nm, and 1552 nm, enabling independent detection of each FBG signal through dedicated photodetectors.

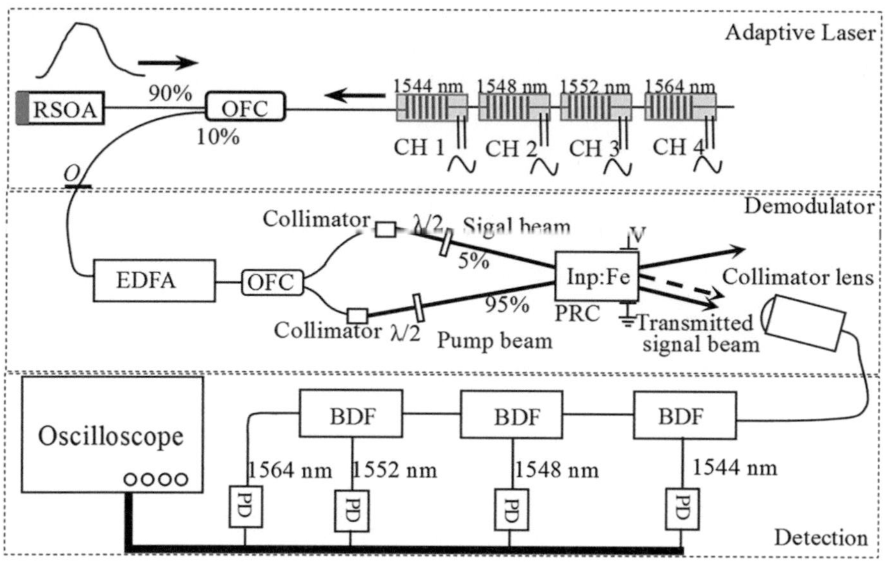

Fig. 18 Experimental configuration system [32]

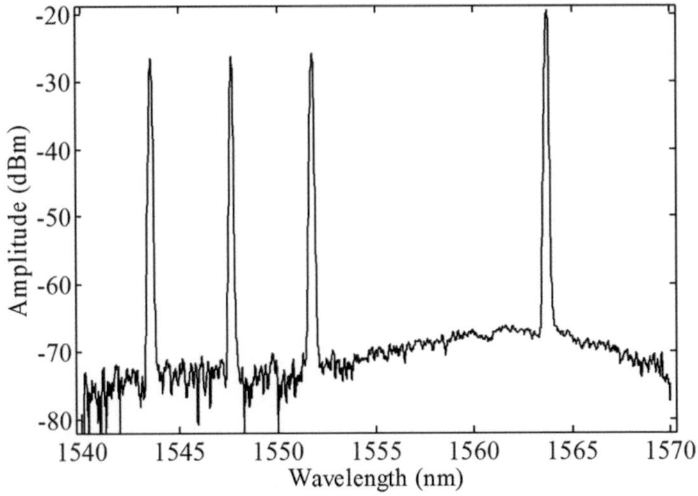

Fig. 19 Output performance of the four-channel RSOA-FBG fiber cavity laser system [32]

A multiplexed system was examined using four fiber Bragg gratings (FBGs), each with a 0.2 nm linewidth, attached to piezoelectric transducer (PZT) stretchers. Different sinusoidal dynamic strains were applied to each FBG: channel 1 (CH1, $\lambda = 1543.57$ nm) received a 4 kHz signal with an amplitude of 5 με, channel 2 (CH2, $\lambda = 1547.64$ nm) was subjected to a 5 kHz signal at the same amplitude, channel 3 (CH3, $\lambda = 1551.76$ nm) experienced a 10 kHz signal, and channel 4 (CH4, $\lambda = 1563.76$ nm) was exposed to a 20 kHz signal, both with 5 με amplitude. The demodulation outcomes for these channels are illustrated in Fig. 20.

To analyze the frequency characteristics of the four channels, a fast Fourier transform (FFT) was utilized, with the resulting spectra depicted in Fig. 21. The data reveals negligible interference between the channels, demonstrating that the integration of a multichannel RSOA-FBG dynamic strain sensing system with a TWM spectral demodulator achieves efficient and adaptable demodulation for multiple FBG-based dynamic strain sensors.

4.6 Aerospace Impact Loading

Under impact loading conditions, a plate produces guided Lamb wave modes, whose acoustic wavelengths are determined by variables including the material properties of the plate, the size of the steel ball, and the speed of impact. In their study, Kirikera et al. [26] employed FBG sensors to capture Lamb waves induced by steel ball impacts and successfully estimated the impact position using an adaptive TWM-FBG dynamic strain sensing system.

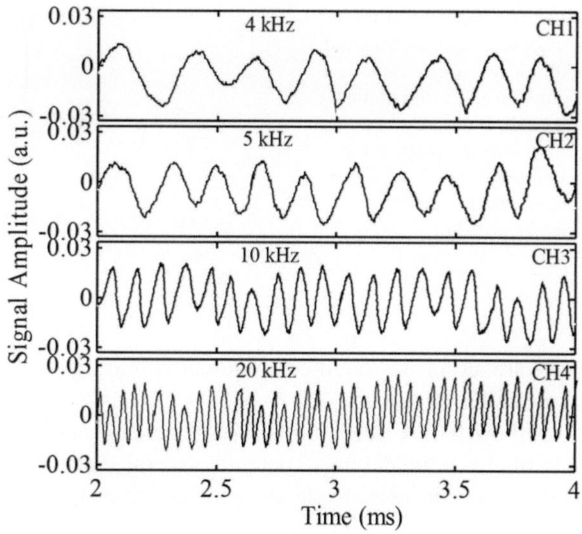

Fig. 20 Simultaneous signal demodulation. Each sensor is subjected to dynamic strains: CH1 at 4 kHz and 5 με, CH2 at 5 kHz and 5 με, CH3 at 10 kHz and 5 με, and CH4 at 20 kHz and 5 με [32]

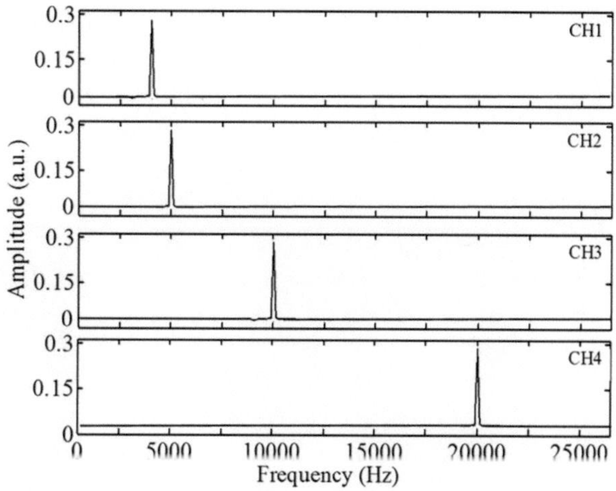

Fig. 21 Fourier spectra of the four-channel signals detected in Fig. 20, obtained using a photorefractive TWM interferometer-based wavelength demodulator [32]

For these tests, a fiber Bragg grating (FBG) sensor with a reflective spectrum linewidth of 0.2 nm was attached to an aluminum plate, as shown in Fig. 22. The separation between the FBG sensor € and the PZT transducer (f) was approximately 5 mm. Impact events were simulated by propelling steel balls at different speeds, while the PZT sensor acted as a trigger to monitor the generated surface waves.

The time-domain signals were analyzed using wavelet transforms. The results demonstrate that the FBG sensor captures acoustic energy up to around 400 kHz, while the PZT transducer detects frequencies extending to approximately 450 kHz, as depicted in Fig. 23c. Theoretical analysis of Lamb waves, performed with COMSOL

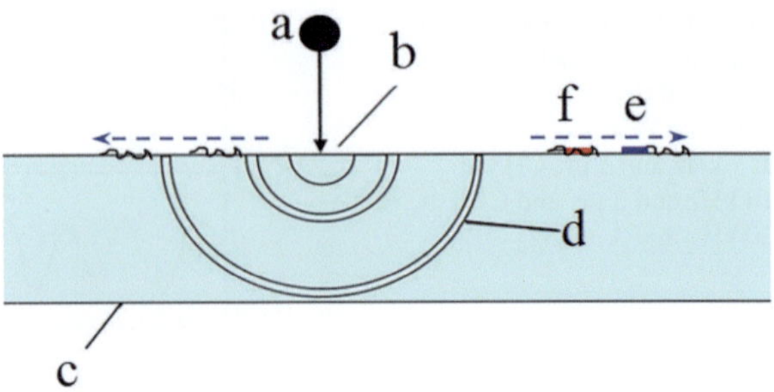

Fig. 22 Schematic of the impact loading experimental setup. Key components include: **a** a 1.5 mm diameter steel ball, **b** the impact point, **c** an aluminum plate, **d** the generated wave, **€** the FBG sensor, and **f** the PZT transducer [33]

software, confirms that the experimental findings correspond to the fundamental A0 mode of Lamb waves.

Although the FBG exhibits slightly lower sensitivity compared to PZTs, it remains suitable for practical use. There are several aspects require further exploration, such as the FBG sensor's attachment method, sensing range, and orientation. The overall system performance, including the demodulator's sensitivity, the signal-to-noise ratio of the captured signal, and the responsiveness, will ultimately determine the threshold for detecting minimal dynamic strain.

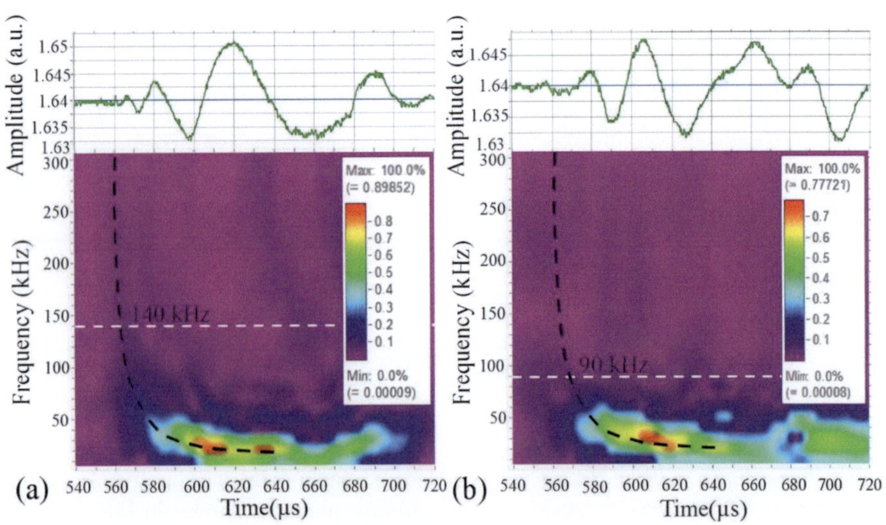

Fig. 23 Dynamic strain performance of the FBG and PZT sensors: **a** time-domain characteristics; **b**, **c** time–frequency responses [33]

5 Challenges and Future Trends

Optical fiber-based multi-physical sensors capable of simultaneous strain, temperature, and humidity monitoring have captured 0.05 μm deformations and 0.1 °C thermal variations in C919 composite wing fatigue tests. The first issue is how to design multi-physical sensors integrated within single fiber for multi-parameter co-sensing. The second issue is extreme environmental adaptation. In cryogenic testing facilities at the Wenchang Space Launch Center, novel fiber-optic strain sensors are being integrated into the Chang'e-8 lunar probe's landing gear. These sensors demonstrate operational stability across temperature ranges from -190 °C to 120 °C, enabling real-time micro-strain measurements during lunar touchdown. In hypersonic research, sapphire fiber sensors with Al_2O_3/SiC composite coatings (developed via chemical vapor deposition) maintain 0.01% measurement accuracy after 1000 thermal cycles up to 2000 °C. These advancements directly support thermal protection system monitoring for hypersonic vehicles. Next issue is fiber material as the material is mostly based on silica. How to improve or reduce thermal sensitivity for specific applications is important to extreme environmental sensing. For example, Harbin Institute of Technology researchers are advancing hollow-core photonic crystal fibers with reduced thermal sensitivity (0.05 pm/°C), enabling precision environmental compensation for spacecraft instrumentation. It has developed carbon nanotube-coated fiber sensors demonstrating 1-year stability at 1000 °C for rocket engine combustion chamber monitoring. The evolution of optical sensors in aerospace SHM is characterized by material science breakthroughs, multi-physical parameter integration, and system miniaturization. Future advancements will focus on extending operational limits in extreme environments, optimizing multi-physics sensing algorithms, and realizing fully integrated photonic SHM systems. These developments are critical for ensuring the safety and reliability of next-generation aerospace vehicles in deep-space exploration and hypersonic applications.

References

1. K. Ranasinghe, R. Sabatini, A. Gardi, S. Bijjahalli, R. Kapoor, T. Fahey, K. Thangavel, Advances in Integrated System Health Management for mission-essential and safety-critical aerospace applications. Prog. Aerosp. Sci. **128**, 100758 (2022)
2. C. Tang, G.Y. Tian, K. Li, R. Sutthaweekul, J. Wu, Smart compressed sensing for online evaluation of CFRP structure integrity. IEEE Trans. Industr. Electron. **64**(12), 9608–9617 (2017)
3. O. Ahmed, X. Wang, M.V. Tran, M.Z. Ismadi, Advancements in fiber-reinforced polymer composite materials damage detection methods: Towards achieving energy-efficient SHM systems. Compos. B Eng. **223**, 109136 (2021)
4. Y. Dong, F. Ansari, Non-destructive testing and evaluation (NDT/NDE) of civil structures rehabilitated using fiber reinforced polymer (FRP) composites, in *Service life estimation and extension of civil engineering structures*. (Woodhead Publishing, 2011), pp. 193–222
5. J. Chen, Z. Yu, H. Jin, Nondestructive testing and evaluation techniques of defects in fiber-reinforced polymer composites: a review. Front. Mater. **9**, 986645 (2022)

6. H. Rocha, C. Semprimoschnig, J.P. Nunes, Sensors for process and structural health monitoring of aerospace composites: A review. Eng. Struct. **237**, 112231 (2021)
7. R.O. Ogunleye, S. Rusnáková, J. Javořík, M. Žaludek, B. Kotlánová, Advanced sensors and sensing systems for structural health monitoring in aerospace composites. Adv. Eng. Mater. **26**(22), 2401745 (2024)
8. M. Philibert, K. Yao, M. Gresil, C. Soutis, Lamb waves-based technologies for structural health monitoring of composite structures for aircraft applications. Eur. J. Mater. **2**(1), 436–474 (2022)
9. M. Abbas, M. Shafiee, Structural health monitoring (SHM) and determination of surface defects in large metallic structures using ultrasonic guided waves. Sensors **18**(11), 3958 (2018)
10. L. Capineri, A. Bulletti, Ultrasonic guided-waves sensors and integrated structural health monitoring systems for impact detection and localization: a review. Sensors **21**(9), 2929 (2021)
11. V. Janapati, F. Kopsaftopoulos, F. Li, S.J. Lee, F.K. Chang, Damage detection sensitivity characterization of acousto-ultrasound-based structural health monitoring techniques. Struct. Health Monit. **15**(2), 143–161 (2016)
12. H. Zhang, J. Si, H. Zhang, H. Wei, Y. Chen, W. Zhu, K. Xu, Q. Zhu, Ultrasonic detection of wrinkles in composites with gradual phase shift migration. Ultrasonics **148**, 107557 (2025)
13. L.-M. Mao, *Fibre-optic ring laser sensing system for ultrasonic detection.* Chongqing University of Technology (2020)
14. A. Roy, A.L. Chakraborty, C.K. Jha, Fiber Bragg grating interrogation using wavelength modulated tunable distributed feedback lasers and a fibre-optic Mach-Zehnder interferometer. Appl. Opt. **56**, 3562–3569 (2017)
15. G. Xue, L. Gong, X. Fang, X. Hu, Measurement accuracy of FBG used as a surface-bonded strain sensor installed by adhesive. Appl. Opt. **57**, 2939–2946 (2018)
16. T. Erdogan, Fiber grating spectra[J]. Light. Technol. **15**(8), 1277–1294 (1997)
17. C. Wu, F. Wu, Fibre grating sensing principle and application, vol. 3, pp. 2–5. (National Defence Industry Press, Beijing, 2011)
18. Y. Rao, Y. Wang, T. Zhu, Principles and applications of optical fibre grating, vol. 136–140, pp. 168–176. (Science Press, Beijing 2006)
19. P.R. Morkel, R.I. Laming, Theoretical modeling of erbium-doped fiber amplifiers with excited-state absorption. Opt. Lett. **14**(19), 1062–1064 (1989)
20. M. Li, K. Liu, W. Jing, G. Peng, Fiber ring laser intra-cavity absorption spectroscopy for gas sensing: analysis and experiment. J. Opt. Soc. Korea **14**(1), 14–21 (2010)
21. Y. Zhang, M. Zhang, W. Jin, H.L. Ho, M.S. Demokan, X.H. Fang, B. Culshaw, G. Stewart, Investigation of erbium-doped fiber laser intra-cavity absorption sensor for gas detection. Opt. Commun. **232**(1–6), 435–441 (2004)
22. S.A. Gebrewold, L. Marazzi, P. Parolari, R. Brenot, S.P. Dúill, R. Bonjour, D. Hillerkuss, C. Hafner, J. Leuthold, Reflective-SOA fiber cavity laser as directly modulated WDM-PON colorless transmitter. IEEE J. Sel. Top. Quant. **20**(5), 503–511 (2014)
23. Z. Vujičić, R.P. Dionísio, A. Shahpari, N.B. Pavlović, A. Teixeira, Efficient dynamic modeling of the reflective semiconductor optical amplifier. IEEE J. Sel. Top. Quant. **19**(5), 1–10 (2013)
24. N.G. Nilsson, Empirical approximations for the Fermi energy in a semiconductor with parabolic bands. Appl. Phys. Lett. **33**(7), 653–654 (1978)
25. A.M. De. Melo, K. Petermann, On the amplified spontaneous emission noise modeling of semiconductor optical amplifiers. Opt. Commun. **181**(18), 4598–4605 (2008)
26. G.R. Kirikera, O. Balogun, S. Krishnaswamy, Adaptive fiber bragg grating sensor network for structural health monitoring: Applications to impact monitoring. Struct. Health Monit. **10**, 5–16 (2010)
27. Y. Qiao, Y. Zhou, S. Krishnaswamy, Adaptive demodulation of dynamic signals from fiber Bragg gratings using two-wave mixing technology. Appl. Opt. **45**, 5132–5142 (2006)
28. C. Davis, D. Robertson, C. Brooks, P. Norman, C. Rosalie, N. Rajic, Reduced length fibre Bragg gratings for high frequency acoustic sensing. Meas. Sci. Technol. **25**, 125105 (2014)
29. H. Wei, C. Tao, S. Krishnaswamy, Fiber Bragg grating dynamic strain sensor using an adaptive reflective semiconductor optical amplifier source. Appl. Opt. **55**(10), 2752–2759 (2016)

30. P. Delaye, Detection of ultrasonic motion of a scattering surface by photorefractive Inp: Fe under an applied dc field. Opt. Soc. Am. B **14**(7), 1723–1734 (1997)
31. C. Yeh, M. Lin, C. Lee, S. Chi, Fiber Bragg grating-based multiplexed sensing system employing fiber laser scheme with semiconductor optical amplifier. Jpn. J. Appl. Phys. **44**(9A), 6590–6592 (2005)
32. H. Wei, S. Krishnaswamy, Comparative assessment of erbium fiber ring lasers and reflective SOA linear lasers for fiber Bragg grating dynamic strain sensing. Appl. Opt. **56**(13), 3867 (2017)
33. H. Wei, C. Tao, Y. Zhu, S. Krishnaswamy, Fiber Bragg grating dynamic strain sensor using an adaptive reflective semiconductor optical amplifier source. Appl. Opt. **55**(10), 2752 (2016)

FBG Sensor Applications in Aerospace Engineering

V. R. Balaji, M. A. Ibrar Jahan, Gaurav Kumar Bharti, Archana Yadav, Srujana, and Rajini V. Honnungar

Abstract Fiber Bragg grating (FBG) sensors have emerged as a preferred choice in various industries, including telecommunications, civil, aerospace, and automotive, for monitoring various parameters due to the advancements in fiber optic technology. They possess exceptional qualities, including outstanding sensitivity, flexibility, lightweight, multiplexing, and immunity to electromagnetic interference. FBG sensors are an exemplary choice for aerospace engineering applications that require remote detection, high precision, and lightweight design. These sensors leverage changes in fiber properties caused by perturbations like polarization, strain or temperature etc. When these perturbations occur, they either alter grating pitch or modal index, resulting in a relative resonance wavelength shift. Measuring this shift enables the detection and quantification of physical changes, yielding essential data for structural integrity evaluations. This chapter covers the utilization of FBG sensors for monitoring the structural health in aircrafts/ Avionics industry. It also covers the case studies that emphasize using FBG for Structural Health Monitoring (SHM) in aircraft, showcasing their efficacy in practical situations. It also illustrates how SHM can contribute to safer, more efficient operations in aerospace engineering.

V. R. Balaji
Centre for Healthcare Advancement, Innovative and Research, Vellore Institute of Technology, Chennai, Chennai, India

M. A. Ibrar Jahan (✉) · R. V. Honnungar
Department of Electronics and Communication Engineering, RNS Institute of Technology, VTU, Bengaluru, India
e-mail: ibrarjahan.ec@gmail.com

G. K. Bharti
School of Electrical, Electronics & Communication Engineering, Galgotias University, Noida, Uttar Pradesh, India
e-mail: gauravkumarbharti7@gmail.com

A. Yadav
Department of Electronics and Communication Engineering, Integral University, Lucknow, India

Srujana
Department of Electronics and Communication Engineering, IEEE Member, Bengaluru, India

C. Marques et al. (eds.), *Advanced Optical Sensors for Aerospace Applications*,
Progress in Optical Science and Photonics 37,
https://doi.org/10.1007/978-981-95-1626-1_4

Keywords FBG · Sensor · Aircraft · Health · Monitoring

1 Introduction

Extensive, intricate, and expensive engineering buildings are primarily designed for longevity. This is particularly recognised for civil engineering structures as well as Ships, Heavy equipment/Machinery, Aeroplanes and Trains. Several among these components are utilised for degradation. An engineering structure is inherently intended to endure degradation for a certain foreseeable lifespan. As the lifetime extends, it turns out to be very challenging to forecast. Such a structure may consequently needs a plan for maintenance/reparability. Historically, this has frequently been insufficiently addressed and may still be the case now. The rationale is that most limited-life goods are seen as having a longer lifespan after their design is finalised. An excellent instance is aviation vehicles. Aircraft, historically constructed entirely from metals, are engineered for a predetermined number of flying hours, determined by typical usage in relation to fatigue. Individuals familiar with fatigue testing understand that the fatigue life of a certain material exhibits considerable variability. A score of two in life is typically considered low. Given that aeroplane structures, and maybe others, are engineered for a low chance of fracture, it is improbable that a component would shatter after meeting its design specifications, unless it was intentionally constructed to be fail-safe Owing to dispersion, it may persist for up until it has already or maybe longer [1]. The inquiry thus emerges: depending on what conditions can we sustain the use of this building beyond its original intended lifespan?

Flight hours, typically employed as a metric for aircraft utilisation and longevity, may not be the only criteria. An aeroplane is frequently used outside the circumstances under which it was originally constructed. This is evident in the military aviation sector. Can the aircraft be utilised until the designated flying hour objective has been met? Initially, the response is: No! One reason is that environmental loads contribute to corrosion, which is challenging to forecast and measure, necessitating determination by inspection. Exceeding the design objective in the context of an operating life update necessitates a significantly revised corrosion in section strategy. A further rationale for not immediately prolonging the operational lifespan of an undamaged structure is that the operating circumstances may differ from those present at the time of the structure's original design. An aircraft, particularly in the military industry, may be required to transport various payloads (e.g., diverse weaponry) and operate within varied flight envelopes owing to advanced flight control systems. Can the existing framework endure this? A reconsideration of the structure is necessary to address this question. Regardless of the supplementary work needed to prolong the operational lifespan of a fading infrastructure, it is prudent to do this form of study, particularly when an infrastructure is expensive. This is particularly accurate for aeroplanes, as the incidents of September 11 have not significantly altered the circumstances. Aircraft have become integral to public transit and defence. Innumerable aircrafts, numbering in tens of thousands have been manufactured throughout the years, with a large number of them continuing to be in operation. Experiences incurred with these flying machines, have revealed that there has been an escalating demand for inspection and maintenance, increasing year after year. It was analyzed that an EF-111A, a fighter plane, required roughly 2200 manhours for inspection and maintenance alone, just for a single aircraft; a full fleet would require many times that number. In 1996, the observed figure rose substantially to requiring manhours exceeding 8000 annually per Aeroplane. Often, magnitude of such undertaking is significantly influenced by the aircraft's usage, complexity, and dimensions. Kudva said that some members of 1993 were inspected solely, with costs of 88, 4, and 29 USD for every hour flown by an F-18 and a T-38, respectively [2]. Table 1 depicts the military ageing aircraft overview.

Any design that is damage-tolerant permits the development of a damage. This can be accomplished by either allowing a fissure to develop to a specific length, at which point, the fissure halted either with the use of a crack-stopper or when the very component fractures, which would transfer the forces acting on it to another part. However, damage tolerance can also be established by presuming that a fracture is present at a location that is difficult to inspect and determining the extent to which the deformity is permitted to expand before it becomes critical.

In all the instances, a fatigue timeline calculation must be conducted, regardless of whether the design is damage-tolerant or safe. This necessitates a load spectrum, 'Kt' which is a factor taking into consideration the stress concentration, a curve (S–N curve) mapping the fatigue timeline, and the stress concentration in the grooves need to be described. The Fatigue timeline curve is experimentally determined through the notched modules using corresponding value of the stress concentration factor (Kt) or through data recording the fatigue, retrieved from unnotched specimens with use of the relationship between notch and strain [3]. The Palmgren–Miner damage accumulation rule is now employed to calculate fatigue life damage for each cycle as expressed in Eq. (1)

Table 1 Overview of ageing military aircraft

Aircraft type	Built year	Total service	Life extension
Boeing B-52H Stratofortress	Mid 50	94	2045
Boeing 707/C-137/C-18/KE-3	Mid 60	97	2030
Boeing F/A-18 Hornet/Super Hornet	1978	1762	2019
BAE Hawk/Boeing T-45 Goshawk	1976	641	–
Northrop T-38	1959	685	2040
Panavia Tornado	1974	746	2018
MiG-21 (incl. licences)	Mid 50	3324	–
L-39/L-39 Albatros/L-159	Mid 70	2233	–
BAE Hawk/Boeing T-45 Goshawk	1976	641	–

$$D = \sum_{i=1}^{n} \frac{1}{N_i} \tag{1}$$

The cumulative damage, the amount of flight cycles or load cycles that were taken into consideration with respect to Load Sequence, threshold number of cycles that a component could withstand with a given constant load and the load sequence under the given load at a constant amplitude load sequence are denoted by the letters D, n, and N respectively. Every time D records the value as 1, the component is regarded as 'unusable'. After that, the component has to be removed in order to implement the safe life design requirements. Fatigue testing reveals considerable variability in fatigue life among specimens subjected to identical loads, typically ranging from a factor of 2 to 4. To mitigate the risk of failure, safety factors are implemented concerning fatigue lives, rendering fatigue damage improbable. Consequently, virtually all components loaded to their calculated fatigue life retain an indeterminate yet potentially significant residual fatigue life. Thus, damage tolerance design, which specifically addresses slow crack growth, is the approach that enables the utilisation of such residual Fatigue life and such a design reliant upon the assessment of lifespan of crack propagation using the principles of fracture mechanics. It begins with the assumption of a starting crack, representing the least crack length that an inspector can consistently detect. The length of the crack is contingent upon its concealment and its existing length at the time of reliable detection. A first fracture inside any sophisticated multi-lap joint could thus end up being significantly longer when compared to the initially observed crack in an aperture-present basic sheet. Protocols for identifying first fractures have been established and are accessible to the designer [4]. The fracture propagation life H is now estimated from the initial crack length to a crack length that the component can still endure. Executing this computation necessitates a stress intensity factor. Data of Crack Propagation and a law of crack propagation, the most basic of which is the Paris law, expressed as Eq. (2).

$$\frac{da}{dN} = C_p \times (\Delta K)m_p \tag{2}$$

where da/dN, Cp, mp, and ΔK represent propagation of crack every cycle, constant of Propagation Behaviour and Constant of stress intensity factor. The stress intensity factor (ΔK) has been ascertained using Eq. (3).

$$\Delta K = \Delta\sigma \times \sqrt{\pi a} \tag{3}$$

Consider an infinite plate, where a crack is subjected to uniaxial stress, Δσ denotes the stress range and a represents the crack length. The remaining two constants, Cp and mp, can be obtained from databases. A fatigue test is conducted on a fractured specimen subjected to a constant stress amplitude Δσ, during which the length of the crack is monitored periodically. In conjunction with Eq. (3), the da/dN against the stress intensity factor (ΔK) for each crack propagation interval is ascertained using

Eq. (4). The resultant connection may be linearised, yielding in

$$\log\left(\frac{da}{dN}\right) = m_p \log(\Delta K) + \log(C_p) \tag{4}$$

Which facilitates the determination of the constants Cp and mp, and is essentially the logarithm of Eq. (2). The findings Derived through tests where $\Delta\sigma$ is varied consistently align over a same region in the fracture propagation curve. Indeed, other variables affect fracture propagation behaviour, including mean stress, ambient conditions, and load sequence effects. The calculation of crack propagation life H until critical failure is only enabled by the availability of multiple factors like propagation curve for the metal used, the Load sequence, the intensity stress factor. After every H/2, the component undergoes re-inspection, ensuring the prevention of catastrophic fracture. If no fracture shows up, the same process repeats. Upon discovery of the break, the component must be fixed. So long no fissure is detected, the iterations repeat. This is referred to as Damage-Tolerant Design. Damage-tolerant architecture is demonstrably beneficial. Such a system enables the use of the material's potential until the component fails. Consequently, components need not be removed upon reaching the initially established fracture propagation and fatigue life thresholds. They can remain till the occurrence of a crack initiation. The start of a crack in any component or a scattered crack behaviour is no longer a cataclysmic event. Components may either be permitted to remain in operation for extended periods or, due to heightened demand in aircraft, be manufactured with reduced weight [5].

1.1 Monitoring Operational Loads

Aircraft, particularly military aircraft, are frequently utilised in a manner that diverges from their original design specifications. The good scenario indicates that the aircraft has utilised less lifespan and incurred less damage than expected, enabling extensions of operating life at reduced costs. The negative example encompasses instances where aircraft have been utilised beyond their original design specifications. This condition can arise when pilots consistently test the limits of an aircraft, therefore substantially altering its performance characteristics in relation to the initial design concept. Extreme alignments can be safely used for a duration when combined with continuous load monitoring. A comparable scenario of surpassing the initial load spectrum may arise with novel payload types (recon equipment or weaponry), necessitating at the very least, another assessment of fatigue life in the concerned aircraft. As an aircraft ages, it becomes increasingly challenging to ascertain its actual remaining service life. The challenge may be mitigated by incorporating into the aircraft, a monitoring system observing and flagging the operational loads [6].

Primary objective of monitoring the structure of anything, really is to assess to what extent has the structure deteriorated? The theory behind safe life design, defines damage as an entity that cannot be quantitatively assessed using conventional testing methods that are not destructive. The fault or damage denoted as D in a component of the structure is assessed using Palmgren–Miner's method Eq. (5), which calculates the sum of the damage accumulated after every singular load cycle, where Di is summation of the inverse of Ni which is defined as the number of load cycles that the component in question can withstand.

$$D = \sum_i D_i = \sum_i \frac{1}{N_i} \tag{5}$$

$$N_i = f(S_i) \tag{6}$$

$$S_i = f(Loads) \tag{7}$$

Stress S is a deciding factor for the number of endurance cycles N_i in Eq. (6), Stress in turn is influenced by loads as shown in Eq. (7), so establishing a rigorous link between loads, their sequence, and resultant damage. Nonetheless, other impurities and presumptions made while linearising the cumulative damage contribute to the aforementioned variability, complicating monitoring efforts. This tendency in design is addressed using a scatter factor, which effectively shifts the fatigue life curve into an area characterised by a low chance of breakage. This method of monitoring the load is being used for many years and is subjected to continuous refinement. The initial methods relied solely on the aeroplane's vertical acceleration, denoted as nz, multiplied by the mass of the aircraft to calculate the force [7].

1.2 Concepts of Aircraft Fleet Management

The prolongation of aircraft operating lifespan substantially influences fleet management and maintenance. Provided that the operating lifespan is brief and the loading circumstances are well delineated, the monitoring of aircraft utilisation may frequently be confined to the observation of average flying data, like flight cycles which is the cycle of one take off till landing. However, in a serviceable but considerably old aircraft, responses to two critical issues become considerably important:

What is the extent of aircraft damage to date? What load spectrum is the aeroplane now experiencing? For aeroplanes lacking onboard load or damage monitoring systems, responses to the first inquiry are often confined to flying cycles and become exceedingly ambitious when a more detailed assessment of damage is sought. Such precise information can only be provided based on many assumptions concerning multiple parameters including but not limited to load spectrum, a comprehensive fatigue life assessment, that are compared to the results of a rigorous non-destructive testing (NDT) examination of all components of the aircraft that are damage critical. The NDT results must be representative and able to be stored in a databank. Before proceeding, the aircraft's loading model is necessary, necessitating the involvement of the aircraft manufacturer.

Given the prevalence of the loads model, the answer to the second question is rendered indolent, somewhat. In order to track the present load spectrum, it is necessary to build a load monitoring system in an aircraft. In conjunction with the calculation of fatigue life, these actual load spectra can be incorporated into the assessments, facilitating continuous monitoring of fatigue life throughout the remaining operational lifespan of the aircraft, while acknowledging the reservations associated with estimating the initial fatigue life. Figure 1 shows the aircraft structural management concept [8].

Each aircraft is equipped with a data gathering equipment that primarily captures a strain (load) sequence or pertinent flying information. The data is subsequently transmitted to a data collection station on the ground, which immediately relays the information either to a Maintenance Coordination Centre if the loads conform to the design spectrum, or to a Fatigue Analysis Centre in instances of load exceeding the design thresholds. The Fatigue Analysis Centre assesses the potential impacts of load exceedance on the integrity of the aircraft's structure and communicates revised repair protocols and plans to the Maintenance Coordination Centre.

A detailed examination of the actions of the maintenance coordination centre is presented.

The inputs received include the load sequence, updates on fatigue-critical locations, damage index rates, and anticipated operational availability from fleet management, which collectively inform the maintenance plan and timeline for the aircraft's utilisation. These actions represent a first foray into the extensively debated domain

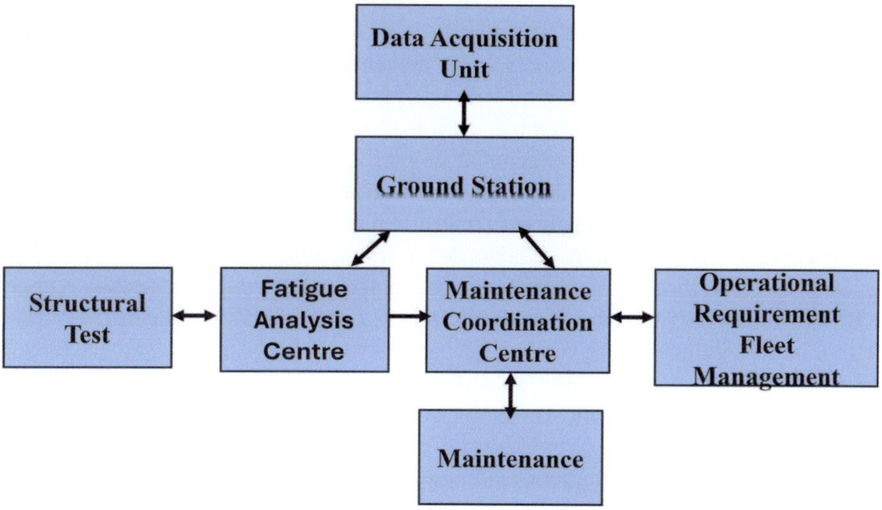

Fig. 1 Structural management concept

of predictive conservation, wherein early data may be utilised to forecast the occurrence of thresholds pertaining to damages, and various maintenance tasks can be consolidated to reduce downtime. In addition to enhanced management of fatigue life consumption and improved upkeep oversight, the benefits of an operational loads monitoring system encompass a deeper comprehension for pilots about the effects of their various flight tactics.

A significant issue in the management of ageing aircraft intended for a safe life arises after the designated lifespan has lapsed. Despite the absence of cracks in the aircraft, it is prohibited from flying based on design considerations. This is where damage monitoring approaches are utilised. Damage monitoring kits capable of continuous monitoring can be installed in sites where the fatigue life has numerically expired. The concept of safe life must now be reinterpreted to resemble the fatigue life of a minuscule fracture, which may be detected by the damage monitoring system without jeopardising the aircraft's structural integrity. This idea may also be used to longer fractures, provided that all security and reliability standards are maintained. Meeting these standards necessitates a comprehensive understanding of the structure and effective evaluation techniques. The benefits of implementing 'flying by inspection' in safe life structures remain mostly unexplored and are a subject of research that may become significantly pertinent when operating ageing fleets.

The anticipated benefits of this type of damage monitoring are numerous. Their scope encompasses not only a permissible extension of the operational lifespan of the aircraft but also a substantial decrease in requirement for inspection, particularly for parts situated in remote and inaccessible locations on the aircraft, where the primary obstacle is the disassembly and reassembly of various components that obstruct access to the respective component.

The principles discussed thus far mostly pertain to fatigue cracking, however they may be readily applied to other degradation modes, including corrosion. The issue of corrosion is being addressed via corrosion prevention strategies. The plan primarily outlines the schedule and location for aircraft inspections, as well as the necessary actions to be implemented upon the detection of corrosion. These procedures pertain to repair techniques, as corrosion must be eradicated in accordance with current requirements upon detection. Improved environmental monitoring methods, along with an enhanced comprehension of corrosion mechanisms, may eventually enable the acceptance of corrosion as a distinct kind of damage.

Another topic to address is the examination of non-metallic materials, namely fiber-reinforced polymer composites. Currently, components fabricated from these materials are engineered for a safe lifespan. Nevertheless, some designers remain sceptical about these 'new' materials concerning aspects like little perceptible impact damages. This uncertainty mostly stems from their ambiguity about load spectrum assumptions, since impact loads become far more pertinent than traditional Inertial Loads. A distinct category of load assessment is to monitor the impact load; a contributing factor to the ongoing development of such methodologies globally. Input information pertaining to cracks, de-laminations and corrosions in composites is now a prominent topic in the field of Predictive Maintenance. Predictive Maintenance, in conjunction with constant monitoring of load and damage, will be crucial for efficient maintenance, enabling the consolidation of numerous maintenance operations and enhancing fleet management [9].

1.3 Traditional Non-destructive Testing

Aircraft need extensive examination, particularly when heavily loaded, as seen with military aircraft during wartime operations or as they age. Aircraft built for damage tolerance need inspection at clearly specified periods. An outline of these periods for a civilian aviation vehicle and the required actions has been depicted in Table 2.

Various monitoring approaches are now employed on buildings.

The most straightforward and commonly employed method is visual examination, utilised primarily for assessing incidental damage and exterior degradation caused by deterioration or wear (exampli gratia: aerodynamic or hydrodynamic profiles, kinds, etc.). Visual inspection is considered for crack monitoring only in very restricted instances, for example, in huge steel damage tolerant structures with large load redundancy, particularly in the offshore industry.

Table 2 Inspection protocols in damage-tolerant civil aircraft design

Definition/Remarks	Interval	Check
Conduct a walkaround to identify apparent inconsistencies (system checks to be addressed by Flight Crew-FCOM)	36 Hours	Daily
Conduct a daily inspection of the landing gear and hydraulic bay with the landing gear door open	8 Calendar days	Weekly
Detailed visual examination from the ground with access to landing gear and hydraulic compartments. Basic operational assessments from the cockpit	300–400 FH	A
Servicing hyd, APU, Power Plant, ldg, lube Maintenance of hydraulic systems, auxiliary power units, power plants, landing gear, and lubrication systems	600–800 FH	2A
Conduct a general visual check from the ground with flaps, slats, and spoilers extended, and engine cowlings opened	1200–1600 FH	4A
Thorough visual examination (external) and internal via expedited access. Various operational testing of the system, lubrication, and replenishment	2500–3500 FH	C
Regarding C, it involves deeper interior penetration by expedited access	5000-7000FH	2C
Regarding C and 2C, but with increased interior penetration in deeper zones. Comprehensive visual examination of regions susceptible to corrosion. Operational assessments	10,000–14,000 FH	4C
Comprehensive visual examination of gasoline tanks and other hard-to-reach locations. Sampling as necessary. Operational assessments	20,000–28,000 FH	8C

A very straightforward method for identifying delamination in composites is the tap test, which involves systematically tapping the structure with an item (e.g., a hammer) and analysing the resultant sound. Upon alteration of this sound, readily discernible to the human ear, damage must be presumed, necessitating a comprehensive study utilising known instrumented NDT procedure.

The most often employed instrumented non-destructive testing technique is ultrasonic testing. Ultrasonics, which examines variations in a propagating audile wave, may be fundamentally used in any material, provided that a discernible differential in sound velocity exists between the solid material and the damage. The eddy current technique is a widely utilised non-destructive testing method, however restricted to electrically conductive materials, mostly applied to metallic parts. Eddy current is somewhat hyperacute to fracture development in metallic constructions and is thus utilised as a supplementary method to findings acquired using ultrasonics.

Additional techniques employed just in particular instances encompass Acoustic Emission, X- Radiation, computed Tomography, Shearography, and Thermography. Acoustic Emission is a well-established passive method, 'listens' for sounds generated by impending harm. It is highly beneficial in benchmark assessments, when extreme loads are exerted, and fault is permissible, as well as in generally 'clean' environments, such as laboratories, where operating noise is predominantly minimised. During in-service operations, the limits of such a technique may be readily apparent. X-Ray and computed tomography are procedures that, as a result of their substantial equipment costs, need to be restricted to components that have high performance and are critical. Applications encompass rotor blades of a Helicopter, high-speed turbine blades, Propellers and discs for jet engines, as well as cast rims for vehicles. Shearography and thermography are developing approaches now in development that provide comprehensive monitoring, necessitating specific monitoring instruments such as cameras coupled with subsequent information processing. Currently, these procedures remain confined to the somewhat 'sterile' laboratory setting.

1.4 Ageing Aircraft Specifications and Lessons

Significant accomplishments include a comprehensive evaluation of operating loads, enhanced comprehension of stress intensity and concentration in the areas of notches and fissures, and increased crack detectability by NDT (Non-Destructive Test) procedures. This resulted in an extension of the fatigue life from 1290 Flight Cycles of the Comet to over 80,000 Flight Cycles achieved by the Aloha Airlines' Boeing 737–200 before it experienced a severe accident in 1988, attributed to what was later identified as multi-site damage (MSD). This disaster has significantly highlighted the importance of direct damage monitoring for ageing aeroplanes. Prior to the Aloha Airlines incident, cracks identified by NDT were thought to appear singularly at a limited number of sites; nevertheless, it has now become evident that the emergence of contiguous cracking from several fatigue-critical points, for instance rivet holes, is far more probable in an old airliner. The close proximity and increased compactness of flaws dramatically expedite fissure spread. The detection of a single crack, measuring 1 mm, on a single side versus conjointed sides of the fastener has a substantial impact. In the former scenario, it remains a 1 mm fracture, however in the later scenario, it extends to a crack of 2 × 1 mm plus the diameter of the rivet, potentially totalling around 10 mm in length. If extensive fractures form at various rivets, the crack propagation inside the structure deteriorates, perhaps leading to a significantly diminished crack propagation lifespan.

MSD has specially launched many initiatives concerning aged aircraft, which encompass enhanced comprehension of areas susceptible to MSD (Multi Site Damage), the impact of MSD on cleft proliferation behaviour, and specialised MSD inspection protocols. To identify the positions of accurate impairment criticality in a hoary plane, it's essential to collect information from multiple narks, particularly the Major Airframe Fatigue Test (MAFT), various reviews from on-the-job to maintenance yard level, and potentially stripping analyses conducted post-MAFT or on well-used components. Various aircraft types must be assessed about damage-critical places, types of damage, loading circumstances, and the incidence of damage occurrence. This information must subsequently be mapped and saved in a database.

1.5 Emerging Structural Health Monitoring Technologies

Emerging Technologies constitutes a broad domain. Several of these issues have already been resolved, including the computer simulation environment. Nonetheless, significant developments are occurring in the hardware sector. Integrating developing technologies into ageing engineering systems may appear unusual; yet, when juxtaposed with the medical profession, the application of advanced technology for the well-being of the elderly reveals minimal distinction.

One of the developing technologies extensively examined in relation to infrastructural surveillance in intelligent tools and assemblies. Smart constituents and edifices refer to the incorporation of sensor technologies and potentially engagement equipment into the assembly or material, wherein, in the event of triggering, the sensory and triggering tools are interconnected through a regulator. This sensor-actuator-controller integration will be able to occur at both apparent and microscopic levels. This design philosophy applied to damage monitoring entails the integration, or preferably the adaptation, of sensors and potentially actuators into the structural component, so making Non-Destructive Testing (NDT) an inherent aspect of the structure.

In the context of Structural Health Monitoring (SHM), the primary focus is on the sensing capabilities inside intelligent assembly technology. The two occurrences to be observed include loads and damage, respectively.

2 Principles of Aircraft Structural Design

The design of an aeroplane or any complicated technical structure is a continuous procedure with iteration. The process typically starts with the specifications articulated by the prospective customer, succeeded by a design idea. This design concept must be analysed, maybe leading to revisions based on client requirements. This procedure is frequently referred to as the design wheel. Upon the availability of a conceptual design, greater focus is placed on factors such as aerodynamics, weight, and propulsion within the first layout. Only in the subsequent phase termed 'updated layout' do structural design, landing gear (for aircraft), cost, and other factors become pertinent. Consequently, solid-state structural considerations are addressed at a rather advanced stage in the design process. The primary objective is to ensure the structural integrity and performance are maintained throughout the system's lifespan. This necessitates a comprehensive understanding of prior experience alongside robust instruments to assess future performance. Aircraft structures are engineered to endure their maximum loading conditions. The circumstances often differ across various aircraft components. Components subjected primarily to compressive and/or shear stress predominantly fail due to buckling, whereas components primarily enduring tensile pressure have greater limitations concerning fatigue and residual strength. Designing of aircraft for stability is straightforward issue; however, designing for fatigue is a bigger difficulty due to time-dependent effects and the limited predictability of mechanical and environmental loading influences. Fatigue and damage tolerance are key considerations during the aircraft design process and during its subsequent operational lifespan. These variables can only be sufficiently addressed when loading circumstances, stress distributions, and material characteristics are understood. This raises a matter of design philosophy, primarily about a component's inspectability throughout service, its repairability or replaceability in the event of damage, and its criticality regarding the potential loss of the aircraft due to component failure. Whether this criterion is met largely depends on the use of safe-life or damage-tolerant principles.

A safe life indicates that the structure is engineered to endure a certain fatigue life—typically expressed in flight hours—under a specific load spectrum without necessitating any inspections. Upon reaching the fatigue life and provided the load spectrum remains unexceeded, the component must be replaced regardless of the absence of damage to date. If loads beyond the loading spectrum, the component must be reevaluated. This may be confined to a numerical assessment alone, leading to a diminished permissible fatigue life or necessitating inspections by non-destructive testing methods [10].

3 Fiber Optic Based Sensors

The use of fiber optic sensors (FOS) for health monitoring of civil and aeronautical structures is currently developing as a pivotal technology. Fiber optic sensors have several benefits and are potentially ideal for intelligent sensing and health monitoring applications. Fiber optic sensors possess a compact form factor and exhibit exceptional sensitivity. FOS exhibits resistance to corrosion and fatigue and is unable to initiate flames or explosions. FOS in composite constructions enables the development of an integrated sensor system. They do not require protection against lightning. They may be affixed at the surface or integrated into sophisticated composites, concrete, or even low-melting alloys. Passive optical sensors linked to control systems (computers) using optical fiber telemetry/communication systems may perform sensing, control, and monitoring activities over the same fiber, hence enhancing system compactness and efficiency. An FOS functions as a transducer, converting measurands such as temperature, stress, strain, rotation, or electric and magnetic currents into a proportional alteration in optical radiation. Light is defined by amplitude (intensity), phase, frequency, and polarisation; hence, any of these factors may vary due to perturbation. The efficacy of the fiber optic sensor is contingent upon the extent of this change and our capacity to monitor and quantify it with dependability and precision. Among the several fiber optic sensor types, interferometric sensors have been extensively researched for their capacity to detect and quantify minute variations. FOS can be constructed utilising any of the changeable characteristics of light: amplitude, phase, frequency, or polarisation. This interferometric phase change-measuring device is frequently favoured due to its great sensitivity and inherent versatility [11].

3.1 Fiber Optic Technology for Load Monitoring

Aircraft loads are measured using flying characteristics or standard electrical strain gauges. The first necessitates a comprehensive description of the load modalities, ideally in virtual reality, whilst the latter demands considerable extra wiring, contingent upon the number of sensors to be incorporated. The quantity of sensors can grow quite considerable, particularly with test aircraft. In the long run, additional load or strain sensors may be required for usage in typically flying aircraft. Fiber optic sensors provide a formidable alternative to conventional electrical strain gauges. Benefits of fiber optic sensors include light-weight, completely passive architectural plans, low power usage, dislike to electromagnetic interference, high bandwidth, high sensitivity, synergy with optical data processing and transmission, long lifespans, and cost-effectiveness. Disadvantages arise from the necessity of integrating optical fibers into the material and positioning them according to significant stresses and strains to ensure the acquisition of valid data. Fiber optic Fabry–Perot interferometer devices have demonstrated efficacy in sensing strain and stress waves generated by acoustic emissions. Their incorporation into composite materials does not impair mechanical qualities, provided the proportion of optical fibers remains very low relative to the other fiber constituents. Interferometric setups can be significantly affected by environmental factors. One of the causes behind this that fiber Bragg grating sensors are under growing research. Although running efficiently under low-frequency and steady loads, FBG sensors, in conjunction with the requisite laser and data processing equipment, are currently undergoing development for employment in the high-frequency acoustic spectrum [12].

3.2 Fiber Bragg Grating Sensor Principle

Fiber Bragg Grating (FBG) is a passive optical component integrated into a single-mode fiber (SMF) section. Its fundamental principle is Fresnel reflection. When light from an optical source passes through the optical fiber, it interacts with the FBG section. The FBG segment comprises periodic variation of low and high refractive indices at a designated location within the core. The periodic structure of FBG produces constructive interference for the specific wavelength of light based on the Bragg reflection law. The wavelength that fulfils the criterion is referred to as the Bragg wavelength as shown in Fig. 2. This wavelength reflects, and the rest will pass through with minimum attenuation [13]. The Bragg condition is given by $\lambda_B = 2n_{eff}\Lambda$ where it is clearly presented in Chapter 2.

Fiber grating constitutes the fundamental concept of optical diffraction gratings. The basic grating Eq. (8) specifies the angle of incidence Θ_1 of the light wave on the FBG.

$$n \sin\theta_2 = n \sin\theta_1 + m\frac{\lambda}{\Lambda} \tag{8}$$

The θ_2 is the angle at which a fiber grating most effectively links light between two modes. Figure 3 shows the diffraction of the incident light wave in the gratings.

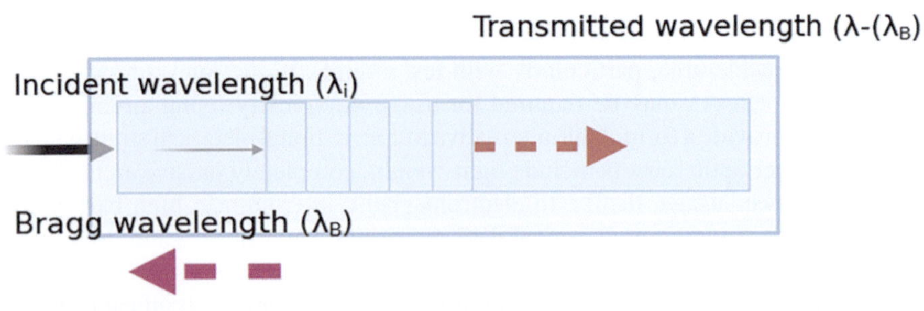

Fig. 2 Schematic view of FBG

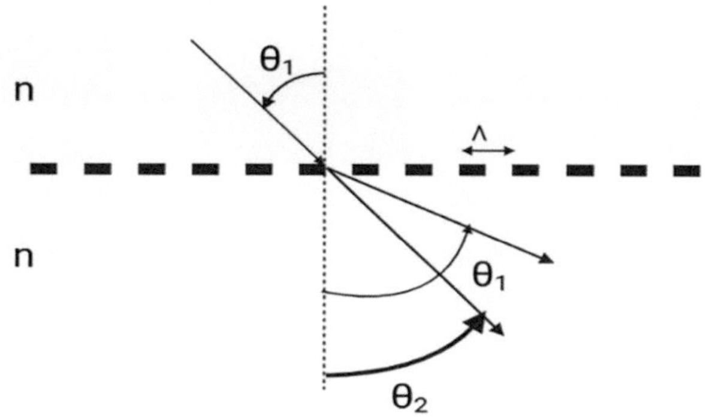

Fig. 3 Schematic of incident light wave diffraction in grating

Aircrafts are often exposed to a myriad of environmental and functional stress that can lead to physical deformation and structural disruptions. Structural health monitoring (SHM) paves a way for continuous monitoring of operational components. SHM not only allows to monitor, maintain and indicate damage in the critical components but also helps in formulating design for a hassle free, lighter and a safer aircraft. Fiber Bragg grating (FBG) sensors are pioneer in structural health monitoring segment due to their stress/strain sensitivity and also they can directly be implemented in sensing the temperature and strain where the temperature and strain variation in the vicinity of the FBG sensors can be monitored with the help of shift in the Bragg wavelength. Due to their highly sensitive response for temperature and strain variations in the atmosphere, they make one of the best suitable SHM sensors for the aviation industries [9].

FBGs for SHMs are selected based on their ability to sense temperature, vibrations, strain and displacement etc. Apart from this appropriate sensor selection also depends on the FBG sensor working mechanism, structure geometry, or it can be application specific, for example: Combination sensor (Temperature – Strain), Embedded Sensors, Multiple- Point Sensing Arrays, Surface mounted Sensors and Distributed sensors. The FGB classification based on Grating type, Refractive index profile and Sensing parameters are shown in Figs. 4, 5 and 6 respectively.

4 Case Studies for Monitoring Aircraft Structural Health

Modern aeroplanes' main flying structures use a considerable amount of complex composite materials. For example, composite materials have largely replaced conventional aluminium alloys in aeroplane structural components, accounting for over half of the total. Modern and future aeroplanes rely on the science of composite materials to reduce certification, operating, and manufacturing costs while increasing fuel efficiency and reducing emissions.

There will likely be a considerable improvement over aircraft at the turn of the century in terms of fuel efficiency, safety, impact on the environment, and noise level.

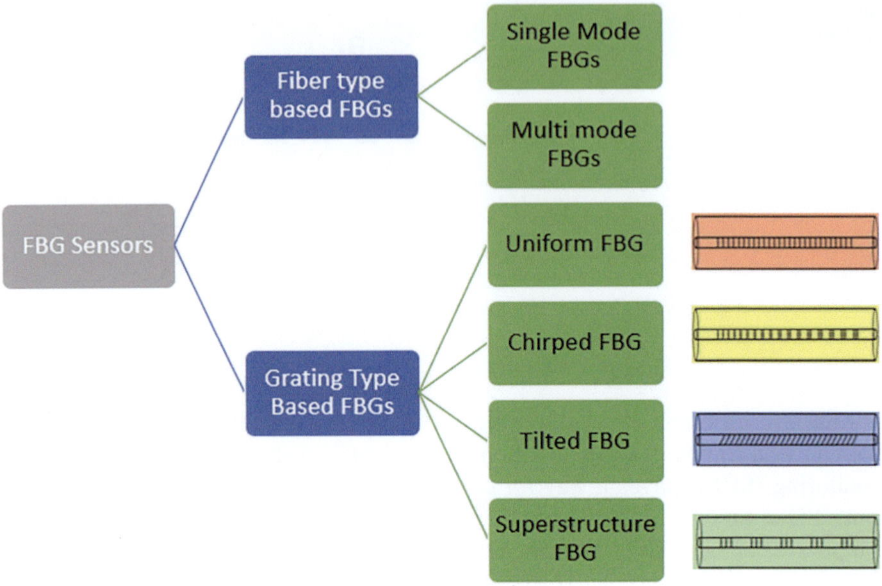

Fig. 4 FBG sensor classification based on grating type

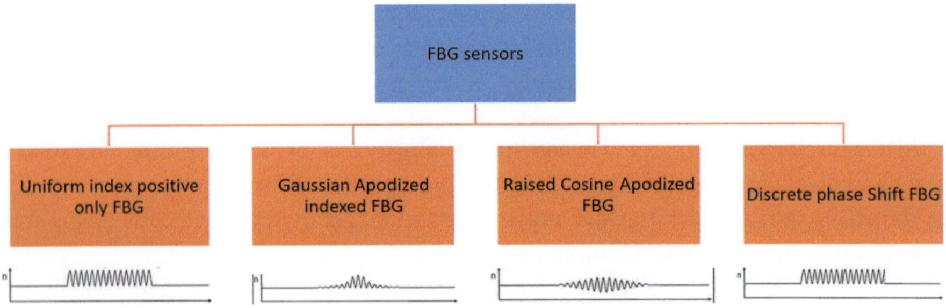

Fig. 5 FBG sensor classification based on refractive index profile

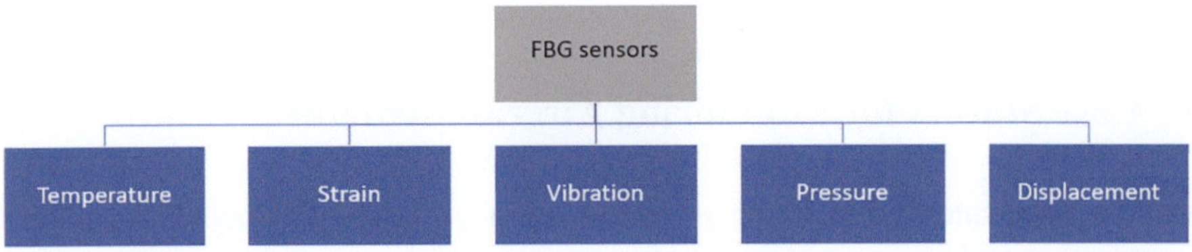

Fig. 6 Sensing parameter based FBG sensor

Clear goals are set, such as reducing CO_2 emissions by fifty percent, NOx emissions by eighty percent, accident rates by eighty percent and time to market by fifty percent. Composites are an essential component in the construction of lightweight, high-strength aircraft structures that significantly cut emissions and fuel consumption without compromising flight safety. However, issues with safety and longevity are more pressing with composites than with metallic equivalents because of their increased complexity. Due to their anisotropy and the fact that they are made up of separate components (matrix and fibers), they behave somewhat erratically when it comes to the start and development of damage, assessing the structure's strength and longevity. The effect of defects, especially impact-related ones, on the durability and composite construction strength and critical damage magnitude, have been the focus of a significant amount of empirical investigation [14, 15].

4.1 *FBG for Monitoring the Wing Load*

The arrangement of the system for measuring the load on composite aircraft wings using fiber Bragg grating sensors is shown in Fig. 7. Leveraging the benefits of integrated FBG sensors for enhanced surveillance while reducing structural disruption, the suggested solution Incorporated fiber Bragg grating sensors into the primary wing component of a composite aircraft.

Installing a monitoring and technology for the processing of data for the composite aircraft allowed for comprehensive investigations of its structural strength and operational management by supplying the entire structural loading history and local strain of critical structures. The CTLS-ELA composite aircraft was selected on which FBG sensor array was deployed to monitor the system during experimentation. The plane has a two seater with a high wing and cantilever wings. The regular empty weight is 325 kg and the maximum takeoff weight is limited to 700 kg. The aeroplane has

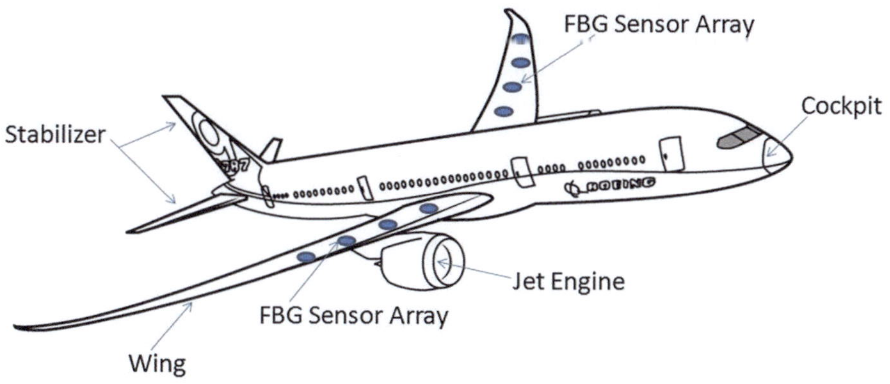

Fig. 7 Schematic view of aircraft wings embedded with FBG sensors

winglets attached to its rectangular wings. Composite materials were used to make nearly every part of the aircraft. This includes the main wings, horizontal and vertical stabilisers, landing gear, and propellers. The main wing structures, which provide most of the lift force, were chosen as the structures to be monitored by the FBG sensor-based operational load system. The test-bed aircraft's main wing structures are a composite spar, upper and lower skins, and a skin. The wing's spar is a spanwise-oriented beam that bears the bending strain caused by lift. The spar belts, which are located at the spar's flanges, are the areas where the bending stress is highest in the spar. As a result, the spar belt was equipped with FBG sensors.

Strain gauges measuring around 12 mm in length were manufactured by FBG Korea, Inc. of South Korea and coated with acrylate. In this setup, six FBG strain sensors were positioned. Starting 500 mm from the wing root, the first sensor of the FBG sensor array was placed on the spar belt. The optical fiber's departure point from the adhesive compound combination at the wing root received extra care. Before applying the adhesive compound mixture and spar, the fiber at the root was fastened to the inner wing skin surface, away from the spar belt, to reduce the likelihood of fracture. A commercial-grade ARINC 801 compatible cable was used to link the FBG sensor array optical fiber line termination. Commercial FBG temperature sensors were incorporated as an auxiliary measure to reduce the impact of temperature variations on the FBG sensors' wavelength fluctuations that occur during flight. Attached to the inner bottom skin 2500 mm from the wing root, one FBG temperature sensor was installed in each wing. Positions of the FBG sensors, which began 500 mm from the base of the wing, were incorporated into the spar web and lower spar belt. Interwoven with the spar web and lower spar belt was the multiplexed, single-fiber, bare FBG sensor array. After the spar belt was cemented to the bottom wing skin structure, the FBG sensor array was attached to it. Then, the adhesion compound mixture and the spar were applied. The composite wings structures were cured after the FBG sensor arrays were installed both on the left and right wing sections. The reflected spectra were measured using an interrogator after the implantable FBG sensor arrays had cured to ensure their integrity. The interrogation along with data processing equipment for in-flight data gathering was housed in the luggage compartment of the test-bed composite aircraft. The onboard system for data processing and interrogation consisted of the DAPS module and the BIG-SM04 module. By applying calibration loads to the structure, the ground load calibration test was carried out to determine the link between strain distribution and wing load. In order to measure the local strain caused by bending stress, the researchers attached FBG sensors to the lower spar belt in a span-wise orientation. This made the sensors sensitive to the bending moment acting on the wing structure. Subsequently, calibration loads were used to establish a link between the strain responses of the implanted FBG sensors and the bending moment at the root. Using the aerodynamic data for the wing's lift distribution, the bending moment was then converted to wing lift. In order to confirm the results of the ground load calibration test and to take into consideration factors that were not included in that test, the calibration flight test was carried out. The majority of the flight test for calibration was spent cruising. The wing's overall lift equals its weight. This led to comparing the expected wing load with the aircraft weight just before each flight to ensure accuracy [16, 17].

4.2 *FBG for Monitoring Tail Boom Structure of Aircraft*

As part of this study, a number of FBG sensors were mounted on the testbed aircraft's tail boom structure, and structural data was recorded while in flight using an on-board measurement system. Figure 8 illustrates the pictorial representation of the aircraft tail boom structure.

The ground system then analysed and examined the in-flight data in conjunction with flying characteristics. An airborne part of the system gathers data while in flight, and a terrestrial part processes and stores that data afterward. The HUMS was selected for testing on the CTLS-ELA aircraft. Two people may be comfortably accommodated on board this high-wing monoplane, which has cantilevered wings. Although the maximum takeoff weight is limited to 600 kg, the usual empty weight is 326 kg. A rectangular wing with winglets is a characteristic of the testbed aeroplane. The fuselage, main wings, landing gear, propellers, horizontal stabiliser, and vertical stabiliser are all made of composite material, which is what makes the testbed aircraft unique. For the purpose of measuring strain while in flight, the tail boom construction was used. Aeroplanes have a structure called the tail boom that connects the fuselage to the tail wing. All of the forces acting on the tail wing are transmitted to the tail boom. The foremost stresses on the tail boom come from the bending of vertical and horizontal axes caused by landing impacts and control surface operations. As part of this study, many sensors were mounted on the tail boom of the aircraft to measure and assess strain changes that occurred while in flight. After the aircraft was finished being built, FBG sensors were installed inside of the tail boom. An adhesive was

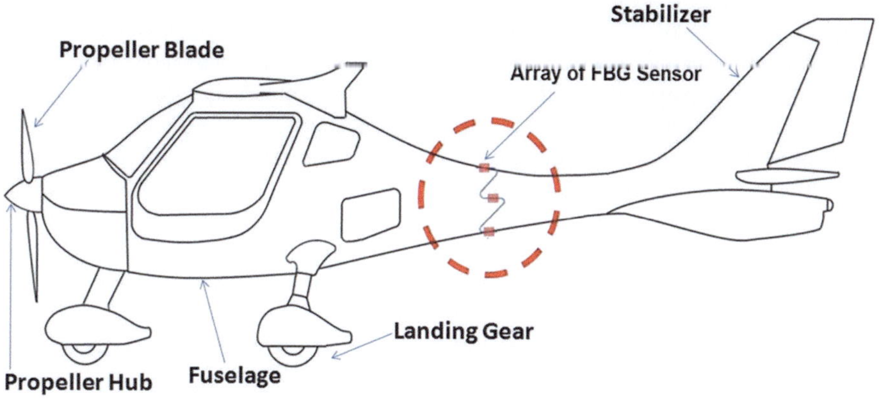

Fig. 8 Schematic view of aircraft tail boom structure

used to secure the FBG sensors. The four strain sensors and one temperature sensor made up an FBG sensor array. Along the longitudinal axis and the circumferential axis within the respective cross-section of the fuselage, the array of FBG sensor was positioned 2300 mm from the end of the tail boom. To track the strain caused by the bending loads applied by the forces acting on the tail part of the wing, the longitudinal axis was equipped with FBG strain sensors. The FBG sensor locations were established by referring to the frontal view from the testbed aircraft's cockpit. Vertical bending strain was measured by attaching the FBG strain sensors to the top and bottom, while horizontal bending strain was assessed by attaching them on both the left and right sides. The FBG strain sensor was put above the temperature sensor, which was placed next to it. The testbed aircraft, which operated normally inflight conditions, was shown to have successfully implemented the HUMS during the flying demonstration. Additionally, the testbed aircraft's in-flight data showed commendable results that were in line with the loading and flying conditions during takeoff, cruising, turning manoeuvres, and landing [18, 19].

4.3 Deformation Shape Prediction Using FBG

The ability to accurately forecast deformation forms is a crucial component of structural integrity surveillance and management for aeroplanes. Predicting wing deflection profiles is one method that may be utilised for the purpose of "in-flight" monitoring of lightweight aeroelastic systems. These systems include flexible aircraft wings and unmanned air vehicles (UAVs), and commercial aircraft respectively. As an alternative to electrical strain gauges, there is the possibility of employing fiber Bragg gratings for the purpose of monitoring strain within these components for the purpose of shape-detection algorithms. In 2003 the "in-flight" disintegration caused in NASA's Helios wing and the realisation of part of the catastrophe was perhaps attributed to severe wing disfigurement of the NASA's 'Dryden Flight Research Centre' (DFRC) acknowledged the fact that it was important to build a shape-detection system that was dependable. Composite materials were used for the majority of the construction of the Helios Prototype, which was a powered by solar energy "the atmosphere satellite" having a wingspan of 75.29 m, which was more than that of the Boeing 747. The epoxy and carbon fiber were used in the construction of the core tubular wing spar as well as the wing ribs themselves. Due to which, the DFRC proposed a unique monitoring procedure employing FBG sensors for the collecting the strain data in real time, as well as a computational efficacious algorithm that is capable of accurately calculating the deformation field based on the data that was acquired [20, 21]. Figure 9 shows the structure of aircraft.

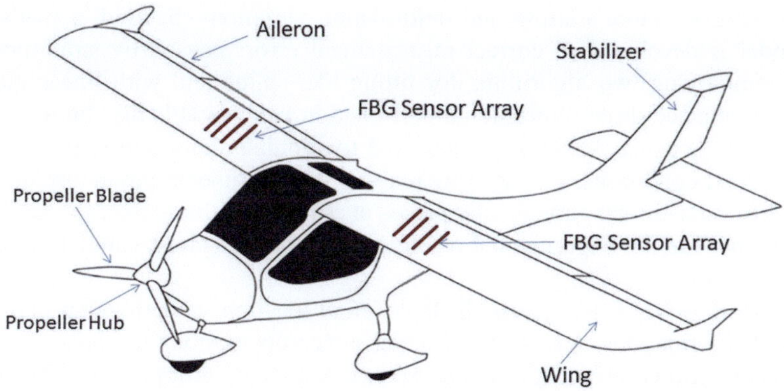

Fig. 9 Schematic view of aircraft structure

4.4 FBG-Based Sensor Array for SHM in Aircraft

This work reports a novel approach to monitoring aircraft structural integrity using a Fiber Bragg Grating (FBG) sensing network. Traditional electronic sensors face limitations due to electromagnetic interference, mechanical constraints, and harsh environmental conditions, which make them less suitable for aerospace applications. This study focuses on monitoring aircraft wings, which are prone to deformation due to self-weight, aerodynamic forces, and environmental stresses, potentially leading to structural failures. The system utilizes an FBG array-based sensing network to provide accurate strain measurements and three-dimensional structural reconstructions, as shown in Fig. 10, ensuring real-time monitoring and enhanced aircraft safety.

The study employs a three-dimensional position correction algorithm to improve strain measurement accuracy. The methodology consists of three key components. First, the FBG sensing principle measures wavelength shifts corresponding to the

Fig. 10 FBG sensor array in aircraft

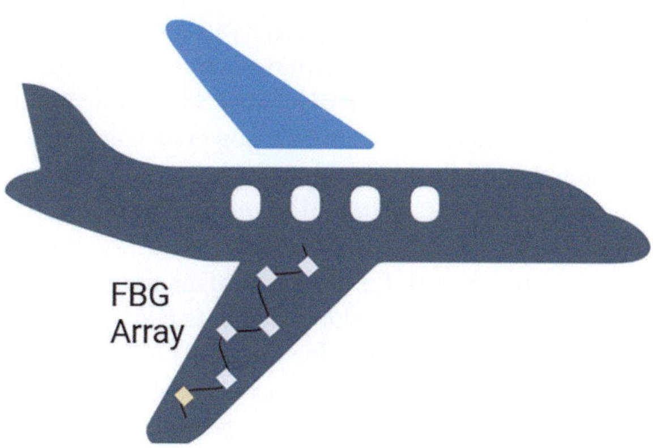

aircraft wing structure's stress and deformation variations. Second, a mathematical model is developed to correct measurement errors caused by strain-position offsets, employing two algorithms for fitting like spline and with linear interpolations to enhance three-dimensional deformation analysis. Finally, finite element analysis (FEA) using ANSYS is conducted to simulate deformation trends under different load conditions, followed by real-world experiments on an aircraft wing model. FBG sensors are strategically placed at critical locations based on strain field distributions, and the experimental results are compared against simulation data to validate accuracy.

A network of six FBG sensors is deployed at locations experiencing maximum structural deformations. A real-time sensing and compensation mechanism is developed to correct measurement errors caused by nonlinear deformations. The system undergoes experiments under different load conditions (0-100N) to evaluate accuracy, demonstrating superior performance over traditional methods. The study compares the bitone and spline interpolation methods with linear interpolation, proving that the former achieves higher measurement precision in deformation reconstruction. Additionally, a handheld optical measuring instrument (Metrascan) is used for 3D position verification, further validating the system's reliability.

The interpolation techniques of spline and bitone achieve relative errors below 4.96%, with an absolute error level not more than 0.014 mm and an average relative error below approx. 3.47%, outperforming traditional linear interpolation by reducing error by 5.84%. The system provides real-time visualization of three-dimensional deformation patterns, enabling accurate structural analysis. Its stability and robustness make it highly suitable for aerospace applications, where real-time monitoring is crucial for safety [22, 23].

4.5 Design and Implementation of SHM in Aircraft Using FBG Grating Array (FBGA)

Overview: often the aircraft wings endure stress and are prone to deformation due to load, turbulence and atmospheric variations. The damage in the wings can be fatal and hence it is very crucial to have continuous structural health monitoring to control the damage. The outline FBG based SHM is given in Fig. 11.

FBG grating sensors can be a powerful solution for early damage detection when embedded in the aircraft wings. The strain variation within the wings can then be collected using data acquisition models and then the deformation parameters were calculated based on the strain displacement. Liu et.al proposed an FBG grating array strain sensor network for the SHM of an aircraft wings where a 3-dimensional stress correction algorithm was used for data acquisition and the deformation parameters errors were solved. Errors due to strain offset were analysed at the measured points. In general, the FBG grating sensors are embedded along the Ribs and Spars of the aircraft wings. With this, it very feasible to identify damages in the structure at

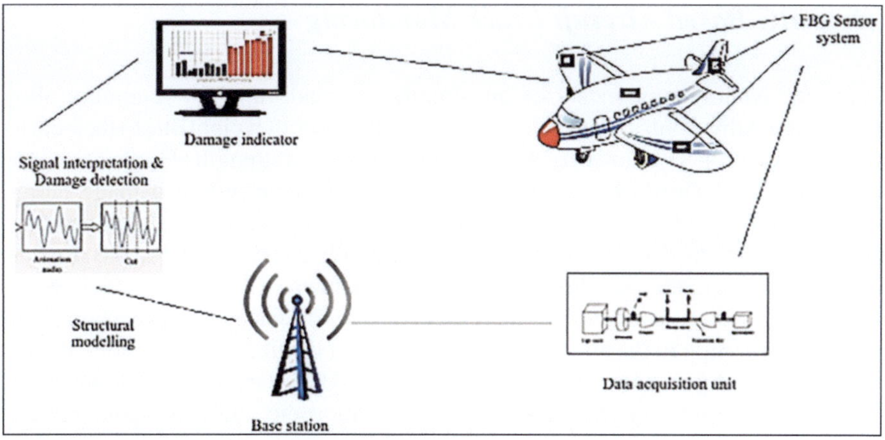

Fig. 11 FBG sensor-based aircraft SHM system

quicker pace as the FBG deals with the optical data when compared to the conventional damage detection methodologies thus promotes early damage detection and maintenance [24, 25].

4.6 *FBG Based Aircraft Landing Gears Load Monitoring for Hard Landing Detection*

Overview: overall safety of an aircraft during the takeoff and landing depends on the design and strength of the landing gears as they are the ones that handle the operational load and can anytime offer resistance to operational load during normal as well as in extreme conditions.Hard landing can be referred as an unusual contact of an aircraft with the runway resulting in a vigorous impact which can damage the landing gears or the entire aircraft rotor system. At times the damage may be fatal and hence it is very crucial to examine and service the landing gears before every flight. Here FBG Array paired with accelerometers and strain gauges that are embedded in the landing gears to monitor stress to identify any damages; to monitor vibrations and the accelerometer to monitor the system dynamics [26]. With this FBG sensors powered predictive analysis with real time monitoring and maintenance reduces the possibility of landing gear failure and promotes safety. Furthermore, the FBG sensor based SHM assures lowers operational costs and offers promising maintenance schedules which indeed increases the aircraft's lifespan.

4.7 FBG Based Aircraft Crack Monitoring

Overview: Major components of an aircraft are made up of Aluminium alloy. Aluminium Alloy contributes more than 80% of the aircraft weight due to their corrosion resistant property. However Aluminium may lose its strength at higher temperatures [20, 27, 28]. During the operation of an aircraft, it has to endure, damage, fatigue and alternating loads. Many structural variations such as slots, pins, and apertures may observe cracks due to strain which can eventually cause damage in the structure thereby increasing the possibility of serious breakdown of an entire vehicle which can be fatal. Thus it becomes extremely crucial to continuously observe/examine the metal configuration in the aircraft to detect and avoid inevitable destruction. Modern detection techniques such as acoustic emission (AE) is majorly implemented to detect the structure abnormalities, crack propogation etc., they are susceptible to noise and interference as well as they require heavy transmitter which makes them less feasible for continuous monitoring as it is very challenging when it comes to accuracy in locating the damage.

Metal fatigue is majorly happens due to periodic loads. Initially the damage initiates in the materials beneath the loads and after a period of time the damage/crack which eventually breaks. With periodical loading the damage length increases after reaching one point the component fails (fracture). In order to avoid component failure FBG are deployed near the cracks to analyse the prorogation with the help of real time strain monitoring. FBGS are more accurate as they immune to noise and hence are more suitable for SHM [29].

4.8 FBG Fault Diagnosis in Aeroplanes Using Advanced Wavelength Demodulation and Anomaly Index

The study reports a fault detection apparatus for FBG sensing systems used in aircraft SHM. FBG sensors have high precision, small size, and immunity to EMI (electromagnetic interference), making them ideal for real-time monitoring of vibration, strain and temperature in aircraft structures. However, the faults in FBG sensors occur due to aging, installation errors, or environmental conditions that may compromise data accuracy. This work reports a hardware-based fault detection device and an efficient fault detection algorithm to ensure rapid and accurate fault diagnosis in FBG sensor networks.

The FBG sensing system integrates SDM (Space Division Multiplexing) and WDM (wavelength division multiplexing technologies to improve sensor distribution and signal transmission. 25 FBG sensors are placed at significant locations on the aircraft, including the wings, fuselage, and tail. The fault detection hardware includes a power supply, an Industrial Personal Computer (IPC), communication interface, an FBG demodulator and a temperature control module. A significant innovation is using Volume Phase Grating (VPG) and Gaussian Curve Fitting, which enhances wavelength demodulation accuracy, allowing for precise monitoring of sensor performance.

To detect faults efficiently, the study introduces an anomaly index based on standard deviation (σ), signal length (L), Light intensity (I), and energy value in the temporal domain (P) are the four essential signal characteristics. By implementing loop iteration method, the accumulation of center values of these features are identified, and the Euclidean distance across the feature points is used to classify sensors as healthy, abnormal, or faulty. A simulation experiment was conducted to validate the system, where five common FBG sensor faults, optical cable breakage, excessive bending, excessive shock, power failure, and sensor detachment were artificially introduced. The fault detection instrument successfully identified faulty sensors in all cases. A real-world airplane test confirmed the method's effectiveness, accurately detecting two abnormal FBG and three faulty sensors. The reported work offers several advantages, like an automated and fast fault detection approach that significantly reduces maintenance costs and downtime compared to manual fault-checking. The integration of WDM and SDM ensures a scalable and lightweight sensor network. Additionally, the Gaussian curve fitting technique improves the accuracy of wavelength demodulation, further enhancing the system's reliability. However, the method could be improved by scaling to more extensive sensor networks, incorporating machine learning for noise reduction, and conducting long-term durability assessments under extreme environmental conditions [22, 30–32].

5 Conclusion

This chapter emphasises the application of fiber optic sensors for structural health monitoring (SHM) in aircraft, illustrated via several case studies. Advancements in fiber optic technology have enabled the utilisation of FBG sensors for the monitoring of various parameters in aircraft. They possess exceptional sensitivity, flexibility, lightweight characteristics, multiplexing capabilities, and a remarkable resistance to electromagnetic interference (EMI). The FBG sensors are optimal for aerospace engineering applications requiring lightweight, high-precision, and remote detecting capabilities.

References

1. K. Diamanti, C. Soutis, Structural health monitoring techniques for aircraft composite structures. Prog. Aerosp. Sci. **46**(8), 342–352 (2010)
2. J.L. Rudd, Durability and damage tolerance, in *Aging Combat Aircraft Fleets—Long Term Applications* (1996)
3. Y. Caron, Y. Richard, CF-188 fatigue life management program. *NATO publication RTO-MP-7, AC/323 (AVT) TP/4, Paper, 4* (1998)

4. J.B. Ihn, F.K. Chang, H. Speckmann, Built-in diagnostics for monitoring crack growth in aircraft structures. Key Eng. Mater. **204**, 299–308 (2001)
5. C. Boller, W.J. Staszewski, Aircraft structural health and usage monitoring. Health Monit. Aerosp. Struct. 29–73 (2004)
6. D. Broek, *The Practical Use of Fracture Mechanics.* (Springer Science & Business Media) (2012)
7. S. Beard, F.K. Chang, Active damage detection in filament wound composite tubes using built-in sensors and actuators. J. Intell. Mater. Syst. Struct. **8**(10), 891–897 (1997)
8. C. Boller, Encyclopedia of structural health monitoring (2009)
9. K. Chandler, S. Ferguson, T. Graver, A. Csipkes, A. Mendez, On-line structural health and fire monitoring of a composite personal aircraft using an FBG sensing system, in *Smart Sensor Phenomena, Technology, Networks, and Systems 2008*, vol. 6933, (SPIE, 2008, April), pp. 130–135
10. D. Barazanchy, M. Martinez, B. Rocha, M. Yanishevsky, A hybrid structural health monitoring system for the detection and localization of damage in composite structures. J. Sens. **2014**(1), 109403 (2014)
11. B. Culshaw, A. Kersey, Fiber-optic sensing: a historical perspective. J. Light. Technol. **26**(9), 1064–1078 (2008)
12. H. Guo, G. Xiao, N. Mrad, J. Yao, Fiber optic sensors for structural health monitoring of air platforms. Sensors **11**(4), 3687–3705 (2011)
13. K.O. Hill, G. Meltz, Fiber Bragg grating technology fundamentals and overview. J. Light. Technol. **15**(8), 1263–1276 (1997)
14. H. Kwon, Y. Park, J.H. Kim, C.G. Kim, Embedded fiber Bragg grating sensor–based wing load monitoring system for composite aircraft. Struct. Health Monit. **18**(4), 1337–1351 (2019)
15. H. Kwon, Y. Park, C. Shin, J.H. Kim, C.G. Kim, In-flight strain monitoring of aircraft tail boom structure using a fiber Bragg grating sensor based health and usage monitoring system. Int. J. Aeronaut. Space Sci. **22**(3), 567–577 (2021)
16. J. Frieden, J. Cugnoni, J. Botsis, T. Gmür, D. Ćorić, High-speed internal strain measurements in composite structures under dynamic load using embedded FBG sensors. Compos. Struct. **92**(8), 1905–1912 (2010)
17. G. Luyckx, E. Voet, N. Lammens, J. Degrieck, Strain measurements of composite laminates with embedded fiber Bragg gratings: criticism and opportunities for research. Sensors **11**(1), 384–408 (2010)
18. W.L. Ko, W.L. Richards, V.T. Tran, *Displacement theories for in-flight deformed shape predictions of aerospace structures* (No. H-2652) (2007)
19. J. Bakalyar, C. Jutte, Validation tests of fiber optic strain-based operational shape and load measurements, in *53rd AIAA/ASME/ASCE/AHS/ASC Structures, Structural Dynamics and Materials Conference 20th AIAA/ASME/AHS Adaptive Structures Conference 14th AIAA*, p. 1904 (2012)
20. R. Di Sante, Fiber optic sensors for structural health monitoring of aircraft composite structures: recent advances and applications. Sensors **15**(8), 18666–18713 (2015)
21. C. Du, D. Kong, C. Xu, Development of a fault detection instrument for fiber Bragg grating sensing system on airplane. Micromachines **13**(6), 882 (2022)
22. R.O. Ogunleye, S. Rusnáková, J. Javořík, M. Žaludek, B. Kotlánová, Advanced sensors and sensing systems for structural health monitoring in aerospace composites. Adv. Eng. Mater. **26**(22), 2401745 (2024)
23. Z. Liu, T. Shao, L. Yin, C. Liu, Local structural health monitoring system in aircraft based on fiber Bragg grating array. Results Opt. **11**, 100393 (2023)
24. A. Othonos, K. Kalli, G.E. Kohnke, Fiber Bragg gratings: fundamentals and applications in telecommunications and sensing. Phys. Today **53**(5), 61–62 (2000)
25. M.H. Yassin, M.H. Farhat, R. Soleimanpour, M. Nahas, Fiber Bragg grating (FBG)-based sensors: a review of technology and recent applications in structural health monitoring (SHM) of civil engineering structures. Discov. Civ. Eng. **1**(1), 1–39 (2024)

26. G.C. Kahandawa, J. Epaarachchi, H. Wang, K.T. Lau, Use of FBG sensors for SHM in aerospace structures. Photonic Sens. **2**, 203–214 (2012)
27. Z. Ma, X. Chen, Fiber Bragg gratings sensors for aircraft wing shape measurement: recent applications and technical analysis. Sensors **19**(1), 55 (2018)
28. D. Howe, Aircraft loading and structural layout (2004)
29. A. Iele, M. Leone, M. Consales, G.V. Persiano, A. Brindisi, S. Ameduri, A. Concilio, M. Ciminello, A. Apicella, F. Bocchetto, A. Cusano, Load monitoring of aircraft landing gears using fiber optic sensors. Sens. Actuators A Phys. **281**, 31–41 (2018)
30. R.J. Sun, L.M. Sun, Z. Sun, Application of FBG sensing technologies to large bridge structural health monitoring. J. Tongji Univ. **36**(2), 149 (2008)
31. N. Godin, P. Reynaud, G. Fantozzi, Challenges and limitations in the identification of acoustic emission signature of damage mechanisms in composites materials. Appl. Sci. **8**(8), 1267 (2018)
32. Q. Liu, M. Sun, H. Qin, K. Wang, H. Sun, Y. Wang, X. Qing, Embedded piezoresistive sensor network for multi-failure mode monitoring in CFRP bolted joints. IEEE Sens. J. (2024)

Optical Sensors for Aircraft Structural Health Monitoring

Arnaldo Leal-Junior and Wilfried Blanc

Abstract This chapter presents an overview of distributed and quasi-distributed optical fiber sensing technologies for structural health monitoring. In this case, the main focus is on the evaluation of the structural integrity of aircraft structures and composites. The Fiber Bragg Gratings (FBGs) quasi-distributed sensing approach is presented as a feasible and straightforward method to measure multiple discrete regions along the aircraft with high resolution and small cross-sensitivity, whereas optical frequency domain reflectometry (OFDR) approaches can be employed for distributed sensing along the optical fiber with sub-millimeter spatial resolution. In addition, this chapter also focuses on the use of new high-scattering optical fibers in which the core of such fibers is doped with engineered nanoparticles to increase the backscattering. Such approaches enable not only to extend the performance of current OFDR systems, but also to enable the development of new low-cost distributed sensing approaches using transmission and reflection analysis.

Keywords Fiber bragg gratings · Optical frequency-domain reflectometry · Transmission-reflection analysis · Nanoparticles-doped optical fibers

1 Introduction

The new approaches to structural health monitoring (SHM) use sensor-acquired data to analyze and evaluate the performance of structures [1]. By identifying, locating, and assessing damage, SHM ensures greater control and safety, allowing preventive measures to maintain structural efficiency [2]. Over time, the functionality of civil and

A. Leal-Junior (✉)
Mechanical Engineering Department, Federal University of Espirito Santo, Brazil and Department of Cybernetics and Biomedical Engineering, VSB–Technical University of Ostrava, Ostrava, Czech Republic
e-mail: arnaldo.leal@ufes.br

W. Blanc
Université Côte d'Azur, CNRS, INPHYNI, Nice, France
e-mail: wilfried.blanc@univ-cotedazur.fr

C. Marques et al. (eds.), *Advanced Optical Sensors for Aerospace Applications*, Progress in Optical Science and Photonics 37,
https://doi.org/10.1007/978-981-95-1626-1_5

mechanical structures can deteriorate due to factors like corrosion, fluctuating load conditions, and the development of cracks [3]. Damage that accumulates throughout the service life of a structure can create significant operational hazards, compromising both safety and reliability. In addition, unforeseen events that were not considered during the design process may cause catastrophic structural failure [4].

To ensure structural integrity and safety, structures of different nature can integrate structural health monitoring systems (SHM), which aim to automate continuous monitoring, inspection, and damage detection processes with minimal human intervention [5]. SHM systems play a critical role in ensuring real-time and online detection of defects, as well as monitoring key parameters such as strain, stress, and temperature, thereby enabling optimal maintenance strategies that enhance safety, reliability, and the service life of structures [6]. An SHM system generally comprises three key elements: a sensor network, a data processing unit (which handles data collection, transmission, and storage), and an evaluation system used for assessing structural integrity. The sensor network plays a crucial role in ensuring stable, precise, and dependable monitoring, as it detects a range of physical and chemical indicators related to the health of the structure [7].

Traditional SHM relies on periodic inspections by maintenance personnel, which are neither real-time nor human-dependent, often delaying the detection of critical issues such as pipeline leakages or bridge faults, thus increasing potential losses [8]. The experimental investigation quantifies the structural condition through on-site testing, which are classified as nondestructive (NDT) or minor destructive test, where the latter is generally limited in many applications for real-time monitoring [5]. The tests are broadly classified into static and dynamic types and may be performed periodically or continuously. Static tests include techniques such as infrared thermography, X-ray imaging, tomography, ultrasound/sonic testing [9]. However, dynamic tests encompass methods such as ambient vibration testing, radar or laser interferometry, and seismic interferometry [10]. In dynamic testing approaches and real-time monitoring systems, sensors are classified into hardware-based and software-based SHM systems [9]. Hardware-based methods include technologies such as visual inspection, acoustic or vibration signature detection, optical fiber sensors, ground penetration radar, and acceleration gradient methods, each offering unique capabilities for assessing structural integrity and health, as summarized in [5].

Optical sensors are highly adaptable and capable of measuring a wide range of parameters, making them popular across diverse fields, from biomedical to industrial applications [11]. Known for their intrinsic safety, corrosion resistance, and ability to perform well in challenging environments [12, 13], these sensors offer substantial benefits. Their lightweight and compact design facilitates seamless integration into structures or embedding within materials such as concrete and composites, as shown in [14]. Additionally, the multiplexing ability of optical fiber sensors makes them well suited for distributed structural monitoring [14]. These sensors can track not only strain and deflection, but also vibration and acceleration, providing a comprehensive solution for monitoring structural health.

2 A Comprehensive Review of Optical Fiber Sensors for Aircraft Structural Health Monitoring

In recent decades, composite materials have seen widespread adoption in the aerospace industry, especially in advanced aircraft models like the Boeing 787 and Airbus A350, which are composed of approximately 50% composite materials [15], has contributed significantly to advances in the field of SHM for composite structures. This advancement is the result of extensive research focused on creating materials with outstanding mechanical properties, including a high strength-to-weight ratio and superior corrosion resistance [16]. The adoption of composite structures in aircrafts has also helped reduce operational costs by lowering fuel consumption and minimizing maintenance requirements. However, despite these advantages, the use of composites presents challenges, particularly in the inspection and monitoring of structural damage [17]. This damage often occurs internally, making it difficult to detect during routine maintenance protocols that rely on visual inspection of external layers [18]. These challenges highlight the necessity for advanced SHM methods to maintain the safety and reliability of composite structures in aircraft [19].

In the aerospace sector, the use of SHM is still in its early stages [20], with the integration into composite structures proving especially difficult. For effective aircraft maintenance in real-world conditions, a thorough damage assessment is crucial across all four SHM levels, rather than depending on a single level alone (see Fig. 1). These levels include: damage detection, damage localization, identification of the damage type, and assessment of the damage severity. Each of these steps is crucial for ensuring the safety, longevity, and performance of composite structures in aircraft [21].

The successful implementation of SHM in aircraft depends on two essential factors: it must be a cost-effective solution for airline operators, minimizing economic losses due to downtime, and it must deliver accurate, reliable data on the structural integrity, thus improving safety and preventing potential failures. To address these needs, both industry and academia have been advancing a variety of SHM technologies [22], with the most promising approaches including acoustic emission techniques, micro-electromechanical systems (MEMS), electrical strain gauges, crack wires, optical technologies, and comparative vacuum monitoring [23]. Among these, optical technologies have emerged as the most promising for future aircraft airframes, particularly in the context of SHM.

Environmental factors, such as temperature, humidity, and pH levels, can significantly contribute to increased maintenance costs and aircraft downtime due to corrosion in various components [24]. As a result, environmental analysis is crucial for detecting structural issues linked to environmental influences, particularly corrosion of the aircraft's structure and parts [18]. Corrosion refers to the electrochemical degradation of metal structures, caused by chemical reactions between the material and its surroundings [25]. The harsh environmental conditions experienced during aircraft operations–such as freezing temperatures combined with intense heat in specific areas–expose the aircraft's structure and components to a

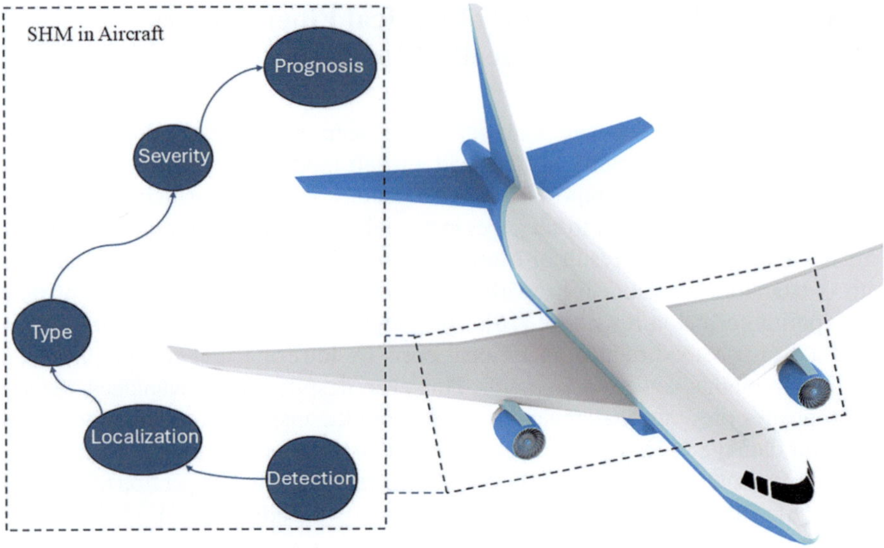

Fig. 1 SHM overview in aircrafts

variety of corrosion processes [18]. This situation complicates the material selection process, requiring designs to factor in corrosion resistance [24]. However, balancing corrosion resistance with other essential performance characteristics like stiffness, weight, and strength presents a multi-objective optimization challenge, where it may not be feasible to achieve all desired material properties with a single material choice.

To address this issue, corrosion detection in structures and components has been actively investigated [24]. Typically, detection methods are based on vision systems, which have limitations when it comes to analyzing difficult-to-reach areas. As a result, the use of environmental sensors for indirect corrosion detection has gained attention, where local or environmental conditions are monitored to correlate with corrosion parameters such as location, time, and rates [26]. In this context, monitoring environmental parameters, such as temperature, humidity, and concentrations of chemical compounds, can be linked to corrosion, helping to detect and estimate it early [27]. Despite the impact of atmospheric pollutants and compounds, such as acid sulfates, acid chlorides, and sea salts carried by moisture, temperature and moisture detection remain critical factors in corrosion analysis [24]. In dynamic moisture measurement, the time during which atmospheric conditions allow surface moisture formation is termed wetness, and it serves as a key factor in the initiation and growth of corrosion [27]. Additionally, temperature plays an essential role in corrosion analysis, as it not only affects the onset of corrosion but also influences its type [24]. Several temperature thresholds have been explored in the literature to correlate with different types of corrosion, underscoring the importance of continuous temperature monitoring in aircraft structures and components, particularly those exposed to moisture or atmospheric pollutants [24].

Optical fiber sensors (OFS) are increasingly recognized by both the scientific and industrial sectors, as well as end-users, for their significant potential in providing continuous, real-time monitoring of aircraft structures across a variety of applications [28]. These applications span from flight environment and navigation sensors to monitoring the pilot's health. Additionally, the integration of OFS into composite materials during the layup process offers the distinct benefit of enabling continuous monitoring of these structures throughout their entire lifecycle. This integration enhances safety, cost-effectiveness, reliability, and extends the operational lifespan of the structures. Moreover, OFS are renowned for their beneficial properties, such as immunity to electromagnetic interference, compact size, multiplexing capabilities, passive operation, biocompatibility, and chemical stability [29].

The advantages of using optical fiber sensors (OFS) for data measurement and reporting are numerous. These include reducing the overall weight of the aircraft by replacing bulky systems, cutting costs (as OFS can be more cost-efficient than other types, thanks to features like multiplexing and multipoint capabilities) [29], and providing predictive data that can enhance fuel efficiency during flights. OFS offer capabilities and measurements that traditional components cannot achieve [28]. Moreover, OFS can be installed more quickly compared to alternative systems. Due to the nature of optical fibers, multiple sensors can be integrated into a single fiber, simplifying the installation process by requiring only one fiber to be attached to the aircraft structure and linked to the data acquisition system.

OFS can be categorized based on their spatial capabilities, which determine whether they provide point-based measurements or monitor a wider area along the fiber's length [30]. This classification consists of three main types: point sensors, quasi-distributed sensors, and distributed sensors [31]. Point sensors deliver localized readings of the measured physical parameter, whereas distributed sensors offer measurements over a range, with values corresponding to specific positions along the fiber [32]. For distributed sensing, backscattering-based technologies are typically employed, with the primary methods being Optical Time Domain Reflectometry (OTDR) and Optical Frequency Domain Reflectometry (OFDR) [33].

Optical fiber-based sensors have emerged as a leading research focus within the sensor community, owing to their numerous benefits, including compact size, immunity to electromagnetic interference, passive operation, multiplexing capabilities, chemical stability, and biocompatibility [34]. These benefits have led to a wide range of applications in different fields, industry [35], structural health monitoring [36], biochemistry [37], and medicine [38]. Optical fiber sensors are highly versatile, with many different approaches utilized over the years, including intensity variation [39], fluorescence/absorbance [40], long-period gratings [41], fiber Bragg gratings (FBGs) [42], nonuniform gratings [43], nonlinear effects [44], specklegrams [45], interferometers [46], and surface plasmon resonance [47]. One of the key benefits of optical fiber sensors is their ability to be integrated into both rigid and flexible structures, thanks to their inherent properties like flexibility, compactness, and chemical stability [48]. They can also be integrated with various dyes [49] and dopants [50]. This ability to integrate different materials is particularly important in the development of humidity sensors using silica optical fibers, as silica is not inherently sensitive to humidity or moisture absorption [51].

Another approach to improve Rayleigh backscattering (RBS) power involves integrating oxide nanoparticles into silica optical fibers, known as dielectric nanoparticle (DNP) doped fibers [52]. These fibers, enriched with DNPs, offer a promising solution to improve spatial resolution in systems based on time-resolved acoustics (TRA) [53]. This advancement is demonstrated in the work of Silveira et al. [54], where the use of a fiber doped with Mg-silicate nanoparticles led to a significant increase in spatial resolution, allowing measurements in the millimeter range.

The low spatial resolution in standard single-mode fibers (SMFs) is primarily due to their inherently low Rayleigh scattering coefficient, which results in weak backscattered optical signals. However, a recent innovation in optical fiber technology, presented in a study [55], introduces a novel fiber featuring a core made of silica and oxide nanoparticles with sizes below 200 nm. Such specialized optical fiber exhibits a significant increase in Rayleigh scattering compared to standard SMFs, leading to enhanced backscattered optical power.

3 Point and Distributed Optical Fiber Sensors

3.1 Sensors Based on Intensity Variation

Generally, intensity variation-based sensors function by detecting fluctuations in optical power–either transmitted or reflected–along an optical fiber as their sensing method [56]. These sensors often rely on optical fiber attenuation mechanisms and their relationship with environmental changes and mechanical stresses to estimate a particular parameter [57]. Among the different sensor types, intensity variation sensors typically depend on macrobending losses in the optical fiber and the coupling of light between an optical source and the waveguide. These sensors have been widely applied in various sensing applications, as summarized in [58]. They generally utilize multimode polymer optical fibers (POFs) and take advantage of radiation losses induced by macrobending. When the fiber is connected to a rotating joint, changes in the joint angle lead to variations in output power, as bending in one direction reduces the output power. As a result, these sensors can effectively measure strain or stress in different structures and applications [59].

A typical macrobending-based POF sensor is composed of an optical fiber with a modified lateral section, often achieved through material removal to form a D-shaped fiber, which creates a sensitive zone that improves both sensitivity and the linearity of signal attenuation under bending. A standard POF consists of three layers: the core, the cladding, and the coating, all with a circular cross section. The core, which has a higher refractive index than the cladding, is mainly responsible for transmitting the optical signal, while the coating offers mechanical and chemical protection. Light losses occur due to absorption and frustrated total internal reflection when the fiber contacts a surface [58]. When the fiber is bent, the incident angle increases, causing variations in the transmitted signal. In the sensitive zone, where the cladding has been removed, radiation losses increase substantially [39]. Additionally, surface scattering contributes to these losses due to the coupling between higher and lower guided modes [60]. It's important to note that the sensor's sensitivity and linearity are closely linked to the length and depth of the material removed from the optical fiber [61].

Another sensing principle used in intensity variation-based sensors is the light coupling principle, which depends on the alignment differences between two optical fibers. POFs with a 1 mm core diameter are commonly used due to their larger acceptance cone, which is determined by the numerical aperture. This feature allows for a broader measurement range [62]. A typical setup includes two fibers: one, referred to as the "illuminated" fiber, is linked to the light source, while the other, called the "nonilluminated" fiber, is connected to the photodetector. The illuminated fiber emits a cone of light, called the acceptance cone. When the input of the non-illuminated fiber falls within the boundaries of this cone, light is transmitted and the output power is measured by the photodetector. Furthermore, any variations in fiber alignment, whether angular, axial, or lateral, affect the coupling efficiency, which can be detected and quantified by the photodetector.

As intensity-variation-based sensors are generally sensitive to bending along the optical fibers, such a sensor system is generally used for curvature and slope detections, where such applications are closely related to SHM of different structures [63]. It is also worth noting that the light coupling principle was already employed in measuring deflection in structural elements, an important parameter for the SHM approach in structural elements [64]. In addition, different configurations for displacement monitoring were proposed [65], where such configurations show promising aspects of the wide measurement range that can be readily employed in SHM applications.

In recent years, the main application of intensity variation-based optical fiber sensors is related to portable and wearable technologies [66], since the key advantages of this sensor system are related to its low cost and simple signal acquisition, making it suitable for wearable sensor systems. However, in aerospace applications, portability is generally not a limiting factor and the low cost of the intensity variation sensors does not compensate for their smaller accuracy (due to the sensitivity to deviations in optical source power) when compared with wavelength encoded data (such as interferometers and grating-based sensors). For this reason, there are only a few demonstrations of intensity variation sensors in SHM monitoring in aircrafts, including a sensor system for wings monitoring [67] and sensors for composite material monitoring using triboluminescent fibers [68].

3.2 Sensors Based on Interferometry

Various types of interferometric sensors exist, including Mach-Zehnder, Fabry-Perot, Sagnac, and Michelson interferometers, as outlined in [69]. One of the key benefits of this sensing approach is its simple fabrication process and the relatively low cost of the setup, especially when compared to fiber Bragg gratings (FBGs) [70]. Additionally, optical interferometers offer high sensitivity, excellent accuracy, and the ability to sense multiple parameters simultaneously [69].

Fabry-Perot interferometers (FPIs) are known for their high fabrication flexibility and simplicity. The FPI principle relies on two reflective surfaces separated by a fixed distance within another medium, forming a cavity [69]. Typically, the FPI response is measured in reflection rather than transmission, making these sensors ideal for use in harsh environments or areas that are difficult to access, while also allowing for the creation of compact sensing elements. As a result, FPIs have been employed to measure various parameters such as strain [71], temperature [72], pressure [73], liquid level [74], relative humidity [75], and even as an edge filter in low-cost FBG interrogators [76].

One of the key advantages of Fabry-Perot interferometers (FPIs) is the ability to fabricate them using various approaches, as outlined in [77]. Several fabrication techniques are available for creating FPIs, such as using photonic crystal fibers [72], silica tubes [71], reflective surfaces [78], air bubbles in fusion splicing [77], femtosecond laser micromachining [79], and even utilizing the periodic voids created by the catastrophic fuse effect on optical fibers [80]. A recent innovative fabrication method was proposed in [75], which involves using curable ultraviolet (UV) resins to form the FPI cavity. This method offers several benefits, including lower relative costs, no need for specialized equipment, and better control over the cavity's dimensional and geometric parameters, as further discussed in [81].

In general, the high accuracy and sensitivity of FPIs make them suitable for different SHM applications, especially for crack detection [82] and small strain/deflection measurement [72]. However, the necessity of using optical spectrum analyzers in conjunction with the lack of multiplexing capabilities of such sensor approach makes them unfavorable for distributed sensing applications, including SHM in aircrafts, since they generally need multiple sensors distributed along the structural element. However, a parameter related to SHM is the vibration monitoring in the aircraft structure. To that extent, the interferometers can be readily applied for highly sensitive vibration sensing using the so-called optical fiber accelerometers.

Optical fiber-based accelerometers have seen significant advancements and widespread applications, extending beyond navigation systems to structural health monitoring [76] and physiological parameter tracking [83]. Various optical fiber sensor approaches have been explored for their development, as extensively reviewed in [84]. To achieve greater precision in optical fiber-based accelerometers, interferometric designs–particularly those utilizing Fabry-Perot interferometers (FPI)–have been developed [84]. These accelerometers, which rely on an optical cavity with a movable structure, experience changes in cavity length in response to acceleration.

This design allows for sub-nanometer resolution, providing exceptional sensitivity [85]. The integration of micro-optical-electromechanical systems (MOEMS) has led to innovative accelerometer configurations that further improve performance [86]. In this approach, a microscale proof mass and a silicon frame form the accelerometer structure, with displacement and acceleration measurements derived from variations in the spectral features of the cavity. Moreover, the accelerometers parameters, such as proof mass and stiffness, can be adjusted to attain high resonant frequencies and reduce noise. Another MOEMS-based accelerometer, inspired by biological structures, was presented in [87]. In this design, a battery serves as the proof mass, and the sensors are integrated into a transparent, web-like structure, enabling accurate movement analysis and highly sensitive displacement detection. Such accelerometers can be used to assess structural conditions and vibration sensing in different aerospace applications.

It is important to mention that the most widely known application of interferometers in aerospace applications is in navigation systems. In inertial navigation systems, advancements in gyroscope technology and rotation measurements are crucial for the progression of navigation systems [88]. In this context, the rotation rate reflects changes in the system's heading and attitude. Additionally, accelerometers play a vital role in navigation systems by measuring acceleration amplitude and direction. When gyroscopes are used, they are typically mounted on a frame or mechanical structure to measure the angular velocity of the rotating body [89]. Gyroscopes are generally classified into mechanical, optical, and MEMS types [89]. The development of optical gyroscopes has been driven by continuous advancements in optoelectronic technologies [90], particularly with the widespread adoption of optical fiber technologies. A Fiber Optic Gyroscope (FOG) is an interferometric sensor that measures angular velocity using optical fibers, leveraging the Sagnac effect to determine rotational rate. This technology offers significant advantages, such as high sensitivity, resistance to temperature and pressure fluctuations, and immunity to electromagnetic fields [57]. Due to these benefits, fiber optic gyroscopes are commonly employed in general navigation systems, especially in aviation [91] and aerospace [92] applications.

3.3 *Fiber Bragg Gratings Sensors*

Fiber Bragg gratings (FBGs) are formed by periodic modulations of the refractive index of the fiber core [93]. These modulations can be generated through the interference of two laser beams [94], although the grating inscription process in polymers can lead to photodegradation [95]. The extent of this degradation depends on the wavelength and intensity of the light source [96]. In polymer fibers, a 248 nm laser was initially avoided for Fiber Bragg Grating (FBG) inscription due to concerns over ablation [97]. However, the observed ablation was mainly a result of the high energy density used during the inscription process. Research has demonstrated that by using lower energy densities and repetition rates, refractive index modifications can be achieved without causing ablation [98]. For instance, applying an energy density of 40 mJ/cm^2 at a repetition rate 5 Hz allows for controlled refractive index changes. Additionally, 248 nm irradiation has been shown to promote crosslinking between ester side chains in polymer molecules, thereby increasing the refractive index. However, prolonged exposure to 248-nm UV light leads to polymer photodegradation, reducing material density, and decreasing the refractive index over time [96]. Further degradation eventually results in polymer ablation, which occurs more rapidly at higher energy densities.

In [99], a low-energy density and a repetition rate 1 Hz were effectively employed to inscribe Fiber Bragg Gratings (FBGs) in polymer optical fibers, avoiding material ablation and significantly shortening the inscription time. This approach allowed for POFBG inscription in undoped polymethyl methacrylate (PMMA) in just 30 s, much faster than the 7-minute process typically required with a 325-nm continuous laser [100]. Due to this improved efficiency, the same repetition rate was later applied to grating inscription for other sensor technologies. Since the ideal energy density varies according to the polymer's specific characteristics [96], experimental optimization was carried out for different polymer optical fiber (POF) materials.

In FBG-based sensors, the change in the reflected wavelength is measured in relation to the monitored parameter, regardless of the material. The FBG exhibits inherent sensitivity to both temperature and strain, as described by Eq. 1 [34]:

$$\frac{\Delta\lambda_B}{\lambda_B} = (1 - P_e)\epsilon + (\alpha_T + \xi)\Delta T, \tag{1}$$

In this equation, $\Delta\lambda_B$ represents the shift in the Bragg wavelength, λ_B is the Bragg wavelength itself, P_e denotes the photoelastic constant, α is the fiber's thermal expansion coefficient, ξ refers to the thermo-optic coefficient, ΔT is the temperature change, and ϵ stands for the strain. Initially, FBG applications were mainly concentrated on measuring basic parameters such as strain and temperature. However, FBGs can also be integrated into various structures, enabling the detection of additional parameters beyond just strain and temperature, as further discussed below.

FBGs are widely used in strain sensor applications due to their high multiplexing capability and strong linear response to fiber deformation. An FBG is made of an optical fiber in which the refractive index is periodically altered along its length [94]. In the oil and gas sector, strain sensors are vital, especially in structural health monitoring (SHM) systems for refining tanks, storage facilities, and distribution terminals [101].

As outlined in Sect. 1, SHM systems are essential for detecting structural changes or potential damage across various industries, such as civil, mechanical, electrical, and aerospace engineering [102]. The process of developing SHM systems includes several stages, namely sensing, data acquisition, analysis, evaluation, and decision-making [103]. A wide range of sensing technologies are employed in the creation of SHM sensors, including corrosion coupons, electrical resistance sensors, electrochemical sensors, ultrasonic testing sensors, electromagnetic sensors, and optical fiber sensors [104].

FBGs offer significant advantages for strain field measurement in SHM applications, particularly in non-destructive testing for storage tanks. By deploying FBG arrays, it is possible to detect structural defects or weak points within tanks. FBG arrays offer unique technical, economic, and operational advantages when compared to traditional methods like magnetic flux sensors, ultrasonic testing, and acoustic emission techniques, as discussed in [105].

From an economic perspective, the high multiplexing capability of FBGs allows the integration of multiple sensors within a single optical fiber cable [24]. Many commercial FBG demodulators support multiple channels, allowing the deployment of many sensor systems using the same acquisition system, which significantly reduces sensor costs, especially when monitoring entire industrial plants with multiple tanks, as well as in different aerospace structures [15].

Some aircrafts are already in operation with integrated optical fiber sensor networks, which allows in-flight measurements. In 2013, Airbus announced its long-term vision of integrating distributed FBG optical sensors into all new aircraft [106]. Several developed solutions have already reached a relatively mature stage. For instance, FBG-based local damage detection has been successfully demonstrated in composite aircraft components, such as bonded repairs [107]. However, the widespread adoption of these sensing systems is still hindered by factors like sensor performance when embedded, detection capability, maintainability, the availability of low-cost interrogation equipment, and the lack of standardization and a clear certification framework.

A major challenge for multipoint (FBG-based) and distributed detection techniques, especially in damage detection, is developing controlled methods to monitor critical parameters associated with the onset and progression of damage in large structures over time. Achieving this requires sufficient physical and spatial resolution, even when the precise location of the damage is unknown. In parallel, aeronautical companies and organizations are focused on enhancing safety, reliability, and aerodynamic efficiency, while also reducing operating costs in future aircraft [16].

Aircraft health monitoring relies on sensors to evaluate the integrity of advanced materials that will support next-generation airframes. Graphite-reinforced composites are particularly advantageous due to their superior strength and stiffness-to-weight ratios compared to traditional aluminum alloys. FBGs, with their multiplexing capabilities, offer the potential for high-density sensor networks with minimal weight penalties. Various sensors have been developed for monitoring load, strain, and shape, with applications such as wing shape measurement and landing gear monitoring [28]. In addition, optical devices have been reported to monitor the health of composite materials [108].

Graphite-epoxy panels integrated with various types of optical fibers have been manufactured and subjected to thermal and mechanical tests to assess both the composite strength and the durability of the sensor [109]. The experimental results demonstrated that both surface-mounted and embedded optical fibers can effectively measure strain and temperature [108]. When compared to traditional instrumentation, FBGs integrated with composites showed significant potential for monitoring the structural integrity of critical aerospace components. Additionally, fibers were successfully embedded in high-performance graphite/polyimide and graphite/epoxy parts, confirming their suitability for structural applications in aircraft [108]. These experiments also highlighted the use of single-mode optical fibers for measuring strain and temperature in composites. Figure 2 provides an overview of the sensor elements along the optical fiber, which can be used to measure the elongation of aircraft flaps, wing shapes, strain distribution, and structural damage conditions, along with the data collected by a data logger [108]. By measuring and correlating changes in chemical and physical properties, it is possible to better understand composite aging. This information allows for predictive decision making, enabling a more accurate assessment of the health of composite parts and helping to determine the necessary maintenance procedures, thus preventing catastrophic failure.

3.4 Distributed Optical Fiber Sensors

The OTDR (Optical Time Domain Reflectometer) was initially proposed to monitor the health of communication networks. The operating principle involves injecting a short optical pulse into the optical fiber and monitoring its delay via the Rayleigh backscattering (RBS) signal [110]. To increase the RBS power in distributed sensing systems, phase-sensitive optical time-domain reflectometry (phase-OTDR) is commonly used. In this configuration, a coherent optical source (i.e., with a narrow linewidth and long coherence length) is used and connected to a 99:1 optical coupler.

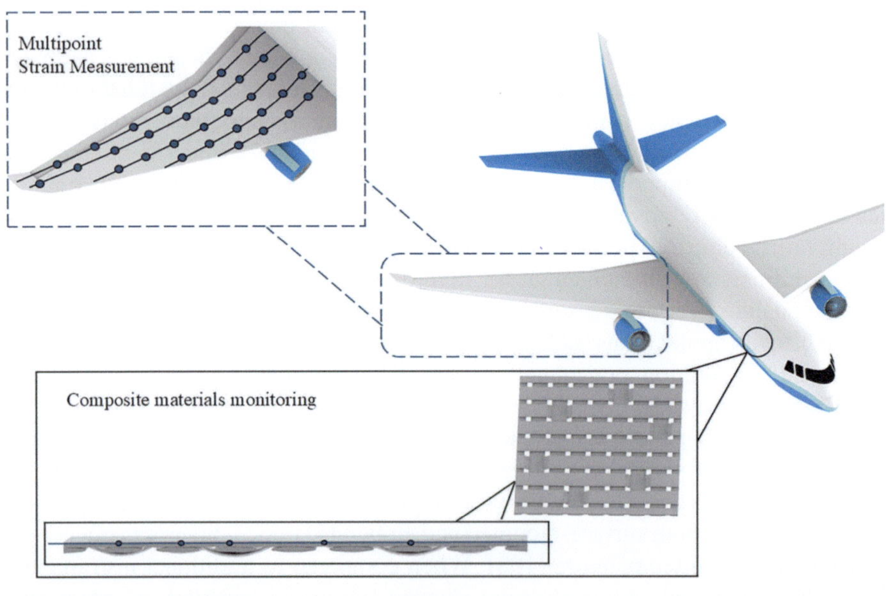

Fig. 2 Overview of FBG applications in aircrafts

The higher-power arm is connected to an electro-optic modulator, which regulates the laser intensity according to the sinusoidal signal produced by an arbitrary waveform generator (AWG), creating an optical carrier with frequency f0. The optical signal is then amplified using an Erbium-Doped Fiber Amplifier (EDFA) and injected into the sensing fiber through an optical circulator. The RBS signal is acquired by a balanced photodetector, with its second input being the output signal from the lower-power arm of the optical coupler. During detection, the beat between both signals is acquired, and the output of the photodetector is connected to an analog-to-digital converter, conditioning the signal for the digital processing stage. Through Fourier and Hilbert transforms, it is possible to demodulate the digital signal and obtain the RBS trace along the sensing fiber, allowing identification of external acoustic vibrations [33].

In contrast, some sensors have already been implemented using optical frequency domain reflectometry (OFDR) techniques. OFDR systems are implemented using a tunable laser with linear frequency scanning. In the configuration that exploits the coherence of light, the optical power of the laser is the input of a MachâŁ"Zehnder interferometer, where the first arm contains the reference signal and the second arm contains the RBS signal from the sensing optical fiber [111]. Both signals are recombined, producing a beat signal with a frequency scan similar to that of the optical source and a time delay proportional to the distance light travels through the fiber [110, 111].

The main advantage of OFDR over OTDR is its high signal-to-noise ratio (SNR) and spatial resolution. However, these systems require optical sources with very precise frequency scanning and are highly sensitive to phase-fading phenomena, which limits their sensing distance and bandwidth [110]. The SNR is defined as the ratio between the RBS power and the noise level, and it is influenced by factors such as the power and pulse width emitted by the laser, as well as the sensitivity of the photodetector [110]. Increasing the pulse power can improve the SNR, but this parameter is limited, as high powers lead to undesirable non-linear phenomena [112]. To increase spatial resolution, it is necessary to narrow the pulse width, which reduces the input optical power, requires higher sensitivity and bandwidth in detection, and consequently raises the noise level, thus reducing the SNR [110]. Finding the optimal balance between spatial resolution and SNR is one of the main challenges faced in the implementation of OTDR systems.

Another relevant factor for the analysis of field-installed sensors is temperature sensitivity, as sensing methodologies are inherently sensitive to temperature variations. In this context, techniques to compensate for the effects of temperature cross-sensitivity are crucial for accurate signal measurement [113]. Various methods related to sensor frequency responses, along with material responses as a function of temperature (since temperature can influence the thermal and mechanical properties of materials [114]), are important. Although there are automatic temperature compensation techniques in commercial distributed sensing equipment, specific tests under cyclic and continuous temperature variations have shown the need for additional signal processing techniques, as conventional methods do not take into account material deterioration and their sensitivity to temperature variations (especially sudden changes in this parameter) [113]. Furthermore, the application in SHM in airframes also requires compensation techniques for deformations in the optical fiber that may be caused by airflow and impacts with objects.

Several techniques for distributed sensing incorporate strategies to increase RBS and, consequently, improve SNR, sensing range, and / or measurement sensitivity [115]. Some commonly adopted strategies include exposing the fiber core to the environment, inscribing ultralong or weak FBGs, and exposing the fiber to ultraviolet light [116]. Another alternative is to use doped optical fibers with nanoparticles [52]. Compared to other special fibers (such as fibers made from alternative glass), these combine the advantages of silica–such as low cost, high durability, and transparency–with the benefits of nanoparticles that act as local scattering centers, increasing RBS [50]. The nanoparticles can contain various alkaline-earth metals, e.g., magnesium, and calcium, which are customized based on the dopant solution content in the manufacturing process, altering the RBS gain and the attenuation coefficient. Moreover, doped fibers can be handled like conventional telecommunications fibers, allowing them to be cleaved and spliced to traditional fibers with a single mode operation.

The features of distributed optical fiber sensors (using both OTDR and OFDR approaches) make them suitable for SHM in different applications, as already summarized in the literature [19]. The possibility of direct measurement of mechanical loads (especially axial deformation [117]) along the optical fiber allows numerous SHM applications in different fields such as geotechnical, pipeline monitoring, civil structures (such as tunnels and bridges) and even railway monitoring for traffic assessment [19]. The versatility of such sensors enable for spatial resolution ranging from millimeters to meters with detection distances that can be as high as in kilometer range. In addition, the gauge length can also be optimized depending on the applications, where sub-centimeter gauge length can be applied [118] as well as meter long gauge length for large area SHM applications.

The use of distributed optical fiber sensors spans a wide range of applications, including those in aerospace [119]. In this context, the primary method involves integrating these sensors into various materials of the aircraft for distributed structural health monitoring (SHM) assessments [120]. To that extent, there is the development of a distributed sensor system to detect disbond damage obtained from the removal of bolts in a wing [121]. Furthermore, the SHM of the cabin was also investigated using distributed optical fiber sensors under different pressurization conditions [122]. The use of distributed acoustic sensors using optical fibers also opens the path for additional applications in SHM that can even affect the airframe design, where it is possible to monitor the entire airframe structure under flight conditions. This possibility was demonstrated in [123] with the use of distributed acoustic sensors in an aircraft model during wind tunnel tests. In order to summarize the applications of distributed optical fiber sensing in aircraft SHM, Fig. 3 presents a schematic overview of the applications discussed.

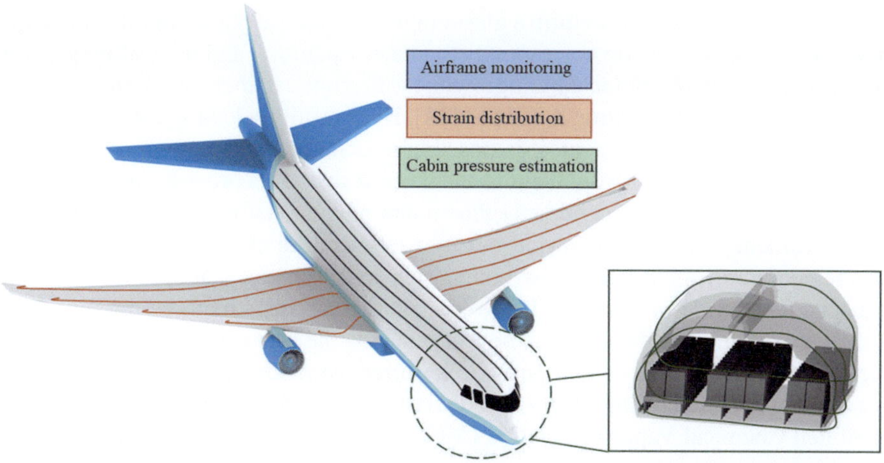

Fig. 3 Overview of distributed sensors using optical fibers for applications in aircraft SHM

4 NP-Doped Optical Fiber Sensors

As mentioned in Sect. 3, the use of distributed optical fiber sensors opens the pathway for multiple applications in aerospace research field. The development of novel distributed optical fiber sensor approaches is closely related to the advances in optical fibers. The development of NP-doped optical fibers marks a significant breakthrough in optical fiber fabrication, greatly advancing distributed sensor systems. Their enhanced backscattering leads to improved sensitivity and resolution, benefiting both OTDR and OFDR approaches.

4.1 NP-Doped Optical Fiber Fabrication

To obtain nanoparticles in optical fibers, two approaches have been addressed. The first, the most widespread, consists in forming the nanoparticles directly in the glass during the manufacturing process by using thermodynamic mechanism such as phase separation or nucleation/growth. The second route consists in embedding already synthesized nanoparticles. These two approaches are now described.

Thermodynamics Approach The thermodynamic approach to form nanoparticles in optical fibers was historically the first used. It is still widely used. When a multicomponent glass is subjected to high temperature, it can, to minimize its energy, cease to be a homogeneous glass to form a glass containing several phases of different compositions. This mechanism corresponds to phase separation and nucleation/growth. Based on the principle of nucleation/growth, the first optical fiber containing nanoparticles was obtained by heating the fiber [124]. By carefully selecting the temperature and duration of heat treatment, it is possible to control the formation of the initial nuclei and subsequently grow them to the desired size. This process corresponds to that used to prepare glass-ceramics (glass containing crystalline phases). For fibers, heat treatment is generally carried out at a temperature close to the glass transition temperature, applied for a few hours. This approach makes it possible to obtain nanocrystals, oxide or fluoride. However, heating the fiber requires removing the polymer cladding and the fiber can be weakened by the heat treatment. To overcome these limitations, it has been proposed to leverage thermodynamic mechanisms during various manufacturing stages, particularly when creating the preform via the Modified Chemical Vapor Deposition (MCVD) process [125]. This industrial technique involves the vapor deposition of a glass layer inside a silica tube, followed by soaking this porous layer in a doping solution containing alkaline earth ions, for example, Mg, Ca, or Sr. During the densification of the porous layer and the collapsing of the tube to form the preform, temperatures exceeding 2000°C are reached. At such high temperatures, the silica glass containing the alkaline earth ions undergoes phase separation, forming nanoparticles. These nanoparticles, composed of alkaline earth-based silicates, have sizes that vary depending on the concentration of alkaline earth ions in the doping solution [55]. This approach can be extended to any ions that trigger phase separation mechanisms in the glass and depend on the involved temperatures.

Doping with Pre-synthesized Nanoparticles The second approach used to obtain nanoparticles in optical fibers consists in doping with already synthesized nanoparticles. The very large choice of commercialized nanoparticles would allow to consider compositions of nanoparticles that cannot be obtained by the thermodynamic approaches described above. The high temperatures involved in the manufacturing process pose a significant challenge to maintaining the integrity of the nanoparticles. An example of this challenge is observed with $YbPO_4$ crystallites dispersed in the doping solution, which were found to remain intact in both the preform and the fiber [126]. This is a rare occurrence, as most nanocrystals typically react with the silica matrix, resulting in the formation of new nanoparticles. For instance, doping with $Y_3Al_5O_{12}$ (YAG) nanocrystals leads to the formation of YPO_4 nanocrystals in the fiber if the core layer contains phosphorus. If the core layer lacks phosphorus, amorphous YAS nanoparticles ($Y_2O_3 - Al_2O_3 - SiO_2$) are formed instead [127, 128]. These examples emphasize the importance of characterizing the nanoparticles within the fiber, as they may not be identical to the pre-synthesized ones. Such characterization still remains a significant challenge, especially for amorphous nanoparticles. However, the development of characterization techniques such as atom probe tomography (APT) now make it possible to achieve a 3D distribution of chemical species with sub-nm resolution [55].

Don't Forget the Drawing Step Optical fibers are obtained by hot drawing of a preform. In the case of telecom fibers, this step can be considered as a homothety operation. But for fibers containing nanoparticles, drawing cannot be considered as a simple change of scale. Indeed, it has been shown that during drawing, the nanoparticles, spherical in the preform, tend to elongate in the fiber along the drawing axis [129]. Depending on the drawing conditions, for example by drawing at a lower temperature, the elongated particles can fragment into spherical daughter particles. This fragmentation mechanism is linked to the Plateau-Rayleigh instability. These mechanical effects have been observed for particles higher than 50 nm. By drawing at a higher temperature, thermodynamic effects can intervene again and lead to the nanoparticles melting, totally or partially [130, 131]. The stretching parameters are therefore also very important to consider when designing the nanoparticles in the fiber.

4.2 NP-Doped Optical Fiber Distributed Sensors

As one of the main advantages of the NP-doped optical fibers is their contrast with telecom fibers, where the latter are designed to minimize possible Rayleigh scattering, since the main aim of these fibers is long-distance communication [132]. For this reason, the backscattering values of around -120 dBm, despite being detected in some commercial equipment, induce a limitation in terms of spatial resolution and sensitivity, since it is close to the equipment detection limits.

To that extent, NP-doped optical fibers have a higher concentration of scattering regions in the fiber core (or around the fiber core), resulting in an increase in backscattering, crucial for higher resolution in OFDR and OTDR systems, which resulted in many applications not only in biological detection [133], but also in shape reconstruction using scattering-level multiplexing (SLMux) systems [134] in which multiple fibers are combined with delay lines for complex 3D shape reconstruction [135].

In addition, the use of NP-doped fibers in conjunction with OFDR approaches resulted in distributed strain detection in different structures [132]. In this case, it is possible to detect the strain distribution along different NP-doped fibers, where the RBS of each fiber can be used to enhance spatial resolution [136]. Furthermore, the integration of NP-doped optical fibers with deep learning techniques can enhance the detection performance, making it possible to provide a 3D shape reconstruction with a single optical fiber cable distributed along the sensing structure [115]. The NP-doped distributed sensing device has significant applications in various fields, including biomedical applications, robotics, environmental monitoring, and, notably, structural health monitoring, with additional potential for detecting defects based on the frequency response of structural elements, particularly in aerospace applications [15].

4.3 Low Cost Approaches for NP-Doped Optical Fiber Sensors

Despite the numerous advantages of distributed optical fiber sensing, there are some limitations that can affect certain sensor applications. One of the primary drawbacks is the generally high cost of these systems. The high cost of distributed optical fiber sensors can be a significant barrier to their use in certain applications, particularly those where cost-effectiveness is critical [137]. This high cost arises from the need for acquisition systems that incorporate bulky and expensive components, such as modulated light sources and, in some cases, microwave photonic circuits [33]. Furthermore, portability of these systems is hindered by the bulkiness of the components used in traditional distributed optical fiber sensor setups [138].

The Transmission-Reflection Analysis (TRA) method offers a cost-effective alternative to conventional distributed optical fiber sensors, particularly those based on OFDR, for sensing mechanical disturbances along optical fiber cables [139]. This approach measures transmitted and backscattered optical powers using an unmodulated continuous wave light source [140], which allows for the use of low-cost light sources, such as light-emitting diodes, in sensor development, reducing system costs and improving portability [141]. Additionally, by utilizing only photodetectors for signal acquisition, the TRA method further contributes to cost savings and enhances portability. However, TRA has a lower spatial resolution (around 1 m) when applied to standard silica fibers [142], due to the weak Rayleigh backscattering signal of these fibers [116]. Moreover, early developments of the TRA approach indicated that it could measure only one mechanical disturbance at a time [140]. The conventional signal analysis in TRA uses the transmitted optical power as a function of the backscattered optical power, with each transmission and reflection pair corresponding to the position of a mechanical disturbance along the fiber. Thus, the TRA method faces two main limitations: (i) poor spatial resolution due to the weak Rayleigh backscattering of standard silica fibers, and (ii) the inability to detect multiple mechanical disturbances simultaneously due to signal processing constraints.

To address the limitations typically associated with the TRA approach, using optical fibers with high scattering properties can significantly enhance spatial resolution. A significant advancement in cost-effective distributed optical fiber sensors is the incorporation of NP-doped optical fibers, which can reduce spatial resolution to just a few centimeters [54]. This innovation has proven to be a breakthrough in improving sensor performance. In a comparative study, it was found that NP-doped optical fibers, with their enhanced backscattering, provided a spatial resolution improvement of three orders of magnitude compared to standard single-mode fibers. Specifically, spatial resolution improved from around 1 m using standard single-mode fiber to approximately 1 cm with magnesium and erbium co-doped fibers [54].

It is important to note that various nanoparticle compositions can be fabricated to tailor fiber properties for different applications [52]. In [50], a study was conducted on the TRA responses of different NP-doped optical fiber compositions, where various NP-doped fibers were subjected to sequential mechanical disturbances, such as macro bending in localized regions of the fiber. Using a common experimental setup, the study compared the spatial resolution, signal-to-noise ratio, accuracy, and attenuation across different fiber compositions. The results highlighted that the size and composition of the nanoparticles directly influenced the sensor's performance, with nanoparticles around 100 nm in size leading to higher attenuation but also increased reflectance (see Fig. 4).

The second limitation of the TRA approach is related to its signal processing method, which relies on a linear regression model using both transmission and reflection optical powers. This approach faces challenges when multiple mechanical disturbances occur simultaneously along the optical fiber cable, as it struggles with the superposition of these disturbances [54]. To address this challenge, a nonlinear model that relates the amplitude and location of mechanical disturbances to transmitted and backscattered optical signals could help detect multiple disturbances and improve overall system performance by simultaneously assessing the amplitude and location of disturbances [143]. In this context, machine learning techniques, particularly neural networks, offer the potential to enhance sensor performance by employing nonlinear models for data analysis. In [143], a feed-forward neural network (FFNN) was proposed to address the limitations of the TRA approach. This FFNN, designed with two hidden layers containing 600 and 300 nodes, was used to classify the location of mechanical disturbances within six predefined regions. Its

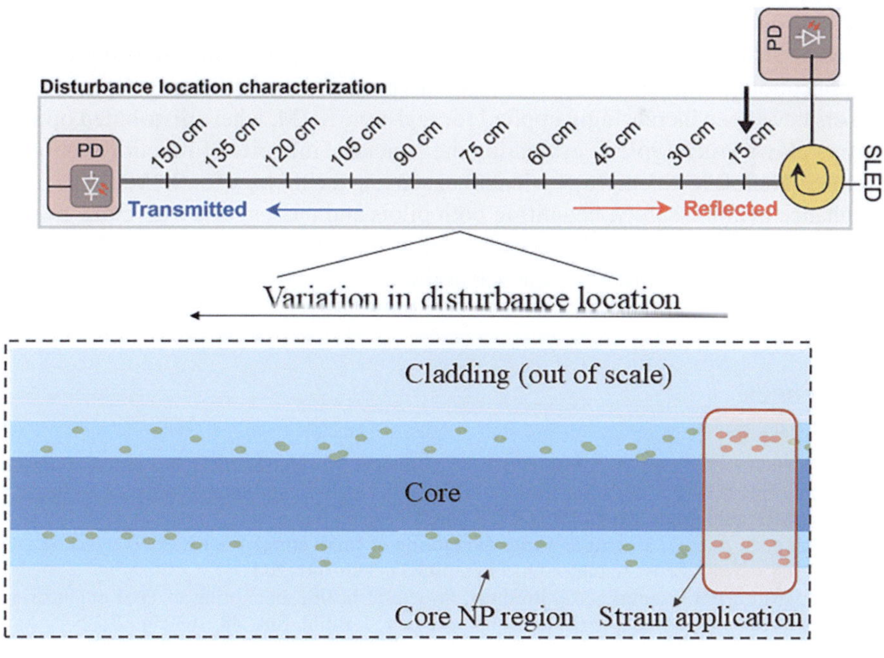

Fig. 4 Schematic representation of NP-doped optical fibers and their application in distributed sensing

objective was to accurately pinpoint the disturbances' positions, even in the presence of multiple disturbances with varying amplitudes occurring simultaneously. The results showed that this approach could successfully detect up to three simultaneous mechanical disturbances along a single NP-doped optical fiber, achieving an impressive accuracy of 99.43% [143].

This development represents a major step forward in cost-effective distributed optical fiber sensing, as it partially addresses the TRA approach's key limitation of being able to analyze only one disturbance at a time. The ability to simultaneously detect multiple disturbances opens up a range of potential applications, particularly in healthcare, where real-time, multi-event detection is often required. For this reason, the TRA-based distributed sensor was applied in distributed strain sensing in different applications using fiber embedded in flexible textiles [53] and in hard floors [144]. In addition, the use of multiple fibers for multiparameter sensing, which can also be used for shape reconstruction in multiple planes [145].

5 Conclusions

In recent decades, the aerospace industry has seen the widespread adoption of composite materials, with modern aircrafts primarily constructed from composites. This shift has greatly advanced the development of SHM for composite aerospace structures and created a demand for new technologies to facilitate this monitoring. Although SHM in the aeronautical and aerospace sectors has made notable progress, its application to composite structures remains particularly challenging. Optical fiber sensors have been increasingly applied for real-time SHM, where distributed optical sensors play a crucial role in evaluating the structural integrity of functional parts in aircrafts. Therefore, to employ optical fiber sensing technology for SHM can be used to enhance aviation safety, benefiting both pilots and passengers, where the growth of the aviation industry will proceed hand in hand with advancements in optical technologies, thanks to their unique capabilities.

References

1. A. Malekloo, E. Ozer, M. AlHamaydeh, M. Girolami, Machine learning and structural health monitoring overview with emerging technology and high-dimensional data source highlights. Struct. Health Monit. **21**(4), 1906–1955 (2022)
2. J. Leng, A. Asundi, Structural health monitoring of smart composite materials by using efpi and fbg sensors. Sens. Actuators A: Phys. **103**(3), 330–340 (2003)
3. M. Mishra, P.B. Lourenço, G. Ramana, Structural health monitoring of civil engineering structures by using the internet of things: a review. J. Build. Eng. **48**, 103954 (2022)
4. H. Qarib, H. Adeli, Recent advances in health monitoring of civil structures. Sci. Iranica **21**(6), 1733–1742 (2014)

5. F.J. Pallarés, M. Betti, G. Bartoli, and L. Pallarés, Structural health monitoring (SHM) and Nondestructive testing (NDT) of slender masonry structures: a practical review. Const. Build. Mater. **297** (2021)
6. M. Sun, W.J. Staszewski, R.N. Swamy, Smart sensing technologies for structural health monitoring of civil engineering structures. Adv. Civ. Eng. 2010 (2010)
7. Z. Liu, Smart sensors for structural health monitoring and nondestructive evaluation. Sensors, **24**(2) (2024)
8. Z.H. Warsi, S.M. Irshad, F. Khan, M.A. Shahbaz, M. Junaid, S.U. Amin, Sensors for structural health monitoring: a review, in *2019 2nd International Conference on Latest Trends in Electrical Engineering and Computing Technologies, INTELLECT 2019*, pp. 1–6 (2019)
9. M. Fawad, M. Salamak, G. Poprawa, K. Koris, M. Jasinski, P. Lazinski, D. Piotrowski, M. Hasnain, M. Gerges, Automation of structural health monitoring (SHM) system of a bridge using BIMification approach and BIM-based finite element model development. Sci. Rep. **13**(1), 1–18, (2023) [Online]. Available: https://doi.org/10.1038/s41598-023-40355-7
10. A. Caprino, S. Puliero, F. Lorenzoni, M. Floris, F. da Porto, Satellite SAR interferometry and on-site traditional SHM to monitor the post-earthquake behavior of the civic tower in L'Aquila (Abruzzo Region, Italy), Remote Sens. **15**(6) (2023)
11. A.G. Mohapatra, A. Khanna, D. Gupta, M. Mohanty, V.H.C. de Albuquerque, An experimental approach to evaluate machine learning models for the estimation of load distribution on suspension bridge using fbg sensors and iot. Comput. Intell. **38**(3), 747–769 (2022)
12. Z. Zhaobo, L. Sheng, W. Yibo, L. Hongnan, Structural health monitoring for bridge crane based on low temperature-sensitivity fbg sensors. Front. Phys. 346 (2021)
13. A.G. Mohapatra, J. Talukdar, T.C. Mishra, S. Anand, A. Jaiswal, A. Khanna, D. Gupta, Fiber bragg grating sensors driven structural health monitoring by using multimedia-enabled iot and big data technology. Multimedia Tools Appl. 1–21 (2022)
14. Y. Fan, M. Kahrizi, Characterization of a fbg strain gage array embedded in composite structure. Sens. Actuators A: Physical **121**(2), 297–305 (2005)
15. C. Marques, A. Leal-Júnior, S. Kumar, Multifunctional integration of optical fibers and nanomaterials for aircraft systems. Materials **16**(4), 1–29 (2023)
16. L. Świech, Calibration of a load measurement system for an unmanned aircraft composite wing based on fibre bragg gratings and electrical strain gauges. Aerospace, **7**(3) (2020)
17. S. Minakuchi, N. Takeda, Recent advancement in optical fiber sensing for aerospace composite structures. Photon. Sens. **3**(4), 345–354 (2013)
18. M. Czaban, Aircraft corrosion–review of corrosion processes and its effects in selected cases. Fatigue Aircraft Struct. **2018**(10), 5–20 (2018)
19. L. Zhu, G. Sun, W. Bao, Z. You, F. Meng, M. Dong, Structural deformation monitoring of flight vehicles based on optical fiber sensing technology: a review and future perspectives. Engineering **16**, 39–55 (2022). [Online]. Available: https://doi.org/10.1016/j.eng.2021.02.022
20. P. Cawley, Structural health monitoring: closing the gap between research and industrial deployment. Struct. Health Monitoring **17**(5), 1225–1244 (2018)
21. C. Boller, State-of-the-art in structural health monitoring for aeronautics. *International Symposium on NDT in Aerospace*, no. January 2008, pp. 1–8, 2008. [Online]. Available: http://www.hf.faa.gov/docs/508/docs/drury_doc.pdf%5Cnhttp://citeseerx.ist.psu.edu/viewdoc/download?doi=10.1.1.151.7689&rep=rep1&type=pdf
22. I. Lopez, N. Sarigul-Klijn, A review of uncertainty in flight vehicle structural damage monitoring, diagnosis and control: challenges and opportunities. Progress in Aerospace Sci. **46**(7): 247–273 (2010). [Online]. Available: https://doi.org/10.1016/j.paerosci.2010.03.003
23. D. Roach, Real time crack detection using mountable Comparative Vacuum monitoring sensors. Smart Struct. Syst. **5**(4), 317–328 (2009)
24. L. Li, M. Chakik, R. Prakash, A review of corrosion in aircraft structures and graphene-based sensors for advanced corrosion monitoring. Sensors **21**(9) (2021)
25. R. Newman, Pitting corrosion of metals. The electrochemical society interface **19**(1), 33–38 (2010). [Online]. Available: https://iopscience.iop.org/article/10.1149/2.F03101if

26. C. Jawalkar, S. Kant, A review on use of aluminium alloys in aircraft components, i-manager's J. Mater. Sci. **3**(3), 33–38 (2015). [Online]. Available: http://www.imanagerpublications.com/article/3673
27. Y. Chen, H. Huang, Y. Zhang, C. Wang, W. Fan, A method of atmospheric corrosion prediction for aircraft structure. Mater. Corros. **70**(1), 79–90 (2019)
28. Z. Ma, X. Chen, Fiber bragg gratings sensors for aircraft wing shape measurement: recent applications and technical analysis. Sens. (Switzerland) **19**(1) (2019)
29. M.J. Nicolas, R.W. Sullivan, W.L. Richards, Large scale applications using FBG sensors: determination of in-flight loads and shape of a composite aircraft wing. Aerospace **3**(3) (2016)
30. A.G. Leal-Junior, C.R. Díaz, C. Marques, M.J. Pontes, A. Frizera, Multiplexing technique for quasi-distributed sensors arrays in polymer optical fiber intensity variation-based sensors. Opt. Laser Technol. **111**, 81–88 (2019). [Online]. Available: https://linkinghub.elsevier.com/retrieve/pii/S0030399218313343
31. C.A.F. Marques, A. Pospori, D. Saez-Rodriguez, K. Nielsen, O. Bang, D.J. Webb, Aviation fuel gauging sensor utilizing multiple diaphragm sensors incorporating polymer optical fiber bragg gratings. IEEE Sens. J. **16**(15), 6122–6129 (2016)
32. A. Ukil, H. Braendle, P. Krippner, Distributed temperature sensing: review of technology and applications. IEEE Sens. J. **12**(5), 885–892 (May2012)
33. Z. Ding, C. Wang, K. Liu, J. Jiang, D. Yang, G. Pan, Z. Pu, T. Liu, Distributed Optical fiber sensors based on optical frequency domain reflectometry: a review. Sensors **18**(4), 1072, 2018. [Online]. Available: http://www.mdpi.com/1424-8220/18/4/1072
34. A. Cusano, A. Cutolo, J. Albert, *Fiber Bragg Grating Sensors: Market Overview and New Perspectives* (Bentham Science Publishers, Potomac, 2009)
35. X. Qiao, Z. Shao, W. Bao, Q. Rong, Fiber bragg grating sensors for the oil industry. Sensors, **17**(3), 429 (2017). [Online]. Available: http://www.mdpi.com/1424-8220/17/3/429
36. D. Kinet, P. Mégret, K. Goossen, L. Qiu, D. Heider, C. Caucheteur, Fiber bragg grating sensors toward structural health monitoring in composite materials: challenges and solutions. Sensors **14**(4), 7394–7419 (2014). [Online]. Available: http://www.mdpi.com/1424-8220/14/4/7394/
37. T. Guo, F. Liu, B.-o. Guan, J. Albert, [INVITED] Tilted fiber grating mechanical and biochemical sensors. Opt. Laser Technol. **78**, 19–33 (2016). [Online]. Available: https://linkinghub.elsevier.com/retrieve/pii/S0030399215303893
38. V. Mishra, N. Singh, U. Tiwari, P. Kapur, Fiber grating sensors in medicine: current and emerging applications. Sens. Actuators, A: Phys. **167**(2), 279–290 (2011). [Online]. Available: https://doi.org/10.1016/j.sna.2011.02.045
39. A. Leal Junior, A. Frizera, M.M.J. Pontes, A.G. Leal-Junior, A. Frizera, M.M.J. Pontes, Analytical model for a polymer optical fiber under dynamic bending. Opt. Laser Technol. **93**, 92–98 (2017). [Online]. Available: https://doi.org/10.1016/j.optlastec.2017.02.009
40. A. Leal-Junior, J. Guo, R. Min, A.J. Fernandes, A. Frizera, C. Marques, Photonic smart bandage for wound healing assessment. Photon. Res. **9**(3), 272 (2021)
41. E. Anemogiannis, S. Member, E.N. Glytsis, Transmission characteristics of long-period fiber gratings having arbitrary azimuthal/radial refractive index variations. J. Lightwave Technol. **21**(1), 218–227 (2003)
42. C.A.F. Marques, R. Min, A. Leal Junior, P. Antunes, A. Fasano, G. Woyessa, K. Nielsen, H.K. Rasmussen, B. Ortega, O. Bang, Fast and stable gratings inscription in POFs made of different materials with pulsed 248 nm KrF laser. Opt. Express, **26**(2), 2013 (2018). [Online]. Available: https://www.osapublishing.org/abstract.cfm?URI=oe-26-2-2013
43. C.A. Marques, P. Antunes, P. Mergo, D.J. Webb, P. Andre, Chirped bragg gratings in PMMA step-index polymer optical fiber. IEEE Photonics Technol. Lett. **29**(6), 500–503 (2017)
44. Y. Mizuno, N. Hayashi, H. Fukuda, K.Y. Song, K. Nakamura, Ultrahigh-speed distributed brillouin reflectometry. Light: Sci. and Appl. **5**(12), e16184 (2016). [Online]. Available: http://www.nature.com/doifinder/10.1038/lsa.2016.184
45. A. Leal-Junior, A. Frizera, C. Marques, M. Pontes, Optical fiber specklegram sensors for mechanical measurements: a review. IEEE Sens. J. 20(2) (2020)

46. A. Pospori, D.J. Webb, Stress sensitivity analysis of optical fiber bragg grating-based fabry–pérot interferometric sensors. J. Lightwave Technol. **35**(13), 2654–2659 (2017). [Online]. Available: http://ieeexplore.ieee.org/document/7906508/
47. C. Leitão, A. Leal-Junior, A.R. Almeida, S.O. Pereira, F.M. Costa, J.L. Pinto, C. Marques, Cortisol AuPd plasmonic unclad POF biosensor. Biotechnol. Rep. **29** (2021)
48. A. Leal-Junior, C. Díaz, A. Frizera, C. Marques, M. Ribeiro, M. Pontes, Simultaneous measurement of pressure and temperature with a single FBG embedded in a polymer diaphragm. Opt. Laser Technol. **112** (2019)
49. J. Guo, M. Niu, C. Yang, Highly flexible and stretchable optical strain sensing for human motion detection. Optica **4**(10), 1285–1288 (2017)
50. L. Avellar, M. Silveira, C.A. Diaz, C. Marques, A. Frizera, W. Blanc, A. Leal-Junior, Transmission-reflection performance analysis using oxide nanoparticle-doped high scattering fibers. IEEE Photon. Technol. Lett. **34**(16), 874–877 (2022)
51. A. Leal-Junior, A. Theodosiou, A. Frizera-Neto, M.J. Pontes, E. Shafir, O. Palchik, N. Tal, S. Zilberman, G. Berkovic, P. Antunes, P. André, K. Kalli, C. Marques, Characterization of a new polymer optical fiber with enhanced sensing capabilities using a Bragg grating, Opt. Lett. **43**(19), 4799–4802 (2018). [Online]. Available: https://www.osapublishing.org/abstract.cfm?URI=ol-43-19-4799
52. W. Blanc, Z. Lu, T. Robine, F. Pigeonneau, C. Molardi, D. Tosi, Nanoparticles in optical fiber, issue and opportunity of light scattering. Opt. Mater. Exp. **12**(7), 2635–2652 (2022)
53. A.G. Leal-Junior, D. Ribeiro, L.M. Avellar, M. Silveira, C.A.R. Diaz, A. Frizera-Neto, W. Blanc, E. Rocon, C. Marques, Wearable and fully-portable smart garment for mechanical perturbation detection with nanoparticles optical fibers. IEEE Sens. J. **21**, 2995–3003. [Online]. Available: https://ieeexplore.ieee.org/document/9197623/
54. M. Silveira, A. Frizera, A. Leal-Junior, D. Ribeiro, C. Marques, W. Blanc, C.A. R. Díaz, Transmission–reflection analysis in high scattering optical fibers: a comparison with single-mode optical fiber. Opt. Fiber Technol. **58**, 102303 (2020). [Online]. Available: https://linkinghub.elsevier.com/retrieve/pii/S1068520020302935
55. W. Blanc, I. Martin, H. Francois-Saint-Cyr, X. Bidault, S. Chaussedent, C. Hombourger, S. Lacomme, P. Le Coustumer, D.R. Neuville, D.J. Larson et al. Compositional changes at the early stages of nanoparticles growth in glasses. J. Phys. Chem. C **123**(47), 29008–29014
56. A. Leal-Junior, J. Silva, L. Macedo, A. Marchesi, S. Morau, J. Valentino, F. Valentim, M. Costa, The role of optical fiber sensors in the new generation of healthcare devices: a review. Sens. Diagnost. **3**(7), 1135–1158 (2024)
57. A.G. Leal-Junior, C.A. Diaz, L.M. Avellar, M.J. Pontes, C. Marques, A. Frizera, Polymer optical fiber sensors in healthcare applications: a comprehensive review. Sensors **19**(14), 3156 (2019). [Online]. Available: https://www.mdpi.com/1424-8220/19/14/3156
58. L. Bilro, N. Alberto, J.L. Pinto, R. Nogueira, Optical sensors based on plastic fibers. Sens. (Switzerland) **12**(9), 12184 12207 (2012)
59. A. Leal-Junior, L. Avellar, J. Jaimes, C. Díaz, W. Dos Santos, A.A. Siqueira, M.J. Pontes, C. Marques, A. Frizera, Polymer optical fiber-based integrated instrumentation in a robot-assisted rehabilitation smart environment: a proof of concept. Sensors (Switzerland) **20**(11), 1–16 (2020)
60. R. Liu, Z. Fu, Y. Zhao, Q. Cao, S. Wang, Operation principle of a bend enhanced curvature optical fiber sensor. IEEE International Conference on Intelligent Robots and Systems, pp. 1966–1971 (2006)
61. A.G. Leal-Junior, A. Frizera, M. José Pontes, Sensitive zone parameters and curvature radius evaluation for polymer optical fiber curvature sensors. Opt. Laser Technol. **100**, pp. 272–281 (2018). [Online]. Available: http://linkinghub.elsevier.com/retrieve/pii/S0030399217309684
62. A. Leal-Junior, A. Frizera, Optical fiber sensors for the next generation of rehabilitation robotics. Elsevier (2022). [Online]. Available: https://linkinghub.elsevier.com/retrieve/pii/C20200021547
63. V. Biazi, L. Avellar, A. Frizera, A. Leal-Junior, Influence of two-plane position and stress on intensity-variation-based sensors: towards shape sensing in polymer optical fibers. Sensors, **21**(23) (2021)

64. P.F.C. Antunes, H. Varum, P.S. Andre, Intensity-encoded polymer optical fiber accelerometer. IEEE Sens. J. **13**(5), 1716–1720 (2013)
65. A. Ghaffar, Q. Li, S. Ali Haider, A. Sun, A.G. Leal-Junior, L. Xu, M. Chhattal, M. Mehdi, A simple and high-resolution POF displacement sensor based on face-coupling method. Measurement: J. Int. Measurement Conf. **187**(March 2021) (2022)
66. L. Avellar, A. Frizera, A. Leal-Junior, POF Smart Pants: a fully portable optical fiber-integrated smart textile for remote monitoring of lower limb biomechanics. Biomed. Opt. Express **14**(7), 3689 (2023)
67. F.M. Reis, P.F. Da Costa Antunes, N.M. Mendes Maia, A.R. Carvalho, P.S. De Brito Andre, Structural health monitoring suitable for airborne components using the speckle pattern in plastic optical fibers. IEEE Sens. J. **17**(15), 4791–4796 (2017)
68. Z. Monette, A.K. Kasar, P.L. Menezes, Advances in triboluminescence and mechanoluminescence. J. Mater. Sci.: Mater. Electron. **30**(22), 19675–19690 (2019). [Online]. Available: https://doi.org/10.1007/s10854-019-02369-8
69. B.H. Lee, Y.H. Kim, K.S. Park, J.B. Eom, M.J. Kim, B.S. Rho, H.Y. Choi, Interferometric fiber optic sensors. Sensors **12**(3), 2467–2486 (2012)
70. C.A. Rodríguez, C.E.S. Castellani, A. Frizera-Neto, M.J. Pontes, M.R.N. Ribeiro, Envelope-based technique for liquid level sensors using an in-line fiber mach-zehnder interferometer. Appl. Opt. **55**(34), 9803 (2016)
71. O. Frazão, J. Kobelke, K. Schuster, C. Aichele, J. Bierlich, J.L. Santos, P. Roriz, K. Wondraczek, M.S. Ferreira, Fabry-perot cavity based on silica tube for strain sensing at high temperatures. Opt. Express **23**(12), 16063 (2015)
72. J. Tian, Y. Jiao, Q. Fu, S. Ji, Z. Li, M. Quan, Y. Yao, A fabry-perot interferometer strain sensor based on concave-core photonic crystal fiber. J. Lightwave Technol. **36**(10), 1952–1958 (2018)
73. J. Zhu, M. Wang, L. Chen, X. Ni, H. Ni, An optical fiber fabry-perot pressure sensor using corrugated diaphragm and angle polished fiber. Opt. Fiber Technol. **34**(1), 42–46 (2017)
74. T. Lü, S. Yang, Extrinsic fabry-perot cavity optical fiber liquid-level sensor. Appl. Opt. **46**(18), 3682 (2007)
75. R. Oliveira, L. Bilro, R. Nogueira, Fabry-pérot cavities based on photopolymerizable resins for sensing applications. Opt. Mater. Express **8**(8), 2208 (2018)
76. C. Díaz, C. Leitão, C. Marques, M. Domingues, N. Alberto, M. Pontes, A. Frizera, M. Ribeiro, P. André, P. Antunes, Low-cost interrogation technique for dynamic measurements with fbg-based devices. Sensors **17**(10), 2414 (2017)
77. M. Rajibul Islam, M. Mahmood Ali, M.H. Lai, K.S. Lim, H. Ahmad, Chronology of fabry-perot interferometer fiber-optic sensors and their applications: a review. Sensors (Switzerland), **14**(4), 7451–7488 (2014)
78. Q. Cui, P. Thakur, C. Rablau, I. Avrutsky, M.M.-C. Cheng, Miniature optical fiber pressure sensor with exfoliated graphene diaphragm. IEEE Sens. J. vol. PP, no. c, pp. 1–1 (2019)
79. A. Theodosiou, X. Hu, C. Caucheteur, K. Kalli, Bragg gratings and fabry-perot cavities in low-loss multimode cytop polymer fibre. IEEE Photon. Technol. Lett. **30**(9), 857–860 (2018)
80. P.F.C. Antunes, M.F.F. Domingues, N.J. Alberto, P.S. André, Optical fiber microcavity strain sensors produced by the catastrophic fuse effect. IEEE Photon. Technol. Lett. **26**(1), 78–81 (2014)
81. R.F. Oliveira, L. Bilro, R. Nogueira, A.M.M. Rocha, Adhesive based fabry-pérot hydrostatic pressure sensor with improved and controlled sensitivity. J. Lightwave Technol. 8724(c), 1–1 (2019)
82. T. Jiao, Z. Zhou, H. Xiao, Investigation into coaxial cable Fabry–Perot interferometers for strain measurement and crack detection in RC structures. Measurement: J. Int. Measurement Conf. **147**, 106873 (2019). [Online]. Available: https://doi.org/10.1016/j.measurement.2019.106873
83. A.G. Leal-Junior, C. R. Díaz, C. Leitão, M. J. Pontes, C. Marques, A. Frizera, Polymer optical fiber-based sensor for simultaneous measurement of breath and heart rate under dynamic movements. Optics Laser Technol. **109**(August 2018), 429–436 (2019). [Online]. Available: https://doi.org/10.1016/j.optlastec.2018.08.036

84. Q.B. Lu, Y.N. Wang, X.X. Wang, Y. Yao, X.W. Wang, W. Huang, Review of micromachined optical accelerometers: from mg to sub-μg. Opto-Electron. Adv. **4**(3), 1–24 (2021)
85. Q. Lu, C. Wang, J. Bai, K. Wang, W. Lian, S. Lou, X. Jiao, G. Yang, Subnanometer resolution displacement sensor based on a grating interferometric cavity with intensity compensation and phase modulation. Appl. Opt. **54**(13), 4188 (2015)
86. Q. Lu, J. Bai, K. Wang, S. He, Design, Optimization, and realization of a High-performance MOEMS accelerometer from a double-device-layer SOI wafer. J. Microelectromech. Syst. **26**(4), 859–869 (2017)
87. A. Leal-Junior, L. Avellar, V. Biazi, M. Simone Soares, A. Frizera, C. Marques, Multifunctional flexible optical waveguide sensor: on the bioinspiration for ultrasensitive sensors development. Opto-Electron. Adv. **5**(10), 1–11 (2022)
88. Z. Wang, G. Wang, S. Kumar, C. Marques, R. Min, X. Li, Recent advancements in resonant fiber optic gyro–a review. IEEE Sens. J. **22**(19), 18240–18252 (2022)
89. V.M. Passaro, A. Cuccovillo, L. Vaiani, M. De Carlo, C.E. Campanella, Gyroscope technology and applications: a review in the industrial perspective. Sensors (Switzerland), **17**(10) (2017)
90. O. Ostroverkhova, Organic optoelectronic materials: mechanisms and applications. Chem. Rev. **116**(22), 13279–13412 (2016)
91. H. Jiao, L. Feng, K. Wang, N. Liu, Z. Yang, Analysis of polarization noise in transmissive single-beam-splitter resonator optic gyro based on hollow-core photonic-crystal fiber. Opt. Express, 25(22), 27806 (2017). [Online]. Available: http://proceedings.spiedigitallibrary.org/proceeding.aspx?articleid=1021977 https://opg.optica.org/abstract.cfm?URI=oe-25-22-27806
92. D. Titterton, J. Weston, Modern inertial navigation technology and its application. Electron. Commun. Eng. J. **12**(2), 49–64 (2000). [Online]. Available: https://digital-library.theiet.org/content/journals/10.1049/ecej_20000202
93. C. Broadway, R. Min, A.G. Leal-Junior, C. Marques, C. Caucheteur, Toward commercial polymer fiber bragg grating sensors: review and applications. J. Lightwave Technol. 37(11), 2605–2615 (2019). [Online]. Available: https://ieeexplore.ieee.org/document/8570755/
94. K.O. Hill, G. Meltz, Fiber bragg grating technology fundementals and overview. IEEE J. Lightwave Technol. **15**(8), 1263–1276 (1997)
95. D. Sáez-Rodríguez, K. Nielsen, O. Bang, D.J. Webb, Photosensitivity mechanism of undoped poly(methyl methacrylate) under uv radiation at 325 nm and its spatial resolution limit. Opt. Lett. **39**(12), 3421–3424 (2014)
96. A. Pospori, C.A.F. Marques, O. Bang, D.J. Webb, P. André, Polymer optical fiber bragg grating inscription with a single uv laser pulse. Opt. Express **25**(8), 9028–9038 (2017)
97. G. Peng, Z. Xiong, P. Chu, Photosensitivity and gratings in dye-doped polymer optical fibers. Opt. Fiber Technol. **5**(2), 242–251 (1999)
98. C. Wochnowski, M.A. Eldin, S. Metev, Uv-laser-assisted degradation of poly(methyl methacrylate). Polymer Degradation Stability **89**(2), 252–264 (2005)
99. R. Oliveira, L. Bilro, R. Nogueira, Bragg gratings in a few mode microstructured polymer optical fiber in less than 30 seconds. Opt. Express **23**(8), 10181 (2015)
100. I.-L. Bundalo, K. Nielsen, C. Markos, O. Bang, Bragg grating writing in pmma microstructured polymer optical fibers in less than 7 minutes. Opt. Express **22**(5), 5270 (2014)
101. A. Leal-Junior, C. Marques, A. Frizera, M. Pontes, Multi-interface level in oil tanks and applications of optical fiber sensors. Opt. Fiber Technol. **40** (2018)
102. C. Farrar, F. Hemez, D. Shunk, D. Stinemates, B. Nadler, A review of structural health monitoring literature: 1996–2001 (2004)
103. H.-P. Chen, Y.-Q. Ni, Structural health monitoring of large civil engineering structures (2018)
104. R.F. Wright, P. Lu, J. Devkota, F. Lu, M. Ziomek-Moroz, P.R. Ohodnicki, Corrosion sensors for structural health monitoring of oil and natural gas infrastructure: a review. Sensors **19**(18) (2019)
105. F. Liu, X. Guo, D. Hu, W. Guo, N. Jin, Comprehensive inspection and evaluation technique for atmospheric storage tanks. Nondestr. Test. Eval. **25**(1), 45–59 (2010)

106. V. Cusati, S. Corcione, V. Memmolo, Impact of structural health monitoring on aircraft operating costs by multidisciplinary analysis. Sensors **21**(20), 1–27 (2021)
107. D. Barazanchy, M. Martinez, B. Rocha, M. Yanishevsky, A hybrid structural health monitoring system for the detection and localization of damage in composite structures. J. Sens. (2014)
108. K. Wood, T. Brown, R. Rogowski, B. Jensen, Fiber optic sensors for health monitoring of morphing airframes: II. Chemical sensing using optical fibers with Bragg gratings. Smart Mater. Struct. **9**(2), 170–174 (2000)
109. S. Mall, S.B. Dosedel, M.W. Holl, The performance of graphite-epoxy composite with embedded optical fibers under compression. Smart Mater. Struct. **5**(2), 209–215 (1996)
110. Z. He, Q. Liu, Optical fiber distributed acoustic sensors: a review. J. Lightwave Technol. **39**(12), 3671–3686 (2021)
111. X. He, X. Xu, M. Zhang, S. Xie, F. Liu, L. Gu, Y. Zhang, Y. Yang, H. Lu, On the phase fading effect in the dual-pulse heterodyne demodulated distributed acoustic sensing system. Optics Express **28**(22), 33433 (2020)
112. A.G. Leal-Junior, C. Marques, Diaphragm-embedded optical fiber sensors: a review and tutorial. IEEE Sens. J. **21**(11), 12719–12733 (2021). [Online]. Available: https://ieeexplore.ieee.org/document/9279330/
113. A. Leal-Junior, A. Frizera, C. Díaz, C. Marques, M. Ribeiro, M.M. J. Pontes, Material features based compensation technique for the temperature effects in a polymer diaphragm-based FBG pressure sensor. Opt. Express, **26**(16), 20590 (2018). [Online]. Available: https://www.osapublishing.org/abstract.cfm?URI=oe-26-16-20590
114. Y. Guo, J. M. Marin, I. Ashry, A. Trichili, M.-N. Havlik, T. K. Ng, C. M. Duarte, B. S. Ooi, Submarine optical fiber communication provides an unrealized deep-sea observation network. Sci. Rep. **13**(1), 1–10 (2023). [Online]. Available: https://doi.org/10.1038/s41598-023-42748-0
115. A. Beisenova, A. Issatayeva, S. Korganbayev, C. Molardi, W. Blanc, D. Tosi, Simultaneous distributed sensing on multiple MgO-doped high scattering fibers by means of scattering-level multiplexing. J. Lightwave Technol. **37**(13), 3413–3421 (2019)
116. S. Loranger, M. Gagné, V. Lambin-Iezzi, R. Kashyap, Rayleigh scatter based order of magnitude increase in distributed temperature and strain sensing by simple UV exposure of optical fibre. Scient. Rep. **5**(June) (2015). [Online]. Available: https://doi.org/10.1038/srep11177
117. H. Murayama, D. Wada, H. Igawa, Structural health monitoring by using fiber-optic distributed strain sensors with high spatial resolution. Photonic Sens. **3**(4), 355–376 (2013)
118. S. Abedin, A.M. Biondi, R. Wu, L. Cao, X. Wang, Structural health monitoring using a new type of distributed fiber optic smart textiles in combination with optical frequency domain reflectometry (OFDR): taking a pedestrian bridge as case study. Sensors **23**(3) (2023)
119. R.O. Ogunleye, S. Rusnáková, J. Javořík, M. Žaludek, B. Kotlánová, Advanced sensors and sensing systems for structural health monitoring in aerospace composites. Adv. Eng. Mater. 2401745 (2024)
120. L.B. Carani, J. Humphrey, M.M. Rahman, O.I. Okoli, Advances in embedded sensor technologies for impact monitoring in composite structures. J. Compos. Sci. **8**(6) (2024)
121. A. Datta, M.J. Augustin, K.M. Gaddikeri, S.R. Viswamurthy, N. Gupta, R.Sundaram, Damage detection in composite aircraft wing-like test-box using distributed fiber optic sensors. Opt. Fiber Technol. **66**(July), 102651 (2021). [Online]. Available: https://doi.org/10.1016/j.yofte.2021.102651
122. N.J. Lawson, R. Correia, S.W. James, M. Partridge, S.E. Staines, J.E. Gautrey, K.P. Garry, J.C. Holt, R.P. Tatam, Development and application of optical fibre strain and pressure sensors for in-flight measurements. Measurement Sci. Technol. **27**(10) (2016)
123. E.G. Bakhoum, C. Zhang, M.H. Cheng, Real time measurement of airplane flutter via distributed acoustic sensing. Aerospace **7**(9), 1–12 (2020)
124. P. Tick, Are low-loss glass-ceramic optical waveguides possible? Optics letters **23**(24), 1904–1905 (1998)
125. W. Blanc, V. Mauroy, L. Nguyen, B. Shivakiran Bhaktha, P. Sebbah, B. P. Pal, B. Dussardier, Fabrication of rare earth-doped transparent glass ceramic optical fibers by modified chemical vapor deposition. J. Am. Ceram. Soc. **94**(8), 2315–2318 (2011)

126. Z. Lu, N. Vakula, M. Ude, M. Cabié, T. Neisius, F. Orange, F. Pigeonneau, L. Petit, W. Blanc, Ybpo4 crystals in as-drawn silica-based optical fibers. Opt. Mater. **138**, 113644 (2023)
127. V. Fuertes, N. Grégoire, P. Labranche, S. Gagnon, N. Hamada, B. Bellanger, Y. Ledemi, S. LaRochelle, Y. Messaddeq, Cubic-shaped and rod-shaped ypo4 nanocrystal-doped optical fibers: implications for next generation of fiber lasers. ACS Appl. Nano Mater. **6**(6), 4337–4348 (2023)
128. T. Cheng, M. Liao, X. Xue, J. Li, W. Gao, X. Li, D. Chen, S. Zheng, Y. Pan, T. Suzuki et al., A silica optical fiber doped with yttrium aluminosilicate nanoparticles for supercontinuum generation. Opt. Mater. **53**, 39–43 (2016)
129. M. Vermillac, J.-F. Lupi, F. Peters, M. Cabie, P. Vennegues, C. Kucera, T. Neisius, J. Ballato, W. Blanc, Fiber-draw-induced elongation and break-up of particles inside the core of a silica-based optical fiber. J. Am. Ceram. Soc. **100**(5), 1814–1819 (2017)
130. Z. Lu, T. Robine, M. Guzik, M. Bellec, D. Tosi, C. Molardi, F. Pigeonneau, W. Blanc, Shaping nanoparticles in optical fibers through thermal engineering, in *Fiber Lasers and Glass Photonics: Materials through Applications III*, vol. 12142 (SPIE, 2022), pp. 81–86
131. V. Fuertes, N. Grégoire, P. Labranche, S. Gagnon, R. Wang, Y. Ledemi, S. LaRochelle, Y. Messaddeq, Engineering nanoparticle features to tune rayleigh scattering in nanoparticles-doped optical fibers. Sci. Rep. **11**(1), 9116 (2021)
132. D. Tosi, C. Molardi, M. Sypabekova, W. Blanc, Enhanced backscattering optical fiber distributed sensors: tutorial and review. IEEE Sens. J. **21**(11), 12667–12678 (2021)
133. M. Sypabekova, S. Korganbayev, W. Blanc, T. Ayupova, A. Bekmurzayeva, M. Shaimerdenova, K. Dukenbayev, C. Molardi, D. Tosi, Fiber optic refractive index sensors through spectral detection of Rayleigh backscattering in a chemically etched MgO-based nanoparticle-doped fiber. Opt. Lett. **43**(24), 5945 (2018)
134. D. Tosi, C. Molardi, W. Blanc, T. Paixão, P. Antunes, C. Marques, Performance analysis of scattering-level multiplexing (SLMux) in distributed fiber-optic backscatter reflectometry physical sensors. Sensors 20(9), 2595 (2020). [Online]. Available: https://www.mdpi.com/1424-8220/20/9/2595
135. A. Issatayeva, A. Amantayeva, W. Blanc, D. Tosi, C. Molardi, Design and analysis of a fiber-optic sensing system for shape reconstruction of a minimally invasive surgical needle. Sci. Rep. **11**(1), 1–12 (2021). [Online]. Available: https://doi.org/10.1038/s41598-021-88117-7
136. W. Blanc, D. Tosi, A. Leal-Junior, M. Ferrari, J. Ballato, Are low- and high-loss glass-ceramic optical fibers possible game changers? Optics Commun. **575**(November), 2025 (2024)
137. N. Arandia, J. I. Garate, J. Mabe, Embedded sensor systems in medical devices: requisites and challenges ahead. Sensors **22**(24), 9917 (2022). [Online]. Available: https://www.mdpi.com/1424-8220/22/24/9917
138. C.A.R. Diaz, A.G. Leal-Junior, L.M. Avellar, P.F.C. Antunes, M.J. Pontes, C.A. Marques, A. Frizera, M.R.N. Ribeiro, Perrogator: a portable energy-efficient interrogator for dynamic monitoring of wavelength-based sensors in wearable applications. Sensors **19**(13), 2962 (2019). [Online]. Available: https://www.mdpi.com/1424-8220/19/13/2962
139. R. López, V. Spirin, S. Miridonov, M. Shlyagin, G. Beltrán, E. Kuzin, Fiber optic distributed sensor for hydrocarbon leak localization based on transmission/reflection measurement. Opt. Laser Technol. **34**(6), 465–469 (2002). [Online]. Available: https://linkinghub.elsevier.com/retrieve/pii/S0030399202000439
140. V.V. Spirin, M.G. Shlyagin, S.V. Miridonov, P.L. Swart, Alarm-condition detection and localization using Rayleigh scattering for a fiber-optic bending sensor with an unmodulated light source. Opt. Commun. **205**(April 2002), 37–41 (2006)
141. V.V. Spirin, Transmission–reflection analysis for localization of temporally successive multi-point perturbations in a distributed fiber-optic loss sensor based on Rayleigh backscattering. Appl. Opt. **42**(7), 1175–1181 (2003)
142. M. Cen, V. Moeyaert, P. Mégret, M. Wuilpart, Localization and quantification of reflective events along an optical fiber using a bi-directional TRA technique. Opt. Express **22**(8), 9839 (2014)

143. L. Avellar, A. Frizera, H. Rocha, M. Silveira, C. Díaz, W. Blanc, C. Marques, A. Leal-Junior, Machine learning-based analysis of multiple simultaneous disturbances applied on a transmission-reflection analysis based distributed sensor using a nanoparticle-doped fiber. Photon. Res. **11**(3), 364 (2023)
144. A. Leal-Junior, L. Avellar, W. Blanc, A. Frizera, C. Marques, Opto-electronic smart home: heterogeneous optical sensors approaches and artificial intelligence for novel paradigms in remote monitoring. IEEE Internet Things J. 1 (2023)
145. A. Leal-Junior, W. Blanc, Temperature-insensitive and cost-effective distributed NP-doped optical fiber sensors. Opt. Fiber Technol. **88**(June),103995 (2024). [Online]. Available: https://doi.org/10.1016/j.yofte.2024.103995

Digital Image Correlation Techniques for Structural Analysis in Aerospace

Łukasz Święch

Abstract The chapter presents the theoretical basis of the digital image correlation method and examples of its application in relation to strength tests. This method is a particularly useful tool for the analysis of load-bearing structures operating under conditions of advanced deformations after loss of stability. In such a situation, the confrontation of the results obtained through experimental research and numerical analyses may be the basis for determining the correctness of the solution. Both the deformation fields on the surface of the tested elements can be compared, as well as selected, representative relations between displacement and load, called equilibrium paths. The results of the research allow us to conclude that correctly combined DIC and FEM results are a highly effective research tool in determining the deformation of structures in the full spectrum of the load realized. Registration of the displacement field over the entire area of the studied structure allows the identification of both the global form of deformation and local effects. Recording can take place with a specific frequency, which, in combination with the information about the load level assigned to each stage, allows for the analysis of the history of deformation of the structure. An unquestionable advantage of the DIC method is also the independence of the possibility of conducting research on the type of material from which the object is made, thanks to which research is not limited only to model tests but can also be carried out in the case of real objects. The tests themselves can be carried out in almost any operating conditions, depending on the configuration of the measuring system. A limitation in the use of digital image correlation systems is the fact that the result of the tests is the field of deformation recorded only on the surface of the tested object, which, except in the case of objects with relatively uncomplicated geometry and constant thickness, makes it impossible to directly determine the state of stress in the tested elements. In such situations, it is necessary to use the results of numerical calculations.

Ł. Święch (✉)
Department of Aerospace Engineering, Faculty of Mechanical Engineering and Aeronautics, Rzeszow University of Technology, Rzeszów, Poland
e-mail: lukasz.swiech@prz.edu.pl

C. Marques et al. (eds.), *Advanced Optical Sensors for Aerospace Applications*, Progress in Optical Science and Photonics 37,
https://doi.org/10.1007/978-981-95-1626-1_6

1 Introduction

A fundamental requirement for the design practice of load-bearing structures is absolute trust in the obtained results. Without diminishing the importance of numerical analysis as a tool of unquestionable effectiveness it is justified, and in some situations essential, to support the design process with a series of experimental studies that significantly increase the reliability of numerical analysis results, utilizing the latest research techniques and well-established traditional methods.

For many years, experimental methods in the mechanics of deformable bodies relied on discrete measurements. These include strain gauge measurements using electro-resistive sensors [46] and fiber optic sensors, which are increasingly used [20, 27, 54].

Strain field measurements were initiated by polarization-optical methods [9]. These studies are conducted on models made of optically active materials. They allow for observing the continuous distribution of optical effects in the form of fringes that correspond to the distribution of the principal strain differences (isochromes) and the main deformation directions (isoclines). The analysis of the three-dimensional state of stress is associated with a more complicated experimental technique than in the case of flat problems [9]. In order to determine the stress at the relevant points of the model, the freezing methods [32, 33] the scattered light method [49, 63] and the integrated elasto-optics method [2, 3].

The development of digital photography technology has also influenced elasto-optical methods, which have evolved into digital elasto-optics, which enable the use of computer algorithms to process and analyze optical effects [36, 40–42].

Due to the fact that conventional strain measurement equipment (strain gauges, extensometers) does not provide information about the distribution of the measured quantities on the surface of the tested element, and elasto-optic measurements are relatively difficult to interpret and limited to materials exhibiting optically active properties, there arose a need to develop new research methods.

Among the range of methods enabling non-contact, field measurements of deformations, the shadow moiré method [39], holographic methods [44] and macular interferometry methods [48] can be distinguished. An overview of the most popular methods can be found in a monograph edited by W. Sharpe [46] and a book on optical measurements edited by Rastogi [43].

Currently, the most dynamically developing group of research methods is based on optical measurements supported by numerical algorithms, enabling unambiguous interpretation of results. A special place among them is occupied by the Digital Image Correlation (DIC) method, which is described on the following pages of this chapter.

2 Digital Image Correlation Method Basics

The increase in computer computing power in experimentation was visible primarily in the possibility of recording field deformations. In 1980, Park [34] presented the possibility of measuring the displacements of points using a photograph of an object with a pattern of spots and analyzing their position changes. Still, the application and accuracy of this method was limited by errors resulting from the decorrelation of individual images. In 1982, Peters and Ransom [37] presented a technique for the digital processing of photographs to determine the displacements of the surface of an element covered by a set of spots illuminated by a laser beam. The concept was to divide the surface of the element into sub-areas, which were compared using a numerical correlation procedure, which made it possible to determine the amount of displacement. Sutton et al. [50] in 1983 developed a numerical algorithm called "the matching process", which became the basis for a two-dimensional version of the Digital Image Correlation (DIC) method. Peters et al. [38] and Chu et al. [8] conducted measurements using the DIC algorithm, proving that it can determine translational and rotational displacements. In the following years, the method was developed by searching for more efficient correlation algorithms [7, 51, 52, 57].

In the 1990s, Luo et al. [25, 26], Kahnjetter [19], and Helm [16] proposed extending the method to include measurements of spatial deformations using a set of two digital cameras. The key to this method is the calibration technique of the measurement system [60, 66].

The DIC method is based on digital processing of the surface image of the tested element recorded during the subsequent stages of deformation growth, which allows the determination of the deformed object's shape and displacement area. The calculation of strain and stress is based on the assumption that the object under study is a continuous medium and that the equations of the theory of elasticity and plasticity can be applied to it [58].

Measurement systems include digital cameras, light sources, and electronic hardware for the acquisition and processing of recorded data. Depending on the measurement system's version, one camera is used to measure in-plane deformation or at least two cameras are used to perform spatial measurements (Fig. 1).

In order to prepare the component for testing, its surface is covered with a stochastic pattern of spots, which allows their location to be unambiguously identified (Fig. 2). The photographs taken are divided into sub-areas, which are usually rectangular or triangular in shape, with a size of dozen by a dozen pixels.

The essence of the method is to find the same area in photographs taken in successive stages of advanced deformation. As a result of the correlation of individual images (deformation stages), it is possible to determine the displacements of the surface points of the element. The displacement area obtained in this way is the basis for determining the deformations, calculated using interpolation functions.

The analyzed image is transformed into a matrix containing 8-bit information in the form of numbers determining the gray level of each pixel of the sub-area (Fig. 3). This number ranges from 0 to 255, with the smallest value represented by black, the

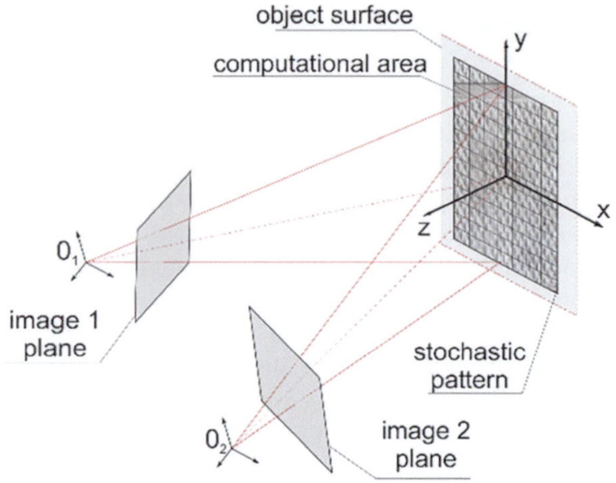

Fig. 1 3D DIC measurement system scheme

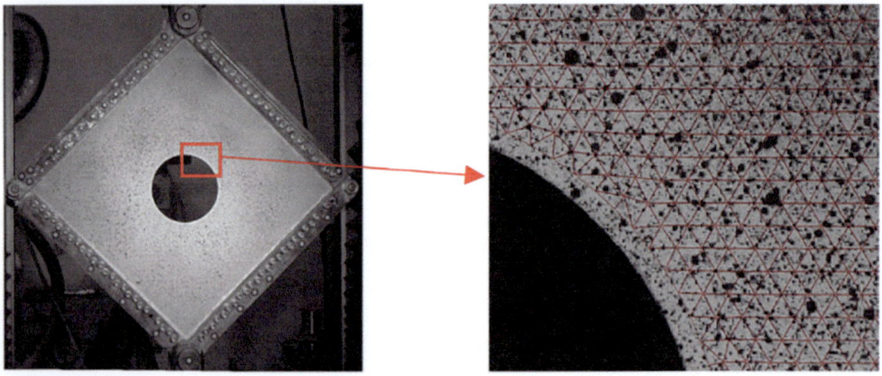

Fig. 2 Test object covered with a pattern of spots and division of the image into control points and calculation sub-areas

largest by white, and all intermediate values corresponding to different levels of gray. The number of combinations of numbers within one sub-area is so large that it allows for unambiguous identification of their location and practical lack of repeatability within the studied area. Due to the fact that the measurement signal is, in fact, the value of the determined gray level of the pixel, the highest possible contrast of the photographs is required, which is obtained by illuminating the surface of the tested element. A change in the illumination intensity during the measurement adversely affects the results.

In order to identify the displacement, consider an area (sub-area) on the surface of an element (Fig. 4) where the location of the midpoint P at coordinates (x, y) is determined. The task of the calculation procedure is to find the same area in the next image, whose midpoint $P\prime$ reaches the coordinates $\left(x_i^*, y_i^*\right)$. Due to the fact that the relative distribution of gray levels of individual pixels in the area does not change

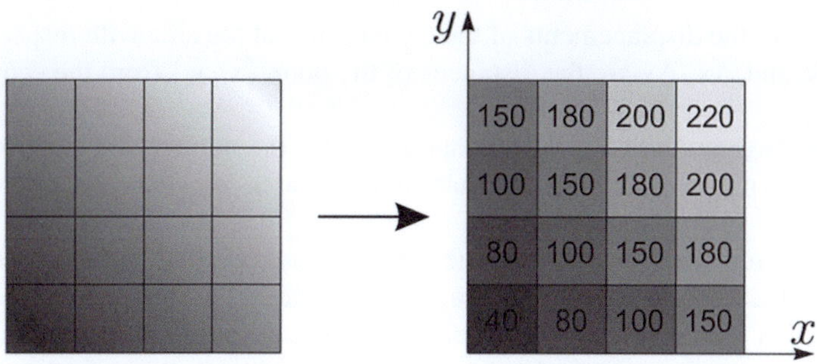

Fig. 3 Image of the computing sub-area and its digital interpretation

with deformation, it is possible to determine the position of any point $Q(x_i, y_i)$ and $Q'(x_i^*, y_i^*)$ using the adopted shape function.

The simplest function for determining translational and rotational displacements as well as normal and tangential deformations results from the expansion into the Taylor series (Eq. 1):

$$
\begin{aligned}
x_i^* &= x_i + u + \frac{\partial u}{\partial x}\Delta x + \frac{\partial u}{\partial y}\Delta y \\
y_i^* &= y_i + v + \frac{\partial v}{\partial x}\Delta x + \frac{\partial v}{\partial y}\Delta y
\end{aligned}
\tag{1}
$$

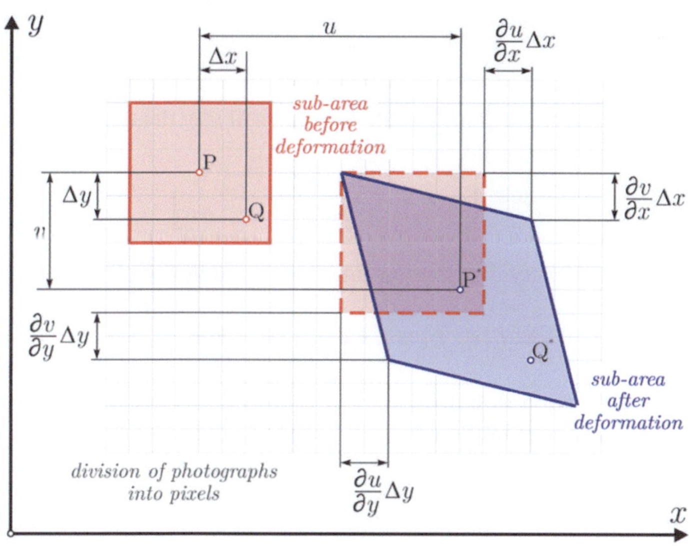

Fig. 4 Description of surface deformations and point displacements

where: u, v are the displacements of the center point of the area with respect to the x and y axes, and Δx, Δy are the distances of the point (x_i, y_i) from the center of the area.

The first two terms of Eq. 1 are responsible for the motion of a rigid body. The following two present affine transformations of the area from the reference state, i.e., deformations.

Each pixel representing a point in the sub-area under consideration is described by a number indicating the level of its gray (light intensity) in the photograph (Fig. 3). Assuming that $f(x_i, y_i)$ is the gray level of the point (x_i, y_i) in the undeformed state, and $g(x_i^*, y_i^*)$ for the point (x_i^*, y_i^*) in the deformed state, theoretically the values of f and g should be identical. In fact, they vary due to the level of lighting, blurring of individual photos, perspective distortion and random measurement noise.

The relationship between them can be defined as:

$$f(x_i, y_i) - e(x_i, y_i) = r_0 + r_1 g\left(x_i^*, y_i^*\right) \tag{2}$$

where r_0 and r_1 are the parameters used to compensate for the gray level resulting from the factors given above (the so-called radiometric transformation), and $e(x_i, y_i)$ is a correction resulting from the estimated value of the measurement noise.

The essence of digital image correlation is to compare sets of numbers between two digital images. If the minimum correlation function is reached, the considered images are characterized by the greatest similarity.

The basic formula for describing correlation is the squared sum of differences:

$$S(p) = \min \sum_{i=1}^{n} (F_i - G_i)^2 \tag{3}$$

where p is denoted by a vector of parameters determining the quality of correlation:

$$p = \left[x, y, u, v, \frac{\partial u}{\partial x}, \frac{\partial u}{\partial y}, \frac{\partial v}{\partial x}, \frac{\partial v}{\partial y}\right] \tag{4}$$

The gray-level intensity distribution functions are denoted as:

$$F_i = f(x_i, y_i) \text{ and } G_i = r_0 + r_1 g\left(x_i^*, y_i^*\right) \tag{5}$$

However, the function described by Eq. 2 does not allow for compensation of the results due to changes in the intensity of illumination and the relative displacement of the light source resulting from the deformation of the tested element. These effects are particularly important for spatial measurements. Therefore, it is necessary to introduce additional optimization criteria that allow obtaining correct results in changing lighting conditions. These changes are taken into account by introducing a G photometric transformation in the form of $\Phi(G) = aG + b$. Where parameter a is the invariant criterion of the illumination scale and b parameter is responsible for the relative displacement of the light source. The correlation function will then take the form:

$$S(p) = \min \sum_{i=1}^{n} (F_i - (aG_i + b))^2 \tag{6}$$

The correlation problem has more unknowns than the equations available. The missing information to get the solution is the gradient of the grayscale distribution between the discrete points represented by pixels.

In 1989, Bruck, et al. [7] proposed the use of the Newton–Raphson method to determine six parameters determining correlation. After determining the first approximation of the values u,v, ∂u/∂x, ∂u/∂y, ∂v/∂x and ∂v/∂y the method is used to find the actual values resulting from minimizing the correlation function S.

Individual changes (increments) of parameters are referred to as:

$$\Delta R_i = -\frac{\nabla(R_i)}{H(R_i)} \tag{7}$$

where R_i is a vector containing the parameter values in the current computational increment:

$$R_i^T = \left(u, v, \frac{\partial u}{\partial x}, \frac{\partial u}{\partial y}, \frac{\partial v}{\partial x}, \frac{\partial v}{\partial y}\right) \tag{8}$$

$\nabla(R_i)$ is a Jacobian containing the derivatives of the correlation function with respect to the parameters:

$$\nabla(R_i)^T = \left(\frac{\partial S}{\partial u}, \frac{\partial S}{\partial s}, \frac{\partial S}{\partial\left(\frac{\partial u}{\partial x}\right)}, \frac{\partial S}{\partial\left(\frac{\partial u}{\partial y}\right)}, \frac{\partial S}{\partial\left(\frac{\partial v}{\partial x}\right)}, \frac{\partial S}{\partial\left(\frac{\partial v}{\partial y}\right)}\right) \tag{9}$$

$H(R_i)$ is a Hesse matrix that contains the second derivatives of the correlation function.

The increments ΔR_i obtained from Eq. 7 are added to the estimated (preliminary) values in successive iterative steps until the convergence of the solution is reached. The criterion of convergence of the solution, and thus stopping the iterative procedure, is adopted by monitoring changes in the displacement level of u and v, usually assuming that the solution converges if these changes do not exceed a 10^{-4} of a pixel. An additional criterion for convergence is the change in the displacement gradient at a level equal to or less than 5×10^{-6}.

The correlation of images allows us to determine the distribution of the degree of gray level of individual points of the body between the successive considered stages of loading, which is equivalent to specifying the average value of displacement. However, such information is incomplete and insufficient to analyze the deformation area of the element because it is discrete. This means that information is received

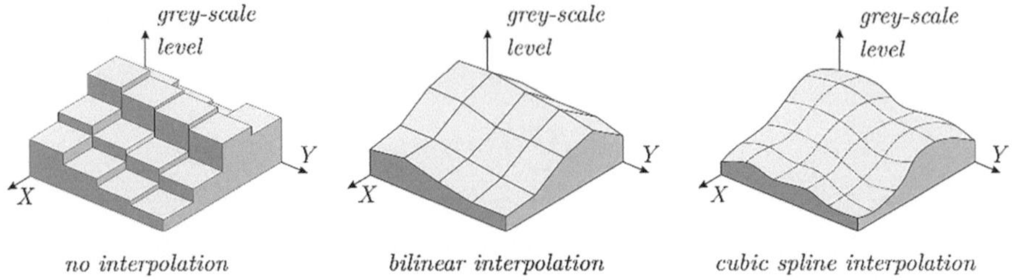

Fig. 5 Example presentation of the measurement area and the result of grayscale level interpolation

only about the grayscale of individual pixels, but it is missing in relation to the nature of the changes in deformation (gray-scale level) between them. For this reason, it is necessary to interpolate the values between adjacent pixels, which allows us to determine the displacements with an accuracy of 1/100th of a pixel. In the initial period of using the method, bilinear or polynomial interpolation was used (Fig. 5).

Higher accuracy can be achieved by using "glued" functions. For example, an interpolation function in the form of a third-degree spline can be written as:

$$g(x, y) = \sum_{m=0}^{3}\sum_{n=0}^{3} \alpha_{mn} x^m y^n \tag{10}$$

where α_{mn} coefficients are determined on the basis of the distribution of gray levels of the points of the analyzed subarea.

The distribution of deformations can be determined by differentiation of the displacement function. However, this operation takes into account the measurement noise contained in the displacement. To eliminate this effect, the displacement function is first smoothed and then differentiated. This makes it possible to increase accuracy. Among the smoothing algorithms, finite element smoothing [28, 52] and thin plate spline smoothing [62] can be distinguished, but there are also different approaches [59].

The use of a single digital camera limits the application of the method to flat problems. To overcome this limitation, a measurement system of at least two cameras is used to observe the object from different angles. This allows us to determine the location of points in a spatial coordinate system. For this purpose, the triangulation process based on the principles of epipolar geometry is used [64].

The triangulation process is preceded by the calibration of the measurement system, which is carried out under the assumption that the cameras are rigidly connected and do not change their position relative to each other. Both cameras observe the same image (calibration element, grid) and take a series of photographs from different angles. The result is a digital representation of the calibration element. It is assumed that one of the cameras provides a reference image. Sixteen parameters are determined as a result of calibration. Six of them are external parameters describing the relative position of the cameras, treated as connected rigid solids, in relation to the calibration element. The remaining five internal parameters for each camera relate the global datum to the coordinate system associated with the camera. These parameters include the scale factors of the photograph, the offset between the origin of the camera's coordinate system and the optical axis, and the radial distortion factor.

During the test, the measuring system must not change its position in relation to the object to be measured. After taking a series of images during the process of loading the structure, numerical procedures are started, the purpose of which is to determine the initial shape of the object and its subsequent deformation.

The process of three-dimensional reconstruction, i.e., determining the spatial location of the points of an object in a reference state, is determined by the correlation of images of an undeformed object between photographs taken by two cameras of the measurement system.

Determination of deformations between successive stages of loading of the object is carried out in the subsequent steps of the numerical procedure, which consist of correlating images between successive stages and between the cameras of the measurement system. After the surface reconstruction process to the 3D system, it is possible to determine the spatial field of displacement by comparing the current and reference stages.

On the basis of the procedures used in the DIC method, a whole group of research methods was created, which enable measurements of the position of points in space, as well as the reconstruction of the three-dimensional geometry of the tested elements.

One of the most common are fringe scanners used to create digital representations of objects of any complex shape [5, 65]. Parallel fringes are projected onto the object under study, the image of which is recorded by a system of stereoscopically arranged digital cameras. Then, the geometry is reproduced using the shadow moiré method [39], but the DIC method is used to identify the position of reference points. Such scanners are primarily used to analyze manufacturing deviations by comparing the measurement results with reference geometry created using CAD computer software. For example, Fig. 6 presents the distribution of manufacturing inaccuracies of the AOS-71 motor glider CFRP fuselage and Fig. 7. shows TWISST-2 unmanned aircraft prepared for measurements and overall deviations of geometry in comparison to the CAD model.

Algorithms identifying the location of points in digital photographs are also used in photogrammetric methods [29]. Taking a series of photos of the tested object with measurement points and calibration objects placed on it allows us to determine the location of points in space.

The use of digital spatial photogrammetry systems also allows us to determine the displacements of measurement points. For this purpose, a series of photos should be taken to determine the location of the points at subsequent levels of static load. As a result, it is possible to determine the displacement vectors of these points. Figure 8 presents the results of the static test of the M3 Pliszka glider in the form of displacement vectors visualized directly on a photograph taken during the measurements and superimposed on the CAD model, thus presenting the distribution of wing deflection

Fig. 6 Measurement of manufacturing geometrical deviations of AOS-71 motor glider [own research]

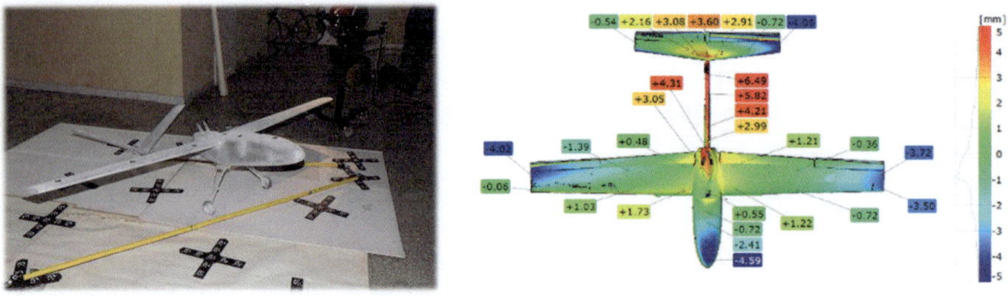

Fig. 7 Measurement of manufacturing deviations of TWISST-2 unmanned aircraft [own research]

under the influence of loads simulating the distribution of aerodynamic forces along the wingspan.

The combination of photogrammetric methods and fringe scanners allows measurements of extensive objects with large dimensions. In this case, the photogrammetric system provides information about the position of the points glued to the object, which then serve as reference points for the fringe scanner. This approach allows for high measurement accuracies and avoids the so-called drift error resulting from the fringe scanner's limited measurement volume (Fig. 9).

3 Application of DIC to Measure Post-Buckling States of Deformations of Thin-Walled Structures

The study of load-bearing structures in post-critical deformation states is a troublesome issue due to the need to take into account nonlinear theories, which, combined with complex geometry, makes it practically impossible to obtain exact solutions achievable only in simple cases [10, 12, 61]. In solving a nonlinear problem, it becomes problematic to interpret the obtained results, sometimes correct from the mathematical point of view in the numerical analysis of FEM but inconsistent with

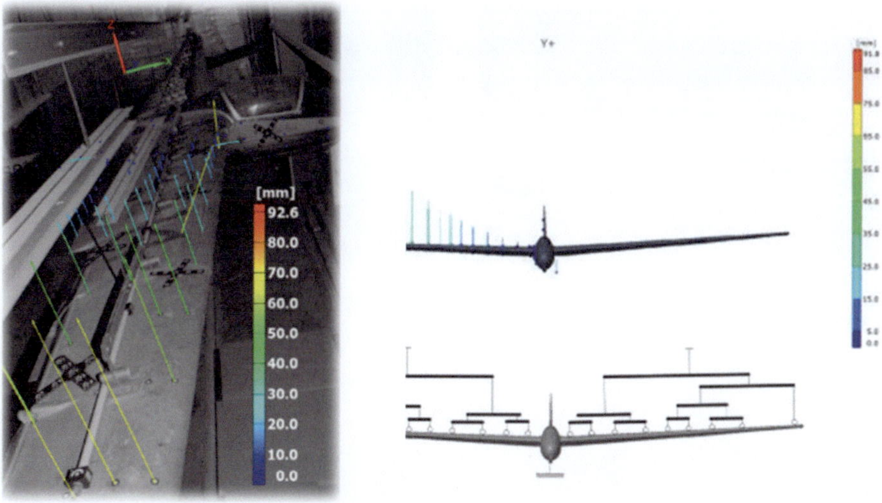

Fig. 8 Result of measurements of deformation of the M3 Pliszka glider during static tests [own research]

the physical course of the phenomenon [4, 6]. In such situations, the choice of the correct solution in many cases requires additional verification experimental studies.

In the general case, obtaining reliable results from numerical analyses and experimental studies of thin-walled structures taking into account critical deformation states can be based on the methodology described below.

The structural testing procedure (Fig. 9) begins with measurements of geometric deviations that occur during manufacturing. The use of a fringe scanner in combination with a digital spatial photogrammetry system allows for the creation of a sufficiently accurate, three-dimensional representation of the object. The resulting model is compared with the geometry created by the CAD software.

Experimental tests are carried out using a testing machine, which ensures control of the level of the given load. The deformation of the structure during the test is recorded by means of a measurement system based on the three-dimensional method of digital image correlation (DIC).

In addition, before loading, the reference stage of the experimental test allows us to determine the initial geometric imperfection of the tested structure, which results both from the method of mounting in the test stand and from performance inaccuracies. This imperfection should be reproduced in the numerical model as the initial geometry of the analyzed system. If this is not possible, i.e., due to excessive complexity, the imperfection can be considered as a perturbation of selected and appropriately scaled buckling modes obtained by linear stability analyses or by loading by external forces in the first step of the numerical analysis.

The results of the experiment are compared with the results of nonlinear numerical analyses. If there is no convergence of the solution or it is considered unsatisfactory,

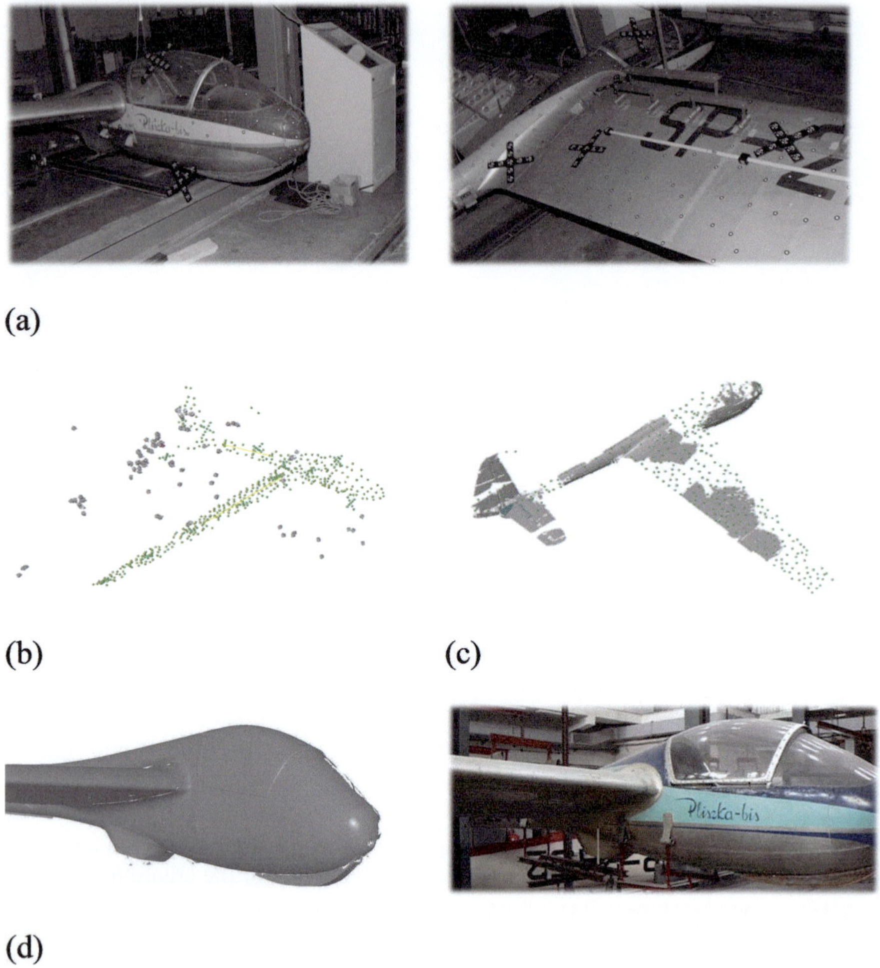

Fig. 9 Measurement of the geometry of the M3 Pliszka glider [own research]: **a** photographs of the glider prepared for measurement, **b** cloud of reference points, **c** fragments of the scanned surface, **d** qualitative comparison of the scan results with a photograph of the real object

the numerical model is modified by correcting the initial imperfection or changing the control parameters, or by choosing a new method for solving the nonlinear problem.

The methodology described above (Fig. 10) allows to obtain the required convergence of the numerical solution in the full range of loads to which the tested structure is subjected.

With reference to practical applications, we will consider the behavior of a thin-walled structure integrally reinforced with ribs in the shape of triangles, in which the load causes a state of advanced deformation. This is one of the well-known methods of integral stiffening, which is becoming increasingly popular in the aerospace industry

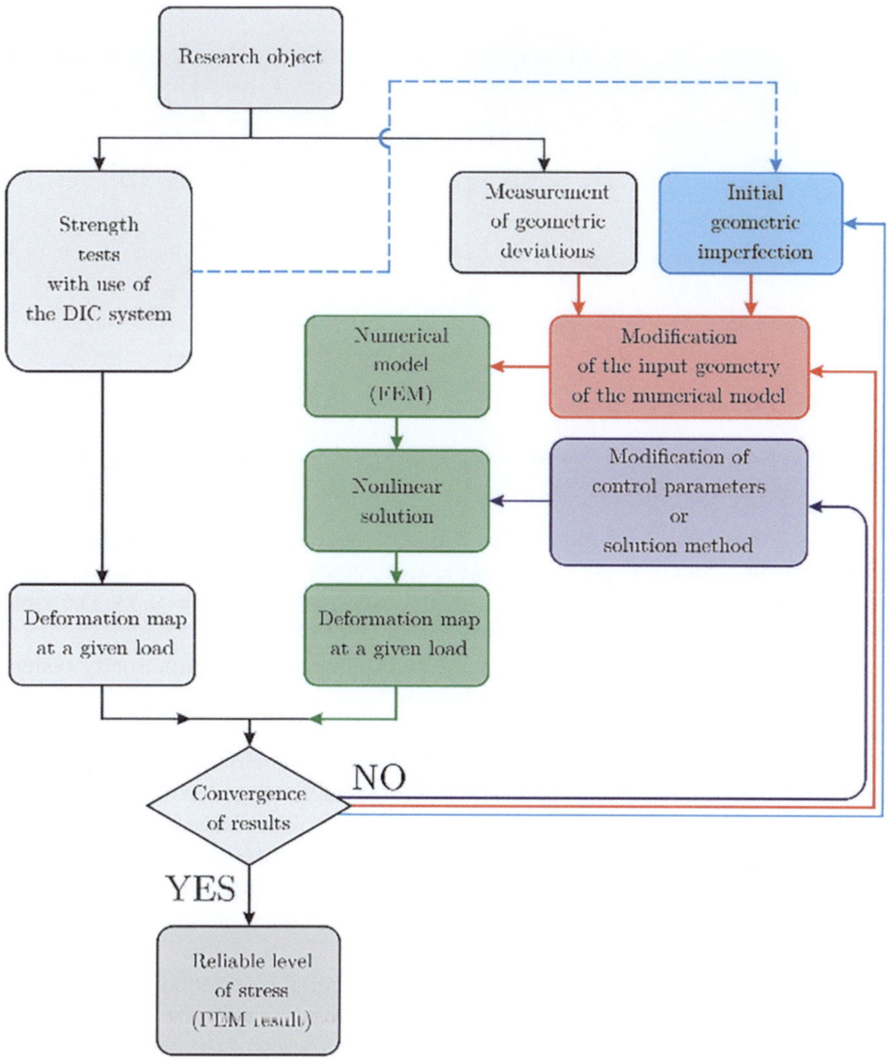

Fig. 10 Procedure for the control of the numerical solution using the results of experimental tests

[30, 53]. This type of solution allows to reduce the weight of the structure and has a positive effect on increasing resistance to local damage [31, 45]

As a test model, the 275 × 275 mm plate (Fig. 11) was made by a CNC machine from a semi-finished product in the form of a 2024 T3 aluminum alloy plate with a thickness of 2 mm. To produce the structure, a specially designed vacuum table was used, ensuring that the appropriate geometric tolerances were maintained. The integral stiffening of the skin was a grid in the form of ribs arranged in the shape of equilateral triangles with dimensions of 45 mm by 90 mm. The rib dimensions were:

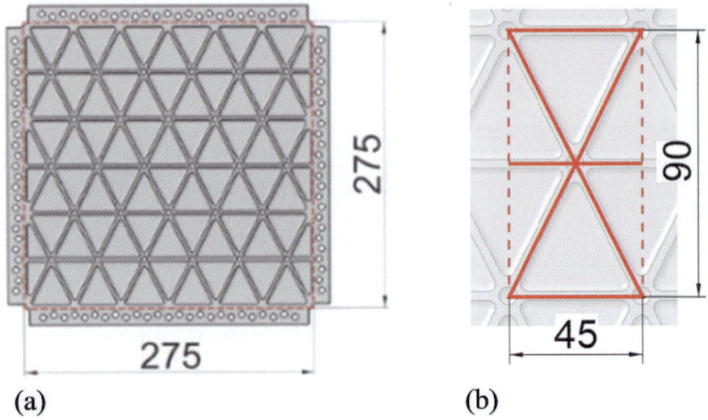

Fig. 11 Geometry of the tested plate: **a** overall dimensions, **b** geometry of the stiffener mesh

thickness 2 mm, height 1.5 mm. Holes with a diameter of 3 mm were drilled in the grid nodes. The thickness of the skin in the area between the stiffeners was 0.5 mm. Young's modulus of material was E = 70 GPa and Poisson ratio ν = 0.33. The yield stress was 345 MPa and the ultimate stress was 488 MPa.

The plate was measured using the GOM TRITOP digital photogrammetry system [14] and the GOM ATOS [38] shadow moiré spatial scanner [15] to identify the manufacturing deviations from the ideal geometry.

As a result of the measurement with the photogrammetric system, the spatial position of 228 points was recorded, which became reference points for 3D scanning. A series of 16 images were taken that allowed the object to be oriented in space (Fig. 12). The result of the measurements with the shadow moiré scanner was a digital representation of the image consisting of 472,827 points, on which 908,173 triangles were stretched to form the surface of the measured element (Figs. 13 and 14).

The result of the measurements in the form of a digital representation of the plate mounted in the frame was compared with the reference geometry (CAD model), which was previously used to program the CNC machining of the plate.

The distribution of geometrical deviations in relation to the geometry of the CAD model is shown in Fig. 15. In the vast majority of the panel surface, the thickness oscillated around 0.53 mm, which corresponds to a deviation of 6% compared to the nominal dimension. In the case of stiffeners, the maximum performance deviation did not exceed 6%, and the "average" height was 2.04 mm, which is a deviation from the nominal dimension of 2%.

During experimental tests, the plate was mounted in a rigid steel frame constituting a four-bar mechanism, providing the possibility of loading with shear load along the edges, inducing a state of pure shear (Fig. 16). The loads were carried out using the Zwick Z050 universal testing machine that allowed to control the level of the load. The skin was loaded with a force of up to 40 kN, which caused the occurrence of

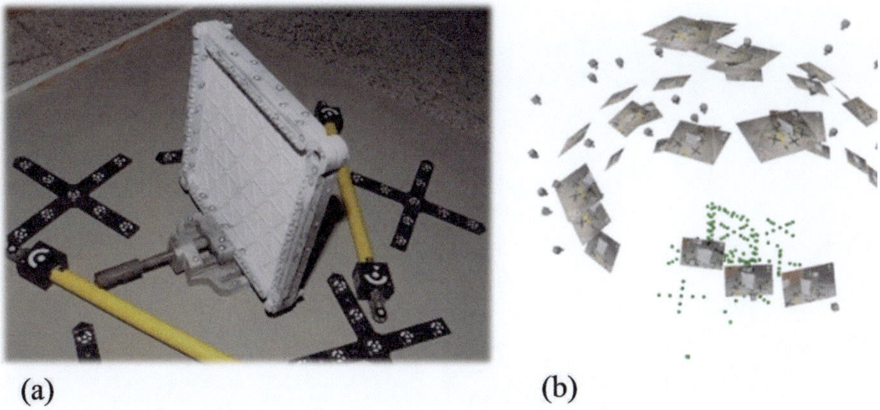

Fig. 12 The plate prepared for photogrammetric tests (**a**) and the measurement result (**b**) with the frames of the photos marked

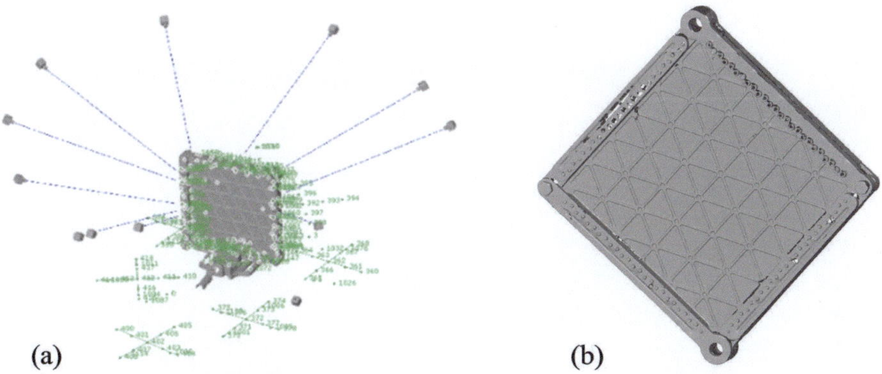

Fig. 13 Result of measurement with a spatial scanner with marked reference points (**a**) and the obtained digital representation of the tested element (**b**)

post-critical deformations, as well as the appearance of plastic deformations of the material of the tested structure.

During loading, the deformation of the unstiffened side of the plate was measured using a three-dimensional digital image correlation (DIC) system. The measurements consisted of taking a series of images of a suitably prepared area at a frequency of 1 Hz. This preparation included covering the measurement surface with a stochastic pattern of black spots. To record the displacements of spots associated with changes in deformations under the influence of an increase in load, two stereometrically arranged digital cameras and a light source were used, which, together with computer software, formed the GOM ARAMIS 5M deformation measurement system based on the digital image correlation method [13].

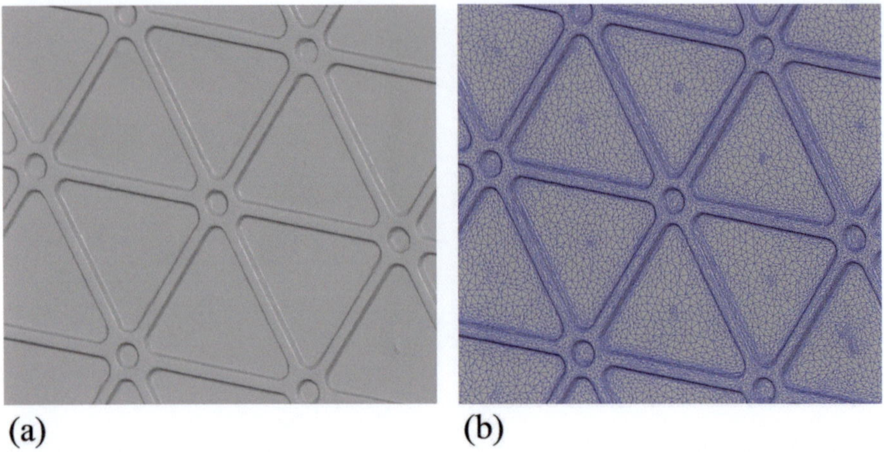

Fig. 14 Details of the surface geometry of the scanned element (**a**) and the mesh representing the shape of the element (**b**)

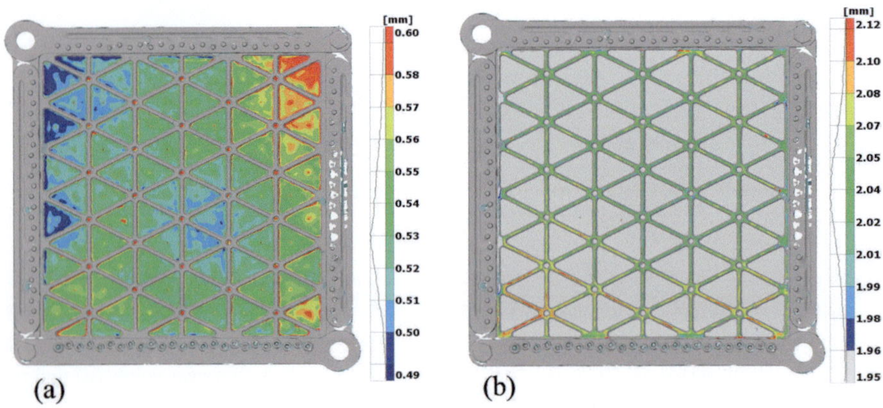

Fig. 15 Distribution of board deviations: **a** plate thickness (nominal value 0.5 mm), **b** height of stiffeners (nominal value 2.0 mm)

The DIC system was connected to the testing machine via a voltage signal, which allowed for the correlation of the load level with the deformation measurement. Figure 17 presents the technical implementation of the test stand.

The results of the tests using the DIC scanner are digital information about the spatial location of the points forming the surface of the tested element in each recorded stage of loading, which results in the creation of a vector field of displacements of the points of the tested object.

In the case under consideration, the measuring volume dimensions were 495 × 435 × 435 mm (H × W × D). Photographs with an image resolution of 2448 by 2050 pixels were divided into 7255 computational subregions (correlation regions)

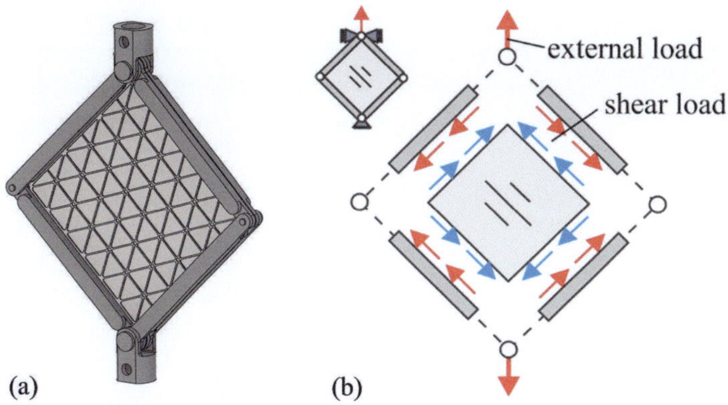

Fig. 16 CAD model of the test stand with plate (**a**) and load diagram (**b**)

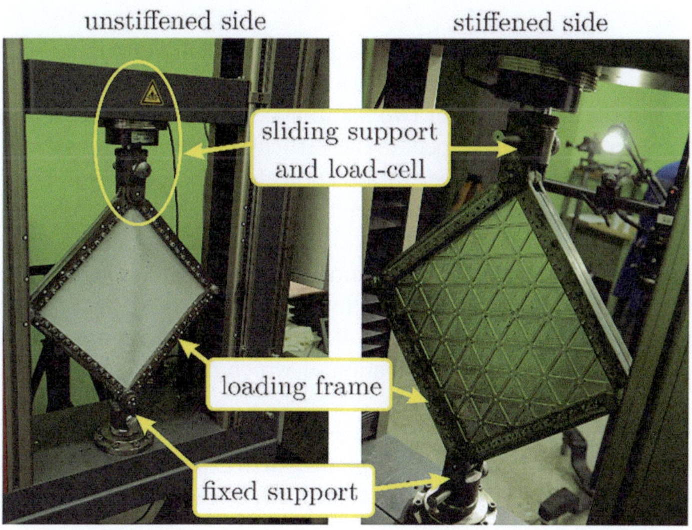

Fig. 17 Test stand prepared for testing—plate mounted in the measuring space of the testing machine

with dimensions of 19 by 19 pixels (Fig. 18). The calibration deviation was 0.040 pixels. The measurement images were taken at a frequency of 1 Hz, and 425 load stages were recorded during the test.

The results obtained with the use of a measurement system based on the DIC method can be analyzed both in the form of graphs determined for specific points, as well as with the use of color maps presenting the field distribution of deformation components determined in relation to any chosen reference system. Figure 19 shows the change of the deflection area of the plate points in the direction normal to its surface in the subsequent loading stages. The global loss of stability manifested in

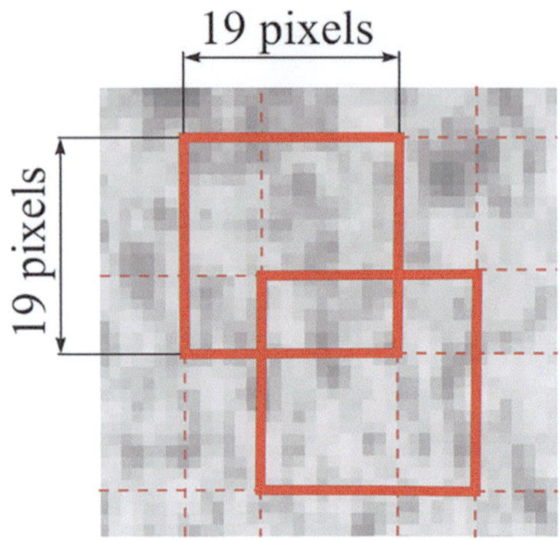

Fig. 18 Division of the board surface into calculation sub-areas in the DIC method

the form of a deflection in the shape of one half-wave in the middle region of the plate geometry. The increase in load led to a deepening of the deflection in the middle of the plate and the appearance of two additional, smaller half-waves of deflection symmetrically arranged with respect to the vertical diagonal of the plate.

Analysis of the deflection of the cross-section along the horizontal diagonal of the plate (relative to the method of attachment during experimental studies) allows for a more accurate trace of the development of the deformation of the system with the increase of load (Fig. 20). As already mentioned, the loss of plate stability manifested itself as one half-wave along the cross-sectional length, whose deflection at maximum load was about 8 mm. At a load of about 40 N/mm, two smaller oppositely directed half-waves began to form, the deformation of which reached a maximum value of 2.5 mm.

Different behavior can be observed in the deflection of the cross-section along the vertical cross-section (in the direction of load input in experimental studies) (Fig. 21). At the maximum load level, a local loss of stability of the covering between the ribs could be observed in terms of plastic deformations.

Based on the deflection history of the characteristic points depending on the load, it is possible to create equilibrium paths that graphically illustrate the changes in the deformation of the structure over the entire load range, taking into account critical and post-critical states.

Figure 22 shows a representative equilibrium path for a point in the center of the principal buckling half-wave with the maximum deflection value. In the case of small load levels, this relationship was linear in nature. With increasing load, a slight change in the angle of inclination of the line can be observed, which means a change in stiffness and a transition to a range of large, i.e., geometrically non-linear, deformations. The lack of rapid change associated with reaching a critical state is caused, among other things, by the geometric imperfection that characterizes the actual research structure.

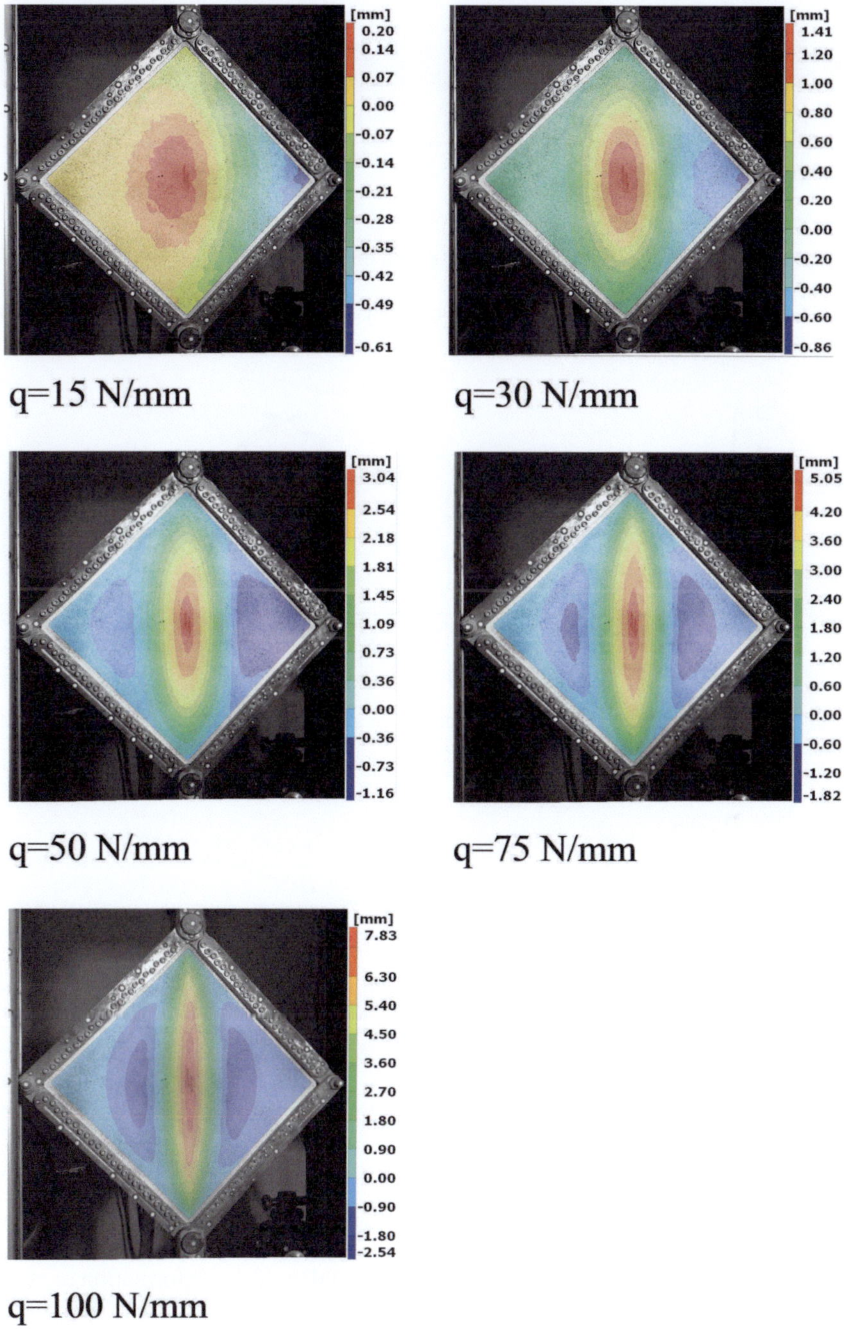

Fig. 19 Deflections of the surface of the stiffened plate for subsequent deformation phases (DIC result)

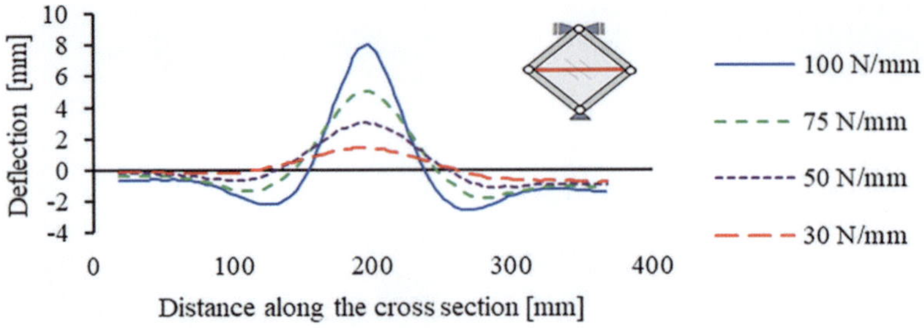

Fig. 20 Change of deflection of a cross-section drawn along the horizontal diagonal of the plate for successive load stages

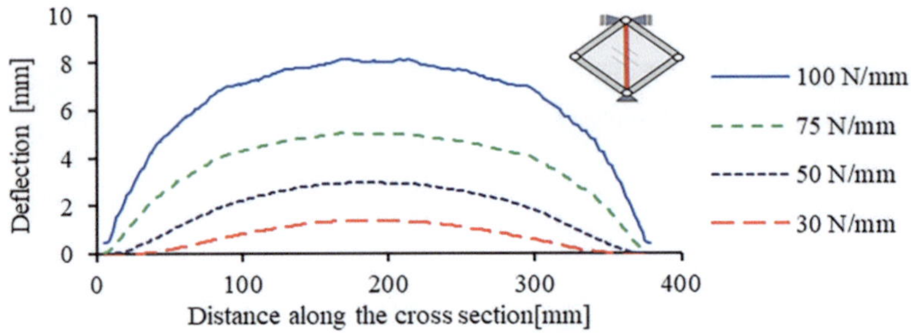

Fig. 21 Change of deflection of a cross-section drawn along the vertical diagonal of the plate for successive load stages

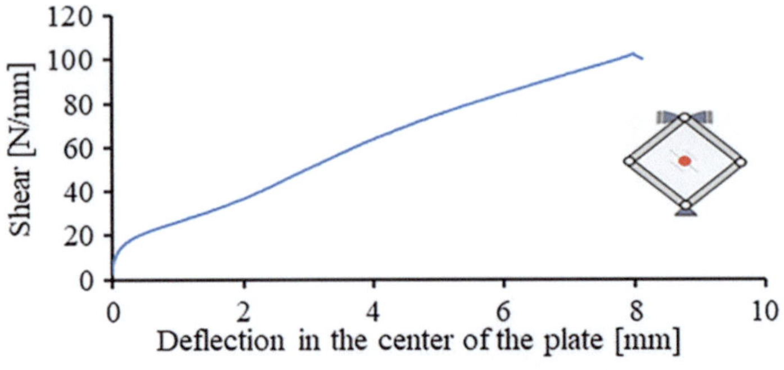

Fig. 22 Equilibrium path in the form of relation between shear load on the edges of the plate and deflection of the point located in the center of the plate

The measurement results in the form of knowledge of the deformation area also allow determining the approximate values of loads causing the loss of stability. In that case. It is necessary to use appropriate approximation methods based on Koiter's methods [21] or the relation of load and deflection (P(w)) or load and square of deflection (P(w2)) at the characteristic point indicating loss of stability [35, 47, 56].

In the case under consideration, pure shear of the plate, the approximate experimental value of the critical load was determined based on the Koiter method (Fig. 23). For this purpose, a fragment of the load–deflection equilibrium path was approximated with a polynomial of the third degree, so that the correlation coefficient R^2 would converge in more than 95% with the form of the original curve in terms of elastic strains. The intersection of the approximation function with the vertical axis of the graph, describing the shear load, allowed determining the approximate value of the critical load of about 15.4 N/mm (Fig. 24). Determination of the critical load value is therefore based on determining its value, under the influence of which the structure experiences a jump from linear to nonlinear geometrically deformations.

As noted above, the purpose of strength tests is to determine whether the structure can safely operate under a given load. The digital image correlation method does not allow to determine the stress distribution, but it provides the basis for determining the correctness of the numerical solution.

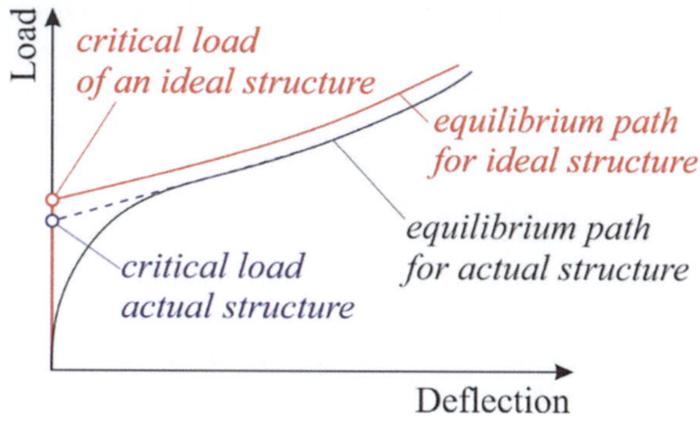

Fig. 23 Schematic method of critical load determination based on experimental results

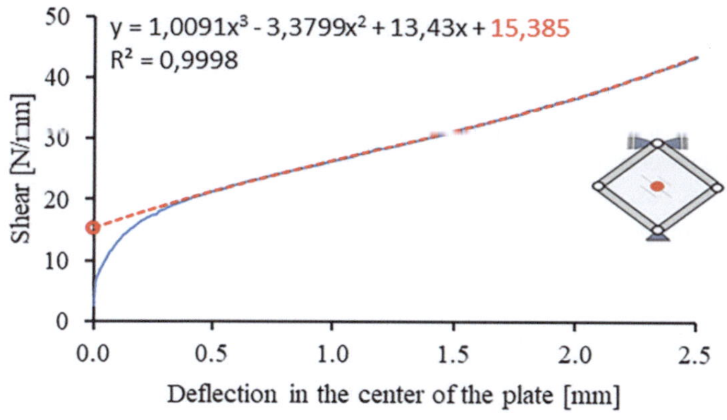

Fig. 24 Determination of the approximate value of critical load based on the results of experimental tests conducted with use of DIC scanner

To determine the level and distribution of stress, the numerical model of the plate with geodetic reinforcements according to the Finite Element Method was prepared in the commercial software ABAQUS 6.13 by Dassault Systèmes [1]

The problem was assumed to be physically and geometrically nonlinear. For the plate material, the elastic–plastic characteristics with reinforcement were adopted.

The analyses in a nonlinear approach were carried out using a modified Newton–Raphson prognostic method with load correction procedures [4, 11]. The parameters that control the advancing of the solution included the initial relative load increment as 0.001 and the maximum as 0.01. Within each increment, the program iterates until the state of static equilibrium of the external load and internal forces is reached with an accuracy of 0.5% and the relative increase of deformation due to loading is checked, which should not be more than 1% in a given increment [55].

The size of each incremental step is automatically selected by a computer algorithm. When the convergence of the solution is achieved in less than five iterations, the next incremental step is increased by 50% of the size. If, on the other hand, the state of static equilibrium is not reached in sixteen successive iterations of the solution of a given increment, the algorithm lowers the value of the load increment by 25% and the calculation is repeated. If convergence cannot be achieved on the next five attempts to reduce the incremental step, the calculation is aborted.

The following results represent only the correct solution to the problem. Detailed analyses taking into account the influence of the modeling method, the size of finite elements, the level of initial geometric imperfection and the influence of the control parameters of the numerical solution on the obtained results can be found in the author's publications [22, 53, 55].

The steel frame in which the structure was mounted was modeled using spatial B31 beam elements with six degrees of freedom in each of the two nodes. The element has a linear shape function and is defined as a Timoshenko beam, which allows the representation of transverse deformations. The corner elements of the frame were reproduced with a HINGE connection [1]. Screw connections between the frame elements and the plate was modeled using MPC TIE elements [1, 6]. In order to reproduce the boundary conditions, a sliding support was used in the upper node of the frame. The lower node was a fixed support.

The plate was modeled using standard four-node (S4R) and three-node (S3R) shell elements based on the Koiter-Sanders theory, with a reduced number of Gaussian integration points [17] and the ability to account for large deformations [18] These elements also have solution control procedures that prevent the occurrence of incorrect forms of deformation such as hourglass and shear locking [24] The mesh elements have been assigned appropriate thicknesses. The adopted method of modeling made it possible to reconstruct the fillets in the corners of the skin, but did not provide the possibility of simulating the fillets in the direction normal to the surface. Schematic diagrams of the adopted finite elements are shown in Fig. 25 while the numerical model of the plate mounted in the frame with the boundary conditions and load is shown in Fig. 26.

The model selected for analysis consisted of 17,947 elements with an average characteristic dimension of 2.5 mm. In this way, a reasonable compromise was ensured

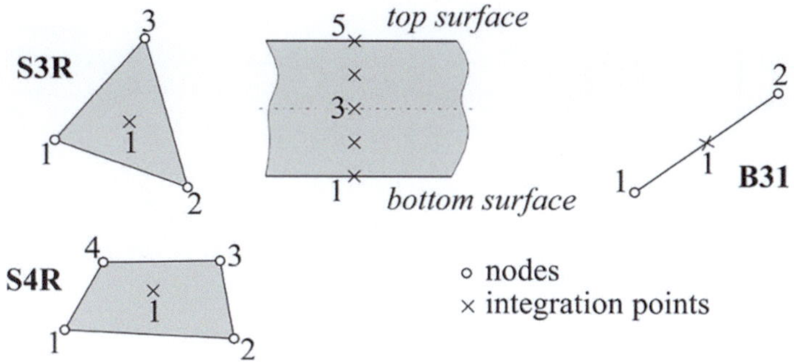

Fig. 25 Finite element diagrams used

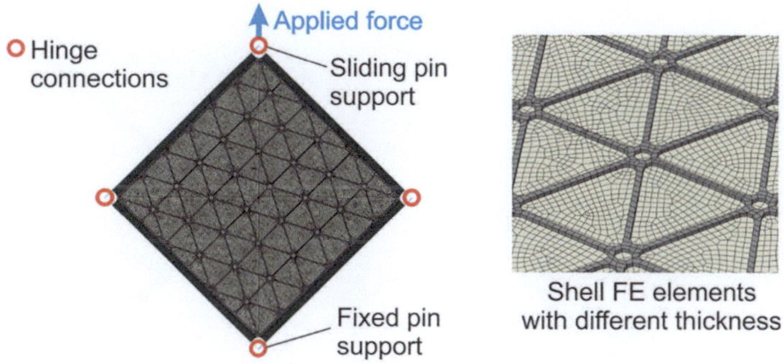

Fig. 26 Plate FE numerical model and boundary conditions

between the accuracy of the results obtained and the time necessary to carry out the calculations.

For comparison purposes, allowing to check the numerical model, curves of the relation of the angle of shear deformation of the plate as a function of acting load were drawn up (Fig. 27). The presented results of numerical calculations show a high agreement with the experiment up to the level of shear load of about 75 N/mm. Above this load value, the FEM calculations showed a tendency to increase the stiffness of the system as the load increased.

Figure 28 shows the results of numerical calculations in the form of transverse deformation areas (deflection) and the distribution of reduced stresses according to the H-M-H hypothesis under conditions of maximum plate loading after loss of stability (102 N/mm).

The assessment of the correctness of numerical modeling of the considered systems is conditioned by the sufficient consistency of the results of numerical analyses with the results of experimental research. The degree of convergence of results seems to be the only rational criterion allowing for similar solutions, especially

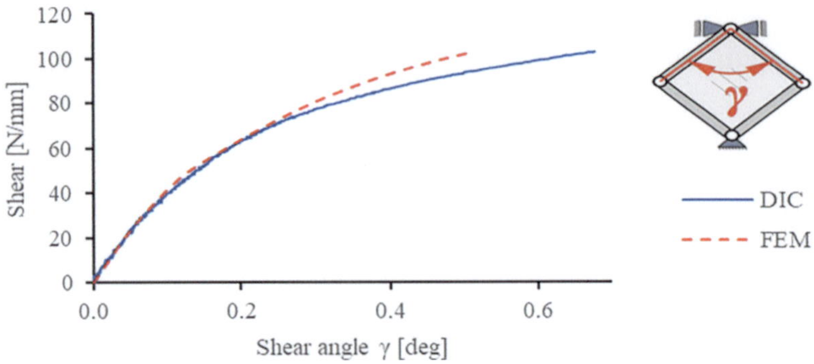

Fig. 27 Comparison of the relationship between the shear load and the angle of the shear γ. FEM analysis and experiment

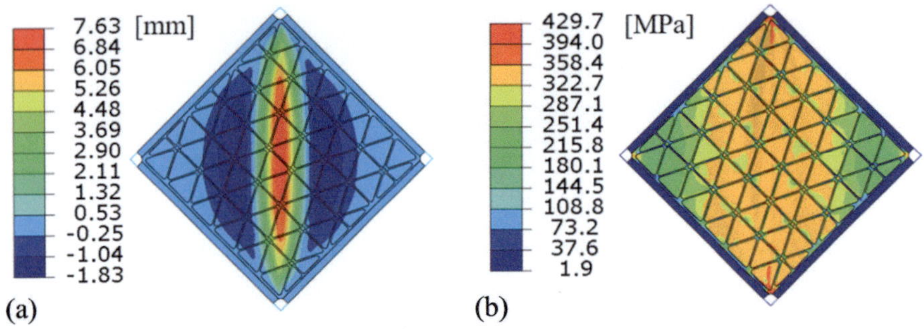

Fig. 28 Deflection area (**a**) and stress distribution according to H-M-H (**b**) for a FEM model. Maximum load level

when the designed system is subjected to loads causing states of advanced deformations. The coherence of both forms of studying the structure is taken up in further considerations.

Considering that only one state of internal forces corresponds to one state of deformation, the criterion for the consistency of the results of numerical analyses with the results of experimental tests of load-bearing structures can only be the consistency of geometry under the conditions of load causing extreme deformations, which is operationally acceptable.

Numerical models with the initial imperfection introduced as the first buckling mode with the magnitude of 1%, 5% and 10% of the plate thickness and for the model with ideal geometry were analyzed.

The problem of the compatibility of the deformation geometry was considered starting from the comparison of the relationship between the deflection of the geometric center of the plate and the load on the structure (Fig. 29). In the phase of deformation advancement below the load of about 85 N/mm, the FEM model retained greater stiffness than that resulting from experimental measurements. Such

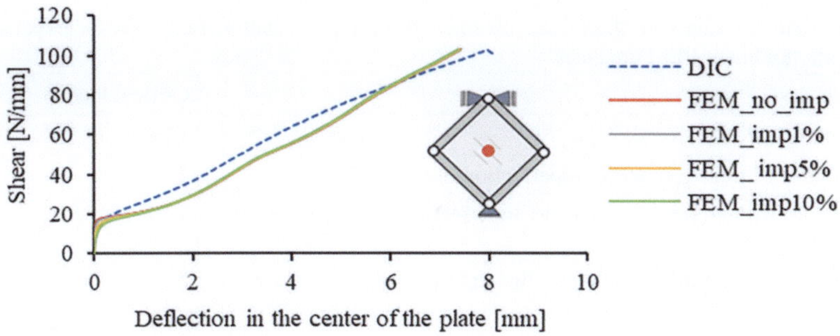

Fig. 29 Comparison of the experimental and numerical equilibrium path for the midpoint of the finned plate (imp. denotes geometric imperfections)

phenomenon can be justified by simplifications of the model geometry (e.g. no fillets between the ribs and the skin) and an oversimplified model of material properties.

The initial geometric imperfection introduced into the FEM model resulted in differences in the behavior of the analyzed structure only for low load levels. Above the flow rate of about 20 N/mm, all equilibrium paths took almost identical shapes (Fig. 30). Numerical calculations have shown that an increase in the value of the initial imperfection leads to a smoother equilibrium path in the range of loads close to loss of stability.

Using the methodology based on the Koiter method, it is also possible to determine the critical load for the cases of numerical analyses carried out in the field of nonlinear deformations. As before, buckling is defined as the deformation of the midpoint of the plate. The results are presented in Table 1.

The deformation and the critical load value resulting from the FEM analyses carried out as an issue for eigenvalues, the so-called linear buckling according to the Lanczos method [23] is presented in Fig. 31. The difference from the results of nonlinear analyses presented in Table 1 is due to the fact that the critical load determined from the linearized analyses refers to a structure with ideal geometry,

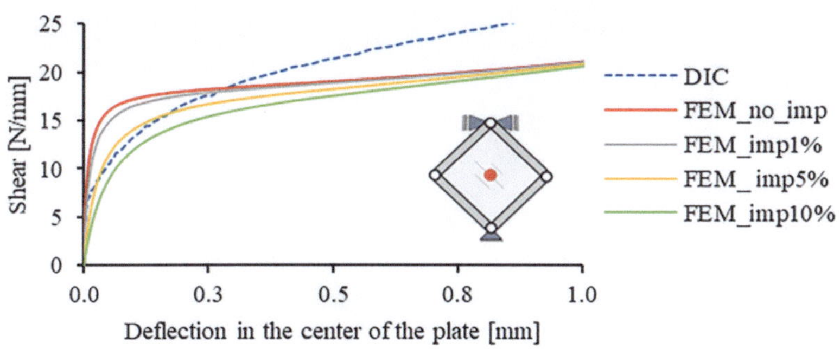

Fig. 30 Comparison of the experimental and numerical equilibrium path for the midpoint of the finned plate. Close-up of the range of loss of stability

Table 1 Critical values of shear load obtained from approximation functions for experimental results and nonlinear FEM analyses

Analysis case	Critical load q_{cr}[N/mm]
Experiment *(DIC)*	15.4
FEM model without geometric imperfection introduced	17.7
FEM model with an initial geometric imperfection of 1% of the plate thickness	17.2
FEM model with an initial geometric imperfection of 5% of the plate thickness	16.6
FEM model with an initial geometric imperfection of 10% of the plate thickness	14.9
FEM—linear buckling	18.3

free from initial imperfections. It should be noted that the critical load level is much higher, which poses a risk when assessing stability based on results obtained from linearized buckling analyses [11].

Figure 32 shows a qualitative comparison of the deformation (deflection) surface for the maximum load level, obtained by numerical analysis and experimental studies. Figure 33 shows the quantitative confrontation of the deflection area obtained under the same loading conditions. It should be noted that despite the differences mentioned above, the general form of the deformation of the numerical model coincides with the result of the experiment.

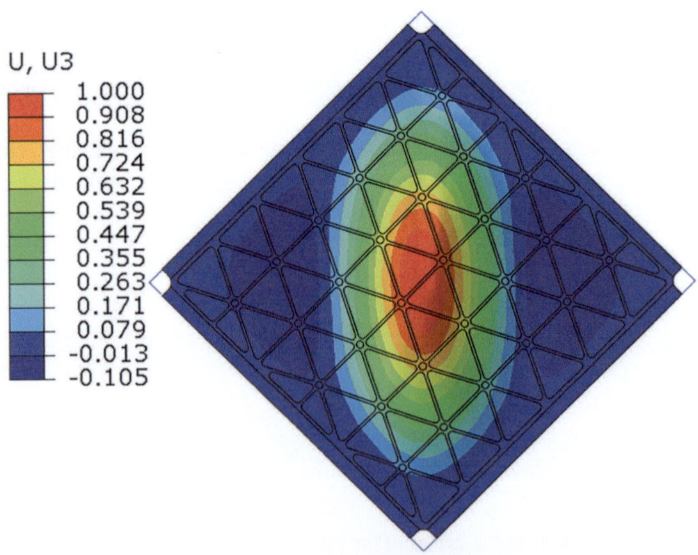

Fig. 31 Normalized deflection area for the first buckling mode of the stiffened plate—FEM result of eigenvalue analysis

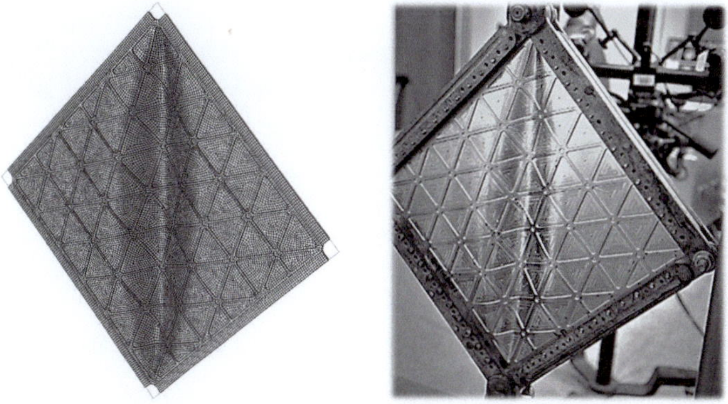

(a) (b)

Fig. 32 Qualitative comparison of the deformation of a plate subjected to the maximum load: **a** result of numerical analyses, **b** experimental result

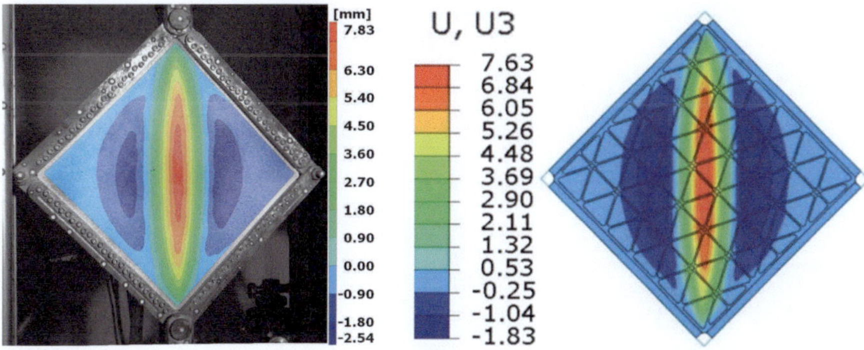

(a) (b)

Fig. 33 Quantitative comparison of plate deformation subjected to maximum load: **a** experimental result (DIC), **b** result of numerical analyses (displacement value in [mm])

The use of a DIC scanner also enabled quantitative comparison of other deformation components. Figures 34 and 35 show the deflection of plate cross-sections along the vertical and horizontal diagonals for the maximum load realized during the experimental tests. The vertical section, in addition to the general deflection, also showed local deformations between the ribs in terms of plastic deformation. This phenomenon is visible in Fig. 34, where the ribs are shown as peaks (hills), and between them deformations can be observed in the form of loss of stability of the skin (Fig. 35).

Figure 36 shows the strain fields in the area of the upper node of the loading frame, recorded by the DIC system and obtained by numerical analysis caused by the action of the maximum load. A comparison of the two results can also be proof of sufficient convergence of the FEM solution in this aspect as well.

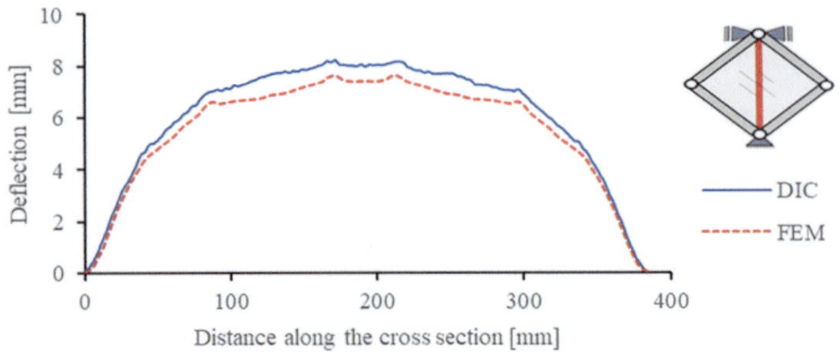

Fig. 34 Deflection of cross-section drawn along the vertical diagonal of the plate subjected to maximum load

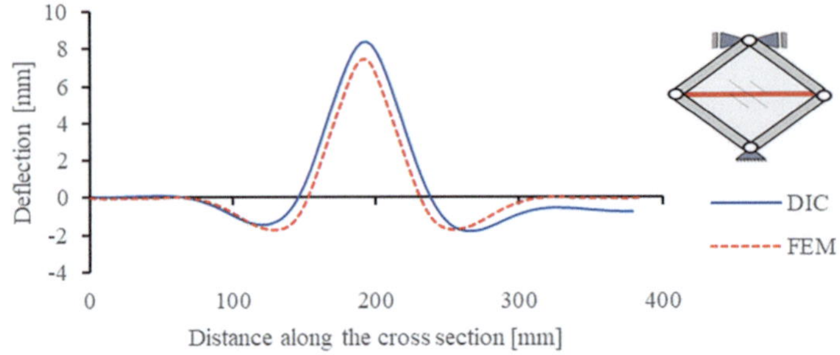

Fig. 35 Deflection of cross-section drawn along the horizontal diagonal of the plate subjected to maximum load

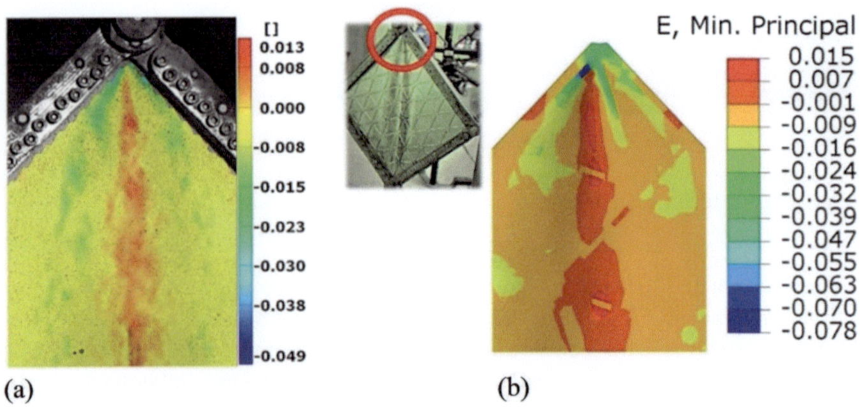

Fig. 36 Comparison of the distribution of the minimum component of the principal strain, **a** experimental result of DIC, **b** result of numerical analyses of FEM

References

1. Abaqus Abaqus, *Theory Guide*. Dassault Systèmes Simulia Corp, Providence, RI
2. H. Aben, Integrated photoelasticity and its applications, in *Static and Dynamic Photoelasticity and Caustics: Recent Developments*. ed. by A. Lagarde (Springer, Vienna, 1987), pp.1–61
3. L. Ainola, H. Aben, On the generalized Wertheim law in integrated photoelasticity. J. Opt. Soc. Am. A JOSAA **25**, 1843–1849 (2008). https://doi.org/10.1364/JOSAA.25.001843
4. T. Belytschko, W.K. Liu, B. Moran, K.I. Elkhodary, *Nonlinear Finite Elements for Continua and Structures*, 2nd edn. (Wiley, Chichester, 2014)
5. P.J. Besl, N.D. McKay, A method for registration of 3-D shapes. IEEE Trans. Pattern Anal. Mach. Intell. **14**, 239–256 (1992). https://doi.org/10.1109/34.121791
6. M.A. Bhatti, *Advanced Topics in Finite Element Analysis of Structures: With Mathematica and MATLAB Computations*. Wiley (2006)
7. H. Bruck, S.R. McNeill, M.A. Sutton, W.H. Peters, Digital image correlation using Newton-Raphson method of partial differential correction. Exp. Mech. **29**, 261–267 (1989)
8. T.C. Chu, W.F. Ranson, M.A. Sutton, Applications of digital-image correlation techniques to experimental mechanics. Experimental Mech. **25**, 232–244 (1985)
9. J.W. Dally, W.F. Riley, *Experimental Stress Analysis*, Subsequent. (McGraw-Hill College, New York, 1991)
10. J.F. Doyle, *Nonlinear Analysis of Thin-Walled Structures* (Springer, New York, 2001)
11. C.A. Felippa, *Nonlinear Finite Element Methods* (ASEN 6107) (2006)
12. Y.B. Fu, R.W. Ogden, *Nonlinear Elasticity: Theory and Applications* (Cambridge University Press, London, 2001)
13. GOM GmbH, *Digital Image Correlation and Strain Computation Basics*; v.2016. GOM GmbH, Braunschweig, Germany (2016)
14. GOM GmbH, *GOM Tritop*; v.2019. GOM GmbH, Braunschweig, Germany (2019)
15. GOM GmbH, *Gom Atos*; v.2019 (2019). GOM GmbH, Braunschweig, Germany
16. J.D. Helm, S.R. McNeill, M.A. Sutton, Improved three-dimensional image correlation for surface displacement measurement. Opt. Eng. **35**, 1911–1920 (1996)
17. T.J.R. Hughes, M. Cohen, M. Haroun, Reduced and selective integration techniques in the finite element analysis of plates. Nucl. Eng. Des. **46**, 203–222 (1978). https://doi.org/10.1016/0029-5493(78)90184-X
18. T.J.R. Hughes, W.K. Liu, Nonlinear finite element analysis of shells-part II. Two-dimensional shells. Comput. Methods Appl. Mech. Eng. **27**, 167–181 (1981). https://doi.org/10.1016/0045-7825(81)90148-1
19. Z.L. Kahnjetter, T.C. Chu, 3-Dimensional displacement measurements using digital image correlation and photogrammic analysis. Exp. Mech. **30**, 10–16 (1990)
20. R. Kashyap, *Fiber Bragg Gratings* (Academic Press, 2009)
21. W.T. Koiter, *The Stability of Elastic Equilibrium*. Stanford University (1970)
22. H. Kopecki, Ł Święch, *Modeling Problems of the Post-critical States of Deformation of Isogrid Plates in the Light of the Preliminary Experimental Investigations* (Zakopane, Poland, 2019), p.020006
23. C. Lanczos, An iteration method for the solution of the eigenvalue problem of linear differential and integral operators. J. Res. Nat. Bur. Stand 255–282 (1950)
24. A. Laulusa, O.A. Bauchau, J.-Y. Choi, V.B.C. Tan, L. Li, Evaluation of some shear deformable shell elements. Int. J. Solids Struct. **43**, 5033–5054 (2006). https://doi.org/10.1016/j.ijsolstr.2005.08.006
25. P. Luo, Y.J. Chao, M.A. Sutton, Application of stereo vision to three-dimensional deformation analyses in fracture experiments. Opt. Eng. **33**, 981–990 (1994)
26. P. Luo, Y.J. Chao, M.A. Sutton, W.H. Peters, Accurate measurement of three-dimensional deformations in deformable and rigid bodies using computer vision. Experimental Mech. **33**, 123–132 (1993)
27. R.M. Measures, S. Abrate, Structural monitoring with fiber optic technology. Appl. Mech. Rev. **55**, B10–B11 (2002). https://doi.org/10.1115/1.1445327

28. L.B. Meng, G.C. Jin, X.F. Yao, Application of iteration and finite element smoothing technique for displacement and strain measurement of digital speckle correlation. Opt. Lasers Eng. **45**, 57–63 (2007)
29. E.M. Mikhail, J.S. Bethel, J.C. McGlone, *Introduction to Modern Photogrammetry* (Wiley, New York, 2001)
30. NASA Marshall Space Flight Center, *Isogrid Design Handbook*. NASA Marshall Space Flight Center, Huntsville, AL, USA (1973)
31. G.I. Nesterenko, *Comparison of Damage Tolerance o Integrally Stiffened and Riveted Structures* (2000)
32. G. Oppel, *The Photoelastic Investigation of Three-dimensional Stress and Strain Conditions* (1937)
33. Z. Orłoś, *Doświadczalna analiza odkształceń i naprężeń* (PWN, Warszawa, 1977)
34. V.J. Park, The range of speckle metrology. Exp. Mech. **20**, 181–191 (1980)
35. M. Paszkiewicz, T. Kubiak, Selected problems concerning determination of the buckling load of channel section beams and columns. Thin-Walled Struct. **93**, 112–121 (2015). https://doi.org/10.1016/j.tws.2015.03.009
36. E.A. Patterson, Digital photoelasticity: principles, practice and potential. Strain **38**, 27–39 (2002). https://doi.org/10.1046/j.0039-2103.2002.00004.x
37. W.H. Peters, W.F. Ranson, Digital imaging techniques in experimental stress analysis. Opt. Eng. **21**, 427–431 (1982)
38. W.H. Peters, W.F. Ranson, M.A. Sutton, T.C. Chu, J. Anderson, Application of digital correlation methods to rigid body mechanics. Opt. Eng. **22**, 738–742 (1983)
39. D. Post, B. Han, P.G. Ifju, Moiré methods for engineering and science—Moiré interferometry and shadow Moiré, in *Photomechanics*. ed. by P.K. Rastogi (Springer, Heidelberg, 2000), pp.151–196
40. K. Ramesh, *Digital Photoelasticity: Advanced Techniques and Applications* (Springer, Berlin, 2000)
41. K. Ramesh, S.K. Mangal, Data acquisition techniques in digital photoelasticity: a review. Opt. Lasers Eng. **30**, 53–75 (1998). https://doi.org/10.1016/S0143-8166(97)00105-X
42. K. Ramesh, S. Sasikumar, Digital photoelasticity: recent developments and diverse applications. Opt. Lasers Eng. **135**, 106186 (2020). https://doi.org/10.1016/j.optlaseng.2020.106186
43. P.K. Rastogi, *Photomechanics* (Springer, Heidelberg, 2000)
44. W. Schumann, J.P. Zürcher, D. Cuche, *Holography and Deformation Analysis* (Springer, Berlin, 1985)
45. M. Shantharaja, K.R. Srinivas, Comparison of riveted panel and integral panel for damage tolerance by using FEA. Int. J. Sci. Res. 426–431 (2014)
46. W.N. Sharpe, *Springer Handbook of Experimental Solid Mechanics* (Springer, New York, 2008)
47. J. Singer, J. Arbocz, T. Weller, *Shell Buckling Experiments* (2008), pp 623–808
48. R.S. Sirohi, *Speckle Metrology* (CRC Press, Boca Raton, 2020)
49. L.S. Srinath, *Scattered Light Photoelasticity* (Tata McGraw-Hill, Delhi, 1983)
50. M. Sutton, W. Wolters, W. Peters, W. Ranson, S. McNeill, Determination of displacements using an improved digital correlation method. Image Vis. Comput. **1**, 133–139 (1983). https://doi.org/10.1016/0262-8856(83)90064-1
51. M.A. Sutton, M. Cheng, W.H. Peters, Y.J. Chao, S.R. McNeill, Application of an optimized digital correlation method to planar deformation analysis. Image Vision Comput. **4**, 143–150 (1986)
52. M.A. Sutton, J.L. Turner, H. Bruck, T.A. Chao, Full-field representation of discretely sampled surface deformation for displacement and strain analysis. Exp. Mech. **31**, 168–177 (1991)
53. Ł. Święch, Experimental and numerical studies of low-profile, triangular grid-stiffened plates subjected to shear load in the post-critical states of deformation. Materials **12**, 3699 (2019). https://doi.org/10.3390/ma12223699
54. Ł. Święch, Calibration of a load measurement system for an unmanned aircraft composite wing based on fibre Bragg gratings and electrical strain gauges. Aerospace **7**, 27 (2020). https://doi.org/10.3390/aerospace7030027

55. Ł. Świech, Komplementarne badania cienkościennych ustrojów nośnych w warunkach dużych deformacji, 1st edn. Oficyna Wydawnicza Politechniki Rzeszowskiej, Rzeszów, eISBN: 978-83-7934-760-5 (2024)
56. A. Teter, Z. Kolakowski, On using load-axial shortening plots to determine the approximate buckling load of short, real angle columns under compression. Compos. Struct. **212**, 175–183 (2019). https://doi.org/10.1016/j.compstruct.2019.01.009
57. Q. Tian, M.N. Huhns, Algorithms for subpixel registration. Comput. Vision Graphics Image Process. **35**, 220–233 (1986). https://doi.org/10.1016/0734-189X(86)90028-9
58. S. Timoshenko, J.M. Gere, Theory of Elastic Stability, 2nd edn. Dover ed. Dover Publications, Mineola, N.Y (2009)
59. W. Tong, Detection of plastic deformation patterns in a binary aluminum alloy. Exp. Mech. **37**, 452–459 (1997)
60. R. Tsai, A versatile camera calibration technique for high-accuracy 3D machine vision metrology using off-the-shelf TV cameras and lenses. IEEE J. Robot. Autom. **3**, 323–344 (1987)
61. Y. Vetyukov, *Nonlinear Mechanics of Thin-Walled Structures* (Springer Vienna, Vienna, 2014)
62. C.C.B. Wang, J.M. Deng, G.A. Ateshian, An automated approach for direct measurement of two-dimensional strain distributions within articular cartilage under unconfined compression. J. Biomech. Eng. Trans. ASME **124**, 557–567 (2002)
63. R. Weller, J.K. Bussey, *Photoelastic Analysis of Three-Dimensional Stress System Using Scattered Light*. Washington DC (1939)
64. G. Xu, Z. Zhang, *Epipolar Geometry in Stereo, Motion and Object Recognition*. Springer Netherlands, Dordrecht (1996)
65. S. Zhang, S.-T. Yau, Three-dimensional shape measurement using a structured light system with dual cameras. Opt. Eng. **47**, 013604 (2008). https://doi.org/10.1117/1.2835686
66. Z. Zhang, A flexible new technique for camera calibration. IEEE Trans. Pattern Anal. Mach. Intell. **22**, 1330–1334 (2000)

Advancements in Optical Sensors for Avionics Systems

Anna Mauro, Chiara Bellezza Prinsi, Aurora Bellone, Massimo Olivero, Alberto Vallan, and Guido Perrone

Abstract Optical sensors are increasingly important in avionics due to their precision, reliability, and ability to operate in extreme conditions. Among them, fiber optic sensors, being all-dielectric structures and not requiring copper wires, stand out for their immunity to electromagnetic interference, making them ideal for applications in which signal integrity is fundamental, as in aircraft environments. Moreover, fiber optic sensors are lightweight and can be easily integrated into structures, in particular composite materials. All this leads to the new paradigm of the "fly-by-light" approach, for which not only command signals are transmitted through optical fibers, but also aircraft control systems rely on optical methods for navigation information and to measure various parameters such as pressure, temperature, and structural deformations. One of the most relevant and already widely deployed applications of photonics in avionics is optical gyroscopes, which are used for inertial navigation, offering reliable and accurate orientation and position data without having to rely on external signals. Another emerging application of optical sensors is in the development of "smart structures", in which they contribute by enabling the detection of the early onset of structural health issues, such as cracks or deformations. These and other new uses of optical sensors are enabling real-time monitoring of aircraft components, enhancing safety while simultaneously reducing downtime and thus operating costs.

Keywords Optical fiber sensors (OFS) · Fiber bragg gratings (FBG) · Structural health monitoring (SHM) · Optical gyroscope · Optical sensors in Aerospace · Optical sensors in Avionics · Sagnac interferometers · Distributed optical sensors

---

A. Mauro · C. Bellezza Prinsi · A. Bellone · M. Olivero · A. Vallan · G. Perrone (✉)
Department of Electronics and Telecommunications, Politecnico Di Torino, Torino, Italy
e-mail: guido.perrone@polito.it

C. Marques et al. (eds.), *Advanced Optical Sensors for Aerospace Applications*,
Progress in Optical Science and Photonics 37,
https://doi.org/10.1007/978-981-95-1626-1_7

1 Introduction

Optical technologies are becoming more and more present on board aircraft: beyond traditional optical instruments and displays, aircraft incorporate many other optical-based systems, especially those that utilize fiber optic technologies. Among them, one of the most widely adopted systems is the Fly-By-Light (FBL) Flight Control System, which is increasingly replacing traditional Fly-By-Wire (FBW) technology.

In the early days of aviation pilots used direct mechanical force on their sticks to move the airplane surfaces. Then, control systems have evolved over the years, first using mechanical and hydraulic systems to steer and maneuver the aircraft beyond the pilot's physical capabilities. Then, in the late 1970s FBW control systems started coming into service, replacing manual flight controls with an electronic interface, so that the controls from the cockpit are converted into electronic signals, which are used by computers to determine how to move the actuators to obtain the desired response at each control surface. FBW technology has brought several advantages, such as lighter weight and easier maintenance, along with some disadvantages, such as susceptibility to electromagnetic interference and high intensity radiated field. The commonly adopted solution is to shield the electrical cables, but this is not always possible, besides for being limited by the additional weight of the cables. A better way to overcome these problems is to use optical fiber cables instead of electrical wires, and to replace electrical sensors with optical sensors [1].

FBL systems employ optical fibers for data transmission, offering many advantages over copper-wire-based systems such as electromagnetic immunity, reduced weight, and enhanced signal integrity, making them the preferred choice for next-generation aircraft. More in general it is possible to say that the well-known advantages of optical fiber in communications also extend to Optical Sensing Networks (OSNs): for instance, thanks to the unique physical and mechanical properties of optical fibers, OSNs are particularly well-suited for hazardous environments, including those that are inflammable, radioactive, or chemically corrosive. As such, fiber optic sensor networks have been used for the real-time measurement of electric and magnetic fields, heat source localization, shock wave detection, battery health monitoring, and in many more applications.

In aerospace, fiber optic sensor networks can play a fundamental role in Structural Health Monitoring (SHM) of aircraft and spacecraft, enabling early damage detection to reduce the risk of catastrophic failures. This application is particularly important because aircraft structures are engineered to be as lightweight as possible, while still maintaining safety margins. Therefore, not only aircraft structures must be intrinsically designed to withstand a wide variety of potential damages, but they also require extensive periodic inspection and maintenance to ensure the structural integrity over time. This approach is time consuming, and, in the end, this translates into additional costs for airlines. In this context, innovation in SHM implies deploying a system of sensors and prediction algorithms, to provide information on the location and evaluation of early signs of load and damage conditions of the aircraft structures. Optical fiber-based monitoring systems can significantly lower maintenance efforts while simultaneously improving overall aircraft safety since Fiber Optic Sensors (FOSs) are well suited for SHM applications because of their intrinsic advantages such as insensitivity to electromagnetic radiation, low mass and small size, high sensitivity and resolution, and, above all, their suitability to be embedded into structures without introducing significant perturbations. The latter aspect is particularly interesting since it enables continuous monitoring and not only during the periodic planned maintenance stops of the aircraft. An optical-based SHM technique can be deployed using two configurations: a combination of single-point sensors to form what it is called a "quasi-distributed network" and truly "distributed sensing", which is particularly suited for very large structures.

So far, the emphasis has been on SHM applications, but optical sensing technologies for aerospace are not limited to this; rather, they are rapidly expanding to encompass all aspects of this sector, including navigation, guidance, and remote sensing applications.

2 Optical Sensors in Aerospace

As highlighted in the previous section, optical sensors are playing an increasingly important role in modern avionics, bringing significant enhancements in aircraft performance, safety, and reliability, thus becoming an essential component of next-generation flight control, navigation, and structural health monitoring systems. Their growing importance is reflected in their market share, which is expanding driven by advancements in fiber optic technology and data processing capabilities [2]. Compared to traditional electronic sensors, optical-based devices offer a unique combination of high sensitivity, immunity to electromagnetic interference, and the ability to operate in harsh environments, making them particularly well-suited for aerospace applications.

Among the various optical sensing technologies, Fiber Optic Sensors (FOSs) stand out as a highly promising solution thanks to their compact size, light weight, minimal invasiveness, and ability to perform distributed sensing over long distances [3, 4]. While their adoption by the aerospace industry is still far from being completed, these sensors have already demonstrated to allow improving the surveillance of critical parameters—in particular, temperature, strain, pressure, and acceleration—in many applications ranging from structural health monitoring, to engine diagnostics, and flight control systems [5–7].

FOS technology has its roots in optical communications, being a sort of spin-off that benefitted from the rapid advancements in the components for high-speed fiber networks. Indeed, the same innovations that have revolutionized telecommunications (e.g., the availability of low-loss optical fibers, high-power laser sources, and advanced photodetectors), have significantly improved the sensitivity, reliability, and cost-effectiveness of fiber optic sensors. In this context, the origins of modern optical sensing can be traced back to the 1970s, a decade marked by two fundamental advancements, such as the development of semiconductor laser diodes capable of operating at room temperature and of low-loss silica optical fibers [8]. These breakthroughs initially revolutionized optical communications, but soon their impact extended to novel compact optical sensing systems. However, the widespread adoption of FOSs faces several challenges, such as how to address some intrinsic cross-sensitivity effects (in particular, strain and temperature) and how to process the huge data volume generated by real-time applications (for example, in case of vibrations, some FOS require processing many optical spectra per second). These limitations can be overcome through the extensive use of data analytics frameworks, taking advantage of Machine Learning (ML) advancements to enhance real-time signal processing, pattern recognition, and noise reduction. Recent studies have pioneered this new combination of optical technologies with Artificial Intelligence (AI), but a lot more work has still to be done before unlocking the full potential of FOSs for continuous and ubiquitous monitoring applications [9].

The key applications of optical sensors in aerospace can be grouped into three categories.

- **Navigation and guidance** [10–15]. Optical sensors are extensively used in precision navigation and positioning for aircraft, satellites, and space probes to ensure stabilization and autonomous trajectory control. The most important are Ring Laser Gyroscopes (RLGs), Fiber-Optic Gyroscopes (FOGs), and LiDAR (Light Detection and Ranging). RLGs are laser-based inertial navigation systems that exploit the Sagnac effect in a ring laser cavity to measure angular velocity. FOGs too rely on the Sagnac effect, but they measure rotational velocity by detecting phase shifts in counter-propagating light waves within an optical fiber coil. These gyroscopes are compact, lightweight, and immune to electromagnetic interference, making them ideal for aircraft, spacecraft, and satellite navigation systems. For instance, they are also used in long-duration missions, such in interplanetary probes and autonomous spacecraft (e.g., NASA's Perseverance rover). LiDAR—the "Optical RADAR"—is an instrument that emits laser pulses instead of radiowave pulses and measure the time it takes for the reflected light to return, creating high-resolution 3D maps of terrain and obstacles. LiDARs are used to detect wind shear and wake turbulence, for terrain mapping (e.g., NASA's Mars Reconnaissance Orbiter), and for autonomous landing.
- **Structural Health Monitoring** (SHM) [16–20]. SHM is a pillar of modern engineering, with application principles that encompass civil and industrial sectors: in the aerospace field it is essential to improve the safety, reliability, and longevity of aircraft. Optical sensors, particularly FOSs, have revolutionized SHM by enabling real-time, high-resolution monitoring of key parameters such as strain, vibrations, and temperature variations, allowing the early identification of potential failures. This is because optical fibers are extremely lightweight and can be embedded into structures without adding significant weight or perturbing their operations. Therefore, optical fiber-based sensors can be integrated into aircraft wings or fuselage to detect material fatigue, enabling predictive maintenance scheduling. Particularly interesting is the use of SHM solutions to detect subtle impact damages in composite airframes, which are becoming the standard in new aircraft models

because their reduced weight allows increasing the fuel efficiency. In the specific case of spacecraft, FOSs can be used to monitor temperature fluctuations and structural stress during launch, orbit, and re-entry. The most suitable FOSs for SHM are Fiber Bragg Gratings (FBGs) and scattering-based distributed sensing systems. FBGs are the most widespread type of fiber optic sensors for their unique combination of sensitivity and robustness to noise. They are point sensors but tens of them can be easily multiplexed along the same fiber to form multi-point quasi-distributed sensing networks, with spatial resolution (i.e., the separation between sensing points) and gauge length that can vary depending on the spacing between two consecutive FBGs and the total number of FBGs. For example, in presence of large temperature gradients over short distances, the spatial resolution can be in the order 1–2 mm, for total distances that are in the 5–10 cm range [21, 22]. Alternatively, especially in large-scale structural monitoring (e.g., monitoring stress distribution along an entire wing or fuselage), truly distributed sensing systems can be used. These are techniques that relate the quantity under measurement with the variations in the scattered light along an optical fiber, enabling the accurate reconstruction of deformation and temperature profiles over very long distances. The most common instruments exploit the Rayleigh backscattering, typically evaluated using Optical Time Domain Reflectometry (OTDR) or Optical Frequency Domain Reflectometry (OFDR) methods. Both quasi-distributed and distributed sensor networks can be embedded within structural components, particularly in carbon fiber composites, to create "smart structures". These are engineered materials that integrate self-monitoring capabilities to adapt performance in real-time depending on external stimuli such as temperature variations, mechanical stress, etc. Despite these are relative new applications for aerospace sector, in the literature are reported examples of smart structures already utilized to detect early signs of damage, material fatigue, or impact stress in fuel tanks and aircraft wings.

- **Remote environmental sensing** [23–27]. Optical sensors, usually deployed on board satellites or specialized airplanes, are commonly used in Earth and other planet observation, for example for the continuous monitoring of the composition of the atmosphere or of extreme weather events. Among the most advanced optical techniques in remote sensing is hyperspectral imaging, which captures data across hundreds of narrow spectral bands instead of the limited bands used in traditional imaging. This approach enhances the ability to distinguish between materials with subtle spectral differences. Instruments such as NASA's Hyperion Hyperspectral Imager and ESA's Sentinel-2 MSI utilize this technology to monitor land use, ocean ecosystems, and environmental changes. An important subcategory of these sensors includes optical spectrometers designed for planetary exploration. Installed on spacecraft, these instruments analyze the composition of the atmosphere or of the surface in search of potential signs of habitability. For instance, spectrometers aboard Mars orbiters and rovers have allowed the detection of water-related minerals, the evaluation of atmospheric methane levels, and the study of the planet's seasonal variations.

Aerospace applications impose stringent requirements, since components and systems—including optical devices—are exposed to harsh environments but must maintain high reliability and performance over long periods. Consider for example the case of space applications for their extreme conditions. Although the specific set of requirements depends on the mission type (e.g., launch methodology, orbit altitude, interplanetary travel, deep space), major space agencies such as ESA and NASA impose strict performance criteria for photonic devices, including:

- Operational reliability for 10–20 years.
- Radiation resistance to prevent performance degradation in outer space.
- Broad temperature tolerance, typically ranging at least from −45 °C to 85 °C.
- Resistance to extreme vibrations and shocks occurring during launch.
- Vacuum compatibility, with low contamination and minimal outgassing, preventing optical degradation.

Fiber optic systems can be designed to meet these requirements; however, several challenges must be addressed to enable their broader adoption. For example, fiber interconnection represents a critical aspect, as space-qualified connectors are costly and prone to failure. An optimized approach involves pre-manufactured sensor arrays, onsite fiber splicing for long sections, and connectors only at fiber endpoints to minimize potential failures. Materials and packaging have limitations too. For start, laser diodes emitting around 1550 nm—the most common wavelength for sensing, an of optical communication systems—are made of alloys of indium phosphide and face reliability risks due to indium creep, which can lead to catastrophic device failure under extreme temperatures. Additionally, high vacuum environments can alter the chemical and physical properties of optical components, leading to performance degradation. Another problem comes from contamination risks due to possible outgassing of epoxies used in optical assemblies. Therefore, to ensure long-term reliable operation of optical components in the very challenging aerospace field, continuous advancements in materials, fiber integration techniques, and contamination-resistant packaging are essential.

3 Types of Optical Sensors

As mentioned in the previous section, the aerospace sector relies on different optical sensing technologies to support fundamental applications, including navigation, structural health monitoring, environmental sensing, and communication. The range of optical sensors is too broad to be covered in a single chapter; therefore, despite their importance, image-based sensors, such as cameras, will not be considered. Instead, the focus will be on optical sensors implemented using free-space discrete components or optical fibers, with emphasis on the latter, the so-called Fiber Optic Sensors (FOSs). FOSs are particularly attractive for aerospace applications due to their unique advantages such as compactness, lightweight design, immunity to electromagnetic interference, and capability for distributed sensing.

Optical sensors can be broadly categorized based on their working principles, so on how the quantity under measure interacts with light. Interestingly, their underlying principle of operation is exactly the opposite of that of optical communication systems: while optical communications strive to transmit signals with minimal distortion and interference, optical sensing fundamentally depends on the intentional perturbations of the optical signal caused by the quantity being measured. This peculiarity allows optical sensors to be designed exploiting variations in any property of an optical wave, such as amplitude modulation, phase modulation, polarization changes, and wavelength conversion.

- **Amplitude modulation** is characteristic of **intensity-based** (sometimes also called "intensiometric") sensors, which operate by detecting modifications in the intensity of light due to absorption, scattering, or reflection. Despite all optical sensors ultimately rely on intensity detection at the receiving end, the term "intensity-based" specifically refers to those sensors in which the quantity to be measured directly modulates the received power. The elementary scheme of intensity-based sensors is illustrated in Fig. 1, in which the quantity to be measured directly influences the propagation loss along the optical path between a light source and a receiver, leading to a corresponding change in the received power. For example, a simple displacement sensor can be built with this scheme since the received power decreases as the separation between source and receiver increases due to diffraction effects.

This approach is the simplest and most cost-effective optical solution because it requires only a light source (usually a Superluminescent LED, SLED, or a laser, typically a semiconductor laser diode, for more demanding applications), the transmission medium (which can be air or most commonly an optical fiber) that serves also as the transducer, and a photodetector at the receiving end. The straightforward design makes these sensors an attractive choice, although the simplicity comes at the expense of very poor selectivity, since any type of perturbation that affects the received power, including fluctuations in the source emissions, is indistinguishable from the actual quantity under measurement. Moreover, even when implemented using optical fibers, this sensing approach does not allow multiplexing along the same fiber, meaning that each sensor in a network must be individually connected

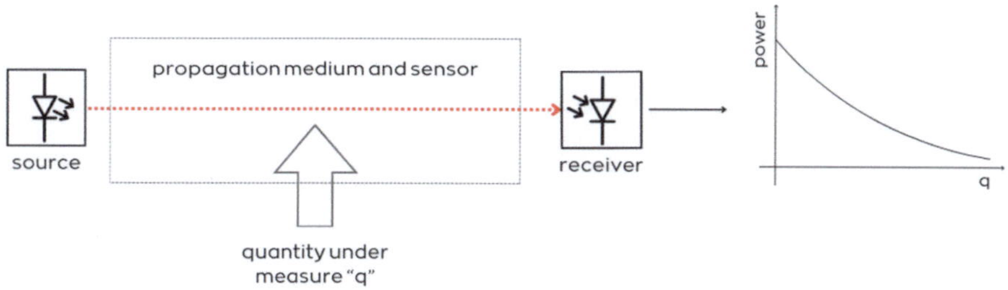

Fig. 1 Schematic of an intensity-based optical sensor

to a dedicated receiver via its own fiber. In other words, it is not possible to have multiple sensors sharing the same optical fiber for signal transmission. This limitation significantly increases the complexity, cost, and weight of multi-point sensing networks, making it less efficient compared to other optical approaches that allow Wavelength-Division Multiplexing (WDM) techniques for which multiple sensors share a single fiber optic link. A partial exception to this limitation is the case of fully distributed intensity-based FOSs that utilize Time-Division Multiplexing (TDM), such as those implemented through the already mentioned OFDR. However, these methods require sophisticated interrogation units and advanced signal processing techniques, so that they can no longer be considered a low-cost solution, limiting their practicality. Intensity-based sensors can be implemented in two configurations, namely transmission and reflection modes. In transmission mode the light source and detector are positioned at opposite ends of the transmission medium (say, the optical fiber), as in Fig. 1. In reflection mode the source and detector are located on the same side and, as the name implies, the signal proportional to quantity under measure must be somehow reflected to reach the receiver, crossing twice the sensing region.

- **Phase modulation** is characteristic of **interferometric** sensors, which rely on the interference between two co- or counter-propagating waves that undergo different phase changes. The measured physical quantity (e.g., strain, pressure, temperature) induces a phase shift in one of the two optical signals, causing modifications in the interference pattern. An example is sketched in Fig. 2. These sensors are typically characterized by highly sensitivity and thus are widely used in precision measurements. Moreover, unlike intensity-based sensors, they are less affected by variations in optical power due to source fluctuations or other types of losses. Finally, when implemented in optical fibers, some form of multiplexing is possible, enabling the development of quasi-distributed sensing networks. For example, FBGs,[1] which rely on interference effects generated by periodic refractive index modulation in the fiber core, seamlessly support WDM. This is because in FBGs the quantity under measurement is encoded into the spectral response and multiple gratings with different spectral behavior can be inscribed along a single fiber, enabling the efficient implementation of quasi-distributed sensing networks. Other types of interferometric sensors, such as Fabry-Pérot-based fiber sensors, also allow for some degree of multiplexing, though with greater limitations because of their broad spectral response. Nonetheless, with advanced signal

[1] FBGs are often regarded as a standalone category of fiber optic sensors due to their widespread use and distinct working principle. However, within the conceptual framework adopted in this chapter, they can be classified as a special type of interferometric sensor. This classification stems from the fact that FBGs operate based on interference effects created by periodic refractive index variations along the fiber core. These periodic structures cause partial reflections at each grating point, leading to the formation of an interference pattern that determines the spectral response, with a reflection peak at the so-called Bragg wavelength. Since their measurement principle relies on detecting shifts in reflected wavelengths, FBGs share fundamental characteristics with Fabry-Pérot and other interferometric sensors, making them functionally similar in terms of how they respond to strain, temperature, and pressure variations.

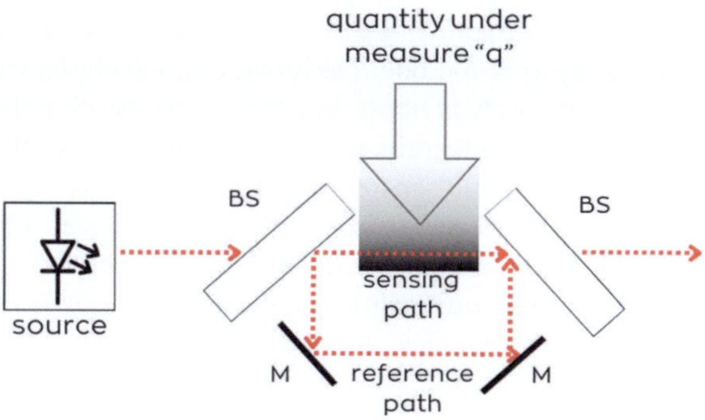

Fig. 2 Schematic of a possible implementations of an interferometric optical sensor using discrete free-space components (in particular, a Mach–Zehnder interferometer); BS stands for "beam splitter", while M for "mirror". The output is an interferogram, also known as "bull eye"

processing techniques, it is still possible to implement multi-point sensing configurations, albeit with reduced scalability compared to WDM-based FBG networks. There are, however, also downsides in interferometric sensors. To start, they require complex interrogation setups, and this makes these sensors more expensive than intensity-based sensors. Then, their high sensitivity makes their response also very subjected to external factors like temperature variations and mechanical vibrations, which can introduce unwanted phase noise. This implies that, especially for free-space implementation (e.g., the Mach–Zehnder interferometer sketched in Fig. 2) precise optical alignment is required and this can be challenging in dynamic environments, such as those typical of aerospace applications.

- **Polarization change** is characteristic of **polarimetric** sensors, which exploit the changes in the polarization state of light as it propagates through an optical medium. The measured physical parameter (e.g., stress, strain, temperature, or magnetic fields) induces birefringence or alters the polarization state, which can then be analyzed using polarimeters or other detection techniques. These sensors are typically used to measure electric and magnetic field intensities, or high currents. They can also be used to measure deformations, especially in the fiber-optic implementation, but in this case, it is difficult to locate the position of the applied force. The main advantage of polarimetric sensors is the high achievable sensitivity. However, they also suffer from poor selectivity because of the cross-sensitivity with any perturbation that induces unwanted polarization changes, such as temperature fluctuations, mechanical stress, or fiber bending (in case of optical fiber-based devices). Moreover, they require polarization stable signal sources, which means more expensive components (e.g., polarization maintaining fibers and polarization controllers), careful calibration and suitable data processing algorithms to reduce the unavoidable fluctuations in applications that require long-term monitoring. Therefore, such sensors in general are best suited for measuring vibrations rather than quasi-static deformation. A possible schematic representation of this system closely resembles that of Fig. 1, with the difference that the receiver must incorporate a polarization-sensitive element, such as a polarizer (called in this case "analyzer"). This component converts changes in polarization state into intensity variations, which are then detected and transformed into an electrical signal by a photodiode. The most common implementation of polarimetric sensors is based on optical fibers, in which micro-bends induce birefringence, allowing the measurement of strain, pressure, or temperature changes. However, like intensity-based fiber optic sensors, polarimetric sensors typically cannot be multiplexed, meaning that each sensor requires a dedicated fiber connection to its receiver, limiting their scalability in large sensing networks.

- **Wavelength conversion** is typical of inelastic scattering approaches or of spectroscopic techniques. The working principle of these sensors is the detection of the changes in the wavelength (or frequency) of light induced by the interaction between the incident optical signal and the measured physical parameter. This can be due to nonlinear optical effects (such as Raman or Brillouin scattering), fluorescence, or shifts in the absorption/emission spectrum of specific molecules called "fluorophores". For example, fluorescence can be exploited to develop optical thermometers that do not suffer from cross-sensitivity with strain (differently for example of FBGs). The main advantage of this category of sensors is their intrinsic multiplexing capability for which multiple sensors can be interrogated simultaneously, particularly in fiber optic architectures. Actually, nonlinear scattering in optical fiber is exploited for fully distributed sensing. However, these sensors require quite expensive equipment, such as spectrometers or wavelength analyzers, increasing system complexity and cost. Moreover, in the case of nonlinear wavelength conversion, quite high optical power levels and long integration times are necessary, limiting their practical utilization in many cases.

These four categories can be implemented exploiting several specific approaches, the most common being those in the following.

- **Absorption sensors**, the prototype of the intensity-based category, operate by detecting changes in received signal intensity due to the absorption of specific wavelengths by the target substance [28]. This principle is particularly suitable for evaluating the concentration of specific gases, such as carbon dioxide (CO_2) or methane (CH_4) [29]. In the aerospace sector, absorption sensors are used in atmospheric monitoring, fuel quality assessment, and safety systems. For instance, in life-support systems air composition within aircraft and spacecraft cabins can be continuously monitored through CO_2- and O_2-specific absorption sensors [30]. Absorption sensors are also employed in fire and smoke detection systems to generate early warnings by identifying combustion-related gases [31]. Furthermore, infrared absorption sensors are utilized to detect contaminants in fuel or changes in the combustion products, ensuring optimal engine performance [32]. Beyond onboard applications, absorption at specific wavelengths is exploited in Earth observation missions for climate research and atmospheric studies. Examples are satellites such as NASA's Orbiting Carbon Observatory-2 (OCO-2), which has specific instruments to measure global CO_2 concentrations [33].

- **Scattering sensors**, another type of intensity-based optical sensors, rely on power variations at the receiver caused by scattering phenomena. When implemented as free-space discrete-component sensors, they detect light scattered by macroscopic particles, enabling the evaluation of particle concentrations in aerosols or fluids, such as those due to combustion particulate [34–36]. These sensors are also employed for early smoke detection in aircraft cabins and cargo compartments [37, 38] and for identifying ice crystal formations [39]. In fiber optic implementations changes in a specific type of intrinsic scattering, the Rayleigh scattering, are at the roots of distributed sensing systems used for real-time monitoring of strain and temperature variations in aircraft and spacecraft structures [40, 41]. Rayleigh scattering arises from sub-wavelength inhomogeneities intrinsic to the structure of any matter, including the material constituting the core of the fiber, and it is sensitive to local deformations of the fiber and to its heating or cooling.
- **Reflective sensors** are intensity-based sensors too, in which the quantity under measurement (e.g., a distance) is determined by changes in the collected light reflected from a surface [42, 43]. Given the working principle, these sensors are particularly suited for non-contact measurements [44]. When implemented with fiber optic technology, they typically make use of large core fibers or of fiber bundles to improve light collecting capabilities [45, 46]. In aviation, they can be used to verify the correct deployment of landing gear and to monitor deflections and deformations in real-time, such as of wings or of the turbines in the engines. In space applications, optical reflective sensors can be used to assist in precise alignment and docking with other spacecraft for servicing or refueling.
- **Evanescent wave sensors** are intensity-based sensors that exploit the interaction of the portion of the field that extends beyond the core of an optical waveguide (e.g., an optical fiber or a planar waveguide)—the so-called "evanescent field"—with the surrounding medium. Sensors implemented according to this approach are highly sensitive to variations in the refractive index of the surrounding medium, which in turn is related to changes in its chemical composition. Additionally, upon proper functionalization, they can be used for the selective detection of specific biomolecules. These sensors are widely used in chemical and biosensing applications, such as detecting specific gases, monitoring environmental pollutants, and performing real-time biomedical diagnostics [47, 48]. A notable variation of this approach is Surface Plasmon Resonance (SPR), in which the sensitivity is enhanced by exploiting the resonant oscillations of conduction electrons at the interface between a metal and a dielectric medium [49, 50]. SPR-based sensors are extensively used in clinical diagnostics due to their ability to provide label-free and highly sensitive detection. This sensor category may not seem to have an immediate and direct use in aerospace; however, measuring chemical reactions and other physical properties in harsh environment, can be relevant also for aerospace applications [51]. For instance, they can be employed for detecting gas leakages. Indeed, since these sensors can detect trace amounts of hazardous gases such as hydrogen, oxygen, and Volatile Organic Compounds (VOCs), they can be used to monitor fuel tanks, measure fuel composition, and detect impurities in

cryogenic liquids used in rocket propulsion [52]. Besides, they can also monitor air quality, humidity, and chemical contaminants in aircraft and spacecraft.

- **Interferometric sensors**, such as Michelson, Fabry-Pérot, Mach–Zehnder, and Sagnac interferometers [53–55]. The **Michelson Interferometer** (MI) has probably the simpler geometry, in which a beam splitter is used to divide light into two perpendicular arms: a sensing and a reference arm. Then, two beams are reflected by mirrors and recombine producing an interference pattern that depend on the optical path difference [56–58]. It has applications in precision metrology, gravitational wave detection (as in the LIGO experiment), and astronomical interferometry [59, 60]. In aerospace, it is utilized for precision displacement measurement and atmospheric studies [61–64]. The **Fabry-Pérot Interferometer** (FPI) consists of two parallel, highly reflective mirrors forming an optical cavity. When a broadband signal is shined on the interferometer, light undergoes multiple reflections between these mirrors, creating a wavelength dependent interference pattern that presents maximum transmission when the cavity length is multiple of half-wavelength [65, 66]. Depending on the mirror reflectivity, it is possible to obtain a very high spectral selectivity (high finesse FPI). FPIs are widely used in high-resolution spectroscopy, laser wavelength stabilization, and in optical filters. Sensing applications typically exploit the dependency of cavity optical length with strain and temperature. In aerospace, FPIs can be integrated in Doppler wind lidar systems and atmospheric gas sensing [67]. Other applications reported in the literature include the analysis of rocket engine emissions [68], hypersonic force measurements [69], and high-resolution spectroscopy [70]. The **Mach–Zehnder Interferometer** (MZI), one of the most widespread interferometers, has a structure similar to the MI, for which an incoming beam is split into two paths, the sensing and the reference arm, respectively. The two beams travel through different optical media (in particular, the sensing beam is subjected to phase variations induced by the quantity under measurement) before recombining through another beam splitter. Any phase difference introduced between the two beams results in interference fringes that can eventually be converted into changes in the received power [71, 72]. MZIs are extensively used in optical sensing, telecommunications, and quantum optics. In aerospace, it finds applications in fiber optic gyroscopes, turbulence detection, and structural health monitoring [73–75]. The **Sagnac Interferometer** (SI) is a ring interferometer in which light travels in opposite directions along a closed-loop path [76, 77]. Rotation of the interferometer causes a phase shift due to the Sagnac effect, generating interference fringes and thus changes in the received power in dependence of the angular velocity. This principle is the foundation of Fiber-Optic Gyroscopes (FOGs) and Ring Laser Gyroscopes (RLGs), both of which are essential for inertial navigation in aircraft, spacecraft, and satellites [78–80]. SI has also been used for the detection of gravitational waves [81].
- **Fiber Bragg Grating** (FBG) **sensors**, among the most widely used fiber optic sensing technologies, rely on periodic variations in the refractive index of an optical fiber core, which causes the selective reflection of specific wavelengths (the Bragg wavelength). Any external perturbation—such as strain, temperature,

pressure, or mechanical deformation—induces a shift in the Bragg wavelength [82, 83]. FBGs are known for their high sensitivity, multiplexing capability, and versatility, for which have become the gold standard in fiber optic sensing [84]. In aerospace applications they can be embedded in aircraft fuselages, wings, and spacecraft components to monitor strain, thermal expansion, and structural integrity [85–87]. Another application is in rocket engines and fuel tanks to detect pressure variations and prevent failures. Given the importance, a specific section of this chapter will be devoted to a more in-depth analysis of FBG-based sensors.

- **Polarimetric sensors**, which measure the variations in the polarization state of light, are used in imaging, navigation, and environmental sensing [88]. Indeed, by analyzing changes in polarization caused by surface reflections, atmospheric scattering, or structural deformations, these sensors are employed from long time in satellites and airborne platforms to improve cloud characterization, ocean surface monitoring, and vegetation analysis [89–91]. Fiber polarimetric sensors can be used in aerospace SHM to detect stress-induced birefringence in aircraft materials, providing insights into structural integrity and early signs of fatigue or damage [92, 93].
- **Luminescence and fluorescence-based sensors** relate the quantity under measurement to the fluorescent emission from target molecules when excited by a specific light source. They are widely used in biomedical diagnostics to identify markers and in chemical detection, but they can also find applications in the aerospace field; examples are the measurement of molecular oxygen (both in fuel tanks and as pressure-sensitive paints) and the monitoring of corrosion and of degradation of hydraulic fluids [94, 95].
- **Raman and spectroscopic sensors** analyze molecular vibrations to identify chemical compounds [96, 97]. In the aerospace and space exploration sector they are fundamental for the analysis of mineral compositions on planetary surfaces, as in the NASA's Perseverance rover in combination with fluorescence sensors [98, 99] or in other rovers [100]. Other applications include the monitoring of gases [101] and of biomolecules [102], which may also serve as indicators of the health of astronauts [103].
- **LiDAR** (from Light Detection and Ranging), or LIDAR, often described as the "optical RADAR", is an advanced remote sensing technology that utilizes near-infrared laser pulses to measure distances with exceptional accuracy by analyzing the time delay between emission and reflection [104, 105]. LIDARs are used to generate high-resolution 3D maps, detect objects, and provide real-time spatial analysis. The shorter wavelength used in LIDAR compared to RADAR, enables superior resolution, allowing precise object recognition and terrain mapping with centimeter-level accuracy. However, the effective range of LIDARs is typically limited to a few kilometers, whereas RADARs can detect objects hundreds of kilometers away. Moreover, LIDAR signals are more susceptible to atmospheric interference. For example, fog, rain, or smoke can scatter or absorb the laser pulses, reducing the detection capabilities. LIDAR technology is increasingly becoming popular because it is used in autonomous vehicles for detecting pedestrians, vehicles, and road conditions [106]. In aviation, LIDARs are used to detect

the presence of microbursts and turbulences. They are also employed in aircraft and spacecraft for precision landing guidance, altitude measurement, and obstacle detection [107–111]. For example, NASA employs LIDAR instruments for planetary surface mapping, as in the Mars Perseverance Rover's hazard detection and terrain analysis systems.

Given their importance in modern sensing applications, the following sections focus on a more detailed description of the most widely used fiber-optic sensors in aerospace. Before delving into specific sensor types, an introduction to optical fibers is presented, briefly explaining their structure and principles of operation to better understand their advantages in various sensing applications.

4 Fiber Optic Sensors

4.1 Optical Fibers

Optical fibers are cylindrical dielectric waveguides operating at optical frequencies. In general, a waveguide is a structure that confines electromagnetic waves to a defined region of space, and it is used to transmit energy with minimal loss along a certain direction, conventionally indicated as the z-axis. In the microwave domain, this term typically refers to a hollow metallic pipe, whereas at optical frequencies, waveguides take the form of miniaturized, all-dielectric rods. Most common optical fibers have a cross-sectional structure composed of two concentric layers, as shown in Fig. 3:

- the "**core**", the inner region where light is confined and guided;
- the "**cladding**", the outer layer that surrounds the core, controlling the optical confinement and preventing the interaction of the guided waves with the surrounding media.

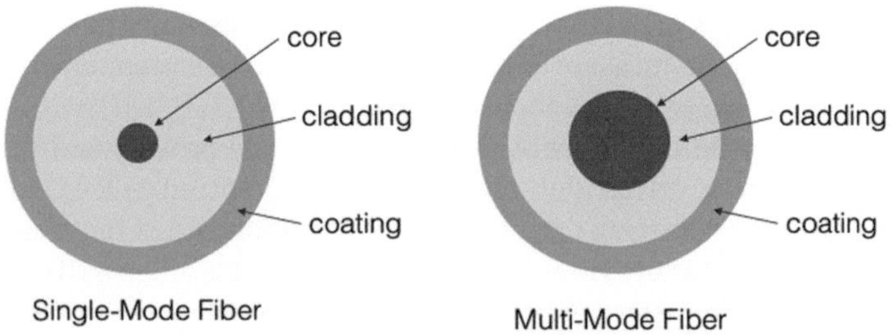

Fig. 3 Sketch of the cross-section of single-mode (left) and multi-mode (right) optical fibers (dimensions not in scale)

Common fibers have both the core and cladding typically made of silica, although entirely polymer-based fibers, known as Plastic Optical Fibers (POFs), are also available. These two layers are further coated with one or more protective layers, usually made of acrylate or polyimide, to improve durability and environmental resistance. The confinement of light within the core of an optical fiber is obtained through Total Internal Reflection (TIR) at the core-cladding interface and this requires that the core has a slightly higher refractive index than the cladding. However, while necessary, TIR alone is not sufficient to ensure guided propagation. Indeed, the propagation in a waveguide—thus including optical fibers—is commonly studied as the combination of particular field distributions called **guided modes** (or simply modes), which have the property of maintaining their transverse shape along the fiber's axis during propagation, while undergoing a phase shift to account for the delay due to the traveling time. Guided modes must satisfy not only the TIR condition but also a phase-matching condition, which results in a discrete set of supported modes. Modes form an orthogonal basis that can be used to express any field distribution entering the fiber. Besides for guided modes, there are **radiation** (or radiated) **modes**, which do not remain confined within the fiber and instead gradually lose power as they propagate. These modes can arise due to several factors, including:

- Poor coupling between the source radiation pattern and the fiber's guided mode spectrum.
- Discontinuities, misalignments, or inhomogeneities within the fiber.
- Excessively sharp bends, which can cause mode leakage.

A generic input optical field can be expressed as a combination of both guided and radiation modes.

Depending on the wavelength, core diameter, and refractive index contrast, an optical fiber may support either a single guided mode or multiple modes:

- **Single-Mode Fibers (SMFs)**: they support only one guided mode, the fundamental mode, making them more critical to excite (for instance, they require laser diodes and not LEDs) and to use (in terms of mechanical alignment constraints) but necessary for some highly accurate sensing techniques, such as those based on FBGs.
- **Multi-Mode Fibers (MMFs)**: they support multiple guided modes (i.e., the fundamental and many higher order modes), typically in the order of hundreds, which makes light coupling much easier—thus less expensive—since the larger number of supported field distributions allows describing almost any input field profile. They are ideal for intensity-based approaches but cannot be used for high-precision sensing applications, which typically exploit the interaction with the fundamental mode only and thus require SMFs.

Standard telecom-grade optical fibers, which are also those used for fiber optic sensing, are intended to work around 1300 nm and 1550 nm (respectively the so-called second and third optical communication windows) and have core diameter around 9 μm for SMFs and around 50 μm for MMFs. For both types the cladding diameter is 125 μm, while the diameter of the first protective coating is 250 μm.

Fiber optic technology provides an exceptional platform for developing sensors capable of detecting virtually any physical or chemical quantity [4, 112].

Broadly speaking, FOSs can be distinguished between intrinsic and extrinsic types. In intrinsic FOSs, a specific section of the fiber serves as the sensing element, meaning that the measured quantity directly affects the fiber's optical properties, such as intensity, phase, wavelength, or polarization. Conversely, in extrinsic FOSs, the fiber acts as a conduit, transmitting light to and from a miniature, non-fiber-based optical component, which modulates one or more properties of light in response to the measured parameter. In other words, intrinsic sensors respond to environmental changes by altering light propagation within the fiber, whereas extrinsic sensors rely on external optical transducers to modulate the optical signal. For example, it is known that in optical fibers with loops, modes, especially higher-order ones, experience propagation losses that depend on the radius of curvature. This principle can be exploited to design an intrinsic intensity-based FOS for pressure sensing, as schematically illustrated in Fig. 4. When pressure is applied, it induces slight bending in the fiber, altering the curvature radius and increasing bend-induced optical losses, with a subsequent attenuation in the transmitted signal.

Since the fiber remains continuous and does not require additional external components, intrinsic fiber FOSs are more robust, reliable, and resistant to harsh environments. Moreover, they can be typically assembled in compact packages that makes them less invasive and well-suited to be integrated withing structures. However, this approach also presents limitations, as the range of sensing techniques available for intrinsic sensors is narrower compared to extrinsic FOSs, which can incorporate specialized external sensing elements to enhance performance and versatility. Indeed, in extrinsic sensors, the transducer can be optimized exploiting all the possible sensing technologies and the fiber acts as a mere replacement for copper wires in cases where these are cannot be used or are strongly limited, such as due to long distances or in presence of strong electromagnetic fields. An example of an extrinsic FOS is a fiber optic pressure sensor in which the optical fiber transmits light

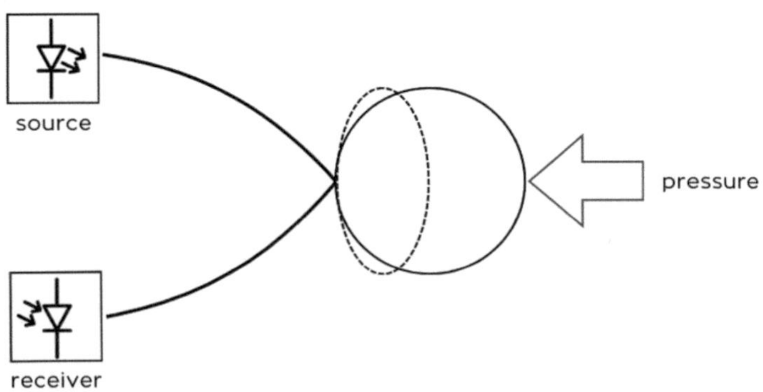

Fig. 4 Sketch of an intrinsic intensity-based FOS for pressure sensing exploiting bending loss (effect not in scale)

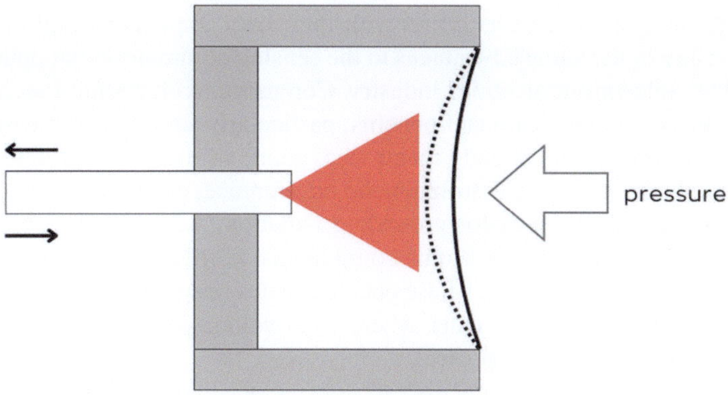

Fig. 5 Sketch of an extrinsic intensity-based FOS for pressure sensing

to and from a diaphragm that deforms in response to pressure changes, altering the intensity or phase of the reflected light. A sketch of such sensors is reported in Fig. 5.

Another example of an extrinsic fiber optic sensor is a high-voltage sensor based on the Electro-Optic (EO) Pockels effect. This requires a crystal exhibiting electro-optic properties, such as lithium niobate ($LiNbO_3$) or potassium dihydrogen phosphate (KDP), in which a change in birefringence occurs when exposed to an electric field. A laser beam passes through the electro-optic crystal, and its polarization state is altered proportionally to the applied voltage; then, an optical fiber is used to transmit the modulated light signal to a remote detection unit, where it is analyzed using a polarizer or an interferometer to determine the change of the received power corresponding to the applied voltage. This technology is particularly suited for monitoring high-voltage cables because it is a non-contact technique with an intrinsic electrical isolation from the measurement system.

Although classification may vary, some also consider electro-mechanical devices equipped with an electrical-to-optical signal conversion unit as a form of extrinsic fiber optic sensors; however, this classification is questionable since there is an optical fiber, but the sensing principle has nothing to do with optics and the fiber merely serves as a transmission medium to replace copper wires.

4.2 Advantages of Fiber Optic Sensors

Fiber optic sensing has many advantages over traditional methods, including high resistance to harsh environments, intrinsic impossibility to start fires, immunity to electrical disturbances, and long-distance transmission capabilities. These properties make fiber optics a valuable tool in virtually all application domains, such as structural health monitoring, industrial process control, oil and gas exploration, biomedical engineering, and many more.

Although fiber-optic sensors come with their own implementation challenges, they offer many of the same advantages to the sensing community that optical fibers provide to the telecommunications industry. Compared to electronic-based sensors, fiber-optic sensors have significant benefits, particularly for detailed measurements in harsh environments. As already mentioned, being all-dielectric structures, FOSs are intrinsically immune to electromagnetic interference, electrical noise, and short circuits, eliminating the need for grounding isolation. Additionally, optical fibers allow remotely locating any electronic components, keeping them away from high-temperature or dangerous areas. These characteristics make FOSs particularly well-suited for hazardous environments where high-voltage electricity or flammable materials, such as jet fuel, are present.

Other key advantages of optical fiber sensors include their small size—accounting for minimally invasive impact–, no need for electrical power at remote locations, and, for some of them, the ability to multiplex sensors along a single fiber. For instance, a single optical fiber can integrate hundreds of Bragg gratings, allowing multi-point or multi-parameter measurements. Similarly for Rayleigh or Brillouin scattering-based sensing systems.

Multiplexing can be achieved through WDM, TDM, or a combination of both. Due to their inherent immunity to electromagnetic interference and non-conductive nature, fiber-optic sensors are particularly well-suited for hazardous environments where high-voltage electricity or flammable materials, such as jet fuel, are present.

4.3 *Fiber Bragg Gratings*

Fiber Bragg gratings are the most widely used type of fiber optic sensors due to their exceptional combination of robustness, sensitivity, and versatility. An FBG is a narrowband reflector formed by inducing a permanent periodic modulation in the refractive index of the core of a single-mode fiber, as sketched on the right part of Fig. 6.

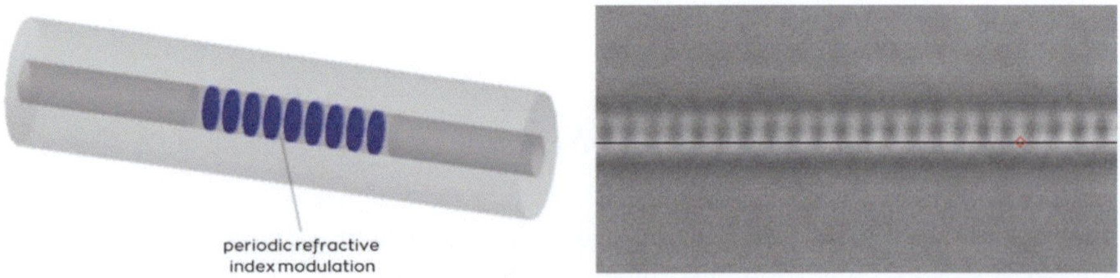

Fig. 6 Schematic representation of a fiber Bragg grating (left) and its realization through point-by-point direct writing using a femtosecond laser (right)

Fig. 7 A photograph of a femtosecond laser direct writing setup, where the laser beam is precisely focused onto the optical fiber, which is positioned using high-accuracy translation stages

Traditionally, this modulation was achieved by exposing a special photosensitive optical fiber to intense ultraviolet light through a phase mask, following a preparatory treatment to enhance photosensitivity [113–115]. However, modern fabrication techniques increasingly favor femtosecond (fs) laser inscription, which is effective in any type of fiber without requiring additional treatments [116, 117]. Furthermore, fs-laser direct writing (Fig. 7) enables the fabrication of FBGs without the need for a phase mask, using the Point-by-Point or Line-by-Line approach, as shown in the right part of Fig. 6. These methods offer greater flexibility, allowing different periodicities simply by adjusting the spacing between inscribed points or lines. Recent advancements in femtosecond laser technology have also enabled the fabrication of FBGs that can withstand extreme conditions. For example, Type-II FBGs fabricated in conventional single-mode fibers have demonstrated long-term thermal stability at 1000 °C for several hundred hours [118].

From a formal point of view, according to the Coupled Mode Theory (CMT), the periodic refractive index modulation in the core induces a "counter-directional coupling", for which part of the energy associated to the incident (only) mode of the single-mode fiber is transferred to the corresponding backward propagating mode [119, 120]. This coupling is more effective for wavelengths around the Bragg wavelength (λ_B), which is a particular value that satisfies a specific relation between the modal effective refractive index (n_{eff}) and the grating modulation period (Λ):

$$\lambda_B = 2\, n_{\text{eff}}\, \Lambda \qquad (1)$$

The FBG structure has therefore a spectral behavior similar to that of a wavelength-selective mirror with a notch-like spectral response, reflecting a narrow spectral portion around the Bragg wavelength while allowing all other wavelengths to pass through.

Both the modal effective refractive index and the grating modulation period in Eq. 1 are influenced by temperature and strain applied to the optical fiber. Consequently, the Bragg wavelength becomes inherently sensitive to these external factors. For linear variations it is possible to relate the shift in the Bragg wavelength with strain and temperature as:

$$\frac{\Delta\lambda_B}{\lambda_B} = (1 - p_e)\varepsilon + \alpha_t \Delta T \tag{2}$$

where $\Delta\lambda_B$ is the shift in Bragg wavelength, λ_B is the Bragg wavelength, p_e is the photo-elastic coefficient of the fiber, ε is the applied strain, α_t is the thermo-optic coefficient, and ΔT is the temperature variation. For standard telecommunication grade silicate fibers, the variation of the Bragg wavelength with strain and temperature is around 1 pm/με and 10 pm/°C, respectively. Equation 2 shows that FBGs are simultaneously sensitive to both mechanical strain and temperature variations, requiring proper compensation techniques (e.g., using reference gratings) to discriminate between the two quantities.

FBGs are intrinsic sensors, meaning that the quantity being measured (such as strain or temperature) directly modifies the fiber's optical properties. Since FBG sensors are integrated into the fiber core, they eliminate problems associated with mechanical alignments; then, since the quantity under measure is encoded into wavelength and not amplitude or phase, they are less sensitive to intensity variations, ensuring higher stability and accuracy even in presence of quite relevant source fluctuations. The downside of wavelength-encoded sensing is that the interrogation system must be capable of detecting very small wavelength shifts with high precision. Several techniques can achieve this, but the two most common approaches use:

- Broadband light source, typically a Superluminescent Light-Emitting Diode (SLED), combined with a spectrometer or an Optical Spectrum Analyzer (OSA) as receiver to examine the reflected spectrum. When implemented with a compact, CMOS-based spectrometer, this approach is cost-effective but suffers from limited spectral resolution. Alternatively, the use of an OSA improves the resolution but at the cost of increased complexity, higher expenses, and slower measurement speeds. The latter drawback is particularly critical because it makes this configuration practically unsuitable for dynamic sensing applications.
- Tunable laser source, used to sweeps a defined wavelength range, while a photodetector records the reflected intensity. This method offers superior resolution and faster response times, making it particularly well-suited for dynamic sensing applications where real-time monitoring is essential.

Multiple FBG-based sensors can coexist in the same fiber, as shown in Fig. 8, exploiting the possibility to inscribe sections with different grating periods along a single optical fiber, obtaining an array of independent FBGs with different Bragg wavelengths.

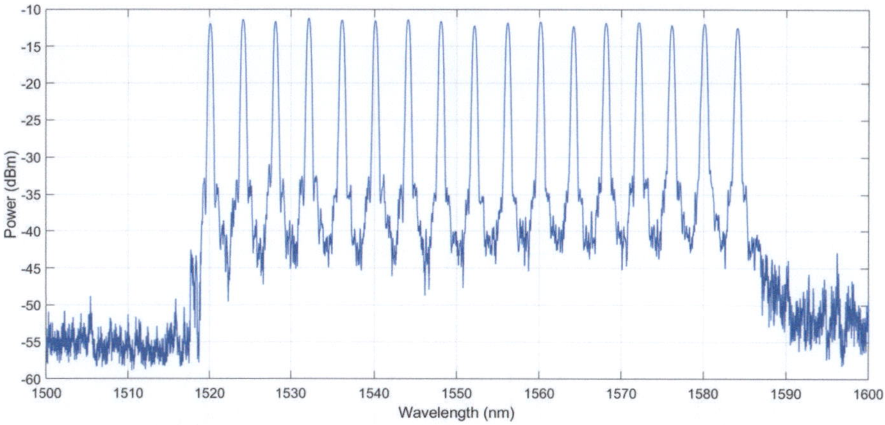

Fig. 8 Example of the spectral response of multiplexed FBGs. Each peak corresponds to an FBG with different periodicity, leading to a different Bragg wavelength

This way, an FBG-based multi-point quasi-distributed sensor network can be implemented, allowing the simultaneous monitoring of multiple parameters along an extended structure. This significantly reduces cabling complexity, weight, and installation costs compared to traditional electrical or single-point fiber optic sensors.

4.4 Distributed FOS

The multiplexing capability of FBGs can be exploited for quasi-distributed sensing. However, truly fully distributed sensing (often identified with the acronym DOFS, from Distributed OFS) can be obtained only with scattering-based fiber sensors. This category of sensors comprises devices exploiting Rayleigh, Brillouin, or Raman scattering. All these approaches share the same underlying principle for which a short light pulse is injected into a single-mode fiber and backscattered signals are detected with a time delay corresponding to their generation point, which in turn is related to the position of the strain or temperature perturbation. However, the three mentioned approach differ for the intensity and frequency of the backscattered signals. In more detail, as schematically illustrated in Fig. 9, Rayleigh scattering produces a signal at the same frequency as the incident beam. In contrast, both Brillouin and Raman scattering generate two distinct components—the "Stokes" and "anti-Stokes" components—which exhibit characteristic positive and negative frequency shifts relative to the incident light. The frequency shift is on the order of a few gigahertz for Brillouin scattering and several terahertz for Raman scattering. The exact spectral position and/or intensity of these components depend on the fiber's operating conditions, making them useful for detecting and quantifying external perturbations such as strain, temperature, or mechanical stress. In particular, the

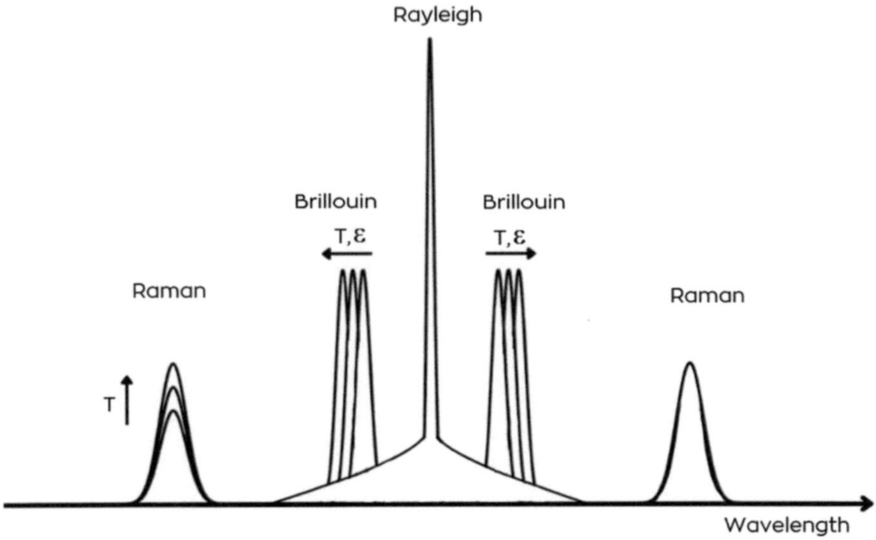

Fig. 9 Schematic representation spectral position of Rayleigh, Brillouin, and Raman scattered signals

spectral shift of Stokes and anti-Stokes Brillouin scattered components is dependent on both local temperature and strain, while the intensity of the anti-Stokes Raman scattered component is dependent on the temperature only and the Stokes Raman scattered component is practically constant with temperature.

Rayleigh scattering:

Rayleigh scattering falls under the category of elastic scattering, meaning that the scattered light retains the same frequency as the incident light. It occurs due to microscopic fluctuations in the refractive index on a scale much smaller than the optical wavelength. When an optical wave encounters these irregularities, it scatters in all directions, with a small portion recaptured by the fiber and propagating back toward the source. Instrumentation based on Rayleigh scattering can be broadly classified into two categories based on their operational mode: Optical Time-Domain Reflectometry[2] (**OTDR**)—including its phase-sensitive variant (**Φ-OTDR**)—and Optical Frequency-Domain Reflectometry (**OFDR**).

In an OTDR, a light pulse is launched into one end of the fiber, generating a weak but continuous backscattered signal as it propagates. This backscattering results from unavoidable microscopic inhomogeneities within the fiber. The back-reflected power, known as the "Rayleigh signature", reaches the receiver at different times, proportional to the distance traveled through the propagation speed in the fiber. Due to fiber attenuation, this signal exhibits an exponential decay with distance. The Rayleigh

[2] The R in the acronyms OTDR and OFDR may stand for "Reflectometry" if referring to the physical approach or for "Reflectometer" if referring to the instrument that implements the approach.

signature is unique to a specific fiber under given operating conditions. However, when external perturbations—such as mechanical deformations, temperature variations, or strain—occur along the fiber, they introduce additional scattering contributions. These disturbances alter the Rayleigh signature, generating localized variations in the backscattered signal. Since the propagation speed in the fiber is known, the time delay between the emitted pulse and the received backscatter allows precise spatial localization of the perturbation. This process, illustrated in Fig. 10, forms the basis of distributed fiber optic sensing. While OTDRs can cover distances of hundreds of kilometers, they have relatively poor spatial resolution (typically in the order of tens of meters). Due to this limitation, they are primarily used to detect discontinuities in optical communication networks rather than for sensing applications.

A more advanced variant, Φ-OTDR (Phase-Sensitive OTDR), enables distributed vibration sensing, also known as Distributed Acoustic Sensing (DAS) [121]. Φ-OTDR employs coherent light pulses from a narrow-linewidth laser. Unlike conventional OTDR, which relies on intensity variations, in Φ-OTDR when an external perturbation (such as mechanical vibration or strain) affects the fiber, it induces a phase shift in the backscattered signal. This shift is detected using heterodyne detection or interferometric techniques, allowing precise measurement of the disturbance. The key advantage of Φ-OTDR is its significantly higher sensitivity and improved spatial resolution, often well below 1 m, over a range of several kilometers.

As the name suggests, OFDR operates in the frequency domain [122, 123]. In a basic OFDR setup, a tunable laser source linearly sweeps its optical carrier frequency over time. The frequency-modulated continuous-wave signal (probe signal) is then split into two paths: the test arm, which includes the sensing fiber, and the reference arm, where a reflector (local oscillator) is placed. The Rayleigh-backscattered test signal coherently interferes with the reference signal at a coupler. This interference is then converted into the electrical domain by a photodetector, yielding beat signals that contain amplitude and phase information about the reflection sites in the sensor fiber. OFDR offers exceptionally high spatial resolution, reaching the hundreds-of-micrometers scale, with a sensing range of several tens of meters. This level of precision makes it a strong competitor to quasi-distributed sensing networks based on FBGs, with the advantage of not requiring any modification to the optical fiber. However, to achieve these resolutions OFDR typically requires several seconds per

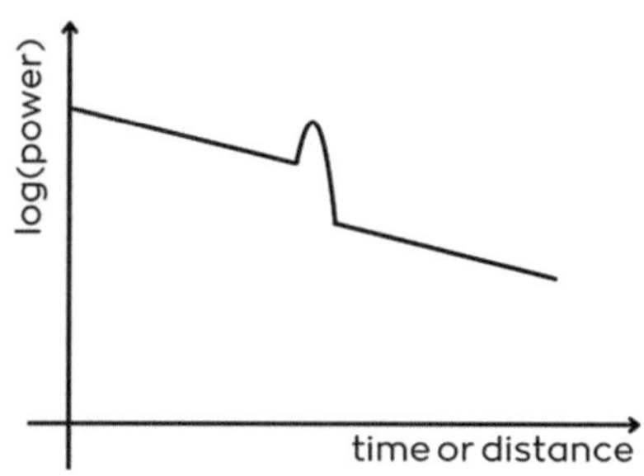

Fig. 10 Schematic representation of an OTDR trace (not in scale) with highlighted a contribution due to a perturbation along the fiber. The slope of the trace is non-zero due to the fiber attenuation

measurement because of the long integration time, making it unsuitable for detecting high-frequency perturbations such as vibrations.

Brillouin scattering:

This phenomenon falls under the inelastic scattering types because it involves a wavelength shift (or frequency shift) in the scattered light. It originates from the interaction between light and acoustic waves within the optical fiber, which are induced by light propagation. The Brillouin frequency shift is highly sensitive to temperature and strain, making it a powerful technique for distributed sensing applications [124–127]. Brillouin Optical Time-Domain Reflectometry (**BOTDR**) and Brillouin Optical Time-Domain Analysis (**BOTDA**) are two widely common implementations. Although these two approaches share similarities, they differ in their operational configurations. The BOTDR is a one-end access system that uses a single laser, injected into one end of the sensing fiber. Since it requires access to only one fiber end it is more practical for real-world applications and is particularly useful for detecting strain and temperature variations over long distances. On the contrary, the BOTDA is a two-end access system that exploit two counter-propagating lasers, injected from both ends of the fiber. It achieves higher measurement accuracy, and a larger sensing range compared to BOTDR. However, it cannot determine the exact breakpoint location in case of fiber failure.

BOTDR, with a sensing range extending tens of kilometers and a spatial resolution of approximately 1 m, is ideal for continuous monitoring of critical infrastructures, such as tunnels, dams, and large-scale industrial facilities.

Raman scattering:

This type of scattering also falls under inelastic scattering, as it involves a wavelength shift in the scattered light. Raman scattering is caused by the interaction of light with molecular vibrations in the fiber, resulting in frequency shifts that are exclusively temperature-dependent [97, 128]. This lack of cross-sensitivity to strain makes Raman OTDR particularly advantageous for temperature monitoring applications, such as in industrial process and fire detection systems. However, Raman scattering requires a higher power threshold than Brillouin scattering, making it less commonly used in distributed sensing applications. Despite this, its performance in terms of spatial resolution and operational range is comparable to that of Brillouin-based sensing systems.

4.5 Sagnac Interferometers

Sagnac interferometers are based on the homonymous Sagnac effect, named after the French physicist Georges Sagnac, who first demonstrated this phenomenon in 1913. Sagnac conducted his experiment in an attempt to confirm the existence of the aether, a hypothetical medium once believed to be necessary for light propagation. Although the concept of aether was already considered obsolete at the time, his

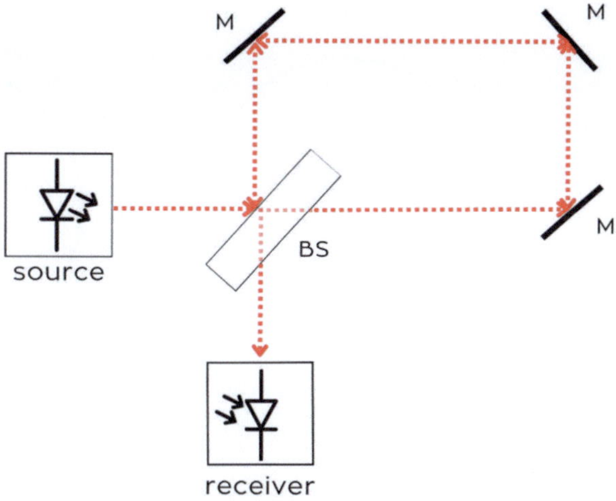

Fig. 11 Schematic representation of a Sagnac interferometer; BS stands for "beam splitter", while M for "mirror"

work has profound implications not only in modern physics, but also in modern devices used for navigation [76, 77]. Among the various types of interferometers (e.g., Michelson, Fabry-Pérot, and Mach–Zehnder), the Sagnac interferometer is less commonly known. However, as mentioned, it plays a critical role in modern navigation systems, particularly in gyroscopes, which are fundamental instruments for inertial navigation in aerospace, marine, and defense applications.

A Sagnac interferometer consists of a closed optical path in the form of a ring or loop, as schematically illustrated in Fig. 11. A beam of light is split into two components, which then travel in opposite directions along the same closed-loop path. Upon returning to the point of entry, the two beams interfere, producing an interference pattern.

The key principle behind the Sagnac effect is that the relative phase shift between the two counter-propagating beams depends on the angular velocity of the system. When the interferometer is stationary with respect to an inertial (non-rotating) reference frame, both beams take the same amount of time to traverse the loop. As a result, their phase difference remains zero, and no shift is observed in the interference pattern. However, when the interferometer is rotating, one beam effectively travels a longer optical path than the other. This can be visualized considering for simplicity an ideal circular light path in a rotating interferometer. An observer in an inertial (non-rotating) reference frame sees light entering the interferometer at point A (Fig. 12-left) and traveling in opposite directions along the loop at the same speed (e.g., the speed of light in vacuum). However, due to the interferometer's rotation, the observer now sees that by the time the light completes its journey around the loop, the beam splitter has moved to a new position, B (Fig. 12-right). As a result, the co-rotating wave (moving in the same direction as the interferometer's rotation) must travel a longer path than a single loop to reach the detector, while the counter-rotating wave (moving against the rotation) follows a shorter path than one full loop. This difference in optical path length introduces a phase shift between the two waves. In

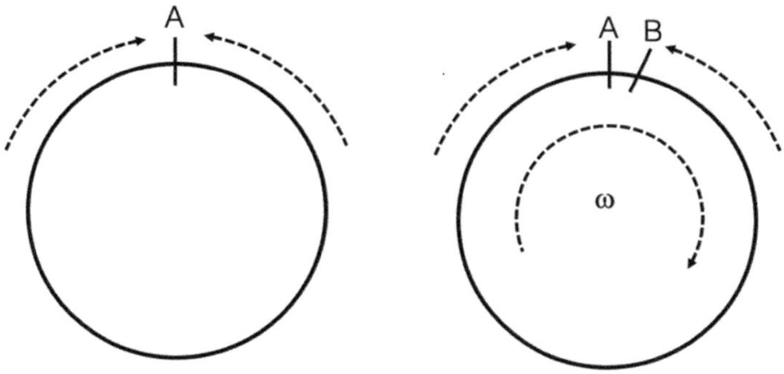

Fig. 12 Sagnac effect in a vacuum considering an ideal circular path: system at rest (left) and rotating (right)

other words, since the loop is moving, the beam traveling in the same direction as the rotation has a longer path to complete a full circuit, while the counter-propagating beam has a shorter path. This difference in path length introduces a phase shift $\Delta\phi$ between the two beams, which is directly proportional to the angular velocity w of the system:

$$\Delta\phi \propto \frac{8\pi}{\lambda c}\,\omega A \tag{3}$$

being A the area of the loop.

This property justifies why Sagnac interferometers are at the foundation of FOGs. However, a more rigorous approach would be needed to fully explain the behavior of FOGs, considering the laws of electromagnetism in a rotating frame and solve the corresponding propagation equations.

5 Selected Applications of FOSs in Aviation

5.1 Structural Health Monitoring

The combination of minimal invasiveness, resilience in harsh environments (including flammable and hazardous conditions), and multiplexing capability makes FOSs, particularly FBGs and DAS, an optimal solution for implementing Condition-Based Maintenance (CBM) programs in aviation. For instance, by integrating FBGs directly into aircraft components, especially composite materials, the real-time monitoring of stress, strain, temperature, and structural integrity without adding significant weight or complexity becomes possible.

Increasingly, aircraft designers have been turning to composites to make their vehicles lighter, more fuel-efficient, and more comfortable for passengers. Composites offer a remarkably high stiffness-to-weight ratio, due to the mechanical properties of the fibers (usually carbon or glass) embedded in a polymeric matrix. Other advantages of these materials include but are not limited to corrosion resistance, noise reduction, improved fatigue life and design flexibility.

While older models such as the Boeing 777 and Airbus A380 already used composite materials on a small scale, the first airliner to benefit from having an airframe that consisted primarily of composite materials was the Boeing 787 Dreamliner family. These structures increased the fuel efficiency by 20% compared to the Boeing 767 series, its predecessor. A direct competitor of the Boeing 787, the Airbus A350 also makes extensive use of composite materials. Both aircraft witnessed a reduction in net fuel usage and minimized maintenance costs thanks to these innovative materials. In the future, new aircraft series will be made up of a much higher percentage of carbon fiber composites. However, these materials are anisotropic and comprise of different materials for their matrix and fibers, making them more complex than traditional metal alloys, and resulting in unpredictable behavior concerning the propagation of damage and the residual life and strength of the structure, prompting durability and safety concerns.

The integration of optical fiber sensors into composite materials during the lay-up process enables real-time monitoring of structural integrity throughout the entire life cycle of composite components. This enhances safety, reliability, and operational lifespan, ultimately improving cost efficiency. Additionally, embedding the sensing fiber within the composite structure protects it from external environmental factors, preventing detachment or damage.

Currently, fiber optic sensors can be installed using two primary methods: surface mounting and implanting. Surface-mounted sensors are widely used for short-term applications, such as temporary testing, but their long-term reliability is limited due to aging adhesives, metal welding degradation, and the risk of detachment. In contrast, implanted sensors, embedded within the composite material during manufacturing, can be employed to form a quasi-distributed or fully distributed sensing network, offering besides superior durability and reliability, improved measurement performance.

In quasi-distributed sensing systems, multiple FBG sensors inscribed along a single fiber enable distributed strain measurements, which can be used to detect crack propagation and structural changes. FBG sensors also provide a viable alternative to traditional strain gauges, offering advantages such as:

- Multiplexing capability, allowing hundreds of sensors to be interrogated simultaneously by a single multi-channel interrogator.
- Simultaneous measurement of multiple physical parameters, including temperature, strain, and pressure.
- High speed rate interrogation to enable dynamic strain measurements.
- Reduction in cabling complexity, as opposed to conventional electronic sensors that require extensive wiring and signal management.

Beyond FBG-based systems, scattering-based techniques in embedded standard telecom fibers, such as Rayleigh and Brillouin scattering, enable fully distributed sensing, where the entire fiber acts as a sensor with strain resolution at the centimeter or sub-centimeter scale. These systems allow for large-scale monitoring of fatigue analysis, although typically limited to low frequency deformations for the reasons mentioned in the previous sections. In both cases the versatility of quasi-distributed or fully distributed embedded monitoring systems makes them ideal for real-time fatigue monitoring and early damage detection in aerospace structures.

Fatigue is a major concern in aircraft structures, developing progressively as a result of repeated flight cycles. It begins with microscopic cracks in components subjected to cyclic loading, which grow over time until they pose a structural risk. Key contributors to fatigue accumulation include: (i) surface loading during flight maneuvers; (ii) pressurization and depressurization cycles of the fuselage; (iii) cyclic bending of metal structures; (iv) residual stress, surface treatments, and corrosion effects. Fatigue cracks can appear in any area exposed to concentrated cyclic stresses, making continuous monitoring essential for components prone to structural failure, particularly in high-speed moving parts.

The damage-tolerant design philosophy adopted in modern aircraft assumes that microscopic defects exist from the beginning of a component's operational life. Fracture mechanics techniques are employed to assess whether these defects will cause only minor damage—such as small cracks or localized corrosion—before scheduled periodic inspections can address them. This approach relies on the use of high-tenacity materials and Non-Destructive Testing (NDT) methods to track progressive damage. However, current NDT inspections require aircraft to be taken out of service, with manual checks performed either visually or using bulky, point-by-point scanning instruments. In many cases, inspections necessitate disassembly, increasing the risk of unintended structural damage. Additionally, these procedures require highly skilled personnel, adding to maintenance costs and complexity.

Particularly attractive is the use of FBGs for the detection of Barely Visible Impact Damage (BVID), which is a critical concern in aerospace engineering, particularly in aircraft and spacecraft structures that use composite materials [129–131]. Indeed, unlike metallic components, which often show clear signs of deformation upon impact, composites can sustain internal delamination and matrix cracking with little to no visible surface indication. Particularly, in high-stress components like wings, fuselage panels, and landing gear structures, even minor undetected damage can be catastrophic under extreme operational conditions. This makes BVID highly dangerous, as it can compromise structural integrity without immediate detection since BVID can progress into more severe damage under cyclic loading, leading to unexpected failures in flight. Traditional visual inspections are ineffective, as BVID often occurs beneath the surface. FBGs can constitute an effective solution for the early warning of BVID, much more effective than other non-destructive testing techniques, such as ultrasonic C-scan because the can provide real-time data during the operation of the aircraft. The interest for the early detection of BVID is also for spacecrafts as micrometeoroid and orbital debris impacts can cause BVID in shielding and panels, jeopardizing mission success.

In monitoring aircraft composite structures, the detection and localization of impacts is of the highest importance, especially for internal sub-structures where damages are undetectable during visual inspections. Most of the FBGs applications are to measure quasi-static strain or temperature variations. In contrast, SHM of aircraft structures requires scan rate capability of hundreds to thousands of Hertz to capture higher frequency phenomena such as impact damage. Indeed, while interrogation at a rate of ~ 500 Hz can capture information of low-velocity impact events, recent works have demonstrated that update rates of 20–100 kHz allow a resolution of features that occur on transient time scales as short as 10 μs [132, 133]. A multipoint sensing network is essential not only for detecting damage occurring at different locations on a structure but also for accurately pinpointing the impact site. Indeed, by interpolating data from multiple sensors thanks to the different arrival times of mechanical waves generated by an impact event, the exact impact location can be predicted with high precision, enabling targeted inspections during routine maintenance.

Another critical aspect of SHM in aviation is the continuous assessment of aircraft engine conditions. In this case too optical-based technologies provide highly reliable, real-time monitoring of key parameters such as vibration, temperature, pressure, and rotational speed, helping to enhance engine performance, safety, and predictive maintenance strategies. It has already been mentioned that fs-written FBGs can withstand very high temperatures, such as those found in the turbines, and thus can serve as a robust sensing solution. However, they can also be complemented by other optical sensing approaches. One of the most promising techniques for temperature measurement in turbine engines is the use of thermographic phosphors [134, 135]. These rare-earth-doped ceramic materials emit light when excited by short-wavelength radiation (typically ultraviolet or blue light). The emitted fluorescence has characteristics such as lifetime, spectral shift, and brightness, which are temperature-dependent, allowing for accurate non-contact temperature sensing in extreme environments. Being ceramic materials, thermographic phosphors have demonstrated their potential for high-temperature measurement on Ceramic-Matrix Composites (CMCs) and Thermal Barrier Coatings (TBCs) used in next-generation jet engines [136–139]. Their ability to withstand extreme thermal conditions makes them ideal for monitoring critical turbine components, where conventional electronic sensors would fail due to harsh temperatures and mechanical stresses.

A third aspect of SHM is Tip Clearance (TC) monitoring of turbine blades are difficult to measure in real time. Environmental temperatures can get very high, more than 1100 °C, and space is extremely limited. A very high resolution is required, in the order of ~ 10 μm for lengths of several millimeters, and data acquisition is also a challenge due to the high velocity of the blade tip. FBGs are not suited for this application, but other optical sensing techniques respond well to these challenges, particularly intensity-based or interferometric reflective sensors [140, 141]. These methods use a multimode fiber to collect the light reflected from the blade, which depends on the distance between the fiber tip and the target.

5.2 *Temperature Sensing for Detecting Icing*

Aircraft icing poses significant risks to flight safety, as it alters aerodynamic properties, increasing drag, reducing lift, and potentially leading to premature stall. These effects require greater pilot awareness and adaptive control strategies to mitigate the impact. During flight, aircraft surfaces can reach temperatures well below −50 °C, while airborne moisture levels can be high. These conditions contribute to rapid ice formation, particularly when Supercooled Large Droplets (SLD) are present. SLD ice accretion has been a factor in numerous aviation incidents, even on aircraft equipped with Ice Protection Systems (IPS). Although modern IPS—such as thermal anti-icing systems using engine bleed air—are effective, their operation reduces engine efficiency and increases fuel consumption. Preventive activation further affects operating costs and emissions. A more efficient approach involves optimizing IPS activation using real-time ice detection systems. Advanced ice sensors provide crucial data on ice formation dynamics, allowing for smarter IPS management that enhances flight safety and energy efficiency.

Ice detection methods are categorized into direct and indirect sensing approaches. Direct sensing methods measure physical ice accretion properties, while indirect methods detect the presence of liquid water in the atmosphere. Industry standards such as SAE AS5498 and EUROCAE ED-103 outline the requirements for in-flight icing detection systems. In particular, the latter establishes sensor development protocols from design to certification, including wind tunnel and flight testing.

A promising innovation in this field is the use of FBG-based thermometers to detect ice formation through temperature variations caused by latent heat release during freezing [142–145]. FBG-based solutions offer advantages over traditional thermal sensors, including faster response times, low weight, and immunity to electromagnetic interference. However, challenges remain in quantifying Liquid Water Content (LWC) due to variable in-flight icing conditions. Nevertheless, FBG sensors have already demonstrated their potential for ice detection in aviation. As already pointed out in the preceding section, FBG sensors have the unique property of being seamlessly embedded into carbon fiber composite materials, conforming to the geometry of aircraft components to detect temperature variations and ice accretion. For example, an anti-icing system for carbon fiber wing leading edges, incorporating FBG sensors to monitor temperature variations and trigger heating when necessary was tested on flat and leading-edge profiles, confirming long-term stability over 1000 heating and cooling cycles [146]. Results demonstrated that FBG sensors accurately monitored temperature and activated the system without compromising structural integrity. A similar approach was also studied and validated during 40 flight hours on the SAFIRE ATR42 aircraft in the framework of the European funded SENS4ICE project [147].

In addition to FBG arrays, fully distributed sensing techniques, such as OFDR, have been explored for developing anti-icing systems, albeit mainly wind turbine blades [148]; nevertheless, the underlying principles are relevant for aviation applications as well.

Beyond quasi-distributed and fully distributed sensing approaches, localized intensity-based fiber optic sensors have also been investigated for ice detection. For instance, one study explored the use of fiber bundles to analyze the optical properties of ice accretion on an aircraft wing's leading edge inside an icing wind tunnel [149]. In this setup, the ice sensor illuminates the ice from within the wing through a transparent window and detects the scattered and reflected light from the micro-bubbles and ice surfaces. By analyzing the unique optical intensity signatures, which correspond to optical diffusion patterns, researchers could effectively differentiate between various types of ice forming on the wing.

5.3 Gyroscopes

Knowing the initial orientation and position of a vehicle, the integration of acceleration and angular velocity measurements allows for precise determination of its trajectory, attitude, and vertical alignment relative to the Earth's surface. This makes gyroscopes—specifically optical gyroscopes—indispensable in avionics, where they provide highly accurate angular velocity measurements for navigation, stabilization, and control. In particular, optical gyroscopes are key components of Inertial Navigation Systems (INS), as they function independently of Global Navigation Satellite Systems (GNSS). This autonomy makes them essential for aircraft, spacecraft, submarines, and military applications, where GPS-denied environments or jamming threats can compromise satellite-based navigation. Optical gyroscopes are the only way known of measuring rotation that involves fundamentally zero moving parts.

Optical gyroscopes are all based on the Sagnac effect, but can be broadly classified into three categories, each offering distinct advantages tailored to different operational requirements: Ring Laser Gyroscopes (RLGs), Fiber Optic Gyroscopes (FOGs), and Integrated Optic Gyroscopes (IOGs).

RLGs operate using two counter-propagating laser beams in a rather short laser loop cavity, measuring light frequency difference in the two directions [12, 150]. In a typical implementation, an RLG is a He–Ne laser cavity with triangular configuration, in which two of the mirrors are quasi-totally reflecting, while the output mirror has a small transmissivity. This configuration enhances the Sagnac effect since light recirculates many times around the resonant cavity. The two counterpropagating beams, picked through the output mirror, have equal wavelengths in steady condition, since the cavity length is the same in both directions.

When a rotation is applied to the system, there is a small difference of cavity lengths due to the Sagnac effect, which yields a frequency difference between the two output beams. This frequency difference is measured by combining the two output beams to get interferences. Since the beams have different frequencies, their phase difference varies, and this makes the intensity of interference intensity modulated at the beat frequency. RLG technology has reached maturity and today the ring-laser gyroscopes represent cornerstone technology, especially in aeronautics. However, still a well-known problem of RLGs is the mode locking between the counterpropagating beams, mainly related to backscattering of the mirrors, which limits the lower bound sensitivity. This phenomenon is mitigated by technological improvements such as dithering of the mirrors [151]. Other limitations of RLGs include the reliability due to gas leakage and long-term drifts.

FOGs, sort of passive forms of RLGs, utilize a much longer coiled optical fiber so that even if the footprint is small, it corresponds to a giant loop [152, 153]. FOGs can be implemented in the resonant arrangement (R-FOG) or as interferometric (I-FOG). The R-FOG is conceptually similar to the RLG, but in this case a passive ring is used, instead of an active cavity, and it is fed by an external source in both directions. The fiber can be coiled, hence the R-FOG provides compactness, but it requires a very narrow source and a corresponding long coherence length to exploits its potential. Instead of measuring the difference between resonant frequencies of the two counter-propagating beams, it is possible to exploit the setup just as a two-beams interferometer which creates a standing wave along the path. This is the principle of operation of the I-FOG. FOGs offer lighter weight, lower power consumption, higher sensitivity, and improved shock resistance compared to RLGs.

IOGs, also called Chip-Scale Gyroscopes, are miniaturized gyroscopes based on photonic integrated circuits. They enable low-cost, compact solutions for small satellites (CubeSats), autonomous systems, and wearable navigation devices.

In order to better understand the key features and performance of the different optical gyroscopes, Table 1 reports a qualitative comparison at glance.

From a quantitative point of view, the performance of optical gyroscopes are measured in terms of:

- short-term noise, measured as power spectral density in $°/\sqrt{h}$
- long-term drift, measured as standard deviation of the drift in °/h

Table 1 Comparison of optical gyroscopes in avionics

Type	Precision	Size/Weight	Reliability	Cost	Applications
I-FOG	High	Large	High	Moderate-high	Aircraft, spacecraft, INS
R-FOG	High	Compact	High	Moderate-high	UAVs, compact navigation systems
RLG	Very High	Moderate	Very high	High	Commercial/ military aircraft, spacecraft
PIC Gyroscope	High (future)	Very Small	High (future)	Low (future)	Emerging applications, next-gen systems

- scale factor, yielding information on the accuracy of measurement of high rotation rates, expressed as ppm or %

Upon these parameters, five grades of performance are classified and summarized in Table 2.

Since 2022, FOGs have surpassed one million units produced, capturing an estimated 40% of the tactical-grade market (as defined in Table 2), compared to 30% for Ring Laser Gyroscopes (RLGs). In the intermediate-grade market (0.01 °/h to 1°/h), FOGs hold approximately 30%, while RLGs remain dominant with 50%. FOGs have also begun to penetrate the high-accuracy navigation-grade market (0.001 °/h to 0.01 °/h) with a 20% share, though RLGs still lead with 65%. Conversely, in the industrial-grade segment (10°/h to 100°/h), both FOGs and RLGs hold a small share (<10%), facing stiff competition from MEMS (Micro-Electro-Mechanical Systems) gyroscopes, which offer higher compactness and lower cost. MEMS-based solutions are rapidly evolving and are particularly attractive for cost-sensitive applications.

The required sensitivity of a gyroscope in aviation depends on its specific application and aircraft type. However, for most aviation-grade gyroscopes, the sensitivity typically falls within the range of 0.01 °/s to 0.1 °/s. This level of precision is critical for accurate navigation and control, ensuring the detection of even minor changes in angular velocity. High sensitivity is especially crucial for instruments such as the attitude indicator and heading indicator, which rely on precise angular measurements to provide accurate orientation and positioning information to pilots. Advanced gyroscopic technologies, including FOGs and RLGs, continue to be the preferred choice in high-precision aerospace and defense applications, despite emerging competition from MEMS-based systems in certain segments.

The research to improve optical gyroscopes and, in particular, FOGs is very active and includes novel setups and interrogation schemes, novel fibers (such as photonic crystal fibers), and many others [154–157].

Table 2 Classification of gyroscopes by performances

	Short-term drift	Long-term drift	Scale factor accuracy
Rate grade	$> 0.5°/\sqrt{h}$	10 to 1000°/h	0.1% to 1%
Tactical grade	0.5 to $0.05°/\sqrt{h}$	1 to 10°/h	100 to 1000 ppm
Intermediate grade	0.05 to $0.005°/\sqrt{h}$	$\leq 1°/h$	10 to 100 ppm
Inertial grade	$< 0.005°/\sqrt{h}$	$< 0.001°/h$	5 ppm
Strategic grade	$< 0.01°/\sqrt{h}$	$< 0.0003°/h$	1 ppm

6 Conclusions

Optical sensors, particularly fiber optic sensors, are playing an increasingly critical role in modern avionics, offering unparalleled advantages in terms of precision, reliability, and immunity to electromagnetic interference. Their inherent lightweight nature, multiplexing capability, and ability to function in extreme environments make them indispensable for a wide range of aerospace applications, from flight control and navigation to structural health monitoring, and environmental sensing.

One of the most transformative implementations of optical sensing in aviation is the Fly-By-Light technology, which enhances flight control systems by replacing traditional electronic sensors and wiring with fiber optic sensors and optical fibers, respectively. This innovation not only improves signal integrity and reduces weight but also mitigates the risks associated with electromagnetic interference. Additionally, fiber optic gyroscopes and ring laser gyroscopes have revolutionized inertial navigation systems, providing highly accurate and reliable orientation data without reliance on external signals, a crucial feature for both aircraft and spacecraft.

Beyond navigation, optical sensors are fundamental for structural health monitoring, enabling real-time monitoring of stress, strain, temperature variations, and potential structural failures in aircraft and spacecraft components. The ability to embed fiber Bragg gratings into composite materials allows for the development of "smart structures" that continuously self-monitor their condition, reducing maintenance costs and improving safety. Alternatively, embedded fibers can be used for implementing distributed sensing approaches, such as Rayleigh, Brillouin, and Raman scattering-based techniques, further extend monitoring capabilities over large areas, proving invaluable for damage detection in lightweight composite airframes.

In addition to structural health monitoring, optical sensing technologies are transforming aerospace environmental monitoring and remote sensing. LiDAR systems provide high-resolution 3D terrain mapping and obstacle detection, while spectroscopic sensors, such as Raman and fluorescence-based systems, play a crucial role in atmospheric analysis and planetary exploration.

Despite their numerous advantages, the widespread adoption of optical sensors in avionics still faces some challenges. These include cross-sensitivity effects, high data processing requirements, and the need for cost-effective interrogation systems. However, advancements in machine learning and artificial intelligence algorithms are progressively addressing these challenges, improving data interpretation and enabling real-time decision-making for complex aerospace applications.

Looking ahead, continued research and development in optical sensing technologies will further enhance aviation safety, efficiency, and performance. As the aerospace industry moves toward increasingly autonomous and energy-efficient systems, optical sensors will remain at the forefront of innovation, shaping the future of aviation and space exploration.

References

1. R.P.G. Collinson, Fly-by-wire, in *Introduction to Avionics Systems* (Springer, Cham, 2023).
2. *Aerospace sensors market forecasts to 2030—global analysis by type, platform, connectivity, application, end user and by geography*. Stratistics Market Research Consulting (2024)
3. C. Lopatin, Opto-mechanical fiber optic sensors, in *Aerospace Applications of Optical Fiber Mechanical Sensors* (Butterworth-Heinemann, 2018), pp. 237–262
4. D. Tosi, G. Perrone, *Fiber-optic sensors for biomedical applications* (Artech House, 2018).
5. I. McKenzie, S. Ibrahim, E. Haddad, S. Abad, A. Hurni, L.K. Cheng, Fiber optic sensing in spacecraft engineering: an historical perspective from the European Space Agency. Front. Phys. **9**, 71944 (2021)
6. C. Marques, A. Leal-Júnior, S. Kumar, Multifunctional integration of optical fibers and nanomaterials for aircraft systems. Materials **16**, 1433 (2023)
7. A. Cutolo et al., Innovative photonic sensors for safety and security, part II: Aerospace and submarine applications. Sensors **23**, 2417 (2023)
8. G.P. Agrawal, Optical communication: its history and recent progress, in *Optics in Our Time* (Springer, Cham, 2016).
9. A. Venketeswaran et al., Recent advances in machine learning for fiber optic sensor applications. Adv. Intell. Syst. **4**, 2100067 (2022)
10. J.M. Rebordão, Space optical navigation techniques: an overview. Proc. SPIE **8785**, 87850J (2013)
11. D.W. Heckman, M. Baretela, Interferometric fiber optic gyro technology (IFOG). IEEE Aerosp. Electron. Syst. Mag. **15**, 23–28 (2000)
12. K.N. Chopra, Ring laser gyroscopes, in *Optoelectronic Gyroscopes. Progress in Optical Science and Photonics* (Springer Singapore, 2021).
13. H. Xu, L. Wang, Y. Zu, W. Gou, Y. Hu, Application and development of fiber optic gyroscope inertial navigation system in underground space. Sensors **23**, 5627 (2023)
14. J.A. Christian, S. Cryan, A survey of LIDAR technology and its use in spacecraft relative navigation, in *AIAA Guidance, Navigation, and Control (GNC) Conference*, AIAA 2013-4641 (2013)
15. T.P. Setterfield, R.A. Hewitt, P.-T. Chen, A.T. Espinoza, N. Trawny, A. Katake, LiDAR-inertial based navigation and mapping for precision landing, in *2021 IEEE Aerospace Conference* (2021), pp. 1–19
16. R. Di Sante, Fibre optic sensors for structural health monitoring of aircraft composite structures: recent advances and applications. Sensors **15**, 18666–18713 (2015)
17. G.C. Kahandawa et al., Use of FBG sensors for SHM in aerospace structures. Photonic Sens. **2**, 203–214 (2012)
18. H. Rocha, C. Semprimoschnig, J.P. Nunes, Sensors for process and structural health monitoring of aerospace composites: a review. Eng. Structures **237**, 112231 (2021)
19. R.O. Ogunleye, S. Rusnáková, J. Javořík, M. Žaludek, B. Kotlánová, Advanced sensors and sensing systems for structural health monitoring in aerospace composites. Adv. Eng. Mater. **26**, 2401745 (2024)
20. A.E.S. Nosseir, E.A. Slejko, A. Cervone, C.J. Oton, F. Di Pasquale, Composite structures with embedded fiber optic sensors: a smart propellant tank for future spacecraft applications. Acta Astronaut. **223**, 144–158 (2024)
21. R. Gassino, G. Perrone, A. Vallan, Temperature monitoring with fiber Bragg grating sensors in nonuniform conditions. IEEE Trans. Instrum. Meas. **69**, 1336–1343 (2020)
22. A. Beccaria et al., Temperature monitoring of tumor hyperthermal treatments with optical fibers: comparison of distributed and quasi-distributed techniques. Opt. Fiber Technol. **60**, 102340 (2020)
23. M.J. Persky, A review of spaceborne infrared Fourier transform spectrometers for remote sensing. Rev. Sci. Instrum. **66**, 4763–4797 (1995)
24. A.D. Scott, N. Rowlands, A. Bell, Miniature spectrometers for planetary remote sensing, in *Proceedings of SPIE*, vol. 5660 (2004)

25. T. Adão, J. Hruška, L. Pádua, J. Bessa, E. Peres, R. Morais, J.J. Sousa, Hyperspectral imaging: a review on UAV-based sensors, data processing and applications for agriculture and forestry. Remote Sens. **9**, 1110 (2017)
26. S.E. Qian, Overview of hyperspectral imaging remote sensing from satellites, in *Advances in Hyperspectral Image Processing Techniques*, ch.2 (John Wiley & Sons Ltd, 2022), pp. 41–66
27. S.S. Panda, M.N. Rao, P.S. Thenkabail, D. Misra, J.P. Fitzgerald, Remote sensing systems—platforms and sensors—aerial, satellite, UAV, optical, RADAR, and LiDAR, in *Remote Sensing Handbook*, vol. I (CRC Press, 2024)
28. R.K. Jha, Non-dispersive infrared gas sensing technology: a review. IEEE Sens. J. **22**, 6–15 (2022)
29. Z. Du, S. Zhang, J. Li, N. Gao, K. Tong, Mid-infrared tunable laser-based broadband fingerprint absorption spectroscopy for trace gas sensing: a review. Appl. Sci. **9**, 338 (2019)
30. A. Lambrecht, K. Schmitt, 16—mid-infrared gas-sensing systems and applications, in *Mid-infrared Optoelectronics* (Woodhead Publishing, 2020), pp. 661–715
31. J. Liu, X. Zhang, S. Zhu et al., Development and measurement of airborne oxygen sensor for aircraft inerting system based on TDLAS with pressure compensation and 2f/1f normalized method. Infrared Phys. Technol. **131**, 104689 (2023)
32. P. Vozka, G. Kilaz, A review of aviation turbine fuel chemical composition-property relations. Fuel **268**, 117391 (2020)
33. https://ocov2.jpl.nasa.gov/observatory/spacecraft/#:~:text=OCO%2D2%20is%20based%20on,light%20within%20the%20electromagnetic%20spectrum. Last accessed 14 Feb 2025
34. A.R. Jones, Light scattering in combustion, in *Light Scattering Reviews* (Springer Praxis Books, Heidelberg, 2006)
35. D. Feikema, W. Kim, Y. Sivathanu, A small angle scattering sensor system for the characterization of combustion generated particulate, in *AIAA 45th AIAA Aerospace Sciences Meeting and Exhibit*, vol. 982 (2007)
36. A. Ehn, J. Zhu, X. Li, J. Kiefer, Advanced laser-based techniques for gas-phase diagnostics in combustion and aerospace engineering. Appl. Spectrosc. **71**, 341–366 (2017)
37. X. Wang et al., Evaluation of gas and particle sensors for detecting spacecraft-relevant fire emissions. Fire Saf. J. **113**, 102977 (2020)
38. R. Zheng, S. Lu, Z. Shi, C. Li, H. Jia, S. Wang, Research on the aerosol identification method for the fire smoke detection in aircraft cargo compartment. Fire Saf. J. **130**, 103574 (2022)
39. A. Ikiades, Fiber optic ice sensor for measuring ice thickness, type and the freezing fraction on aircraft wings. Aerospace **10** (2023)
40. S. Kreger, B. Soller, D. Gifford, R. Duncan, et al., *Optical backscatter reflectometry as a measurement tool for fiber-optics in avionics and aerospace applications*, SAE Technical Papers, 2007-01-3863 (2007)
41. X. Bao, Y. Wang, "Recent advancements in Rayleigh scattering-based distributed fiber sensors," Adv Devices Instrum., vol. 2021, 2021.
42. G. Perrone, A. Vallan, A low-cost optical sensor for noncontact vibration measurements. IEEE Trans. Instrum. Meas. **58**, 1650–1656 (2009)
43. A. Vallan, M.L. Casalicchio, M. Olivero, G. Perrone, Assessment of a dual-wavelength compensation technique for displacement sensors using plastic optical fibers. IEEE Trans. Instrum. Meas. **61**, 1377–1383 (2012)
44. A. Vallan, M.L. Casalicchio, G. Perrone, Displacement and acceleration measurements in vibration tests using a fiber optic sensor. IEEE Trans. Instrum. Meas. **59**, 1389–1396 (2010)
45. R. Fernández-Bello, J. Amorebieta, J. Beloki, G. Aldabaldetreku, I. García, J. Zubia, G. Durana, Performance comparison of three fibre-based reflective optical sensors for aero engine monitorization. Sensors **19**, 2244 (2019)
46. R. Fernández-Bello, J. Amorebieta, I. García, G. Aldabaldetreku, J. Zubia, G. Durana, Review of a custom-designed optical sensing system for aero-engine applications. Int. J. Turbomach. Propuls. Power **6**, 3 (2021)
47. L. Jiao, N. Zhong, X. Zhao, S. Ma, X. Fu, D. Dong, Recent advances in fiber-optic evanescent wave sensors for monitoring organic and inorganic pollutants in water. TrAC Trends Anal. Chem. **127**, 115892 (2020)

48. M.A. Butt, Dielectric waveguide-based sensors with enhanced evanescent field: unveiling the dynamic interaction with the ambient medium for biosensing and gas-sensing applications—a review. Photonics **11**, 198 (2024)
49. V. Yesudasu, H. Shekhar Pradhan, R. Jasvanthbhai Pandya, Recent progress in surface plasmon resonance based sensors: a comprehensive review. Heliyon **7**, e06321 (2021)
50. A. Philip, A.R. Kumar, The performance enhancement of surface plasmon resonance optical sensors using nanomaterials: a review. Coord. Chem. Rev. **458**, 214424 (2022)
51. M. Wu, S. Pu, C. Zhang, W. Liu, S. Han, J. Fu, Triple-channel optical fiber surface plasmon resonance sensor for simultaneous detection of relative humidity, magnetic field, and temperature. IEEE Sens. J. **24**, 22439–22443 (2024)
52. D. Lyu, Q. Huang, X. Wu, Y. Nie, M. Yang, Optical fiber sensors for water and air quality monitoring: a review. Opt. Eng. **63**, 031004 (2023)
53. M.N. Zahid, J. Jiang, S. Rizvi, Reflectometric and interferometric fiber optic sensor's principles and applications. Front. Optoelectron. **12**, 215–226 (2019)
54. A. Miliou, In-fiber interferometric-based sensors: overview and recent advances. Photonics **8**, 265 (2021)
55. N.F. Noori, T.S. Mansour, A review of recently optical hybrid optical fiber interferometers. J. Opt. (2025)
56. S.J. Wellard, J.G. Blakeley, S. Brown, B.Y. Bartschi, E.R. Huppi, Cryogenic Michelson interferometer on the space shuttle, in *Proceedings of SPIE*, vol. 1765 (1993)
57. B.E. Hines, Optical design issues for the micro-precision interferometer testbed for space-based interferometry, in *Proceedings of SPIE*, vol. 1947 (1993)
58. M. Webb, D. Cordray, S. Cronin, G. Walker, Key elements to successful rapid prototyping and building of a Michelson interferometer for space-based sensing applications, in *Proceedings of SPIE*, vol. 5234 (2004)
59. A. Freise, S. Chelkowski, S. Hild, W. Del Pozzo, A. Perreca, A. Vecchio, Triple Michelson interferometer for a third-generation gravitational wave detector. Class. Quantum Grav. **26**, 085012 (2009)
60. A. Staley et al., Achieving resonance in the advanced LIGO gravitational-wave interferometer. Class. Quantum Grav. **31**, 245010 (2014)
61. H. Tsuda et al., Damage monitoring of carbon-fibre-reinforced plastics using Michelson interferometric fibre-optic sensors. J. Mater. Sci. Lett. **17**, 503–506 (1998)
62. H. Tsuda et al., Response of Michelson interferometric fiber-optic sensors to impact loading. J. Mater. Sci. Lett. **19**, 1–2 (2000)
63. J.S. Ibrahim, J.N. Petzing, J.R. Tyrer, Deformation analysis of aircraft wheels using a speckle shearing interferometer. Proc Inst Mech Eng Part G J Aerosp Eng **218**, 287–295 (2004)
64. G.G. Shepherd, Development of wind measurement systems for future space missions. Acta Astronaut. **115**, 206–217 (2015)
65. M.R. Islam, M.M. Ali, M.H. Lai, K.S. Lim, H. Ahmad, Chronology of Fabry-Perot interferometer fiber-optic sensors and their applications: a review. Sensors **14**, 7451–7488 (2014)
66. C. Zhu et al., Advances in fiber-optic extrinsic Fabry-Perot interferometric physical and mechanical sensors: a review. IEEE Sens. J. **23**, 6406–6426 (2023)
67. H. Xia, D. Sun, Y. Yang, F. Shen, J. Dong, T. Kobayashi, Fabry-Perot interferometer based Mie Doppler lidar for low tropospheric wind observation. Appl. Opt. **46**, 7120–7131 (2007)
68. R. Bickford, G. Madzsar, Fabry-Perot interferometer development for rocket engine plume spectroscopy, in *AIAA 26th Joint Propulsion Conference* (1990)
69. H. Qiu, F. Min, S. Zhong, X. Song, Y. Yang, Hypersonic force measurements using internal balance based on optical micromachined Fabry-Perot interferometry. Rev. Sci. Instrum. **89**, 035004 (2018)
70. S. Löhle, P. Jenniskens, High-resolution spectroscopy of the Hayabusa reentry using a Fabry-Perot interferometer. J. Spacecraft Rockets **51**, 1986–1993 (2014)
71. L. Li, L. Xia, Z. Xie, D. Liu, All-fiber Mach-Zehnder interferometers for sensing applications. Opt. Express **20**, 11109–11120 (2012)

72. S. Xiao, B. Wu, S. Ma, T. Ye, In-fiber Mach–Zehnder interferometers for sensing, in *Advanced Optical and Optoelectronic Fibers* (Springer, Singapore, 2024)
73. C. Lan, W. Zhou, Y. Xie, Detection of ultrasonic stress waves in structures using 3D shaped optic fiber based on a Mach-Zehnder interferometer. Sensors **18**, 1218 (2018)
74. X. Ren, H. Yu, X. Yao, H. Su, P. Hu, Shock boundary layer interaction and aero-optical effects in a transonic flow over hemisphere-on-cylinder turrets. Int. J. Aerosp. Eng. 3397763 (2022)
75. Y.V. Filatov et al., Experimental investigation of an optical resonator gyroscope with a Mach-Zehnder modulator and its sensitive elements. Photonics **10**, 4 (2023)
76. B. Culshaw, The optical fibre Sagnac interferometer: an overview of its principles and applications. Meas. Sci. Technol. **17** (2005)
77. B. Barrett et al., The Sagnac effect: 20 years of development in matter-wave interferometry. Comptes Rendus Physique **15**, 875–883 (2014)
78. H. C. Lefevre, Y. Bourbin, P. Graindorge, H. J. Arditty, Review of fiber optic gyroscopes, in *Proceedings of SPIE*, vol. 0369 (1983)
79. Z. Wang, G. Wang, S. Kumar, C. Marques, R. Min, X. Li, Recent advancements in resonant fiber optic gyro—a review. IEEE Sens. J. **22**, 18240–18252 (2022)
80. E. Udd, FIBER optic sensors based on the Sagnac interferometer and passive ring resonator, in *Fiber Optic Sensors* (Wiley, 2024)
81. K.-X. Sun, M.M. Fejer, E. Gustafson, R.L. Byer, Sagnac interferometer for gravitational-wave detection. Phys. Rev. Lett. **76**, 3053 (1996)
82. J.K. Sahota, N. Gupta, D. Dhawan, Fiber Bragg grating sensors for monitoring of physical parameters: a comprehensive review. Opt. Eng. **59**, 060901 (2020)
83. D. Kinet, P. Mégret, K.W. Goossen, L. Qiu, D. Heider, C. Caucheteur, Fiber Bragg grating sensors toward structural health monitoring in composite materials: challenges and solutions. Sensors **14**, 7394–7419 (2014)
84. S.J. Mihailov, Fiber Bragg grating sensors for harsh environments. Sensors **12**, 1898–1918 (2012)
85. D. Tosi, M. Olivero, G. Perrone, Dynamic strain measurement system with fiber Bragg gratings and noise mitigation techniques. Meas. Sci. Technol. **20**, 065203 (2009)
86. J. Jiang, et al., Development of optical fiber sensing instrument for aviation and aerospace application, in *Proceedings of the SPIE*, vol. 9044 (2013)
87. G. Hegde, S. Asokan, G. Hegde, Fiber Bragg grating sensors for aerospace applications: a review. ISSS J. Micro Smart Syst. **11**, 257–275 (2022)
88. F. Snik, et al., An overview of polarimetric sensing techniques and technology with applications to different research fields, in *Proceedings of the SPIE*, vol. 9099 (2014)
89. R.M. Taylor, M.J. Ranshaw, Coherence multiplexed polarimetric fibre sensor arrays for aerospace applications. Opt. Lasers Eng. **16**, 223–236 (1992)
90. W.-M. Boerner, H. Mott, E. Luneburg, Polarimetry in remote sensing: basic and applied concepts, IGARSS'97, in *1997 IEEE International Geoscience and Remote Sensing Symposium Proceedings*, pp. 1401–1403 (1997)
91. R.J. Peralta, et al., Aerosol polarimetry sensor for the glory mission, in *Proceedings of the SPIE*, vol. 6786 (2007)
92. I. Pitropakis, H. Pfeiffer, M. Weves, The use of a polarimetric fibre sensor to detect impact damage in aircraft composites, in *Emerging Technologies in Non-Destructive Testing V* (Taylor & Francis Group, London, 2012)
93. K. Bednarska et al., Hybrid fiber optic sensor systems in structural health monitoring in aircraft structures. Materials **13**, 2249 (2020)
94. B. Pedras, G. Orellana, M.N. Berberan-Santos, Luminescence-based sensors for aeronautical applications, in *Fluorescence in Industry*, Springer Series on Fluorescence, vol. 18 (Springer, Cham, 2019)
95. D. Attwood, D. Osborne, B. de lacy Costello, N. Ratcliffe, J. Lam, Fibre optic oxygen sensor for long term use in jet fuel, featuring phosphorescent dye chemically bonded to functionalised silica. Sensing Bio-Sensing Res. **26** (2019)

96. A. Orlando, F. Franceschini, C. Muscas, S. Pidkova, M. Bartoli, M. Rovere, A. Tagliaferro, A comprehensive review on Raman spectroscopy applications. Chemosensors **9**, 262 (2021)
97. J. Li, M. Zhang, Physics and applications of Raman distributed optical fiber sensing. Light Sci Appl **11**, 128 (2022)
98. K. Uckert et al., Calibration of the SHERLOC deep ultraviolet fluorescence–Raman spectrometer on the perseverance rover. Appl. Spectrosc. **75**, 763–773 (2021)
99. E.A. Lalla, et al., Raman characterization of the CanMars rover field campaign samples using the Raman laser spectrometer ExoMars simulator: implications for Mars and planetary exploration. Astrobiology **22** (2022)
100. R.S. Jakubek et al., Calibration of Raman bandwidths on the scanning habitable environments with Raman and luminescence for organics and chemicals (SHERLOC) deep ultraviolet Raman and fluorescence instrument aboard the perseverance rover. Appl. Spectrosc. **78**, 993–1008 (2024)
101. S. Yao, M. Zhang, H. Gao, Y. Zhang, J. Liu, Y. Yang, X. Wang, Development of four-channel hydrogen sensor based on stimulate Raman spectroscopy, in *Proceedings of the SPIE*, vol. 12968 (2023)
102. V. Snitka et al., Surface-enhanced Raman scattering sensors for biomedical and molecular detection applications in space. CEAS Space J. **13**, 509–520 (2021)
103. F. Maita, et al., Low cost and label free Raman sensors based on Ag-coated ZnO nanorods for monitoring astronaut's health, in *2023 IEEE 10th International Workshop on Metrology for AeroSpace (MetroAeroSpace)* (2023), pp. 363–367
104. G.G. Gimmestad, D.W. Roberts, *Lidar engineering. Introduction to basic principles* (Cambridge University Press, 2023)
105. N. Li et al., A progress review on solid-state LiDAR and nanophotonics-based LiDAR sensors. Laser Photonics Rev. **16**, 2100511 (2002)
106. P.Y. Leong, N.S. Ahmad, LiDAR-based obstacle avoidance with autonomous vehicles: a comprehensive review. IEEE Access **12**, 164248–164261 (2024)
107. J. Mao et al., The NASA Goddard CO_2 sounder Lidar: 2017 airborne campaign as a demonstration toward a future space mission, in *2021 IEEE International Geoscience and Remote Sensing Symposium IGARSS* (2021), pp. 1673–1676
108. E.J. Welton et al., The NASA micro pulse lidar network (MPLNET): early results from development of diurnal climatologies, in *2021 IEEE International Geoscience and Remote Sensing Symposium IGARSS* (2021), pp. 1432–1433
109. Z. Ren et al., Application of LiDAR survey technology in space launch sites. J. Phys.: Conf. Ser. **2364**, 012017 (2022)
110. Z. Wang, M. Menenti, Challenges and opportunities in Lidar remote sensing. Front. Remote Sens. **2**, 641723 (2021)
111. P.D. Mckee, A survey of autonomous navigation techniques applicable to Lunar surface exploration, in *47th Annual American Astronautical Society (AAS) Guidance, Navigation and Control Conference* (2025)
112. C. Pendão, I. Silva, Optical fiber sensors and sensing networks: overview of the main principles and applications. Sensors **22**, 7554 (2022)
113. V. Mizrahi, P.J. Lemaire, T. Erdogan, W.A. Reed, D.J. DiGiovanni, R.M. Atkins, Ultraviolet laser fabrication of ultrastrong optical fiber gratings and of germania-doped channel waveguides. Appl. Phys. Lett. **63**, 1727–1729 (1993)
114. D.Z. Anderson, V. Mizrahi, T. Erdogan, A.E. White, Phase-mask method for volume manufacturing of fiber phase gratings, in *Conference on Optical Fiber Communication/International Conference on Integrated Optics and Optical Fiber Communication*, vol. 4, PD16 (1993)
115. B.-O. Guan, H.-Y. Tam, X.-M. Tao, X.-Y. Dong, Highly stable fiber Bragg gratings written in hydrogen-loaded fiber. IEEE Photonics Technol. Lett. **12**, 1349–1351 (2000)
116. A. Roberge, S. Loranger, J.S. Boisvert, F. Monet, R. Kashyap, Femtosecond laser direct-writing of high quality first-order Bragg gratings with arbitrary complex apodization by phase modulation. Opt. Express **30**, 30405–30419 (2022)

117. A.G. Kuznetsov, A.V. Dostovalov, A.Y. Kokhanovskiy, S.A. Babin, Multicore fiber lasers with fs-inscribed grating arrays: recent developments and perspectives, in *2024 Photonics & Electromagnetics Research Symposium (PIERS)*, pp. 1–7 (2024)
118. X. Dong, L. Zeng, D. Chu, X. Sun, J. Duan, Enhanced high temperature properties of overexposed FBG fabricated by femtosecond laser. Measurement **195**, 111118 (2022)
119. W.-P. Huang, Coupled-mode theory for optical waveguides: an overview. J. Opt. Soc. Am. A **11**, 963–983 (1994)
120. M. McCall, On the application of coupled mode theory for modeling fiber Bragg gratings. J. Lightwave Technol. **18**, 236 (2000)
121. T.F.B. Marie, Y. Bin, H. Dezhi, A. Bowen, Principle and application state of fully distributed fiber optic vibration detection technology based on Φ-OTDR: a review. IEEE Sens. J. **21**, 16428–16442 (2021)
122. R.G. Duncan, et al., *OFDR-based distributed sensing and fault detection for single-and multi-mode avionics fiber-optics* (2009)
123. Z. Ding et al., Distributed optical fiber sensors based on optical frequency domain reflectometry: a review. Sensors **18**, 1072 (2018)
124. M. DeMerchant, A. Brown, X. Bao, T. Bremner, Structural monitoring by use of a Brillouin distributed sensor. Appl. Opt. **38**, 2755–2759 (1999)
125. L. Thévenaz, Brillouin distributed time-domain sensing in optical fibers: state of the art and perspectives. Front. Optoelectron. China **3**, 13–21 (2010)
126. X. Bao, L. Chen, Recent progress in Brillouin scattering based fiber sensors. Sensors **11**, 4152–4187 (2011)
127. Y. Muanenda, C. J. Oton, F. Di Pasquale, Application of Raman and Brillouin scattering phenomena in distributed optical fiber sensing. Front. Phys. Sec. Optics Photonics **7** (2019)
128. A. Datta, V. Sankar, B. Srinivasan, Challenges and opportunities in distributed anti-Stokes Raman thermometry. ISSS J. Micro Smart Syst. **11**, 179–206 (2022)
129. S. Goossens et al., A global assessment of barely visible impact damage for CFRP sub-components with FBG-based sensors. Compos. Struct. **272**, 114025 (2021)
130. S. Goossens, K. Muñoz, M. Jiménez, M. Mora Mendíaz, F. Berghmans, Barely visible impact damage detection and location on a real scale curved CFRP fuselage panel with optical fibre Bragg grating sensors, in *Procedia Structural Integrity*, vol. 52 (2024), pp. 647–654
131. Z. Tang, X. Gui, W. Guo, Z. Li, Impact damage monitoring of composite materials using optical fiber sensors, in *2024 Global Reliability and Prognostics and Health Management Conference* (2024)
132. W.-S. Yang, H.-D. Le, H.-C. Hsu, C.-W. Wu, C.-C. Chiang, Design and application of FBG shock sensor for high-G shock tester and a wider frequency range. Optik **311**, 171888 (2024)
133. N.R. Marinho, R. Loendersloot, J.W. Wiegman, F. Grooteman, T. Tinga, Evaluating sensor performance for impact identification in composites: a comprehensive comparison of FBGs with PZTs, in *Structural Health Monitoring* (2025)
134. D. Witkowski, D.A. Rothamer, Evaluation of 15-kHz high-speed Pr:YAG phosphor surface thermometry of a thermal barrier coating in a reciprocating IC engine. Proc. Combust. Inst. **39**, 1289–1297 (2023)
135. J.I. Eldridge, J.A. Setlock, K.N. Lee, Temperature sensing to above 1500 °C using Y2SiO5:Er luminescence lifetime thermometry, in *AIP Conference Proceedings* (2024)
136. X. Wang et al., Advances in modifications and high-temperature applications of silicon carbide ceramic matrix composites in aerospace: a focused review. J. Eur. Ceram. Soc. **41**, 4671–4688 (2021)
137. S. Dhanasekar, et al., A comprehensive study of ceramic matrix composites for space applications, in *Advances in Materials Science and Engineering*, 6160591 (2022)
138. H. Feuk, S. Nilsson, M. Richter, Upconversion phosphor thermometry for use in thermal barrier coatings. Meas. Sci. Technol. **34**, 064003 (2023)
139. W. Zhou et al., Novel strategy for thermal evaluation of film-cooled blades using thermographic phosphors at simulated engine conditions. J. Turbomach. **145**, 091009 (2023)

140. I. García et al., An optical fiber bundle sensor for tip clearance and tip timing measurements in a turbine rig. Sensors **13**, 7385–7398 (2013)
141. I. García et al., Different configurations of a reflective intensity-modulated optical sensor to avoid modal noise in tip-clearance measurements. J. Lightwave Technol. **33**, 2663–2669 (2015)
142. M. González del Val, J. Mora Nogués, P. García Gallego, M. Frövel, Icing condition predictions using FBGS. Sensors **21**, 6053 (2021)
143. M. Gonzalez, M. Frövel, Fiber Bragg grating sensors ice detection: methodologies and performance. Sens. Actuators A **346**, 113778 (2022)
144. J. Gao, X. Gui, Z. Li, N. Jiang, A distributed aircraft ice detection method based on φ-OTDR system and FBG array, in *Proceedings of SPIE*, vol. 13231 (2024)
145. X. Gui, F. Zeng, J. Gao, X. Fu, Z. Li, Detection of aircraft wing icing and de-icing by optical fiber sensing with FBG array. Measurement **247**, 116748 (2025)
146. V.R. Marrazzo et al., Study and validation of an electrical anti-ice integrated system for carbon fiber leading edge airplane wing section. IEEE Trans. Instrum. Meas. **73**, 1–9 (2024)
147. M. Gonzalez del Val, M. Frövel, Flight tests results of a fiber Bragg gratings based ice sensor. Cold Reg. Sci. Technol. **224**, 104248 (2024)
148. S. Shajiee, L.Y. Pao, P.N. Wagner, E.D. Moore, R.R. McLeod, Direct ice sensing and localized closed-loop heating for active de-icing of wind turbine blades, in *2013 American Control Conference*, pp. 634–639 (2013)
149. A. Ikiades, Fiber optic ice sensor for measuring ice thickness, type and the freezing fraction on aircraft wings, in *Aerospace* (2023)
150. M. Faucheux, D. Fayoux, J. J. Roland, The ring laser gyro. J. Optics **19** (1988)
151. S. Hu, G. Lu, Z. Fan, H. Luo, Mirrors movement-induced equivalent rotation effect in ring laser gyros. Opt. Express **21**, 14458–14465 (2013)
152. H.J. Arditty, H.C. Lefèvre, Sagnac effect in fiber gyroscopes. Opt. Lett. **6**, 401–403 (1981)
153. H.C. Lefèvre, *The fiber-optic gyroscope* (Artech House, 2022)
154. G. A. Sanders et al., Hollow-core resonator fiber optic gyroscope using nodeless anti-resonant fiber. Opt. Lett. **46** (2021)
155. S. Liu et al., Performance of a resonant fiber-optic gyroscope based on a broadband source. Appl. Opt. **61**, 4971 (2022)
156. S. Liu, L. Liu, J. Hu, Q. Liu, H. Ma, Z. He, Reduction of relative intensity noise in a broadband source-driven RFOG using a high-frequency modulation technique. Opt. Lett. **47** (2022)
157. J. Hu, S. Liu, L. Liu, H. Ma, Closed-loop resonant fiber-optic gyroscope with a broadband light source. J. Lightwave Technol. **41**(18), 6088–6093 (2023)

Optical Sensors for Stellar-Based Satellite Positioning: Precision Navigation Through Single-Star Tracking

Vaishnavi P. Bhat and S. K. Vinay

Abstract The book discusses the latest use of satellite navigation by using optical sensors, which is technology for single-star tracking applied in accurate positioning for space. Conventional satellite navigation depends on signals from terrestrial systems such as GPS; however, satellites are located further and further away from Earth's orbit, a hindrance to this navigation method. To counter this limitation, many researchers and engineers are pursuing a process of celestial-based navigation, using fixed stars for reference points. High-resolution and high-sensitivity optical sensors to faint light sources, in this scenario, offer a stable and predictable reference to satellites in the form of a particular star. Such stellar-based navigation gives precise orientation and positioning, which is of utmost importance for missions into deep space, where traditional earth-linked navigation cannot be used. This chapter addresses the principles underlying stellar-based positioning and the contribution of optical sensors towards realizing and sustaining accuracy by means of single-star tracking. This is achieved through using a recognizable star as an ever constant point of reference in a manner that the satellites are independently positioned to allow realization of operating independence and minimized Earth dependency in regard to positional updatings. The chapter will start by discussing the technical needs of stellar navigation optical sensors. Such technical needs include an elaborately designed sensitive imaging array and star-identification algorithms as well as error correction. The design of optical sensors is specifically for picking up low-light signals and capturing stars that could be millions of miles away through variable space environments where radiation and cosmic dust as well as extreme temperature changes are possible. It depends mostly on the capability of the optical sensor in recognition of the targeted star out of all possible nearby objects and performing adaptations on tracking parameters according to circumstances for its sureness assurance. Advanced algorithms are needed in doing so using star-tracking software, which relies on pattern detection and forecast models to find the target star always. The chapter also talks about the challenges of developing sensors that can withstand the unforgiving nature of space and operate with high dependability for times without upkeep. Some of the leading

V. P. Bhat · S. K. Vinay (✉)
Department of ISE, PESITM, Shivamogga, Karnataka, India
e-mail: vinaysk@pestrust.edu.in

C. Marques et al. (eds.), *Advanced Optical Sensors for Aerospace Applications*,
Progress in Optical Science and Photonics 37,
https://doi.org/10.1007/978-981-95-1626-1_8

technological developments facilitating additional success with single-star satellite navigation tracking involve photodetector sensitivity, optical component miniaturization, and improvements to computational orientation and drift correction modeling. These allow for improved precision in star-based positioning but potentially also create room for multi-star tracking where the use of two or more celestial bodies is implemented to triangulate and further optimize positional accuracy. Recent case studies of missions show how optical sensors are revolutionizing navigation for satellites operating in deep space. These case studies show how single-star tracking could be used to achieve more efficient and effective navigational autonomy so that the orbits of satellites can be adopted, optimized, and adjusted autonomously with minimal possible intervention by Earth itself. For example, there can be cases where one could reach the outer planets or interstellar space through some satellite, wherein real-time guidance from Earth might seem impossible due to an inordinate distance between two bases. The chapter ends by considering the future of stellar navigation in aerospace. Further advances in optical sensor technology, artificial intelligence, and onboard processing may further increase the accuracy and adaptability of single-star tracking. Navigation with a stellar basis will undoubtedly form the core part when advancing into unknown territories for space exploration, which in itself will form a reliable and autonomous self-orienting and self-positioning of satellites. Having both current capabilities and even future possibilities covered under the discussion, this chapter provides an exhaustive overview into the role single-star optical navigation is ready to play in the next generations of satellite missions-safe as well as efficient expansion into the cosmos.

1 Introduction

Satellite navigation has been the backbone of satellite operation and space exploration all along. Satellites have always relied on ground navigation systems such as the Global Positioning System (GPS) to remain well-positioned. However, once space missions go beyond the Earth's orbit in outer space, traditional approaches become less effective due to the decreasing power of the signal and increasing delay time. Against these limitations, satellite navigation has been a feasible option, employing stars as fixed landmarks for orientation and positioning. The topic of this chapter is the use of optical sensors in satellite navigation, as employed through the method of single-star tracking. Optical sensors designed specifically for stellar navigation can provide autonomous space positioning with assured high precision even without signal availability from the Earth. A single star, being tracked and monitored as a reference point, enables satellites to maintain stable navigation without constant ground-based adjustments. This technique is especially important in deep-space missions, where Earth-based real-time guidance is impossible (Fig. 1).

The advent of optical sensors with high sensitivity, powerful star-identification algorithms, and accurate tracking procedures has greatly facilitated the efficiency of stellar navigation. These sensors must operate efficiently in harsh conditions of space

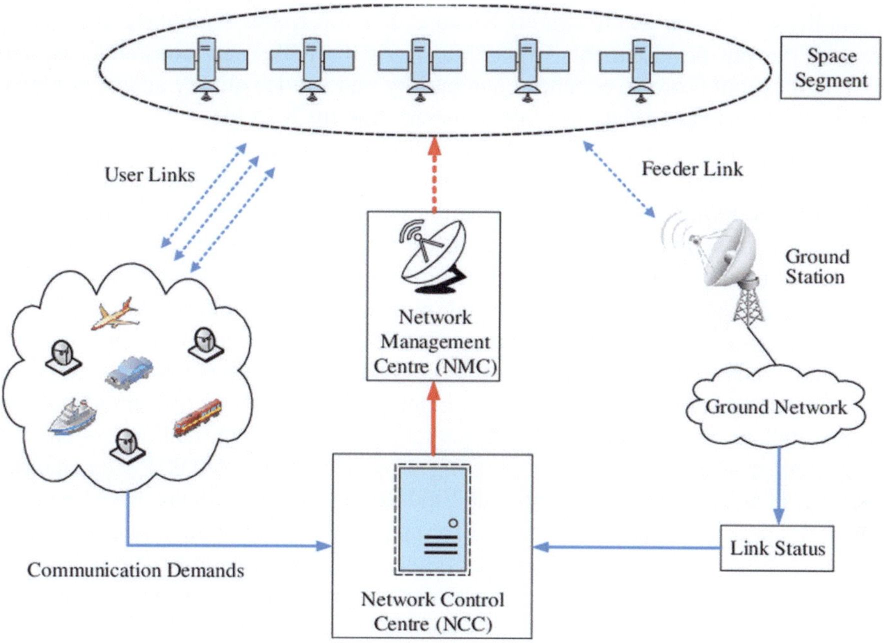

Fig. 1 Diagram of a satellite communication system architecture [1]

environment, such as extremes of temperature, cosmic radiation exposure, and lack of visibility due to dust in space. In a counter to such conditions, scientists and designers came up with strong optical mechanisms that can distinguish targets from background images, detect weak light signals, and continuously adjust tracking parameters. This chapter discusses the science behind single-star tracking, the technological breakthrough that gave rise to its application, and the technical difficulties in designing optical sensors for long-duration spaceflight. It also describes how advances over the last few years—such as novel photodetectors, sub-millimeter-sized optical elements, and predictive models based on machine learning propel autonomous satellite navigation forward. Through the medium of successful flight case histories, this assessment locates growing stellar-based contribution to the reality of autonomous space operation, opening eventually the gateway to more visionary deep-space exploration missions.

Satellite navigation has also been essential for aerospace use, and it has developed dramatically over the last 30 years [2]. Global Navigation Satellite Systems (GNSS) cover different constellations, augmentations, and user equipment [3]. The systems have grown to provide more complex flight phases, such as vertical guidance down to 200 ft [4]. Precision and position accuracy needs create challenges in aerospace applications. Augmentation systems, such as aircraft-based, satellite-based, and ground-based systems, have come together to address these issues [3]. Coupling GNSS with inertial navigation systems (INS) and other sensors has also increased positioning capability [5]. As far as constraints because of radio-frequency interference, there are still current developments in satellite navigation like new constellations and civil signals that promise enhanced availability and service quality and even global coverage of vertical guidance for air transportation by 2025 [4].

Optical sensors are important for satellite navigation and positioning. Space optical sensors can serve both space object observation and satellite navigation [6]. Efficient scheduling of sensors and the employment of multiple sensors with perpendicular lines of sight are advisable to enhance navigation accuracy [6]. Atmospheric refraction impinges on geometric precision in optical satellite positioning, but this may be corrected using calibration based on measured atmospheric parameters and iterative processing [7]. For CubeSats, modulated retroreflector laser links can be designed for high-speed downlinks, taking channel parameters into consideration and suggesting sensing and positioning algorithms [8]. Optical measurements can further enhance the accuracy of satellite tracking by correcting tracking system and orbit prediction errors to achieve pointing error accuracies of 0.001° and satellite tracking accuracies of 0.1–0.2 mrad at 1000 km altitude [9]. Stellar-based navigation and single-star tracking are useful for spacecraft positioning and space debris detection. High-performance star trackers enable autonomous onboard orbit determination without a priori knowledge, which improves deep space mission efficiency and reliability [10]. Single-star tracking can enhance satellite tracking accuracy through system error correction and orbit prediction, with accuracies of 0.001° and 0.1–0.2 mrad for satellites at 1000 km altitude [9]. Space debris detection can be done using star sensors with the aid of autonomous recognition algorithms and local sky reference star image comparison to separate faint, small targets from background stars [11]. Joint estimation of stellar atmospheric refraction and star tracker attitude has been suggested to enhance attitude determination inside the atmosphere with an 80.06% reduction in average reprojection error over conventional approaches [12]. These developments underscore the importance of stellar-based positioning and single-star tracking in contemporary space applications.

2 Principles of Stellar-Based Positioning

Celestial navigation has a proven history and is still essential in life today. Earlier human cultures, including Aboriginal Australians, had a sophisticated understanding of astronomy to be used for navigation, time-keeping, and writing about natural phenomena [13]. This folk knowledge typically existed in oral histories and songlines for the learning of navigational methods. Invention of wood navigational devices during early modern Western civilization (1590–1731) was a serious breakthrough in methods of celestial navigation [14]. Though new technology, like GPS, has come to dominate conventional techniques [15], celestial navigation is still an ongoing field of research. Indeed, fruit flies, Drosophila melanogaster, have been shown in recent times to employ celestial cues like the location of the sun and polarized

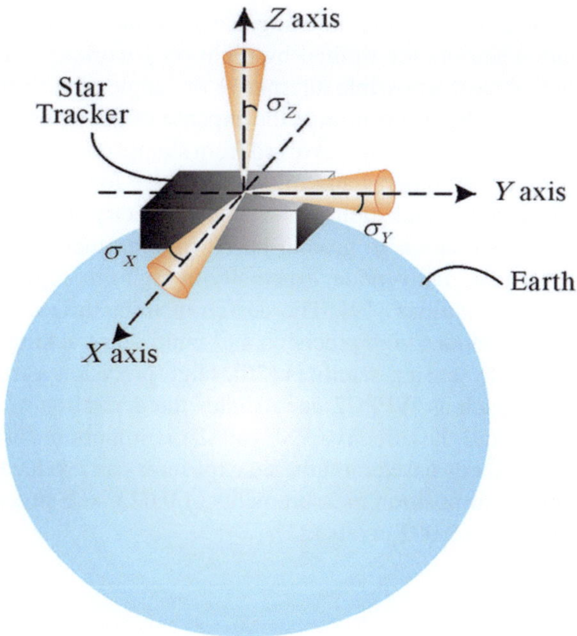

Fig. 2 Pointing and rolling accuracy schematic of the star tracker [17]

light patterns for far-distance navigation, suggesting a shared, ancient capability for celestial navigation in insects [16] (Fig. 2).

Star trackers have evolved from simple pointing devices to sophisticated devices using CCD cameras and microcomputers for accurate attitude determination [18]. Single-mode and multi-mode interferometry techniques have been compared for phase estimation in ground-based interferometry, where single-mode is typically better than multi-mode, especially in high flux regimes and compromised adaptive optics conditions [19]. The design of miniaturized star trackers, such as the HD-1003, provides high precision and multi-star tracking in a small form factor for use in remote sensing satellites [20]. High-precision astrometry with the use of instruments such as WFPC2 necessitates the derivation of a good point-spread function (PSF), and the effective PSF (ePSF) is important for undersampled images. Iterative methods based on dithered exposures can produce very precise ePSFs, allowing single-star position measurements of 0.02 pixels precision and systematic accuracy better than 0.002 pixels [21].

Optical sensors are utilized extensively in star navigation and measuring spacecraft attitude. The most accurate attitude sensor is the star tracker, and it can report local coordinates on the Earth surface as well as 3-axis attitude [22]. Modern image processing methods enable concurrent extraction of stellar and celestial body features with precision, navigation and relative position computation being an easy affair [23]. They opt for the use of internal stellar catalogs for stellar field detection, with computation software to provide computation and photovisualization of the catalog and sensors [24]. The coupling of optical sensors with inertial measurement units is capable of increasing tracking accuracy and efficiency [25]. Although optical navigation sensors are limited by night-only working and requirement for a clear sky, they present a possible alternative or supplement to standard GPS systems that can provide reliable position at the expense of not using satellite data or ground stations [22]. Star trackers have evolved from simple pointing devices to sophisticated devices using CCD cameras and microcomputers for accurate attitude determination [18]. Single-mode and multi-mode interferometry techniques have been compared for phase estimation in ground-based interferometry, where single-mode is typically better than multi-mode, especially in high flux regimes and compromised adaptive optics conditions [19]. The design of miniaturized star trackers, such as the HD-1003, provides high precision and multi-star tracking in a small form factor for use in remote sensing satellites [20]. High-precision astrometry with the use of instruments such as WFPC2 necessitates the derivation of a good point-spread function (PSF), and the effective PSF (ePSF) is important for undersampled images. Iterative methods based on dithered exposures can produce very precise ePSFs, allowing single-star position measurements of 0.02 pixels precision and systematic accuracy better than 0.002 pixels [21].

3 Fundamentals of Optical Sensors for Stellar Tracking

Charge-coupled devices (CCDs) have been the prevalent imaging sensors in astronomy since the 1980s due to their high performance and sensitivity over a wide range of astronomical uses [26]. Nevertheless, complementary metal–oxide–semiconductor (CMOS) sensors are increasingly encroaching on the territory of CCDs in some areas with their low power consumption, price, and on-chip integration capability [27]. CMOS sensors also hold promise in star trackers with the introduction of new techniques such as global-shutter centroiding and adaptive region-of-interest readout that increase signal-to-noise ratio and centroiding accuracy [28]. Even after the introduction of readout noise, high quantum efficiency of the latest CCDs can still qualify them for usage in fringe tracking in optical stellar interferometry [29]. With advancing CMOS technology, customized pixel architecture and manufacturing process are filling up the performance shortfall vis-a-vis CCDs and paving the way for increasing involvement of CMOS sensors in high-end imaging tasks [27] (Fig. 3).

Precise sensor calibration, resolution, and sensitivity are essential in stellar position application. A least squares algorithm is designed for determining satellite attitude, which opened the way for current star tracker algorithms [31]. A thorough exposition of star tracker performance was presented with emphasis on the importance of accurate attitude determination [32]. An all-parameter calibration technique for stellar-inertial navigation systems was presented that estimated biases, scale factors, and inertial measurement unit misalignments as well as star sensor misalignments [33]. The method was able to significantly enhance navigation accuracy in comparison to the conventional approach. A maximum-likelihood sensor array calibration algorithm was presented that corrected unknown gain, phase, and mutual coupling

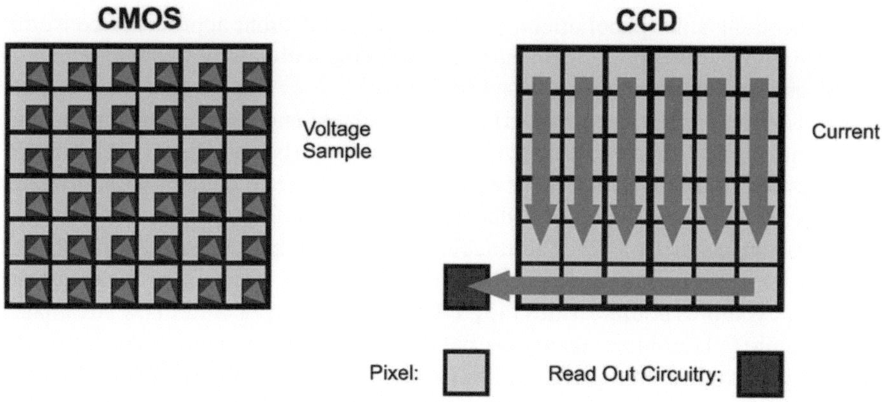

Fig. 3 The comparison of CCD and CMOS sensors [30]

and sensor position errors [34]. These papers together emphasize the significance of accurate calibration methods and error compensation for attaining high-performance stellar positioning systems.

Optical sensors are very essential in spacecraft attitude determination and star tracking. These sensors can perceive different kinds of radiation such as solar, albedo, infrared, and star [35]. In stellar interferometry, noisy sensors such as CCDs are quite appropriate for use in correlation measurement and group delay tracking because they have high quantum efficiency [29]. Low-cost star trackers employing linear optical sensor arrays in conjunction with inertial measurement units have exhibited potential for spatial object tracking at high speeds and high accuracies [25]. Latest advancements in stellar navigation (StarNAV) leverage optical cameras of large field of views to provide measurements of inter-star angles such that velocity is estimated with less precision but with many more measurements. This method has proven to be practicable for utilization in small satellite con8gurations, i.e., 3U CubeSats [36].

4 Design and Functionality of Single-Star Tracking Systems

Design for single-star tracking systems entails a number of approaches and considerations. Independent observatories, such as the Hungarian Automated Telescope (HAT), confirm the viability of automated star tracking for variability surveys and possible exoplanet detection [37]. Solar tracking systems, with shared concepts, can possibly increase photovoltaic panel performance, with two-axis tracking best but more expensive than single-axis systems [38]. High-precision satellite laser ranging receivers such as SLR2000 utilize high-precision tracking capabilities and automatic operation, proving the viability of unbroken round-the-clock 24-h coverage [39]. New ideas, such as the origami-based pneumatic solar tracker for picosatellites, offer light, compact solutions for space missions that utilize bio-inspired mechanisms to provide accurate solar panel orientation [40]. These different approaches constitute the ongoing progress and refinement of star tracking technology in a wide range of distinct astrophysical and space-based projects.

Photosensor oculography (PSOG) is of high spatial and temporal resolution, and sensor tracking and alignment accuracy are key factors to be considered [41]. LiDAR-based systems are also being suggested as an alternative to camera-based systems for object tracking and detection due to the fact that they provide 3D information for better accuracy [42]. The tracking algorithms must simulate a number of sources of errors, including sensor bias, environmental effects, and measurement and target dynamics nonlinearities [43]. PTZ camera networks offer new advantages with dynamic field of view adaptation, but difficult PTZ parameter control and inter-camera synchronization are needed [44]. With such technologies, key parameters for accurate navigation are sensor alignment, focal length, tracking accuracy, and the capacity to adjust to changing environmental conditions and target behaviors (Fig. 4).

Optical system design faces many challenges that need to be resolved in order to provide high performance and reliability. Lens distortion is one of the major concerns that can produce aberrations and compromised image quality in imaging systems. This can be resolved through advanced correction mechanisms, such as aspheric lens elements and computational approaches. Another major challenge is thermal stability, since temperature changes can lead to refractive index variations

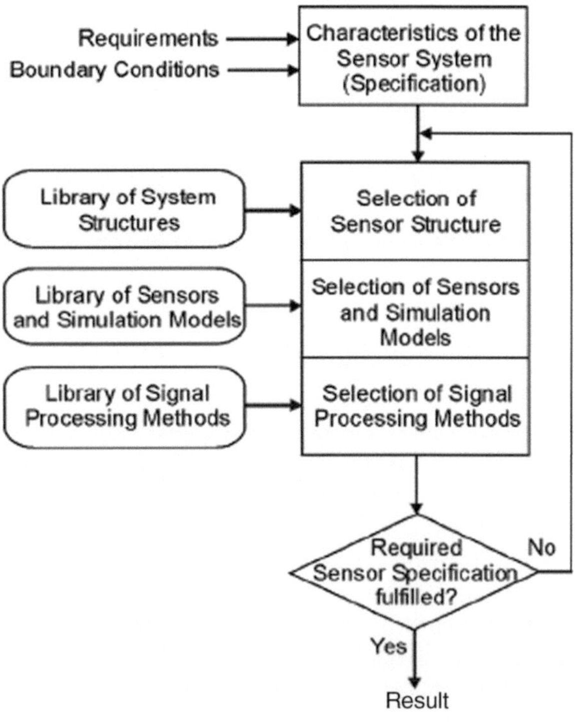

Fig. 4 Design process of sensors and sensor systems [45]

and physical deformation of optical elements. Precision glass molding (PGM) has been recognized as a highly effective method of producing high-quality lenses, especially for high-volume production. This process is strongly sensitive to temperature changes, which can lead to residual stresses and shape distortions and hence impact optical performance and durability [46]. Over the past few years, metalenses have attracted significant interest because of their ability to transcend conventional constraints in optical design. Unlike traditional lenses, which depend on curvature to refract light, metalenses use nanostructures to obtain precise wavefront control. One of the strengths of metalenses is that they can correct chromatic aberration, a typical problem in optical systems. New advances in dispersion engineering approaches and inverse design methods have greatly enhanced their performance and efficiency, positioning them as an auspicious substitute to conventional optics [47]. Thermal stability is especially important in high-power lasers, where even slight thermal fluctuations can cause serious focal shifts and deterioration in performance. To prevent such effects, researchers have used passive compensation methods, including negative dn/dT materials. Through precise optical material selection with reversing thermal expansion behavior, focal shifts are minimized by up to 60%, which provides stable optical performance under fluctuating thermal conditions [48]. The second critical area of optical system design is vibration control, especially in dynamic environment-based systems such as automotive and aerospace systems. Vibration can lead to misalignment, phase errors, and reduced system accuracy. To counteract these problems, MEMS vibratory gyroscopes have been developed with higher robustness to structural and environmental changes. By operating in a stable sense frequency region, the gyroscopes achieve improved bias stability and temperature resistance, and consequently, are used in demanding applications [49].

5 Advancements in Optical Sensing for Precision Navigation

5.1 Recent Developments in Optical Sensor Technology

Recent technological improvements in optical sensor technology have led to unprecedented gains in sensitivity, miniaturization, and processing speed. These improvements have made possible many applications' migration to new levels of performance, ranging from biomedical diagnostics to environmental monitoring and security networks. Progress in laser technology has especially been instrumental in optimizing the functionality of optical sensors by offering higher switching rates, improved eye safety regulation compliance, and higher power outputs. Of the emerging sources of light, photonic crystal surface-emitting lasers (PCSELs) have been extremely promising since they are capable of delivering high-power emission with excellent beam quality, which will be particularly well suited for next-generation optical sensing applications [47].

By far the greatest progress in the field has been the development of quantum biosensing, which dramatically increased sensor sensitivity by exploiting the quantum characteristics of entanglement and superposition. These developments made it possible to create optical biosensors that have the ability to detect biomolecules, pathogens, and toxins selectively at high levels in real time. Filling in the gap between measuring extremely low concentrations accurately, quantum biosensing is now an important tool in biochemical research and medicine diagnosis [50].

In addition, label-free optical micro- and nanosensors are now a cutting-edge technology that can detect single molecules very accurately. The sensors circumvent the use of fluorescent or radioactivity-based labeling and, as an alternative, minimize complexity while increasing detection efficiency. They are used differently in different fields, such as medicine, environmental monitoring, and security, where ultra-sensitivity and real-time monitoring are the main priorities [51].

The quick evolution in high-speed electronic circuits, new signal processing methods, and advanced fabrication techniques allowed for the progressive enhancement of optical sensor systems. These advancements have guaranteed greater quality, reliability, and cost-effectiveness, which has made the utilization of optical sensors more beneficial and economically justifiable in most applications [52]. With the evolution of nanotechnology, there has also been progressive enhancement in the optical sensing capability. Technology such as metasurface has brought about higher beam deflection, larger field-of-view, and dynamic modulation, providing unparalleled control of light-matter interaction. This has opened up new possibilities to investigate ultra-miniature high-performance optical sensors [47]. The quick advancement in high-speed electronic circuits, new signal processing technologies, and advanced fabrication technologies made progressive advancement of optical sensor systems possible. These have provided quality, reliability, and cost-effectiveness, thus making the utilization of optical sensors more rewarding and economically viable in the majority of instances [52]. With advancements in nanotechnology, progressive advancement in the optical sensing ability has been achieved as well. Technology such as metasurface has provided greater beam deflection, greater field-of-view, and dynamic modulation, making unprecedented light-matter interaction control possible. This has opened up new avenues to explore ultra-miniature high-performance optical sensors [47].

5.2 *Role of Sensor Networks and Multi-sensor Fusion*

Multisensor fusion and integration (MFI) is fast-evolving technology, which is significantly contributing to enhanced perception and decision-making across a broad range of applications. Through the integration of data from multiple sensors, MFI avoids the limitations present in each separate sensing system to deliver more accurate, consistent, and reliable output. The method is particularly beneficial in applications where one sensor might be insufficient due to noise, occlusion, or environmental interference [53].

The most significant benefit of MFI is that it combines redundant and complementary data, thus increasing measurement accuracy and reducing uncertainty. Through intelligent data fusion techniques, the system is able to deliver a more accurate description of the observed environment, with high applicability to safety–critical applications. The fusion occurs at multiple levels, such as estimation, classification, and inference, in order to enhance decision-making and enhanced performance [54].

The multi-functionality of MFI has been applied to a broad variety of applications that cut across industrial control, robotics, defense, biomedicine, and transport. In robotic vision, for example, MFI is applied to integrate vision, LiDAR, and inertial sensors to enhance navigation and object recognition. In industrial automation, fusion techniques optimize process control and fault detection in order to optimize efficiency and accuracy. Military systems utilize MFI for battlefield surveillance and target detection, where multiple sensing modalities enhance situational awareness [53, 55] (Fig. 5).

Target tracking is one of the most important uses of MFI where computationally intelligent methods are in demand to develop quality tracking algorithms. With the measurements of multiple sensors, the systems achieve higher accuracy in monitoring moving targets even in changing and adversarial environments. This is specifically advantageous in defense, security, and autonomous navigation of vehicles, for which the success of a mission relies to a great extent on effective tracking [57].

Research in MFI will in the future be aimed at microsensor design, smart sensor design, and adaptive fusion algorithms. These advancements will further improve the performance, efficiency, and effectiveness of multisensor systems, introducing new possibilities of next-generation autonomous systems, medical diagnosis, and intelligent transport networks [53].

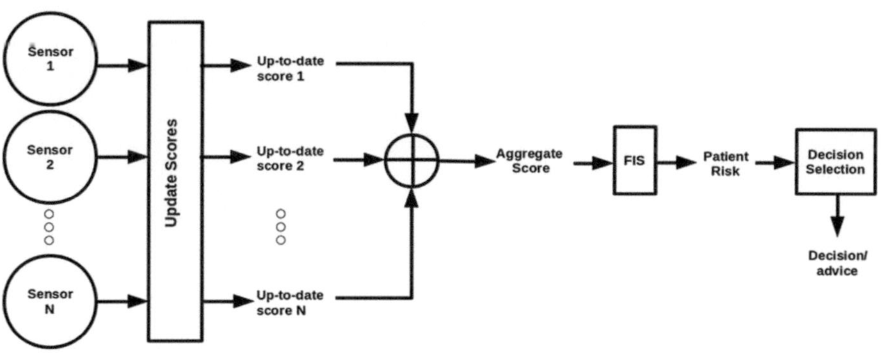

Fig. 5 Architecture of the multi-sensor data fusion model [56]

5.3 *Case Studies of Advanced Optical Sensors*

Advanced optical sensors are crucial in aerospace, where reliability, precision, and longevity are of paramount concern. The sensors find extensive applications in aerospace in structural health monitoring, navigation, and imaging and offer aerospace system safety and efficiency. Structural health monitoring is required in modern aircraft and spacecraft due to the fact that it is possible to monitor stress, fatigue, and damage in composite material. This facilitates predictive maintenance and reduces the possibility of catastrophic failure. The piezoelectric and fiber-optic sensors are emerging as leading technologies in this area, with characteristics of greater sensitivity, resistance to harsh environments, and multiplexing ability that allows multiple sensing points along one optical fiber [58].

Aside from structural health monitoring, emerging sensor technologies are being developed to revolutionize the performance of aerospace engines. Silicon carbide sensors already showed significant promise to be applied in harsh environments, especially in high-temperature propulsion systems. The sensors are very resistant to extremely high pressure and high temperature, thus making them perfect for engine monitoring and flight performance improvement. Their use has the potential to increase fuel efficiency and emissions by allowing real-time data for adaptive control systems [59] (Fig. 6).

Another future development in aerospace technologies is the manufacture of smart structures, which combine actuators with sensors to offer enhanced functionality. The structures are made to dynamically move in accordance with changing states, thus producing real-time adjustments of mechanical as well as optical types. Integrating highly integrated networks of sensors, smart structures offer active vibrational damping, shape morphing, and precise control of the aerodynamic surface, thus offering enhanced overall functioning and responsiveness to aerospace systems [60]. There are sophisticated sensors that form the most critical element of scientific observations and deep-space exploration research in space exploration. Space telescopes, for instance, depend on highly advanced imaging and signal-processing

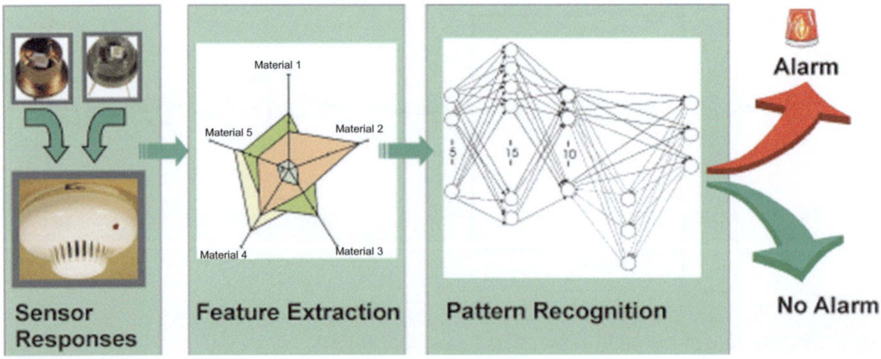

Fig. 6 Future trends in sensor technology [45]

systems for the acquisition of high-resolution astronomical data. The James Webb Space Telescope, which is perhaps the most advanced observatory constructed, utilizes near-infrared detectors and control ASICs to provide high-precision analog-to-digital conversion and data transmission for deep-space exploration. These state-of-the-art components provide high-resolution imaging performance, hence enabling scientists to see the distant-off galaxies, exoplanets, and cosmic phenomena with unprecedented detail [61]. Advancements in optical sensor technology are continually raising the bar in terms of innovation in the aerospace industry, thereby enhancing performance, reliability, and safety. Regardless of whether the application is in structural monitoring, propulsion, smart structures, or space exploration, the innovation is significantly improving stellar positioning accuracy, image performance, and aerospace operations optimization.

6 Data Processing and Error Mitigation Techniques

6.1 Star Detection and Image Processing

Treatment of astronomical images is a basic aspect of obtaining meaningful information from observational data sets; however, it is prone to a sequence of challenges that need to be alleviated to offer precision and reliability. Among these, impulsive noise is a significant problem that may undermine image integrity and hide dim celestial objects. Noise generation is caused by numerous causes and encompasses thermal generation, encounter with cosmic rays, and high readout voltages of imaging detectors. Depending on the consequences, scientists have developed a wide range of filtering algorithms based on the type of astronomical imagery. The performance of the algorithms relies on several parameters, such as stellar profile widths, signal-to-noise ratios, and the stochastic nature of noise in the data set [62].

Aside from noise suppression, source detection is a critical aspect of astronomical image analysis, especially when working with large data sets and weak celestial object sources. Automatic detection algorithms employ a variety of image transformation techniques, as well as detection paradigms, to efficiently separate sources from noise. Such processes allow for the processing of vast data sets gathered from terrestrial as well as extraterrestrial observatories, thereby improving the ability to detect and analyze distant galaxies, stars, and other celestial bodies [63] (Fig. 7).

To maximize data management and operational efficiency, image compression algorithms are used to reduce storage and transmission needs without compromising significant scientific information. Multiresolution methods, including the pyramidal median transform, have been combined with noise reduction algorithms to maintain the photometric and astrometric quality of astronomy images. Through the application of such compression algorithms, scientists are able to maximize data handling processes without compromising significant observational data [65].

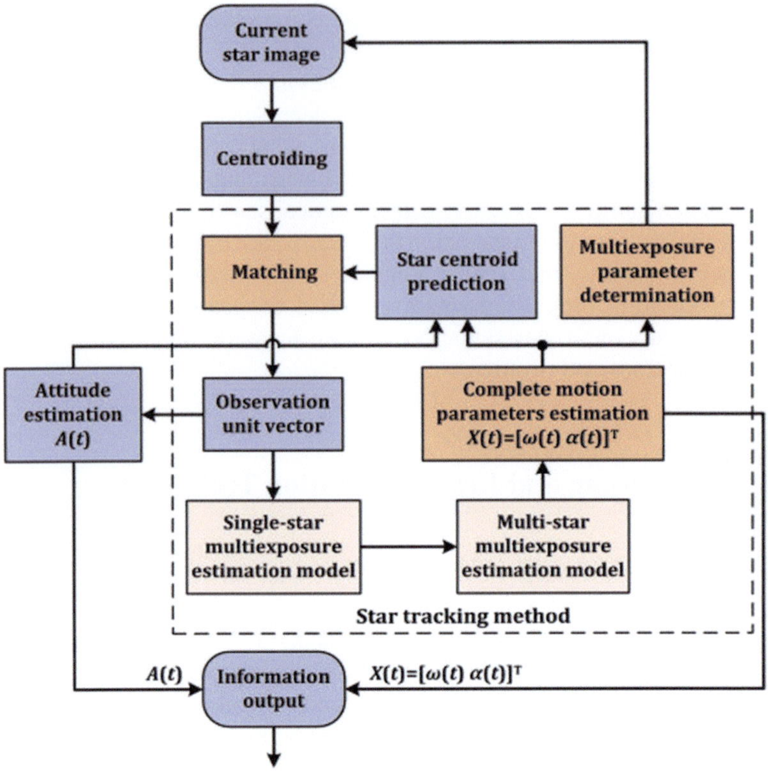

Fig. 7 Flow chart of the star tracking method [64]

Deconvolution is a key aspect of astronomical image processing, particularly in image restoration. The addition of noise, however, makes the deconvolution process complicated and requires special attention in instrumental noise models and the nature of the studied objects in order to provide correct reconstructions. Recent developments in deconvolution methods have seen multiresolution techniques, which improve signal recovery and image acuity. These developments have greatly improved the resolution of one detail in astronomical observations and thus benefit ground-based and space-based telescopes [66]. In general, ongoing development in noise altering, source detection, compression, and deconvolution methods plays a central role in the evolution of astronomical image processing. Through these developments, astronomers are able to retrieve data with high accuracy, enhance the accuracy of observations, and push the frontiers of cosmic research.

6.2 Error Correction and Filtering Techniques

Kalman filtering is a common error correction technique, which has been used in the application of accurate positioning and tracking in many fields. By sensor data fusion and mathematical modeling, the Kalman filter efficiently estimates system parameters, and in the process, removes uncertainties and improves dynamic system reliability.

The method works best in fields like navigation, robotics, and autonomous systems where real-time state estimation is required to ensure maximum performance [67]

Kalman filtering has evolved over the recent years with the evolution of methods that utilize multibody models, thus improving the accuracy of the estimation by adding physical system dynamics to the estimation process. Such models give direct simulation-to-real behavior connections, which enable more accurate monitoring of mechanical systems with complex dynamics. With such methodologies, scientists enhanced motion prediction and control in various fields, for instance, vehicle dynamics and biomechanics [68].

Apart from its common application in positioning, Kalman filtering has been greatly modified for employment in indoor localizations, wherein GPS signals tend to be undependable. For wireless sensor network positioning, new tuning methods to the measurement covariance matrix have been used, rendering the filter maximally sensitive to environmental changes. The inclusion of inertial measurements stabilizes tracking, thus providing enhanced positioning accuracy in sub-meter ranges. Such developments have significantly enhanced indoor localization, thus broadening the scope of applications to include smart buildings, healthcare monitoring, and industrial automation [69].

Kalman filtering continues to be at the core of positioning applications; however, error correction methods are also applied in other areas, notably in computational biology. In the application to the area of next-generation sequencing (NGS), the precise reading of data is dependent on sequencing error correction, which can be caused by events such as polymerase misincorporations and signal noise. Several error-correction frameworks have been used to enhance sequencing data quality, with academic interest in the discovery of standardized tests for comparing studies. These early efforts are directed towards enhancing sequencing accuracy and to identify viable strategies towards future genomic research progress [70]. Generally, the ongoing development of Kalman filtering among other error-correction frameworks is advancing a variety of applications, from navigation and robotics to genomics and data science. Advances foster accuracy, reliability, and efficiency and therefore form essential parts of existing technology.

6.3 Thermal and Environmental Compensation

Optical sensors are used in a broad variety of applications, many of which subject them to extreme environmental conditions that affect their performance and lifespan. These adverse environments—anything from high temperature and cryogenic environments to radiation-laden environments—require the use of sophisticated compensation techniques to ensure accuracy and reliability. In temperature applications at high temperatures, degradation of the sensor can be prevented by proper material selection, doping with rare earth elements, and the use of annealing processes to improve stability and thermal stress resistance. These processes preserve optical properties and structural integrity with exposure to long-term heat [71].

In cryogenic applications, the use of special fiber materials and processes is necessary to avoid mechanical brittleness and signal loss. Severe low temperatures have a tendency to shift refractive indexes and cause microcracking of optical fibers and thus require engineered materials to operate under such stresses without degradation [71]. Furthermore, areas with high irradiation levels in space missions as well as atomic power plants call for high-grade control of both fiber content as well as processes used for preparing them. Radiation-induced attenuation can reduce signal transmission through optical fibers, requiring radiation-hardened materials and coatings to extend sensor life [71].

The thermal lens effect, where temperature variations induce refractive index changes in optical components, leading to beam distortion, is another serious problem in the operation of optical sensors. Passive compensation methods, such as the utilization of materials with a negative thermo-optic coefficient (dn/dT), have been shown to counteract these effects. Experiments have shown that such materials can decrease focal shifts by as much as 60%, thus enhancing system stability in high-power optical systems [48].

Operating and environmental conditions, for example, mechanical loading, temperature changes, and humidity, influence the precision and reliability of Lamb wave-based structural health monitoring systems as well. Changes in signal transmission with environmental conditions can impact such systems used for crack detection and damage to a structure. Various compensatory strategies and baseline-free methods have been developed by researchers to address such problems and provide more stable measurement while limiting environmental effects on system performance [72]. In chemical optosensing, selectivity and sensitivity are a top priority for effective contaminant quantification. Various flow methods and pre-treatment procedures for samples have been employed to ensure maximum detection capacity, such that the sensors record accurately even under complicated chemical matrices.

These developments enable monitoring of pollutants, toxins, and other harmful materials in real time, which enhances environmental security and process control in industries [73]. Generally, compensation technique advancement and material science push the limits of optical sensor performance in hostile environments. Such innovations make optical sensors stronger, precise, and flexible, ensuring their usability in mission-critical applications like aerospace, medicine, environment sensing, and automation.

7 Integration of AI for Enhanced Single-Star Tracking

7.1 Machine Learning for Enhanced Positioning Accuracy

Machine learning (ML) techniques are rapidly becoming relevant for enhancing positioning accuracy in various indoor and outdoor locationing systems. Traditional positioning methods are usually founded on pre-de8ned models and tedious manual calibration, but ML has the potential to offer a data-driven approach that can improve accuracy, learn in different environments, and reduce reliance on human inputs. These benefits have spurred tremendous research and development in ML-based localization techniques for use in smart buildings, autonomous navigation, and location-based services. One of the most important areas where ML has demonstrated immense potential is visible light positioning (VLP), a method that uses LED light sources for indoor localization, especially in Internet of Things (IoT) use cases. Through the use of ML models, scientists have improved 2D and 3D position estimation accuracy, enabling more stable tracking in GPS denied or impossible environments. This enhancement has specific applications in smart homes. Such innovations make optical sensors stronger, precise, and flexible, ensuring their usability in mission-critical applications like aerospace, medicine, environment sensing, and automation. By using ML algorithms, scientists have improved sensor design and created advanced error modeling methods that compensate for drift as well as accumulation of bias with time. Also, ML-based multi-sensor fusion algorithms enhance the robustness of INS, which results in enhanced positioning performance in dynamic conditions like autonomous vehicles and aerospace navigation [74]. For pedestrian localization, ML methods have been used to minimize performance measures other than accuracy. These have served to reduce response times, improve energy efficiency, and enable real-time adaptive localization, thus making them extremely suitable for mobile applications such as navigation assistance, fitness tracking, and emergency response systems. While supervised learning continues to be the prevailing strategy within this field, researchers are now investigating the applications of unsupervised and reinforcement learning strategies that may maximize adaptability while lessening the reliance on an excessive amount of annotated training data [75]. While the many benefits that ML adds to positioning systems are undeniable, various challenges still persist. High computational complexity in ML models may result in high computational resource requirements, which render real-time deployment challenging in resource-limited settings. Moreover, the long training and fine-tuning time for ML models may also act as a bottleneck, especially where large-scale deployment is involved. Progress in overcoming these challenges will necessitate improvements in model optimization, transfer learning, and compact ML frameworks specifically designed for real-time localization [76]. In total, ML-based localization methods are still transforming positioning systems with increased accuracy, flexibility, and efficiency. With advancing research, the incorporation of new ML methods, including self-learning and adaptive models, is likely to further improve localization performance in various applications, ranging from indoor IoT settings to massive-scale autonomous navigation systems.

7.2 *Optimization of Data Processing Through AI*

Artificial intelligence (AI) is revolutionizing the processing and analysis of images in a myriad of applications, especially with remarkable results in medical imaging. AI-based techniques with deep learning models have demonstrated unimaginable potential for image denoising, enhancement, segmentation, and classification. These machine learning techniques consistently outperform traditional image processing techniques through the following capability to automatically learn patterns and characteristics from massive data sets without lowering efficiency or precision, reducing the amount of human work required [77]. In ophthalmology, AI is also contributing significantly to better diagnostic performance and efficient clinical procedures. Artificial intelligence software has been designed to assist in screening, staging, and treatment planning for eye diseases, thus improving quality eye care and improving patient outcomes in general. Computer-based analysis of the retinal photograph has been especially of extreme importance in the diagnosis of early disease in conditions like diabetic retinopathy and age-related macular degeneration, where early treatment is essential to avoid visual impairment. These artificial intelligence-based strategies not only increase the precision of diagnosis but also enable mass screening programs to be conducted, closing gaps in health care [78, 79]. Outside ophthalmology, AI is changing other branches of medicine by making highly complex, labor-intensive tasks automated. In reproductive medicine as well, AI is being sought to be employed for sperm auto-selection, heretofore human eye-based. With very accurate measurement of sperm motility and morphology, AI models can help embryologists choose the best sperm to be used in fertility treatment, which can improve success rates in assisted reproductive technologies [80]. With ongoing developments in AI technology, it is anticipated to further widen its applications in medical imaging with significant advantages such as increased diagnostic accuracy, less subjectivity in evaluation, and increased efficiency in clinical workflow. The implementation of AI-based tools in regular medical practice has the capability to lighten the load of medical professionals while also enabling patients to receive faster and more accurate diagnoses. As research and development keep on evolving, AI will be an incredibly useful resource in most of the medical fields, developing even more advanced, data-based health care solutions.

7.3 Future Trends of AI in Optical Navigation

Technological advancements in artificial intelligence (AI) and optics are driving groundbreaking sensing and navigation technologies that result in extremely intelligent, precise, and responsive systems. Coupling AI with advanced optics is emerging with the new breed of high-performance devices including optical neural networks, intelligent imaging, and smart sensors. Often the more promising of these development areas, metaphotonics involves applying design methodologies through which AI optimizes device functionality, boosts computation efficacy, and makes real-time operation feasible in scenarios like autonomous vehicles and distant monitoring [81]. AI is revolutionizing the design of all-weather, extremely flexible, and intelligent sensor systems in the field of optoelectronic sensing. Such technologies employ next-generation nanomaterials such as 2D materials and quantum dots to provide sensitivity, strength, and tunability under dynamic environmental conditions. AI-based data analysis techniques in integration enable superior object identification, pattern recognition, and real-time decision-making capacity, which are placing the sensors more and more on the forefront in automotive safety systems, surveillance systems, and industrial automation [82]. Indoor location is also progressing at fast speed with the incorporation of artificial-intelligence-based environmental-perception sensors to enhance positioning accuracy and system response. Multi-sensor fusion techniques, which combine data from various sources such as LiDAR, inertial measurement units, and vision-based systems, are also enhancing the reliability of navigation solutions in GPS-denied environments. Also, the convergence of AI with next-generation communication technologies such as 5G and IoT is facilitating real-time, smooth navigation for users in complicated indoor spaces such as smart buildings, airports, and shopping malls [74]. There have also been tremendous improvements in inertial navigation systems (INS), with optical sensing technology improvement being central to the development of smaller, more precise, and highly efficient sensors. Miniature optical gyroscopes and accelerometers are enhancing motion tracking and navigation precision, especially in aerospace, autonomous vehicles, and robots. Besides, emerging technologies like nuclear magnetic resonance (NMR) and cold-atom sensors are also being researched for upcoming navigation technologies, with unprecedented accuracy and stability. To maximize further the performance of low-cost sensors, sophisticated error modeling techniques are being integrated to facilitate more efficient compensation of drift and bias with time, thus enhancing the reliability of these systems [83]. Taken as a whole, these technologies point toward the coming age of wiser, more capable, and extremely accurate sensing and navigation capabilities. As advances in AI propel optical and optoelectronic technology forward, we can look to navigation tools that are not only more robust and efficient but can adapt to an enormous spectrum of dynamic surroundings. These advancements have enormous potential for application in autonomous transportation, military, medical applications, and industrial automation, finally making the world a smarter, more interconnected place.

8 Challenges and Limitations in Stellar-Based Navigation

8.1 Environmental Limitations

Environmental influences have a big impact on the accuracy, reliability, and general efficiency of a host of observation and navigation systems used in a wide range of applications. From transportation surveillance to astronomical observations and biology, conditions in the external environment can pose huge challenges for which sophisticated compensation and mitigation measures must be conceived. For transport, video-based automatic incident detection systems make use of visual information to detect departures from traffic, accidents, and congestion. Weather conditions including shadows, snow, rain, and glare have the potential to significantly interfere with system performance by hiding essential visual information or producing false alarms. To counter these issues, scientists have created compensation techniques, such as adaptive image processing algorithms and deep learning-based enhancement methods, to enhance the resilience of these systems under different weather and illumination conditions [84]. Stellar navigation, employed by some technology and certain animals, also encounters serious environmental challenges. Migrating animals following celestial navigation, for instance, have to extract usable directional information from the stars despite the abstract and constantly variable pattern of faint light points in the night sky. Such organisms depend on sophisticated sensory processing systems to accurately interpret celestial patterns, even when the atmosphere distorts the pattern, or artificial lights pollute the landscape. Understanding these biological navigation systems has enabled autonomous robot navigation and space exploration technology to be improved [85]. In astronomy, the precision of the stellar abundance measurement, a key to understanding the chemical composition, evolution, and formation history of galaxies, can be influenced by environmental factors. The precision of such measurements depends on the utilization of high-resolution spectral data and calibrated standards for consistency between various astronomical surveys. Atmospheric condition variation, equipment, and observational bias can introduce inconsistency in data acquisition, which needs to be strictly calibrated to ensure credible and reproducible outcomes [86].

Technology and astronomy are not the sole fields where biological and laboratory experiments are affected much by environmental factors. Biological and laboratory experiments carried out with lab animals can also be affected by many factors that include temperature, humidity, illumination, and sounds. Uncontrolled factors will add variance to animal response and behavior as well as negate research findings. Thus, a standard and homogeneous environment is pivotal in making biomedical research reliable as well as replicable [87].

8.2 Technical Limitations of Current Sensor Technology

Existing power system sensor technologies are very important in monitoring, control, and protection but suffer from a plethora of limitations making their widespread usage and effectiveness difficult. Electromagnetic coils and current transformers have been used to sense electrical parameters in substations for several decades. Conventional sensors are, however, large and cumbersome, and their deployment is limited to dedicated sites. Their size is a problem in installation, maintenance, and scalability, particularly with the present power grids' requirement for compact and flexible monitoring [88]. To address some of these challenges, optical sensor technologies such as inline-Sagnac current sensors have been suggested as a possible solution. The sensors take advantage of the working principles of fiber-optic sensing to enable high-accuracy measurement with low electromagnetic interference. Nevertheless, even though they possess strengths, they are rather temperature-sensitive with working scale factor ranges being in the 80–200 ppm/°C order. In addition, accurate measurement can only be enabled through highly stable wavelength stability of the light source, which might be challenging to achieve in dynamic grid environments [89]. For residential and commercial power monitoring purposes, smart home sensors have been recognized as an effective solution to track daily energy consumption, enhance safety, and load management optimization. Although the sensors possess promising features, they need additional study for evidence on how well they work, are reliable, and affordable before they can be mass-deployed. Considerations such as cost, user acceptance, and compatibility with existing smart grid infrastructure continue to be the main factors [90]. Yet another newly emerging sensor technology with future application in power systems is the Giant Magneto Resistive (GMR) current sensor. Such sensors are highly sensitive and have rapid response, therefore best suited to applications in smart grids, like instantaneous current measurement and fault identification. But present uses of GMR sensors continue to experience problems regarding stability, linearity, and ruggedness, which require further research and development efforts towards improving their performance and reliability [91]. With the progress of power grids in embracing renewable energy sources, automation, and digitalization, there is a growing need for compact, energy-saving sensors with multi-parameter measurement. The new-generation sensors must not only be smaller and cheaper but also deliver real-time wide-area monitoring in order to facilitate predictive maintenance and grid stability. By leveraging beyond the limitations of existing sensor technologies and benefiting from the advancements in material science, AI-based analytics, and edge computing, new power system monitoring can be more effective, resilient, and attuned to the changing energy landscape [88].

8.3 Current Challenges in Autonomous Stellar Navigation

Autonomous satellite navigation is a dynamic technology that must overcome numerous challenges in an effort to offer secure and efficient fight in more crowded space conditions. There is a growing need for higher levels of automation and autonomy as more satellites are launched into orbit, which requires innovative technologies to support complex orbital behaviors and prevent collisions and operational losses [92]. Artificial Intelligence (AI) and Machine Learning (ML) are currently leading the way in enabling Trusted Autonomous Satellite Operations (TASO). The technologies enhance system performance, flexibility, and decision-making capacity, allowing satellites to dynamically adjust to the environment, maximize trajectory orbits, and automatically diagnose and correct [92, 93]. Nevertheless, incorporation of AI in space systems brings technological and legal issues, i.e., trustworthiness of data, explainability of AI decision-making, and compliance with international space law [93]. Present-day navigation systems in satellites depend on augmentation systems, i.e., space-based or ground-based augmentation services, to render navigation more accurate during phases of combat which are most important. These methods help improve precision in positioning but expose navigation systems to interference generated by radio-frequency interference (RFI), which can lower or even completely disable satellite-based navigation [4]. As space traffic increases and hostile threats such as jamming and spoofing emanate, there is a need to create robust navigation solutions that have lesser dependency on external augmentation systems for preserving functional integrity. Outside of satellite uses, AI-driven navigation techniques are also being researched for small Unmanned Aircraft Systems (sUAS). These techniques utilize sophisticated algorithms and heterogeneous low Size, Weight, Power, and Cost (SWaP-C) sensors to enhance navigation performance in GPS-denied or contested environments [94]. AI-based navigation methods, such as sensor fusion and reinforcement learning, can contribute to improving the resilience and autonomy of sUAS operations and provide safer and more efficient flight through complex airspace. Yet maintaining predictability and determinism of the systems is critical for regulatory approval and meeting stringent safety standards [94]. With advancing autonomous navigation technology, the imperative to cross the challenges of AI implementation, cybersecurity risks, and regulatory guidelines will be all the more pressing. By building AI-driven strategies, enhancing system resilience against jamming, and establishing strict certification standards, future autonomous aerial and satellite navigation will be capable of achieving higher degrees of reliability, safety, and operational efficiency.

9 Future Directions and Trends in Optical Sensing for Satellite Navigation

9.1 Emerging Technologies in Optical Sensors

Recent advances in optical sensing technology have transformed numerous applications by marrying artificial intelligence (AI) and quantum mechanics so that highly smart, precise, and responsive sensing systems have been developed. A few of the most remarkable breakthroughs in this regard are hyperspectral imaging (HSI), where high-resolution spectral information is captured in numerous bands to allow further analysis of more comprehensive data. When combined with AI, HSI significantly enhances data processing, pattern recognition, and interpretation capability. This has been shown to be highly beneficial in a variety of applications, including precision agriculture, where it is employed for crop health monitoring and disease detection, environmental monitoring for detecting pollution and land-use mapping, and medical diagnostics for detecting early diseases and tissue analysis [95].

Another major achievement is the creation of AI-enabled optoelectronic sensors, which are now becoming more advanced, all-weather, and multi-functional systems. These sensors utilize nanomaterials and advanced fabrication techniques to enhance performance, longevity, and efficiency. When enabled with AI, they allow real-time adaptive output, improving autonomous navigation, industrial control, and smart city infrastructure applications. Moreover, their applicability in the Internet of Things (IoT) environment is on the rise, as they enable efficient data exchange and automation between interconnected smart systems [82].

Quantum biosensing has also emerged as a revolutionary technology in biomedicine and environmental sensing. Quantum optical biosensors are highly sensitive because they can detect even very small biological and chemical changes at the molecular level. Such technologies have widespread uses, especially in medical diagnostics, where they enable ultra-sensitive detection of disease, in environmental monitoring for the real-time sensing of pollutants, and in checking food quality for detection of contaminants and controlling safety levels [50]. The quirks of quantum mechanics, for example, entanglement and superposition, have also increased the limits of sensor accuracy and reliability. Further, the merging of AI with flexible sensing technology has led to smart and dynamic sensing systems capable of replicating human sensory potential. With artificial synapses and machine learning included in such devices, they can learn as well as process from external inputs, and thus play a very important part in soft robotics, wearable technology, and monitoring human activity. Such technologies are enabling more human–machine interaction modes as well as more natural interfaces, and applications of which include interactive entertainment, healthcare, and prosthetics [96].

9.2 Improved Sensor Resilience and Efficiency

Recent technological development of sensors has mostly been stimulated by advances in electronic circuits, advanced signal processing technologies, and precise manufacturing. With these technological improvements, highly efficient, reliable, and cheaper sensing systems could be developed to support a multitude of industries and enable them to enhance their decision-making and monitoring real-time within an extensive domain [52]. In the automobile sector, there has been an increasing need for robust spatial perception systems as autonomous and assisted driving technologies are adopted rapidly. Protecting the sensors from both hardware and software failure is a major area of study since even partial system failure may undermine safety. The current advancements underscore sensor redundancy, fault-tolerant design, and AI-assisted error-detecting mechanisms to allow for autonomous vehicles to maintain functional integrity despite uncontrollable environmental and system changes [97]. Advanced hostile-environment sensing technologies are now a necessity for applications demanding durability, accuracy, and reliability in harsh environments. Some of the finest examples of such technologies include Fiber Bragg Grating (FBG) sensors, which are renowned for their outstanding ruggedness, multiplexing, and high sensitivity. These sensors are especially suited for use in high-temperature, radiation-rich, and mechanically demanding environments and are thus a must in aerospace engineering, structural health monitoring, and power systems.

This constant development of FBG sensors stems from the need to further improve their performance. By integrating high-quality materials, researchers aim to improve their mechanical stability and thermal stability, offering long-term reliability in extreme operating conditions. In addition, advancements in thermally stable sensor characteristics are expected to minimize performance degradation from long-term exposure to high temperatures and fluctuating environmental conditions. One of the most significant developments in FBG sensor applications is the integration of IoT functions, which offers real-time remote monitoring and predictive maintenance. Smart sensing technologies facilitate the collection and processing of data from remote or inaccessible locations, making monitoring possible beyond the immediate sensor location. Such innovation enhances operational efficiency, reduces maintenance, and enhances safety through the provision of proactive intervention to structural transformation and environmental hazard [98].

Another area of growing interest is the use of sensor-based material flow characterization (SBMC) in waste management and recycling. The traditional sorting methods are less precise and versatile, but with optical sensor and machine learning algorithm converging, material identification and separation have become far more effective. SBMC not only continues to gain growing significance in sensor-based sorting but also in real-time monitoring of the material flow and automatic process control. The next research phase in this direction will likely employ deep learning methods for better classification, develop low-cost sensor solutions to mass deploy, and tap data streams from field-deployed industrial machinery to improve the sorting and recycling process [99]. These developments all together point towards the revolutionary possibilities of contemporary sensor technologies. With the integration of AI, IoT, and high-end signal processing, industries are able to enhance efficiency, resilience, and automation, opening doors for intelligent and sustainable technological applications in various domains.

9.3 Integration with Other Navigation Systems

The combination of optical sensors with GPS and inertial measurement units (IMUs) has transformed many areas, such as navigation, ecological studies, and remote sensing applications. By integrating Global Navigation Satellite Systems (GNSS), Inertial Navigation Systems (INS), and optical sensors, scientists and engineers can utilize complementary sources of data to enhance positioning accuracy, environmental flexibility, and system reliability. This merging is particularly beneficial in rugged environments, i.e., in Arctic environments, where satellite signals are blocked or erroneous due to adverse weather and geomagnetic disturbances [100] MEMS-based inertial sensors have played a critical role in refining user interfaces and indoor navigation for next-generation consumer products. Their accurate monitoring of motion and gesture sensing have helped greatly towards improving the functionality of smartphone, wearable sensor, and AR system through providing intuitive controls and seamless interactions [101]. These developments have facilitated more responsive game play, accurate motion tracking for health and fitness uses, and greater virtual and augmented reality interactions, whereby MEMS sensors have become unavoidable in present-day smart devices.

Aside from consumer navigation, GPS coupling with other sensing technologies has become imperative in wildlife conservation, behavioral ecology, and ecological monitoring. Accurate tracking animals, consisting of GPS positioning blended with optical and inertial sensors, enables researchers to understand better movement routes, habitat usage, and behavior patterns in varied species, including land animals, birds, and sea creatures [102]. Hybrid systems have improved the monitoring ability of migration patterns, prey-predator relationship, and species adaptation to habitat change, promoting biodiversity conservation activities and environmental research. At forest application scale, sensor fusion has come a long way in forest inventory and resource management. The fusion of LiDAR and optical sensor information has been found to improve precision in forest type classification, species detection, and biomass estimation by as much as 55% over the use of a single sensor [103]. Although these mergers improve overall data quality, they do not translate into meaningful improvement in tree height estimation, which suggests that other sensor technologies may be required in a bid to improve precision. Overall, the integration of optical sensors with GPS and IMUs is a reliable approach to enhancing navigation accuracy, extending the range of research activities, and optimizing data quality in most scientific and industrial applications. As technology advances in the field of sensors, the coming sensor fusion algorithms, real-time processing, and adaptive calibration techniques will provide increasingly accurate positioning, tracking, and environmental monitoring.

10 Case Studies of Stellar-Based Positioning in Aerospace Applications

10.1 Real-World Applications in Aerospace Missions

The latest studies indicate dramatic improvements in satellite positioning technologies, with Global Navigation Satellite Systems (GNSS) at the heart of a broad array of applications. The greatest breakthroughs have been seen in structural health monitoring with GNSS offering high-precision vibration data for buildings, towers, and bridges to enable engineers to identify early signs of structural stress and deformation more accurately [104]. They enable the creation of smart infrastructure systems that are capable of delivering real-time monitoring and predictive maintenance to enhance safety and life expectancy. Due to adverse weather conditions in the Arctic region, standard GNSS constellations offer position accuracy of up to three meters. Due to ionospheric irregularities within the region, adverse weather conditions, and satellite visibility, other positioning solutions such as multi-sensor fusion and ground-based augmentation systems are required to improve further accuracy [105].

Outside Earth-based applications, positioning via satellites has also been proved to work in space missions. The CoRoT satellite mission demonstrated the utility of stellar-based positioning for seismology and exoplanet detection in combination with the circumvention of the limitations of low Earth orbit (LEO) environments, e.g., signal interference and orbital instability [106]. Such success exemplifies the utility of astronomical navigation for deep-space exploration missions to be used in the future. From the viewpoint of signal processing, conventional phase-locked loop (PLL) methods have been employed for GNSS receiver carrier synchronization for many years. Nonetheless, Kalman filter (KF)-based methods have proven to be better alternatives, especially under adverse propagation environments that are often found in aerospace systems. KF techniques enable reliable carrier tracking in GNSS signals and deep-space communication networks, enhancing signal quality and lessening vulnerability to noise and signal degradation [107]. These developments open the door to more robust and accurate satellite navigation systems, which are critical to applications from autonomous spacecraft navigation to high-accuracy scientific missions.

10.2 Effectiveness and Technical Analysis

Ergonomics Interventions: Studies indicate that ergonomics interventions create significant gains in productivity and work safety in the workplace. These interventions decrease the number of musculoskeletal disorders at work, which results in a healthier employee population. The interventions also bring with them very robust economic rewards, with the payback being less than a year because of improved efficiency and decreased absenteeism [108].

Techno-Economic Analysis of Energy Retrofits: A techno-economic model to study energy retrofit technologies reflected significant improvement in performance in the form of reduced energy consumption as well as minimizing carbon footprints. Financial feasibility and sustainability factors involved in the integration of energy-saving programs [109] are primarily the areas under study.

Industrial Benchmarking and Cost-Based Ratios: While in industrial benchmarking, research indicates that low-cost, simple ratios work best for internal performance analysis, especially in cases where price structures are standard and well-established. This method enables organizations to measure efficiency and profitability using the least amount of complexity [110].

Technical Performance During Surgery and Patient Outcomes: In surgical and clinical practice, through the systematic review of 24 studies, there is a proven association that heightened technical performance while carrying out operations contributes directly to patient outcomes improvement in terms of complications and death rate reduction [111]. The evidence identifies the inevitability of technical skills for an improvement in healthcare quality as well as ensuring patients' safety.

10.3 Lessons Learned and Industry Impact

Additive Manufacturing (AM): Additive manufacturing technology is challenged regarding aggregate sizes, nozzle limitations, and product quality. Overcoming these limitations is important for improving material performance and accuracy in manufacturing, which are key fields that need future research and technology development [112].

JPEG 2000 Development: The development of JPEG 2000, the next-generation image compression standard, recognizes collaboration and innovation as drivers towards improved compression processes. Technology innovation as a driver to improved digital imaging capability is cited in this case study [113]. JPEG 2000's advancement over JPEG was facilitated by collaborative efforts in enhancing compression ratios, adding lossless compression, and adding features such as region-of-interest coding. This collaborative work, through the efforts of numerous experts and organizations, illustrates how collective innovation can initiate phenomenal advance in digital imaging technology to enable more efficient image storage and transmission.

Healthcare Data Standards—U.S. Department of Veterans Affairs: The U.S. Department of Veterans Affairs' adoption of healthcare data standards is one instance of the advantages of well-disciplined processes, internal know-how, and coordination of business goals. This is one instance of the maturity of standardization within big companies and strategic planning's value [114].

XML-Based Standards in the United States Home Mortgage Industry: The use of XML-based vertical IS standards can potentially revolutionize the U.S. home mortgage business, with effects depending heavily on implementation strategies. This case study highlights the widespread implications of data exchange standardization on industry processes and efficiency [115].

11 Conclusion

Recent studies have significantly advanced aerospace sensing technologies to the point of revolution in monitoring and commanding vehicles, satellites, and spacecraft in harsh environments. The recent past has witnessed tremendous advancement by scientists and engineers in developing sensors not just durable enough to withstand aerospace environments but also providing reliable real-time data crucial in ensuring safety and optimal performance. The most prominent achievement in this area is the creation of optical fiber sensing (OFS). OFS initially was confined to laboratory research only, but nowadays it has found its application in real-world engineering applications in the form of measuring structural deformation of flight vehicles. Current aircraft and spacecraft are dynamically loaded, subjected to temperature fluctuations, and experience mechanical loads during operation. Fiber-optic sensors integrated into the vehicle structure provide the potential for real-time continuous monitoring of strain, temperature, and other important parameters. This will provide engineers with the opportunity to recognize material fatigue and structural irregularities earlier, prevent catastrophic failure, and extend vehicle operating life. The technological growth from test beds to multi-sensor fusion systems reflects the technological maturity and acceptability of optical fiber sensing for space technology. Together with advances in optical fiber sensing, fiber optic sensor-embedded composite structures have emerged as an exciting trend. Composite materials are extensively used in spacecraft these days because they possess a high strength-to-weight ratio and long life. By integrating fiber optic sensors into these composites, engineers can design "smart" structures that monitor their own health in real time. Smart propellant tanks, for instance, with built-in sensors can provide real-time stress, strain, and temperature readings, ensuring that any emerging issues are identified and addressed before they become significant issues. This type of monitoring is particularly important to long-duration missions, where there are limited maintenance windows and reliability is the biggest issue. Optical sensing technologies have also played an important role in satellite imagery. Perhaps one of the most intriguing advancements in this field is the merging of optical stereo imagery with laser altimeter information. Blending these sources of data improves the geometric positioning quality of satellite imagery, resulting in more precise three-dimensional models of the Earth's surface. These advancements are essential to uses from environmental monitoring and city planning to disaster response and resource prospecting. By combining the advantages of optical and laser-based sensing, hybrid systems can acquire high-resolution topographic data, which improves our capacity to monitor environmental change and react to natural disasters in turn. Beyond the aerospace industry, positioning technologies have advanced considerably in wireless networks. These technologies are most critical for future applications such as extended reality (XR) and smart environments, where the accuracy of positioning data is crucial. By integrating optical sensors with GPS and inertial measurement units, and subsequently augmenting these systems using artificial intelligence and adaptive algorithms, high-precision navigation is today possible in multifaceted outdoor and indoor environments. This intersection not only improves consumer electronics user experience but also paves the way for even more sophisticated applications in robotics, autonomous transportation networks, and smart city infrastructure. Optical interferometry is one area where a lot of development has taken place, especially among the astronomical community. Optical interferometry aligns many telescopes together in a bid to combine their light, producing much higher resolutions than any individual telescope can. This technology has allowed scientists to view celestial objects at never-before-seen resolution, resulting in groundbreaking scientific findings. The technological breakthrough in interferometry has not only assisted us in developing further knowledge regarding the universe, but has also assisted in the development of new optical sensing and imaging techniques that can be used in other industries like aerospace and remote sensing. Spacecraft engineering has in fact been greatly benefited by application of fiber optic sensors. For use in aerospace applications, these sensors are being qualified for operation across a wide range of conditions—from the cryogenic temperatures of missions to deep space to the intense thermal loading of high-speed re-entry through Earth's atmosphere. Their ability to make precise, high-precision measurements in such a wide range of conditions makes them invaluable in the delivery of spacecraft safety and performance. As missions become increasingly ambitious and the environments increasingly hostile, these sensors' role in keeping watch on important parameters increases in significance. Arctic positioning systems demonstrate the versatility of such technologies to special environmental problems. In polar regions, where there are few visible satellites and ionospheric disturbance is widespread, a successful positioning is especially difficult. In spite of these challenges, traditional satellite navigation systems have proven to be able to achieve positioning accuracies of three meters. Such precision is essential for scientific research, navigation, and the safe operation of vehicles in such remote and hostile areas. Another research frontier is aero-optics, which is the study and compensation of optical aberrations caused by aerodynamic flows. Optical systems that are subject to turbulent airflow turbulent boundary layers, separated shear layers, or air flowing over optical turrets—can be severely distorted and reduce image quality and sensor performance. New developments in computation methods allow the engineer to precisely predict such distortions of flows, resulting in the creation of compensating methodologies that ensure optical system integrity regardless of adverse aerodynamic conditions. Combine these developments in optical sensing and positioning technologies and they are giving rise to a new generation of improved monitoring, control, and navigation systems. From structural health monitoring of aerospace vehicles and smart composite materials in space vehicles to high-accuracy satellite imaging and high-reliability positioning systems in hostile environments, these technologies are ushering in safer and more efficient operations. Additionally, the interfacing of these systems with wireless networks and artificial intelligence is providing new avenues for autonomous decision-making and real-time data analysis, so that aerospace missions and applications in related domains in the future are not only innovative but also robust. With ongoing research and development of these technologies, we can expect further development that will expand their uses beyond aerospace to fields such as environmental monitoring, urban planning, and autonomous transport. The ongoing development of optical sensing and positioning technologies promises a future where sophisticated, interconnected technologies not only address the needs of today's challenges but also lay the groundwork for tomorrow's opportunities.

References

1. H. Al-Hraishawi, H. Chougrani, S. Kisseleff, E. Lagunas, S. Chatzinotas, A survey on nongeostationary satellite systems: the communication perspective. IEEE Commun. Surveys Tutorials **25**(1), 101–132 (2023). https://doi.org/10.1109/COMST.2022.3197695
2. G. Chen, F. Xiong, The development of aeronautical mobile satellite services over the past thirty years. IEEE Aerosp. Electron. Syst. Mag. **9**(12), 25–36 (1994). https://doi.org/10.1109/62.334757
3. C.J. Hegarty, E. Chatre, Evolution of the global navigation satellite system (GNSS). Proc. IEEE **96**(12), 1902–1917 (2008). https://doi.org/10.1109/JPROC.2008.2006090
4. J. Blanch, T. Walter, P. Enge, Satellite navigation for aviation in 2025, in *Proceedings of the IEEE*, vol. 100, no. Special Centennial Issue (2012), pp. 1821–1830. https://doi.org/10.1109/JPROC.2012.2190154
5. P.D. Groves, Principles of GNSS, inertial, and multisensor integrated navigation systems, 2nd edition [Book review]. IEEE Aerosp. Electronic Syst. Mag. **30**(2), 26–27 (2015). https://doi.org/10.1109/MAES.2014.14110
6. Y. Hu, X. Zhang, L. Chen, Strategy design and sensor scheduling for optical navigation of low earth orbit satellites. IEEE Sens. J. **18**(23), 9802–9811 (2018). https://doi.org/10.1109/JSEN.2018.2871805
7. Y. Wang, Y. Zhu, M. Wang, S. Jin, Q. Rao, Atmospheric refraction calibration of geometric positioning for optical remote sensing satellite. IEEE Geosci. Remote Sens. Lett. **17**(12), 2130–2134 (2020). https://doi.org/10.1109/LGRS.2019.2963626
8. M. Taghi Dabiri, M. Hasna, S. Althunibat, K. Qaraqe, Modulating retroreflector-based satellite-to-ground optical communications: acquisition, sensing, and positioning. IEEE Trans. Commun. **73**(1), 483–497 (2025). https://doi.org/10.1109/TCOMM.2024.3412779
9. T. Aruga, K. Araki, Improvement of satellite tracking accuracy using optical observations. IEEE Trans. Aerosp. Electron. Syst. **AES-21**(4), 514–521 (1985). https://doi.org/10.1109/TAES.1985.310639

10. M. Betto, J.L. Jørgensen, P.S. Jørgensen, T. Denver, Advanced stellar compass onboard autonomous orbit determination, preliminary performance. Ann. N. Y. Acad. Sci. **1017**(1), 393–407 (2004). https://doi.org/10.1196/annals.1311.022
11. M. Liu, Research on star sensor-based space debris detection and positioning technology. Publ. Astron. Soc. Pac. **135**(1046), 047001 (2023). https://doi.org/10.1088/1538-3873/acc7c9
12. J. Yang, J. Jiang, L. Tian, G. Zhang, Joint estimation of stellar atmospheric refraction and star tracker attitude. IEEE Trans. Instrum. Meas. **71**, 1–12 (2022). https://doi.org/10.1109/TIM.2021.3135343
13. R.P. Norris, Dawes review 5: Australian aboriginal astronomy and navigation. Publ. Astron. Soc. Austral. **33**, e039 (2016). https://doi.org/10.1017/pasa.2016.25
14. J. Bennett, Navigation on wood: wooden navigational instruments, 1590–1731: an analysis of early modern western instruments for celestial navigation, their origins, mathematical concepts, and accuracies. Isis **110**(3), 593–594 (2019). https://doi.org/10.1086/704665
15. N. Ulman, Computers make celestial navigation less celestial? IEEE Aerosp. Electron. Syst. Mag. **4**(3), 39–40 (1989). https://doi.org/10.1109/62.17936
16. T.L. Warren, Y.M. Giraldo, M.H. Dickinson, Celestial navigation in Drosophila. J. Experimental Biol. **222**(Suppl_1) (2019). https://doi.org/10.1242/jeb.186148
17. T. Sun, F. Xing, X. Wang, Z. You, D. Chu, An accuracy measurement method for star trackers based on direct astronomic observation. Sci. Rep. **6**(1), 22593 (2016). https://doi.org/10.1038/srep22593
18. C.C. Liebe, Star trackers for attitude determination. IEEE Aerosp. Electron. Syst. Mag. **10**(6), 10–16 (1995). https://doi.org/10.1109/62.387971
19. E. Tatulli, N. Blind, J.P. Berger, A. Chelli, F. Malbet, Estimating the phase in ground based interferometry: performance comparison between singlemode and multimode schemes. Astron. Astrophys. **524**, A65 (2010). https://doi.org/10.1051/0004-6361/200913356
20. L.W. Cassidy, L. Schlom, *Miniature star tracker for small remote sensing satellites*, in ed. by W.L. Barnes, B.J. Horais (1995), pp. 160–166. https://doi.org/10.1117/12.198942
21. J. Anderson, I.R. King, Toward high-precision astrometry with WFPC2. I. Deriving an accurate point-spread function. Publ. Astron. Soc. Pac. **112**(776), 1360–1382 (2000). https://doi.org/10.1086/316632
22. M.A. Samaan, D. Mortari, J.L. Junkins, Compass star tracker for GPS-like applications. IEEE Trans. Aerosp. Electron. Syst. **44**(4), 1629–1634 (2008). https://doi.org/10.1109/TAES.2008.4667738
23. J. Jiang, H. Wang, G. Zhang, High-accuracy synchronous extraction algorithm of star and celestial body features for optical navigation sensor. IEEE Sens. J. **18**(2), 713–723 (2018). https://doi.org/10.1109/JSEN.2017.2777493
24. K.C. Bürger, F.O. Fialho, C.R.B. Aykroyd, Embedded star catalog calculation tool for autonomous star trackers. Publ. Astronomical Soc. Pacific **136**(10), 104502 (2024). https://doi.org/10.1088/1538-3873/ad7f86
25. A. Kumar, P. Ben-Tzvi, Spatial object tracking system based on linear optical sensor arrays. IEEE Sens. J. **16**(22), 7933–7940 (2016). https://doi.org/10.1109/JSEN.2016.2607120
26. M. Lesser, A summary of charge-coupled devices for astronomy. Publ. Astron. Soc. Pac. **127**(957), 1097–1104 (2015). https://doi.org/10.1086/684054
27. J. Janesick, G. Putnam, Developments and applications of high-performance CCD and CMOS imaging arrays. Annu. Rev. Nucl. Part. Sci. **53**(1), 263–300 (2003). https://doi.org/10.1146/annurev.nucl.53.041002.110431
28. X. Qian, H. Yu, S. Chen, A global-shutter centroiding measurement CMOS image sensor with star region SNR improvement for star trackers. IEEE Trans. Circuits Syst. Video Technol. **26**(8), 1555–1562 (2016). https://doi.org/10.1109/TCSVT.2015.2469091
29. T.A. ten Brummelaar, Correlation measurement and group delay tracking in optical stellar interferometry with a noisy detector. Mon. Not. R. Astron. Soc. **285**(1), 135–150 (1997). https://doi.org/10.1093/mnras/285.1.135
30. K. Micko, P. Papcun, I. Zolotova, Review of IoT sensor systems used for monitoring the road infrastructure. Sensors **23**(9), 4469 (2023). https://doi.org/10.3390/s23094469

31. G. Wahba, A least squares estimate of satellite attitude. SIAM Rev. **7**(3), 409–409 (1965). https://doi.org/10.1137/1007077
32. C.C. Liebe, Accuracy performance of star trackers - a tutorial. IEEE Trans. Aerosp. Electron. Syst. **38**(2), 587–599 (2002). https://doi.org/10.1109/TAES.2002.1008988
33. J. Lu, C. Lei, S. Liang, Y. Yang, An all-parameter system-level calibration for stellar-inertial navigation system on ground. IEEE Trans. Instrum. Meas. **66**(8), 2065–2073 (2017). https://doi.org/10.1109/TIM.2017.2674758
34. B.C. Ng, C.M.S. See, Sensor-array calibration using a maximum-likelihood approach. IEEE Trans. Antennas Propag. **44**(6), 827–835 (1996). https://doi.org/10.1109/8.509886
35. J.R. Coldrick, Optical sensors for spacecraft attitude determination. Opt. Laser Technol. **4**(3), 129–141 (1972). https://doi.org/10.1016/0030-3992(72)90019-9
36. P. McKee, H. Nguyen, M.W. Kudenov, J.A. Christian, StarNAV with a wide field-of-view optical sensor. Acta Astronaut. **197**, 220–234 (2022). https://doi.org/10.1016/j.actaastro.2022.04.027
37. G.Á. Bakos, J. Lázár, I. Papp, P. Sári, E.M. Green, System description and first light curves of the Hungarian automated telescope, an autonomous observatory for variability search. Publ. Astron. Soc. Pac. **114**(799), 974–987 (2002). https://doi.org/10.1086/342382
38. S. Racharla, K. Rajan, Solar tracking system—a review. Int. J. Sustain. Eng. **10**(2), 72–81 (2017). https://doi.org/10.1080/19397038.2016.1267816
39. J.J. Degnan, *Engineering progress on the fully automated photon-counting SLR2000 satellite laser ranging station*, in ed. by U. Schreiber, C. Werner (1999), p. 76. https://doi.org/10.1117/12.373041
40. Q. Qiao, J. Yuan, Y. Shi, X. Ning, F. Wang, Structure, design, and modeling of an origami-inspired pneumatic solar tracking system for the NPU-Phonesat. J. Mech. Robot. **9**(1) (2017). https://doi.org/10.1115/1.4035086
41. I. Rigas, H. Raffle, O.V. Komogortsev, Photosensor oculography: survey and parametric analysis of designs using model-based simulation. IEEE Trans. Hum. Mach. Syst. **48**(6), 670–681 (2018). https://doi.org/10.1109/THMS.2018.2807244
42. M. Hasan et al., LiDAR-based detection, tracking, and property estimation: a contemporary review. Neurocomputing **506**, 393–405 (2022). https://doi.org/10.1016/j.neucom.2022.07.087
43. F. Daum, Tracking and data fusion: handbook of algorithms (Bar-Shalom, Y., et al.: 2011). IEEE Aerosp. Electron. Syst. Mag. **27**(12), 34–35 (2012). https://doi.org/10.1109/MAES.2012.6397662
44. C. Micheloni, B. Rinner, G. Foresti, Video analysis in pan-tilt-zoom camera networks. IEEE Signal Process. Mag. **27**(5), 78–90 (2010). https://doi.org/10.1109/MSP.2010.937333
45. H.-R. Trankler, O. Kanoun, Recent advances in sensor technology, in *IMTC 2001. Proceedings of the 18th IEEE Instrumentation and Measurement Technology Conference. Rediscovering Measurement in the Age of Informatics* (Cat. No.01CH 37188) (IEEE, 2001), pp. 309–316. https://doi.org/10.1109/IMTC.2001.928831
46. L. Zhang, W. Liu, Precision glass molding: toward an optimal fabrication of optical lenses. Front. Mech. Eng. **12**(1), 3–17 (2017). https://doi.org/10.1007/s11465-017-0408-3
47. H. Yu et al., Dispersion engineering of metalenses. Appl. Phys. Lett. **123**(24) (2023). https://doi.org/10.1063/5.0177734
48. M. Stubenvoll, B. Schäfer, K. Mann, Measurement and compensation of laser-induced wavefront deformations and focal shifts in near IR optics. Opt. Express **22**(21), 25385 (2014). https://doi.org/10.1364/OE.22.025385
49. C. Acar, A.R. Schofield, A.A. Trusov, L.E. Costlow, A.M. Shkel, Environmentally robust MEMS vibratory gyroscopes for automotive applications. IEEE Sens. J. **9**(12), 1895–1906 (2009). https://doi.org/10.1109/JSEN.2009.2026466
50. N. Sagar Shrikrishna, R. Sharma, J. Sahoo, A. Kaushik, S. Gandhi, Navigating the landscape of optical biosensors. Chem. Eng. J. **490**, 151661 (2024). https://doi.org/10.1016/j.cej.2024.151661

51. S. Subramanian, H. Wu, T. Constant, J. Xavier, F. Vollmer, Label-free optical single-molecule micro- and nanosensors. Adv. Mater. **30**(51) (2018). https://doi.org/10.1002/adma.201801246
52. O. Kanoun, H.-R. Trankler, Sensor technology advances and future trends. IEEE Trans. Instrum. Meas. **53**(6), 1497–1501 (2004). https://doi.org/10.1109/TIM.2004.834613
53. R.C. Luo, C.-C. Yih, K.L. Su, Multisensor fusion and integration: approaches, applications, and future research directions. IEEE Sens. J. **2**(2), 107–119 (2002). https://doi.org/10.1109/JSEN.2002.1000251
54. R.C. Luo, C.-C. Chang, Multisensor fusion and integration: a review on approaches and its applications in mechatronics. IEEE Trans. Industr. Inform. **8**(1), 49–60 (2012). https://doi.org/10.1109/TII.2011.2173942
55. R.C. Luo, C.C. Chang, C.C. Lai, Multisensor fusion and integration: theories, applications, and its perspectives. IEEE Sens. J. **11**(12), 3122–3138 (2011). https://doi.org/10.1109/JSEN.2011.2166383
56. C. Habib, A. Makhoul, R. Darazi, R. Couturier, Health risk assessment and decision-making for patient monitoring and decision-support using wireless body sensor networks. Information Fusion **47**, 10–22 (2019). https://doi.org/10.1016/j.inffus.2018.06.008
57. D. Smith, S. Singh, Approaches to multisensor data fusion in target tracking: a survey. IEEE Trans. Knowl. Data Eng. **18**(12), 1696–1710 (2006). https://doi.org/10.1109/TKDE.2006.183
58. R.O. Ogunleye, S. Rusnáková, J. Javořík, M. Žaludek, B. Kotlánová, Advanced sensors and sensing systems for structural health monitoring in aerospace composites. Adv. Eng. Mater. **26**(22) (2024). https://doi.org/10.1002/adem.202401745
59. D.G. Senesky, B. Jamshidi, K.B. Cheng, A.P. Pisano, Harsh environment silicon carbide sensors for health and performance monitoring of aerospace systems: a review. IEEE Sens. J. **9**(11), 1472–1478 (2009). https://doi.org/10.1109/JSEN.2009.2026996
60. E.F. Crawley, Intelligent structures for aerospace: a technology overview and assessment. AIAA J. **32**(8) (1994). https://doi.org/10.2514/3.12161
61. M. Loose, State-of-the-art image sensors and signal processing in NASA's space telescopes [In the spotlight]. IEEE Signal Process. Mag. **27**(1), 128–126 (2010). https://doi.org/10.1109/MSP.2009.934924
62. A. Popowicz, A.R. Kurek, T. Blachowicz, V. Orlov, B. Smolka, On the efficiency of techniques for the reduction of impulsive noise in astronomical images. Mon. Not. R. Astron. Soc. **463**(2), 2172–2189 (2016). https://doi.org/10.1093/mnras/stw1983
63. M. Masias, J. Freixenet, X. Lladó, M. Peracaula, A review of source detection approaches in astronomical images. Mon. Not. R. Astron. Soc. **422**(2), 1674–1689 (2012). https://doi.org/10.1111/j.1365-2966.2012.20742.x
64. W. Yu, J. Jiang, G. Zhang, Star tracking method based on multiexposure imaging for intensified star trackers. Appl. Opt. **56**(21), 5961 (2017). https://doi.org/10.1364/AO.56.005961
65. J.-L. Starck, F. Murtagh, B. Pirenne, M. Albrecht, Astronomical image compression based on noise suppression. Publ. Astron. Soc. Pac. **108**, 446 (1996). https://doi.org/10.1086/133746
66. J.L. Starck, E. Pantin, F. Murtagh, Deconvolution in astronomy: a review. Publ. Astron. Soc. Pac. **114**(800), 1051–1069 (2002). https://doi.org/10.1086/342606
67. T. Fossen, T. Pérez, Kalman filtering for positioning and heading control of ships and offshore rigs. IEEE Control. Syst. **29**(6), 32–46 (2009). https://doi.org/10.1109/MCS.2009.934408
68. M.Á. Naya, E. Sanjurjo, A.J. Rodríguez, J. Cuadrado, Kalman filters based on multibody models: linking simulation and real world. A comprehensive review. Multibody Syst. Dyn. **58**(3–4), 479–521 (2023). https://doi.org/10.1007/s11044-023-09893-w
69. A. Correa, M. Barcelo, A. Morell, J.L. Vicario, Enhanced inertial-aided indoor tracking system for wireless sensor networks: a review. IEEE Sens. J. **14**(9), 2921–2929 (2014). https://doi.org/10.1109/JSEN.2014.2325775
70. X. Yang, S.P. Chockalingam, S. Aluru, A survey of error-correction methods for next-generation sequencing. Brief. Bioinform. **14**(1), 56–66 (2013). https://doi.org/10.1093/bib/bbs015
71. Y. Deng, J. Jiang, Optical fiber sensors in extreme temperature and radiation environments: a review. IEEE Sens. J. **22**(14), 13811–13834 (2022). https://doi.org/10.1109/JSEN.2022.3181949

72. R. Gorgin, Y. Luo, Z. Wu, Environmental and operational conditions effects on lamb wave based structural health monitoring systems: a review. Ultrasonics **105**, 106114 (2020). https://doi.org/10.1016/j.ultras.2020.106114
73. E.J. Llorent-Martínez, M.L. Fernández-de Córdova, P. Ortega-Barrales, A. Ruiz-Medina, Analysis of agroalimentary and environmental contaminants using flow-through chemical optosensors. Appl. Spectrosc. Rev. **50**(7), 527–556 (2015). https://doi.org/10.1080/05704928.2015.1027449
74. N. El-Sheimy, Y. Li, Indoor navigation: state of the art and future trends. Satellite Navigation **2**(1), 7 (2021). https://doi.org/10.1186/s43020-021-00041-3
75. V.F. Mirama, L.E. Diez, A. Bahillo, V. Quintero, A survey of machine learning in pedestrian localization systems: applications, open issues and challenges. IEEE Access **9**, 120138–120157 (2021). https://doi.org/10.1109/ACCESS.2021.3108073
76. V.P. Rekkas et al., Artificial intelligence in visible light positioning for indoor IoT: a methodological review. IEEE Open J. Commun. Soc. **4**, 2838–2869 (2023). https://doi.org/10.1109/OJCOMS.2023.3327211
77. R. Archana, P.S.E. Jeevaraj, Deep learning models for digital image processing: a review. Artif. Intell. Rev. **57**(1), 11 (2024). https://doi.org/10.1007/s10462-023-10631-z
78. D. Benet, O.J. Pellicer-Valero, Artificial intelligence: the unstoppable revolution in ophthalmology. Surv. Ophthalmol. **67**(1), 252–270 (2022). https://doi.org/10.1016/j.survophthal.2021.03.003
79. D.S.W. Ting, A.Y. Lee, T.Y. Wong, An ophthalmologist's guide to deciphering studies in artificial intelligence. Ophthalmology **126**(11), 1475–1479 (2019). https://doi.org/10.1016/j.ophtha.2019.09.014
80. P. Cherouveim, C. Velmahos, C.L. Bormann, Artificial intelligence for sperm selection—a systematic review. Fertil. Steril. **120**(1), 24–31 (2023). https://doi.org/10.1016/j.fertnstert.2023.05.157
81. S. Lee, C. Park, J. Rho, Mapping information and light: trends of AI-enabled metaphotonics. Curr. Opin. Solid State Mater. Sci. **29**, 101144 (2024). https://doi.org/10.1016/j.cossms.2024.101144
82. S. Zhang et al., The rise of AI optoelectronic sensors: From nanomaterial synthesis, device design to practical application. Mater. Today Phys. **27**, 100812 (2022). https://doi.org/10.1016/j.mtphys.2022.100812
83. N. El-Sheimy, A. Youssef, Inertial sensors technologies for navigation applications: state of the art and future trends. Satellite Navigation **1**(1), 2 (2020). https://doi.org/10.1186/s43020-019-0001-5
84. M.S. Shehata et al., Video-based automatic incident detection for smart roads: the outdoor environmental challenges regarding false alarms. IEEE Trans. Intell. Transp. Syst. **9**(2), 349–360 (2008). https://doi.org/10.1109/TITS.2008.915644
85. J.J. Foster, J. Smolka, D.-E. Nilsson, M. Dacke, How animals follow the stars. Proc. R. Soc. B Biol. Sci. **285**(1871), 20172322 (2018). https://doi.org/10.1098/rspb.2017.2322
86. P. Jofré, U. Heiter, C. Soubiran, Accuracy and precision of industrial stellar abundances. Annu. Rev. Astron. Astrophys. **57**(1), 571–616 (2019). https://doi.org/10.1146/annurev-astro-091918-104509
87. D.L. Cinti, M.A. Limelin, J. Christian, Induction of liver microsomal mixed-function oxidases by volatile hydrocarbons. Biochem. Pharmacol. **25**(1), 100–103 (1976). https://doi.org/10.1016/0006-2952(76)90180-5
88. A. Swain, E. Abdellatif, A. Mousa, P.W.T. Pong, Sensor technologies for transmission and distribution systems: a review of the latest developments. Energies (Basel) **15**(19), 7339 (2022). https://doi.org/10.3390/en15197339
89. A.H. Rose, F. Ferdous, M.C. Hudson, Optical inline-Sagnac current sensor, Part I: theory. IEEE Trans. Power Deliv. **37**(6), 5385–5393 (2022). https://doi.org/10.1109/TPWRD.2022.3177078
90. D. Ding, R.A. Cooper, P.F. Pasquina, L. Fici-Pasquina, Sensor technology for smart homes. Maturitas **69**(2), 131–136 (2011). https://doi.org/10.1016/j.maturitas.2011.03.016

91. W. Shanxiang, *Review of Current Sensor Technology Based on Giant Magneto Resistive Effect and Possible Applications for Smart Grids*, Power System Technology (2011)
92. K. Thangavel et al., Artificial intelligence for trusted autonomous satellite operations. Prog. Aerosp. Sci. **144**, 100960 (2024). https://doi.org/10.1016/j.paerosci.2023.100960
93. P.A. Oche, G.A. Ewa, N. Ibekwe, Applications and challenges of artificial intelligence in space missions. IEEE Access **12**, 44481–44509 (2024). https://doi.org/10.1109/ACCESS.2021.3132500
94. S. Bijjahalli, R. Sabatini, A. Gardi, Advances in intelligent and autonomous navigation systems for small UAS. Prog. Aerosp. Sci. **115**, 100617 (2020). https://doi.org/10.1016/j.paerosci.2020.100617
95. S.N. Khonina, N.L. Kazanskiy, I.V. Oseledets, A.V. Nikonorov, M.A. Butt, Synergy between artificial intelligence and hyperspectral imagining—a review. Technologies (Basel) **12**(9), 163 (2024). https://doi.org/10.3390/technologies12090163
96. T. Sun et al., Artificial intelligence meets flexible sensors: emerging smart flexible sensing systems driven by machine learning and artificial synapses. Nanomicro. Lett. **16**(1), 14 (2024). https://doi.org/10.1007/s40820-023-01235-x
97. M. Kahlert, H. Peitzmeier, D. Evans, K. Talits, F. Kortmann, C. Tebruegge, Resilience of spatial environment perception toward fully automated driving: a review. IEEE Sens. J. **24**(14), 21801–21812 (2024). https://doi.org/10.1109/JSEN.2024.3375607
98. S.P. Kok, Y.I. Go, X. Wang, M.L.D. Wong, Advances in Fiber Bragg Grating (FBG) sensing: a review of conventional and new approaches and novel sensing materials in harsh and emerging industrial sensing. IEEE Sens. J. **24**(19), 29485–29505 (2024). https://doi.org/10.1109/JSEN.2024.3434351
99. N. Kroell, X. Chen, K. Greiff, A. Feil, Optical sensors and machine learning algorithms in sensor-based material flow characterization for mechanical recycling processes: a systematic literature review. Waste Manage. **149**, 259–290 (2022). https://doi.org/10.1016/j.wasman.2022.05.015
100. L. Ruotsalainen et al., Toward autonomous driving in arctic areas. IEEE Intell. Transp. Syst. Mag. **12**(3), 10–24 (2020). https://doi.org/10.1109/MITS.2020.2994014
101. D.K. Shaeffer, MEMS inertial sensors: a tutorial overview. IEEE Commun. Mag. **51**(4), 100–109 (2013). https://doi.org/10.1109/MCOM.2013.6495768
102. S.M. Tomkiewicz, M.R. Fuller, J.G. Kie, K.K. Bates, Global positioning system and associated technologies in animal behaviour and ecological research. Philos. Trans. R. Soc. B Biol. Sci. **365**(1550), 2163–2176 (2010). https://doi.org/10.1098/rstb.2010.0090
103. C. Xu, J. Morgenroth, B. Manley, Integrating data from discrete return airborne LiDAR and optical sensors to enhance the accuracy of forest description: a review. Curr. Forestry Rep. **1**(3), 206–219 (2015). https://doi.org/10.1007/s40725-015-0019-3
104. J. Yu, X. Meng, B. Yan, B. Xu, Q. Fan, Y. Xie, Global navigation satellite system-based positioning technology for structural health monitoring: a review. Struct. Control Health Monit. **27**(1) (2020). https://doi.org/10.1002/stc.2467
105. A. Yastrebova, M. Hoyhtya, S. Boumard, E.S. Lohan, A. Ometov, Positioning in the Arctic region: state-of-the-art and future perspectives. IEEE Access **9**, 53964–53978 (2021). https://doi.org/10.1109/ACCESS.2021.3069315
106. M. Auvergne et al., The CoRoT satellite in flight: description and performance. Astron. Astrophys. **506**(1), 411–424 (2009). https://doi.org/10.1051/0004-6361/200810860
107. J. Vila-Valls, P. Closas, M. Navarro, C. Fernandez-Prades, Are PLLs dead? A tutorial on kalman filter-based techniques for digital carrier synchronization. IEEE Aerosp. Electron. Syst. Mag. **32**(7), 28–45 (2017). https://doi.org/10.1109/MAES.2017.150260
108. R.W. Goggins, P. Spielholz, G.L. Nothstein, Estimating the effectiveness of ergonomics interventions through case studies: implications for predictive cost-benefit analysis. J. Safety Res. **39**(3), 339–344 (2008). https://doi.org/10.1016/j.jsr.2007.12.006
109. T.G. Ahmed, B. Gudlaugsson, C. Ogwumike, H. Dawood, M. Short, N. Dawood, Evaluation framework for techno-economic analysis of energy system retrofit technologies. Energy Build. **286**, 112967 (2023). https://doi.org/10.1016/j.enbuild.2023.112967

110. M. Schefczyk, Industrial benchmarking: a case study of performance analysis techniques. Int. J. Prod. Econ. **32**(1), 1–11 (1993). https://doi.org/10.1016/0925-5273(93)90003-4
111. A.B. Fecso, P. Szasz, G. Kerezov, T.P. Grantcharov, The effect of technical performance on patient outcomes in surgery. Ann. Surg. **265**(3), 492–501 (2017). https://doi.org/10.1097/SLA.0000000000001959
112. S.M.E. Sepasgozar, A. Shi, L. Yang, S. Shirowzhan, D.J. Edwards, Additive manufacturing applications for industry 4.0: a systematic critical review. Buildings **10**(12), 231 (2020). https://doi.org/10.3390/buildings10120231
113. D.T. Lee, JPEG 2000: retrospective and new developments. Proc. IEEE **93**(1), 32–41 (2005). https://doi.org/10.1109/JPROC.2004.839613
114. O. Bouhaddou et al., Translating standards into practice: experience and lessons learned at the Department of Veterans Affairs. J. Biomed. Inform. **45**(4), 813–823 (2012). https://doi.org/10.1016/j.jbi.2012.01.003
115. R.T. Wigand, C.W. Steinfield, M.L. Markus, Information technology standards choices and industry structure outcomes: the case of the U.S. Home Mortgage Industry. J. Manag. Inf. Syst. **22**(2), 165–191 (2005). https://doi.org/10.1080/07421222.2005.11045843

Laser Spectroscopic Techniques for Aerospace Remote Sensing

Vedika Chavan and V. K. Unnikrishnan

Abstract The chapter describes the applications of remote spectroscopic multi-modal systems using Laser Induced Breakdown Spectroscopy (LIBS), Laser Induced Fluorescence (LIF), and Raman spectroscopy etc. in aerospace. LIBS, LIF, and Raman are remarkably complementary spectroscopic methods employed in laboratories, in-situ, and remote analysis of materials. The whole study and analysis of a sample's properties and their function can be possible only by the utilization of elemental and molecular data simultaneously. This multi-spectral information accessible from a single compact and cost-effective device opens new avenues for comprehensive material characterization of any sample. In essence, the effective combination of these powerful spectroscopic tools can retrieve the identity of a material that is far superior to that monitored when each of these techniques is used individually. The chapter aims to discuss the fundamental aspects, state-of-the-art, and core challenges involved while using these techniques for aerospace applications.

Keywords Remote sensing · Spectroscopy · LIBS · LIF · Raman · Aerospace

1 Introduction

In aerospace, laser spectroscopic remote sensing describes the stand-off analysis of materials, atmospheric conditions, or celestial bodies using such techniques. While investigating aerospace, the specimen cannot be able to bring to the vicinity of the instrument or controlled environment for analysis. Thus, it requires in situ or remote analysis. Remote sensing is the acquisition of information about an object without being in physical contact [1]. The Remote analysis allows for gathering vital data without the need for physical presence or direct contact yielding to risk mitigation in exploring space.

V. Chavan · V. K. Unnikrishnan (✉)
Department of Atomic and Molecular Physics, Manipal Academy of Higher Education, Manipal, Karnataka, India
e-mail: unnikrishnan.vk@manipal.edu

C. Marques et al. (eds.), *Advanced Optical Sensors for Aerospace Applications*,
Progress in Optical Science and Photonics 37,
https://doi.org/10.1007/978-981-95-1626-1_9

The platform to remotely analysis varies depending on altitude, sensor type, and purpose of analysis. The remote platform can broadly be classified into

(a) ground-based
(b) airborne and
(c) spaceborne systems

The ground-based platforms are fixed or mobile stations on Earth's surface. They are used for environmental monitoring, atmospheric studies, etc. The atmospheric boundary layer is the lowest part of the atmosphere. In this region, most of the interaction between the atmosphere and the earth's surface takes place by exchanging heat, momentum, humidity, and aerosols as well as greenhouse and other atmospheric gases. The ground-based remote sensing instruments such as Radio Detection and Ranging (RADAR) wind profilers, microwave radiometers, Doppler wind Light Detection and Ranging (LiDAR), and automatic LiDAR have been utilized for profiling of atmospheric boundary layer [2]. Airborne sensors can be used for aero photography which provides high-resolution mapping along with a precise digital elevation model. The airborne remote sensing platforms include balloons, airplanes, drones, multi-altitude aircraft, etc. These platforms are non-ground-based platforms for early remote sensing works. Balloons are rarely used these days due to lack of stability and the unpredictable course of flight. Currently, the Unmanned Aerial System (UAE) is the emerging trend in remote analysis due to its low cost and ease of control [3, 4]. Airborne platforms work at lower altitudes with more flexibility [5]. Unoccupied Airborne Systems (UAS) provide an exclusive airborne platform that has the potential to report the limitations inherent in both methods of data collection which are manned airborne and spaceborne platform-based data collection via a middle-ground approach [4]. The spaceborne platforms are the ones mounted on on-board spacecraft or satellites. These are used for weather monitoring, communication, navigation, space and planetary exploration. They are not affected by earth's atmosphere. Spaceborne platforms cover large regions from space, by producing reliable data for earth observation [6].

The necessary instrumentation for remote sensing involves an energy source, medium for propagation, specimen under study, signal collection unit, data transmission unit, and processing unit.

There are two types of energy sources: passive and active remote sensing (Fig. 1). Passive remote sensing relies on a natural illumination source (the sun) that emits discontinuous radiation. Some examples of passive remote sensing are photography, radiometers, and infrared imaging. Active remote sensing operates with its energy source for transmitting radiation to the region of interest and recording the reflected or scattered radiation from objects within the range of the system. Active remote sensing works in all weather conditions and dark surroundings whereas passive remote sensing requires natural illumination presence which limits its applications in cloudy and dark conditions.

Remote sensing can further classify based on the spectral acquisition as multispectral remote sensing and hyperspectral remote sensing. Multispectral remote sensing acquires surface information in multiple wavelengths, enabling the classification of

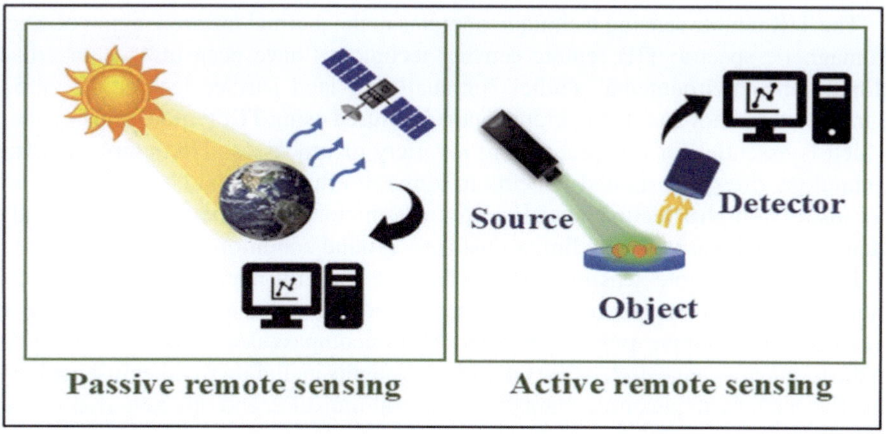

Fig. 1 Illustration of passive and active remote sensing sensor

different types of surfaces due to the property of reflecting light with different intensities and wavelengths. Different spectral behaviour leads to detailed classification. Multispectral remote sensing involves the acquisition of visible, near-infrared, and short-wave infrared images in several broad wavelength bands. A common multispectral imager for monitoring numerous kinds of landscape scale characteristics is NASA's Landsat [7]. Hyperspectral remote sensing, with its high spectral resolution, enables direct identification of individual materials based on their reflectance properties. It allows precise measurement of material spectra, making it possible to detect specific minerals, rocks, soils, and vegetation changes over time with exceptional accuracy [7].

Various approaches have been employed for distant aerospace investigations, including Thermal Infrared (TIR) remote sensing, microwave remote sensing [8], RADAR [9], LiDAR [10, 11], and others. LiDAR is an active sensor-based technique that uses a laser to illuminate the surroundings. LiDAR determines the distance by targeting a surface or object with the source laser and measuring the time required for the reflected light to return to a receive [12].

A LiDAR instrument contains of laser, a scanner, and a receiver. LiDAR technology has applications in disciplines for topographic mapping, forestry, and vegetation analysis for measuring height, urban planning, and infrastructure [13, 14].

The RADAR system utilizes radio waves to determine the distance, direction, and radial velocity of the object to the site. The component of RADAR features a transmitter that produces electromagnetic waves in the radio or microwave region, a transmitting antenna, a receiving antenna, and a processor. It is a classic example of an active remote sensing technique, where radio waves are emitted, and the return signals are analyzed to determine the distance, speed, and characteristics of objects in the surveyed area. The modern application of RADAR technology is highly diverse including weather monitoring, aviation for detecting aircraft, military surveillance applications, air traffic control, etc.

The TIR remote sensing technique operates in the thermal infrared region of electromagnetic spectra. TIR remote sensing techniques have been utilized in urban climate and environmental studies for analysing land surface temperature [15]. Surface energy flows and temperatures are measured using TIR remote sensing data, which is essential for comprehending a variety of processes. TIR remote sensing technology offers substantial benefits in terms of efficiency and cost-effectiveness because of its high detection precision, vast information content, and ability to quickly identify large areas with few limitations from ground conditions [16].

Microwave remote sensing has also become a useful method for tracking soil moisture since it can penetrate surface layers and vegetation and is less sensitive to weather. The technique uses microwaves which encompass electromagnetic radiation in the frequency range of 0.3–300 GHz offers benefits including weather insensitivity and the capacity to penetrate plants, to track soil moisture, and other environmental parameters.

All the remote sensing techniques discussed above have a wide range of applications in aerospace. The major application of these is in mapping, monitoring, imaging, thermal, and depth estimation, which makes them a vivid source of information. The significant limitation of these techniques is the acquisition of relevant information regarding the composition of materials. Thus, for remote material analysis, a different approach is required.

Spectroscopy is a powerful analytical technique that provides insights into the composition, structure, and properties of materials by analysing absorption, emissions, and scattering of radiation by chemical bonds, atoms, or molecules. Since it is non-destructive and has great selectivity and sensitivity, it is a wonderful analytical technique for remote analysis. Numerous spectroscopic techniques are used for material characterization, such as UV–Vis, IR, X-ray diffraction spectroscopy, Nuclear Magnetic resonance (NMR), etc. They all use different approaches and offers insights based on energy shift of atom or molecule. UV–Vis spectroscopy is used to calculate the band gap whereas IR spectroscopy represents the absorption peak corresponding to the frequency of vibrations present between bonds of molecules, and atoms. X-ray Diffraction (XRD) on the other hand provides information on crystal orientation, electron density plots, grain size, crystallinity, and phases with the major advantage of a non-destructive and fast sensing method [17]. However, there are certain limitations associated with these techniques as well. Limitations for UV–Vis spectroscopy are: not all samples absorb wavelengths between 200 and 800 nm and at higher concentrations samples deviate from following Beer-Lambert's law which is the founding principle of UV–Vis spectroscopy [18]. IR and FTIR face challenges due to poor absorbance of IR radiations which leads to noisy data. Also, complex samples like biological or food mixtures being very difficult to analyze due to peak overlapping and inability to detect monoatomic elements, ions etc. Inert gases further limit fictional working of IR and FTIR. XRD often needs Rietveld refinement for the most detailed information like electron density plots, crystalline parameters, miller indices, etc. Also, the sample must be crystalline to obtain information [17, 19].

These laboratory-based spectroscopic techniques provide accurate and precise information in a controlled environment only. The setup is mostly bulky thus making it difficult to use for in situ analysis. In remote material cartelization, it is essential that the instrument is portable to carry it to the sample space.

The remarkable properties of a laser such as coherence, monochromaticity, and directionality make it an optimal fit instrument for long-range sensing, and precision targeting for analysis. Laser spectroscopy techniques enhance conventional optical spectroscopy by providing selective excitation, high spatial, and temporal resolution, portability, and remote detection capabilities. Such advancement in analytical technique extends the application in vast fields. The diverse and uncontrollable conditions such as temperature, radiation, air pressure, etc. in Earth's atmosphere and outer space make remote detection a challenging assignment. Therefore, to match the performance of laboratory systems in uncontrolled conditions, the analytical system needs to have specific attributes.

2 Laser Spectroscopic Techniques in Aerospace

The remote or in-situ spectroscopic analysis system is typically equipped with a laser for efficient detection and analysis of specimens in hostile uncontrolled environment conditions. Some of the standoff spectroscopic techniques are Raman spectroscopy, Laser induced fluorescence (LIF), Laser induced breakdown spectroscopy (LIBS), Infrared spectroscopy (IR), terahertz spectroscopy, Photoacoustic spectroscopy (PAS), etc. [20]. Each technique has a specific methodology and delivers different information depending on the energy transition that takes place in an atom or molecule. In the following section, we discuss about LIBS, Raman, and LIF spectroscopic techniques.

2.1 Laser Induced Breakdown Spectroscopy (LIBS)

LIBS technique provides the elemental insides of the specimen. It depends on the fundamentals of laser ablation and optical emission spectroscopy as shown in Fig. 2. Plasma initiation, production, and decay are the key processes. The high-energy laser pulse causes the formation of plasma containing electrons, ions, and atoms in an excited state. For first few nanoseconds, continuum spectra are obtained due to inverse bremsstrahlung processes, arising from collisions between electrons and photons. Further, when plasma starts to decay and cool down, the electrons, atoms, and ions in the excited energy starts falling into ground states, leading to photon emission with distinct spectral peaks. These spectral peaks are characteristic of the particular emitting atoms and ions. Thus, a sample can be identified by analysing the obtained characteristic emission spectra.

LIBS has remarkable properties such as rapid, in-situ, real-time time analysis and the ability to detect multiple elements without sample pre-treatments or minimal sample preparation. Due to these properties, LIBS is used in various applications

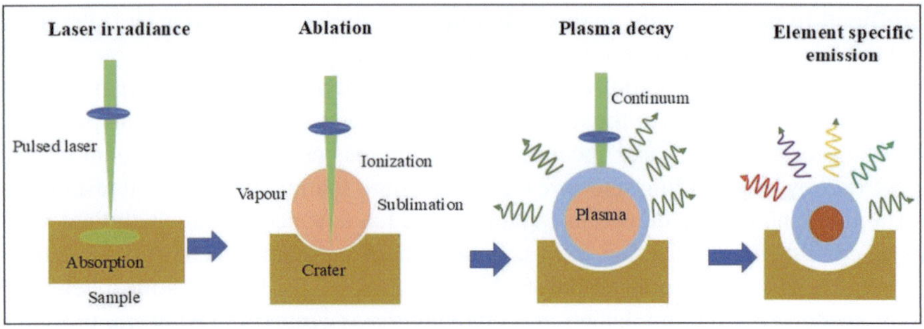

Fig. 2 Schematic of processes occurring during LIBS

such as; environmental monitoring viz. analysis of water, soil, and atmospheric aerosols [21], analysis of materials, liquids, metal, recycling, etc. [22], defence and security applications like explosives detection, identification of unknown substance [23], archaeological and cultural heritages applications viz. identification pigments in paintings, composition analysis of historical artifacts [24], space and planetary exploration for elemental analysis, and many more.

These properties make LIBS an appropriate choice for remote analysis. With LIBS techniques, lighter elements such as H, Be, Li, C, N, O, Na, and Mg, and thin-sample analysis can be detected without the worry of substrate interference. The elimination of sample pre-treatments, and real-time and in-situ analysis make the technique appropriate for remote analysis of samples. This provides a leverage of contactless analysis of unknown specimens in hostile environments. Standoff or remote analysis of the sample by As Is Where Is condition reduces the possible hazards associated with detention of such elements using conventional methods.

Remote or standoff LIBS have been used for the detection of hazardous materials like explosives, biological warfare agents [25], etc. The major advancement of LIBS is its element deletion capabilities in a hostile environment. The very fact that it is not a necessity to bring the sample into the laboratory environment makes the LIBS analysis attractive especially for planetary surface conditions. LIBS has been utilized in several space and planetary missions to analyze the elemental composition of extraterrestrial materials. NASA's Mars Science Laboratory (MSL) rover, Curiosity, with a scientific payload has included a stand-off LIBS spectrometer as part of its ChemCam instrument. ChemCam analyzes rocks and regolith on Mars at 1–7 m distances to identify their chemical makeup [26]. LIBS was deployed on the ISRO Pragyan rover to determine the elemental comprised in lunar regolith and rocks in the Chandrayan-2 mission [8]. These missions highlight the growing interest of remote LIBS in planetary exploration, providing critical data on the elemental composition of celestial bodies.

2.2 *Raman Spectroscopy*

Indian physicist Chandrashekhar Venkata Raman discovered the phenomenon of inelastic scattering of light, known as the Raman effect [27]. This explains the frequency shift of inelastic scattered light. Elastic scattering and inelastic scattering are the two possible results of light scattering off a material. Light scattering on material often results in energy loss which is attributed to inelastic scattering. When there is no loss of energy it is called elastic scattering. This can be further elaborated as: when the energy of scattered light is the same as that of the incident, it is referred to as elastic scattering or Rayleigh scattering. On the other hand, when the scattered light has a different energy than the incident light, it is referred to as inelastic scattering or Raman scattering. This implies that the frequency and wavelength of light that is inelastically scattered will differ from those of the incident light beam. Molecules in a sample undergo continuous motion that creates molecular bonds with continuous vibrations. These vibrations take place at particular frequencies that are particular to the bond type and molecule. Larger molecules with plenty of chemical bonds vibrate at different frequencies due to their varied structures (Fig. 3).

When light interacts with the molecule, some of it is absorbed, and the rest scatters at a different frequency, a phenomenon known as the Raman effect. Thus, identifying these light wavelengths allows us to determine the molecular compounds found within the specimen. The requirement for a molecule to be Raman active is that its polarizability must change during the motion. The resulting scattered light

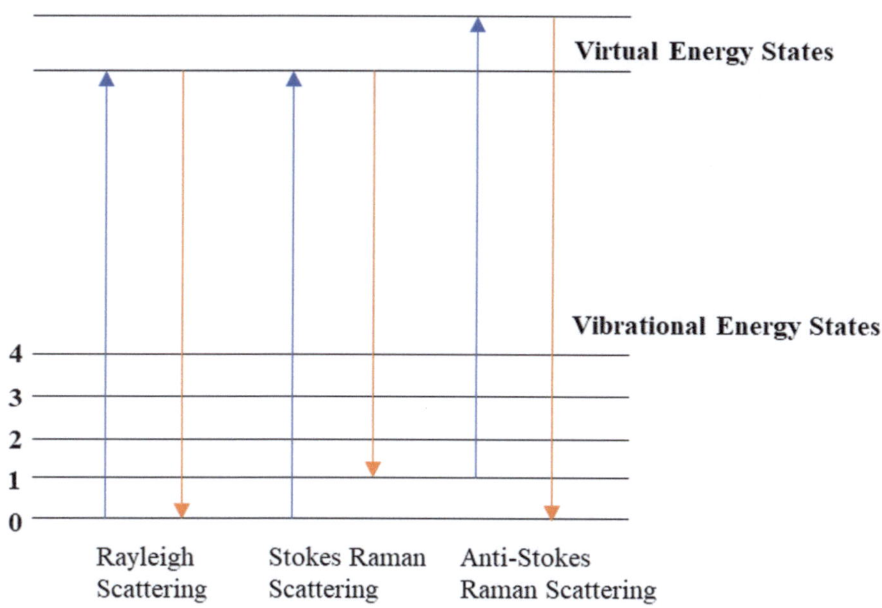

Fig. 3 Schematic of transitions in Raman scattering

in Raman scattering can be photons of either lower or higher frequency than the incident photons. If the resulting photons have a lower frequency, it is called Stokes-Raman scattering, and if they have a higher frequency, it is called anti-Stokes-Raman scattering. In Stokes, scattering arises from the excitation of molecules from a lower energy state to a higher energy state. Thus, the number of Stokes scattering is comparatively higher than that of anti-Stokes scattering. The intensity of Raman scattering is proportional to the magnitude of the change in molecular polarization [28].

Raman spectroscopy detects vibrational transitions within a molecule. Since these transitions are unique to each chemical structure, the resulting Raman spectrum serves as a distinctive molecular fingerprint. Real time monitoring system have employed Raman spectroscopy to identify chemical and biological warfare agents, illicit drugs, and hazardous environmental materials. Moreover, with no requirement for any type of sample preparation in certain applications, it can be used for solid, liquid, or gas forms and it's a non-destructive method. Raman spectroscopy is employed in planetary applications to determine variety of molecular substances, including water, water ice, water-bearing minerals, and organics of special importance to planetary research, can be identified and characterized by the Raman spectroscopy technique [29]. Instruments like Super Cam were used in NASA's Perseverance Rover involves Raman spectroscopy for the identification of minerals and organic compounds on Mars [30].

2.3 *Laser Induced Fluorescence (LIF)*

Luminescence is the emission of light that occurs in an electronically excited state in any substance. The phenomenon of luminescence is classified as fluorescence and phosphorescence. Fluorescence emission occurs in excited singlet states, where the electrons in the excited orbit are coupled to the electrons in the ground state orbit by a contrary spin, resulting in the rapid emission of photons that eventuate while returning to the ground state. This takes place in a few nanoseconds [31]. Whereas in phosphorescence, the emission of light falls within the triplet excited state and the ground state, where the spin of the electron in the excited state is identical, which gives rise to a longer lifetime of milliseconds to seconds. The transition mentioned can be well explained by the Jablonski diagram shown in Fig. 4. The vibrational level of a molecule is represented by horizontal black lines, the bold lines have the lowest energy (ground energy), and the thin ones have the highest vibrational energy, indicating the order of increasing energy in an upward direction. Here, only a few vibrational levels are used to present the continuum. In general, as per the spin-angular momentum configuration, names are assigned to the electronic states. The S denotes the singlet state and the T denotes the triplet state, having total spin angular momentum of 0 and 1, respectively. The S_0, S_1, and S_n represent the singlet ground state, the first excited state, and the nth excited state, respectively. Similarly, T_1 represented the first excited triplet state. The shift among two energy states occurs when a molecule in a ground state absorbs energy equivalent to the energy difference

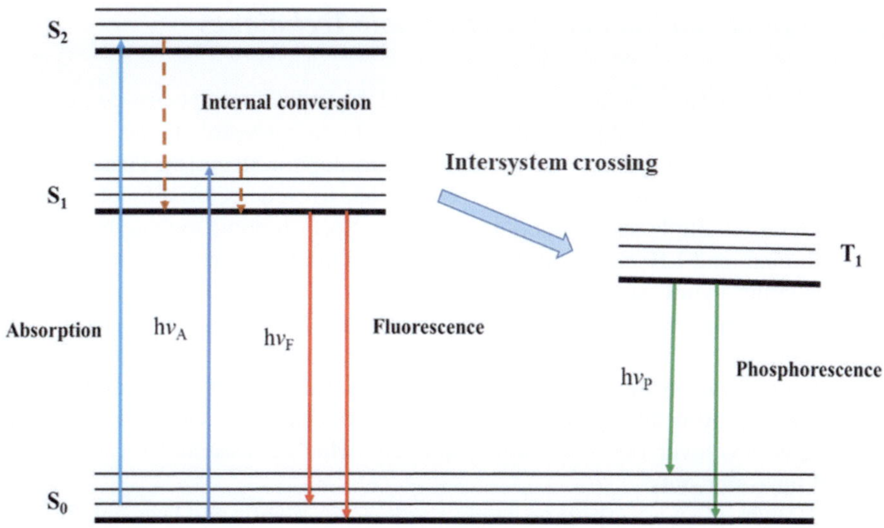

Fig. 4 Jablonski diagram demonstrating the fluorescence process

between its two electronic states and gets excited to the highest vibrational state of the excited electronic state. An excited electron undergoes a non-radiative transition, called an internal conversion process, and relaxes to the lowest vibrational level. The excess vibrational energy is lost within the vibrational levels of the same molecule or to the surrounding molecules. The electrons relax radiatively from S_1 to the vibrational level of the electronic ground state S_0 by emitting photons which results in fluorescence.

Laser-induced fluorescence (LIF) involves the excitation of molecules using a laser beam, which produces fluorescence emission having the characteristic of precise molecular species [32]. The wavelength of the laser source used for LIF setup must match the maximum adsorption wavelength of the target fluorescence molecules so that sufficient excitation to the higher energy level takes place, and then the excited fluorescence molecules return to the ground energy level facilitating fluorescence emission. This technique is mainly used to identify organic substances and biological specimens.

As in the case of aerospace field, Butterworth et al. claim that their results confirm the effective performance of several critical steps for remote and independent operation in novel extraterrestrial environments. The use of LIF techniques can be established as a major improvement in extraterrestrial chemical and biochemical analysis, which is also vital for space exploration [33]. Also, the ability of such analysis systems can be extended to future space missions, mostly to Mars and icy moons like Enceladus and Europa and can also impact environmental monitoring.

Other than space exploration and planetary studies, laser remote spectroscopic systems are used in environmental monitoring, defence and security, archaeological studies, and various other non-reachable sample fields [34–36].

2.4 Multimodal Standoff Spectroscopic Techniques

The standoff spectroscopic system established on individual spectroscopic techniques provides a specific piece of information. In recent years studies have been reported on multimodal or bimodal standoff spectroscopic systems [8]. These systems offer powerful advantages in terms of comprehensive characteristics of material by combining multiple spectroscopic techniques. Spectroscopic techniques such as LIBS, Raman, and LIF share similar instrumentation requirement making them fit for developing hybrid systems as shown in Fig. 5. LIBS requires a higher energy laser pulse for plasma formation whereas Raman and LIF require less energy laser pulse. Therefore, the incident beam energy must be reduced, either through an energy regulator or by increasing the beam spot size on the sample surface. Additionally, a time-gated spectrometer synchronizes data collection with the laser excitation pulse, allowing control over the gate width and delay, which is essential for sequential spectral measurements.

The Raman effect occurs immediately after laser excitation and lasts only briefly, whereas fluorescence involves molecular excitation to higher electronic states, followed by relaxation and subsequent fluorescence emission, classically occurring on the picosecond to nanosecond timescale. Therefore, the optimal sequence for

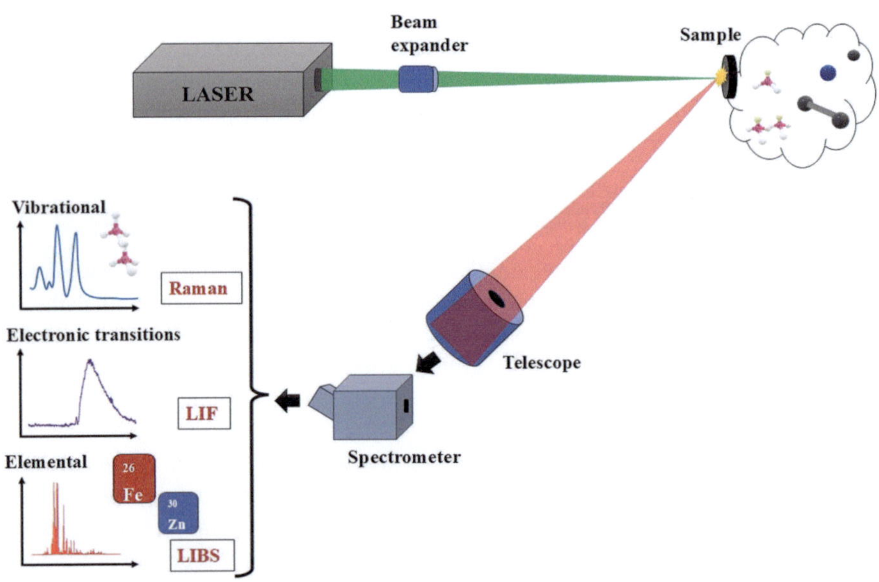

Fig. 5 Illustration of multimodal remote system

spectral acquisition is Raman spectroscopic signals followed by laser-induced fluorescence and laser-induced breakdown spectroscopy respectively. Hybrid spectroscopic systems offer significant advantages by reducing costs, labour, and detection time while providing comprehensive atomic, ionic, and molecular information in a single, efficient setup.

3 Applications of Remote Spectroscopic Techniques in Space Exploration

As discussed above, laser-based remote spectroscopic techniques are powerful methods to perform rapid chemical analysis of samples in As-Is-Where-Is conditions. Their robust data-gathering ability in harsh environments makes them potential tools for exploring space and extraterrestrial substances. The deep vacuum conditions, extreme temperature variation, radiation strengthening, and dust Strome conditions are major environmental challenges associated with planetary and lunar studies [37].

In recent years, LIBS has become a credible technique for space exploration. The Curiosity rover of NASA's Mars Science Laboratory (MSL) was the first on-board application of LIBS [37]. The elementary purpose of the mission is to evaluate the Martian Gale Crater geology and the possibility of supporting life [38]. The ChemCam suit on the rover consists of LIBS instruments to analyze the elemental composition of the Martian surface. They used a laser pulse with a power density greater than 10 MW/mm^2. These pulses focus on the sample (soil or rock) covering a spot of diameter 0.2 to 0.6 mm which was in 7 m of the detection range. A telescope with a 110 mm diameter was employed to detect plasma emissions. The optic fibre coupled the signal to the three-spectrometer, which records the spectra in the wavelength range of 240–850 nm with a resolution of 0.15–0.60 nm [39]. Trace level of boron was detected in calcium sulfate veins [40]. Currently along with the Curiosity rover, the Perseverance rover and Zhurong rover are also equipped with LIBS instrumentation for martial surface exploration using the LIBS technique [41].

The follow up tool to ChemCam, Super Cam, is a component of the scientific payload of NASA's 2020 Mars mission to Jezero Crater. For the first time, a single space exploration tool combined LIBS with Raman spectroscopy. Super Cam is a standoff instrument, designed to be a multi-analytical suite of five coaligned techniques: LIBS, Visible (VIS), and Infrared (IR) Spectroscopy (Vis-IR), Remote Micro-Imaging (RMI) and Time-Resolved Raman and Luminescence (TRR/L) [42].

The first moon mission equipped with the LIBS instrument for lunar elemental analysis was loaded on Indian Space Research Organization's (ISRO) Pragyan rover in Chnadrayan-2 mission [41]. However, due to a harsh landing of lander craft on the lunar surface, the mission was not able to accomplish the possible findings. The latest follow-up lunar mission, Chandramana-3 of ISRO successfully landed near the lunar south pole which has been equipped with low-energy eye-safe LIBS (LE-LIBS). The LE-LIBS setup consists of a low-energy laser of $\leq$ 3 mJ energy having a wavelength of 1535 ± 1 nm. The laser beam divergence is ~ 0.43 mrad and the spot size is ≤ 100 μm. The signal collection unit is a folding geometry comprised of two mirrors and three lenses system. Spectra of the collected signal are obtained by using a combination of Aa holographic reflection grating free from aberration and a Charge Coupled Detector (CCD) [43]. The first LIBS spectra of the moon regolith pointed the existence of O, Si, Fe, Mn, Ti, Al, Ca, and Cr [43].

Around the globe, different space agencies such as NASA, ISRO, CNSA, ESA, and others are trying to acquire insights into the expanded space. Implementing laser-based spectroscopic techniques in the mission provides advancement in the understanding of the extraterrestrial environment.

4 Development in Remote Spectroscopic Analysis

Over the years significant improvement has been made in remote or standoff spectroscopic analysis in terms of advanced equipment, applications, and analytical power. Elementary instrumentation of remote sensors consists of an illumination source, signal collector, data transmission and processing system. The significant improvement in basic instrumentation increases the sensitivity and specificity of the entire system. Recent studies have been reported showing the potential of performing different spectroscopic techniques with one system which enhances the calibre of the system, yielding overall elemental composition and molecular structural information of the specimen under observation. Such a hybrid or multimodal system increases the diversity in data collection abilities of the system. The use of advanced computational techniques enables to retrieve the meaningful insights from complex spectroscopic datasets.

4.1 Basic Instrumentation of Laser Spectroscopic Remote System

The remote spectroscopic system consists of an excitation source, focusing optics, a collection system, a detector, and an analyzer. Terrestrial, planetary, and space explorational studies require a robust source, lasers, due to their high fluence, beam stability, and unidirectionality. They are the most appropriate choices for spectroscopic remote sensing in aerospace. A wide range of lasers including semiconductor lasers, edge-emitting lasers, vertical-cavity surface-emitting lasers, solid-state lasers, etc. were used in space exploration based on the requirements of the mission. Certain lasers, like gas, chemical, and dye lasers, have also been researched for use in space, but they are more challenging to modify for the space environment due to their short lifetime and poor radiation resistance [48]. The parameters associated with lasers such as wavelength, peak power, pulse duration, and repetition rate, are fundamental considerations for laser performance. The laser operation in space is subjected to different limitations compared to its terrestrial applications. For this reason, lasers that are appropriate for space applications must meet a range of requirements, including long lifespan, high efficiency, minimal sensitivity to contamination and optical misalignment, and unattended operation. These fundamental laser parameters significantly influence the quality, accuracy, and type of spectral information.

The laser beam is required to be focused on the distant sample to perform remote measurements this is carried out using focusing optics. The big aperture lenses or the combination of lenses typically called beam expanders are used to focus the laser beam on the targeted sample. The use of a single lens to expand the beam not only fixes the distance of analysis but also the aperture of the lens for far focusing making the instrument handling difficult. The use of beam expanders in the system offers the privilege of changing the focusing distance by adjusting separation distances between the lenses. The beam expander comes in two variants viz. Keplerian and Galilean are shown in Fig. 6. The focal lengths of the primary and secondary lenses are represented by F_1 and F_2 in the figure.

The design of the Keplerian expander comprises a positive short focal length and a long focal length lens as entry and exit lenses respectively, which produces intermediate focus. Whereas the Galilean beam expander comprises a negative lens and a positive lens as entry and exit lenses respectively. The Galilean beam expander shortens the longitudinal dimension and also avoids air breakdown caused due to internal focus under the usage of a high-energy laser source [49]. To reduce different optical abbreviations in the basic combination, and to increase the focusing range and efficiency, multiple lenses are also employed for distant focus. The well-focused beam on the targeted specimen initiates different electronic, vibrational, and rotational transitions which in turn yields the atomic or molecular signal.

Telescopes are often used to collect the signals in standoff spectroscopic systems. Their optical design enables them to gather the maximum amount of signal over a long distance in challenging environments. There are different types of optical combinations in telescopes such as reflector, catadioptric, and refractor. In most of the standoff spectroscopic studies reflecting-type telescopes, a combination of

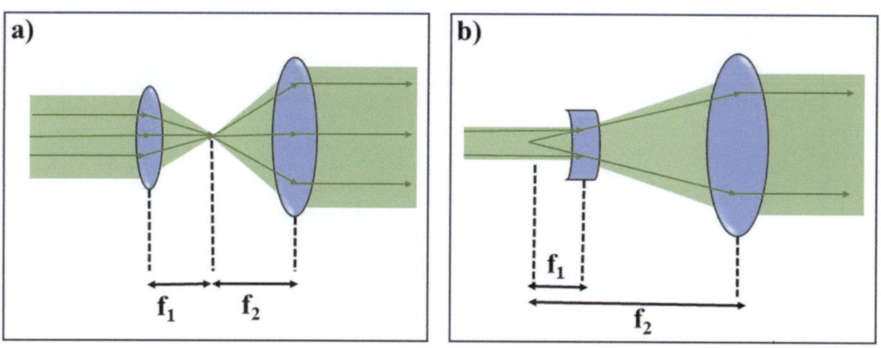

Fig. 6 Schematic of beam expanders: **a** Keplerian beam expander and **b** Galilean beam expander

mirrors, and catadioptric telescopes consisting of lens and mirror combination are used.

Reflecting type Cassegrain telescope are mainly used in standoff spectroscopic systems due to its alignment stability, compact design, and less obstruction in signal coupling as depicted in Fig. 7. The design of the Cassegrain telescope incorporates a parabolic primary mirror with a hole drilled in the centre. A secondary hyperbolic mirror directs the collected signal through the hole in the primary mirror to the eyepiece [50].

Limited studies have also utilized Newtonian telescopes, which use mirrors and lenses to capture and focus signals from remote distances before directing them to the eyepiece. However, this setup can increase system complexity. In contrast, the Cassegrain telescope simplifies the process by focusing the signal at the rear end of the telescope.

The gathered signal is coupled to the spectrometer for further measurement. Czerny Turner and echelle are two commonly used spectroscopic designs (Fig. 8). In Czerny-Turner spectrometer light enters through a slit and is directed through a collimating mirror, followed by a diffraction grating, focussing mirror, and finally reaches to the detector. This design allows for high spectral resolution due to its

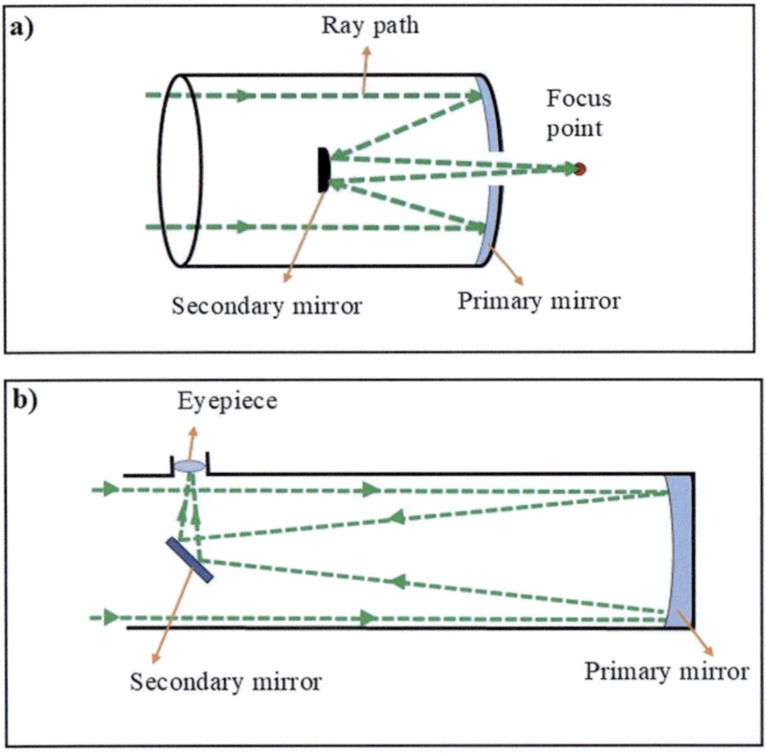

Fig. 7 Schematic of telescopes **a** reflecting type Cassegrain telescope and **b** Newtonian telescope

precise manipulation of light paths. However, it covers a more limited spectral range in a single pass. Therefore, multiple grating or scans are needed for broad-range analysis. On the other hand, the echelle spectrometer employs a diffraction grating with a high blazing angle. To separate overlapping orders produced by the echelle grating, a cross-disperser (like a prism or another grating) is used. This creates a two-dimensional spectrum, with the echelle grating which has dispersing light in one direction and the cross-disperser in the other. Echelle spectrometers can cover a wide spectral range with excellent resolution in one exposure, making them ideal for fast, broad-spectrum analysis.

Both spectrometers often use Intensified Charge-Coupled Device (ICCD) or Charge-Coupled Device (CCD) detectors for optical signal detection. A CCD is a semiconductor device that converts light (photons) into electrical charges, which are then gathered and read out to form an image or spectrum. CCDs are known for their ability to capture detailed images with high spatial resolution and good quantum

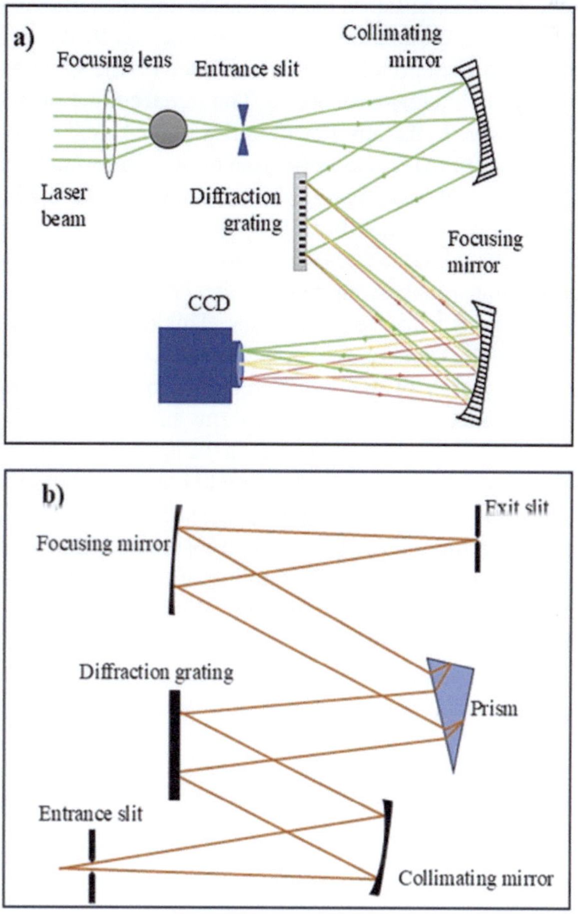

Fig. 8 Internal scheme **a** Czerny turner spectrometer. *Reprinted with permission from {M. Naeem, N.-u.-a. Fatima, M. Hussain, T. Imran, and A. S. Bhatti, "Design simulation of Czerny–Turner configuration-based Raman spectrometer using physical optics propagation algorithm," Optics, vol. 3, no. 1, pp. 1–7, 2022}. Copyright {2022} MDPI* [51] and **b** echelle spectrometer

efficiency, but they have limited sensitivity at very low light levels and slower readout speeds compared to other detector types. In contrast, an ICCD combines a CCD with an image intensifier, which amplifies the incoming light signal before it reaches the CCD, enhancing its sensitivity, especially at low light levels. ICCDs are highly sensitive, capable of detecting single photons, and have fast gating capabilities to record brief light pulses (down to nanoseconds). This makes them ideal for time-resolved studies and real-time imaging in extremely low-light environments.

4.2 Development in System Scheme and Signal Collection Efficiency

The fundamental instrumentation of the standoff spectroscopic system is commonly aligned in the oblique scheme and back collection scheme. In the oblique optical scheme, the excitation source (laser) and the collection optics typically telescope is aligned at a certain angle. As shown in Fig. 9. Whereas in a coaxial optical scheme both the laser and telescope are aligned on the same optical axis. Both schemes have certain advantages over others. The signal collection efficiency is higher in the oblique scheme due to the absence of optics in the signal collection path. However, there is a requirement to refocus collection optics while analysing at different distances, or the sample under investigation has an uneven surface which is normal during stand-off operations. This drawback is resolved in the coaxial scheme. The coaxial scheme is less prone to geometrical distortions. Due to the common optical axis, there is ease in collecting signals from distinct distances. However, the presence of multiple optical components causes signal losses. This can be further improved by reducing optics or using the same optical components for focusing and signal collection.

The quality of the collected signal to the spectrometer significantly impacts the performance of the system in terms of its sensitivity and accuracy. The choice of signal scheme is majorly based on the application and the specific need of the standoff spectroscopic system. In the direct coupling scheme, the collected signal is directly sent to the spectrometer through the slit with precise alignment. This scheme leads to a smaller amount of transmission losses and high signal efficiency but has minimal alignment flexibility. The coupling of the signal can also be done using a fibre optic cable. The fibre-optic coupling allows flexibility in positioning the detector and provides signal protection from environmental factors.

As space samples cannot be brought into the vicinity of the instrument there is a great need to develop a portable, lightweight, stand-off spectroscopic system that can be easily transported or placed on the rover or drone. Such studies are the need of the hour considering the recent quest to explore different regions of the aerospace in depth.

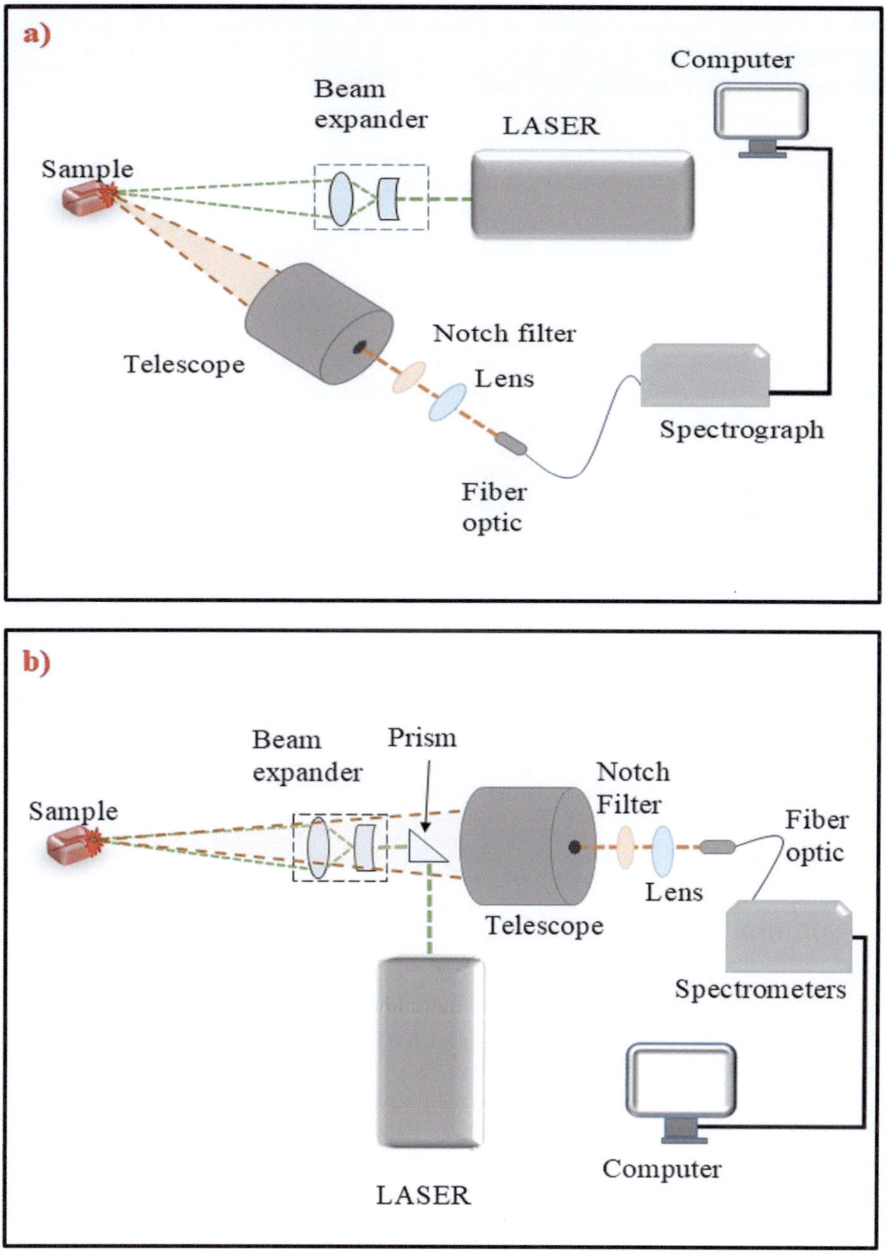

Fig. 9 Optical schemes for standoff spectroscopy system: **a** Oblique scheme and **b** Coaxial scheme

Table 1 Planetary missions exploiting the possibilities of LIBS, Raman, and LIF spectroscopic techniques

Mission	Year	Space agency	Objective	Spectroscopic technique			Ref.
				LIBS	Raman	LIF	
Chandrayan-3	2023	ISRO, India	In-situ analysis of lunar surface composition	✓	×	×	[44]
Tianwen-1	2021	CNSA, China	Exploring geological composition on Mars's surface	✓	×	×	[45]
Chandrayan -2	2019	ISRO, India	Orbital and attempted landing to study the lunar surface	✓	×	×	[42]
Chang'e 7	2024	CNSA, China	Exploration of the lunar surface	×	✓	×	[46]
Mars 2020	2021	NASA, USA	Detailed study of Martian geology and search for signs of the past	✓	✓	(Time-resolved luminescence)	[47]
Mars Science Laboratory	2012	NASA, USA	In-situ analysis of the Martian surface using ChemCam instrument	✓	×	×	[39]

5 Summary and Conclusion

Laser remote spectroscopic techniques have been used for aerospace applications. Lasers' unique properties make them ideal for remote detection. Remote or in-situ spectroscopic systems proved the capability of providing chemical information of samples in harsh, uncontrolled environments providing contactless analysis. Techniques like Raman spectroscopy, LIBS, and LIF identify molecular and elemental compositions. Hybrid standoff spectroscopic systems combine different spectroscopic techniques and yield a wide range of information with a single system. By integrating multiple methods, detection accuracy can be enhanced with reduced costs and improved efficiency in aerospace and other applications. The complementarity of these techniques adds value to interpreting the data meaningfully with a high degree of confidence. There is emerging interest for such compact, low-weight spectroscopy systems which provide portability and in turn, make it easy to install with a drone or rover for space explorations.

References

1. J.B. Campbell, R.H. Wynne, *Introduction to Remote Sensing*. (Guilford Press, 2011), p. 1609181778
2. S. Kotthaus, J.A. Bravo-Aranda, M. Collaud Coen, J.L. Guerrero-Rascado, M.J. Costa, D. Cimini, E.J. O'Connor, M. Hervo, L. Alados-Arboledas, M. Jiménez-Portaz, Atmospheric boundary layer height from ground-based remote sensing: a review of capabilities and limitations. Atmos. Meas. Tech. Discuss. **22**, 1–88 (2022)
3. Z. Zhang, L. Zhu, A review of unmanned aerial vehicle remote sensing: platforms, sensors, data processing methods, and applications. Drones. **7**(6), 398 (2023). https://doi.org/10.3390/drones7060398
4. S. Zhang, S.M. Bogus, C.D. Lippitt, V. Kamat, S. Lee, Implementing remote-sensing methodologies for construction research: an unoccupied airborne system perspective. J. Constr. Eng. Manag. **148**(9), 03122005 (2022). https://doi.org/10.1061/(ASCE)CO.1943-7862.0002347
5. C. Magnard, M. Frioud, D. Small, T. Brehm, H. Essen, E. Meier, Processing of MEMPHIS Ka-band multibaseline interferometric SAR data: From raw data to digital surface models. IEEE J Sel Top. Appl. Earth Observations Remote Sens. **7**(7), 2927–2941 (2014). https://doi.org/10.1109/JSTARS.2014.2315896
6. M.M. Verstraete, D.J. Diner, J.-L. Bézy, Planning for a spaceborne earth observation mission: from user expectations to measurement requirements. Environ Sci Policy **54**, 419–427 (2015). https://doi.org/10.1016/j.envsci.2015.08.005
7. S.S. Kumar, S. Arivazhagan, N. Rangarajan, Remote sensing and GIS applications in environmental sciences–a review. J. Environ. Nanotechnol. **2**(2), 92–101 (2013). https://doi.org/10.13074/jent.2013.06.132025
8. B. Das, P. Rathore, D. Roy, D. Chakraborty, R.S. Jatav, D. Sethi, P. Kumar, Comparison of bagging, boosting and stacking algorithms for surface soil moisture mapping using optical-thermal-microwave remote sensing synergies. CATENA **217**, 106485 (2022). https://doi.org/10.1016/j.catena.2022.106485
9. D. Amitrano, G. Di Martino, R. Guida, P. Iervolino, A. Iodice, M.N. Papa, D. Riccio, G. Ruello, Earth environmental monitoring using multi-temporal synthetic aperture radar: a critical review of selected applications. Remote Sensing **13**(4), 604 (2021). https://doi.org/10.3390/rs13040604
10. T. Fahey, M. Islam, A. Gardi, R. Sabatini. Laser Beam Atmospheric Propagation Modelling for Aerospace LIDAR Applications. Atmosphere, 12,7, 918, 2021 (2021). https://www.mdpi.com/2073-4433/12/7/918
11. N.A. Muhadi, A.F. Abdullah, S.K. Bejo, M.R. Mahadi, A. Mijic, The use of LiDAR-derived DEM in flood applications: a review. Remote Sens. **12**(14), 2308 (2020). https://doi.org/10.3390/rs12142308
12. Li, J. Ibanez-Guzman, Lidar for autonomous driving: the principles, challenges, and trends for automotive lidar and perception systems. IEEE Signal Processing Mag. **37**(4), 50–61 (2020). https://doi.org/10.1109/MSP.2020.2973615
13. M. Beland, G. Parker, B. Sparrow, D. Harding, L. Chasmer, S. Phinn, A. Antonarakis, A. Strahler, On promoting the use of lidar systems in forest ecosystem research. For. Ecol. Manage. **450**, 117484 (2019). https://doi.org/10.1016/j.foreco.2019.117484
14. K. Anderson, S. Hancock, S. Casalegno, A. Griffiths, D. Griffiths, F. Sargent, J. McCallum, D.T.C. Cox, K.J. Gaston, Visualising the urban green volume: Exploring LiDAR voxels with tangible technologies and virtual models. Landsc. Urban Plan. **178**, 248–260 (2018). https://doi.org/10.1016/j.landurbplan.2018.05.024
15. A.M. Coutts, R.J. Harris, T. Phan, S.J. Livesley, N.S. Williams, N.J. Tapper, Thermal infrared remote sensing of urban heat: hotspots, vegetation, and an assessment of techniques for use in urban planning. Remote Sens. Environ. **186**, 637–665 (2016). https://doi.org/10.1016/j.rse.2016.09.007

16. S. Wang, W. Xu, T. Guo., Advances in thermal infrared remote sensing technology for geothermal resource detection. Remote Sens. **16**(10), 1690 (2024). https://www.mdpi.com/2072-4292/16/10/1690
17. A.A. Bunaciu, E.G. UdriŞTioiu, H.Y. Aboul-Enein, X-ray diffraction: instrumentation and applications. Crit. Rev. Anal. Chem. **45**(4), 289–299 (2015). https://doi.org/10.1080/10408347.2014.949616
18. F.S. Rocha, A.J. Gomes, C.N. Lunardi, S. Kaliaguine, G.S. Patience, Experimental methods in chemical engineering: ultraviolet-visible spectroscopy—UV-Vis. Can. J. Chem. Eng. **96**(12), 2512–2517 (2018). https://doi.org/10.1002/cjce.23344
19. G. Kimmel, D. Mogilyanski, Advantages and disadvantages of fast XRPD measurement by using image-plate and rotating anode source Zeitschrift für Kristallographie. Supplements. **30**, 171–176 (2009). https://doi.org/10.1524/zksu.2009.0024
20. L.M. Narlagiri, M. Bharati, R. Beeram, D. Banerjee, V.R. Soma, Recent trends in laser-based standoff detection of hazardous molecules. TrAC, Trends Anal. Chem. **153**, 116645 (2022). https://doi.org/10.1016/j.trac.2022.116645
21. Y. Zhang, T. Zhang, H. Li, Application of laser-induced breakdown spectroscopy (LIBS) in environmental monitoring. Spectrochim. Acta Part B **181**, 106218 (2021). https://doi.org/10.1016/j.sab.2021.106218
22. S. Legnaioli, B. Campanella, F. Poggialini, S. Pagnotta, M.A. Harith, Z.A. Abdel-Salamb, V. Palleschi, Industrial applications of laser-induced breakdown spectroscopy: a review. Anal. Methods **12**(8), 1014–1029 (2020). https://doi.org/10.1039/C9AY02728A
23. F.C. De Lucia, J.L. Gottfried, Laser-induced breakdown spectroscopy for the detection and characterization of explosives. Count. Terrorist Detect. Tech. Explos. **2**, 269–313 (2022). https://doi.org/10.1016/B978-0-444-64104-5.00011-4
24. X. Bai, H. Allègre, M. Gosselin, P. Dillmann, M. Lopez, F. Téreygeol, V. Detalle, Impact of laser-induced breakdown spectroscopy implementation for the quantification of carbon content distribution in archaeological ferrous metals. Spectrochim. Acta Part B **172**, 105964 (2020). https://doi.org/10.1016/j.sab.2020.105964
25. J.L. Gottfried, F.C. De Lucia, C.A. Munson, A.W. Miziolek, Standoff detection of chemical and biological threats using laser-induced breakdown spectroscopy. Appl. Spectrosc. **62**(4), 353–363 (2008)
26. P. Sobron, A. Wang, F. Sobron, Extraction of compositional and hydration information of sulfates from laser-induced plasma spectra recorded under Mars atmospheric conditions—implications for ChemCam investigations on Curiosity rover. Spectrochim. Acta Part B **68**, 1–16 (2012). https://doi.org/10.1016/j.sab.2012.01.002
27. R.S. Das, Y. Agrawal, Raman spectroscopy: recent advancements, techniques and applications. Vib. Spectrosc. **57**(2), 163–176 (2011). https://doi.org/10.1016/j.vibspec.2011.08.003
28. P. Rostron, S. Gaber, D. Gaber, Raman spectroscopy, review. Int. J. Eng. Tech. Res. (IJETR). **6**(1), 50–64 (2026)
29. Y. Dwivedi, Concept and applications of standoff Raman spectroscopy techniques. Mol. Laser Spectrosc. **2**, 483–520 (2020). https://doi.org/10.1016/B978-0-12-818870-5.00014-9
30. R.C. Wiens, S. Maurice, S.H. Robinson, A.E. Nelson, P. Cais, P. Bernardi, R.T. Newell, S. Clegg, S.K. Sharma, S. Storms, The SuperCam instrument suite on the NASA Mars 2020 rover: body unit and combined system tests. Space Sci. Rev. **217**, 1–87 (2021). https://doi.org/10.1007/s11214-020-00777-5
31. R. Lakowicz, Principles of fluorescence spectroscopy **3**, 623–673 (2006). https://doi.org/10.1007/978-0-387-46312-4
32. H. Zeng, *Laser-Induced Fluorescence*. (Springer Science+Business Media New York, 2014). https://doi.org/10.1007/978-3-642-27758-0_793-6
33. A.L. Butterworth, M. Golozar, Z. Estlack, J. McCauley, R.A. Mathies, J. Kim, Integrated high performance microfluidic organic analysis instrument for planetary and space exploration. Lab Chip **24**(9), 2551–2560 (2024). https://doi.org/10.1039/D4LC00012A
34. R. Junjuri, A.P. Gummadi, M.K. Gundawar, Standoff identification of plastic waste using a low-cost compact laser-induced breakdown spectroscopy (LIBS) detection system. Appl. Spectrosc. **78**(10), 1089–1097 (2024). https://doi.org/10.1177/00037028241268348

35. P. Vinod, M.S. Babu, R. Sarathi, N.J. Vasa, S. Kornhuber, Influence of standoff distance and sunlight on detection of pollution deposits on silicone rubber insulators adopting remote LIBS analysis. IEEE Trans. Ind. Appl. **58**(3), 3285–3293 (2022). https://doi.org/10.1109/TIA.2022.3159771
36. A. Abdallah, A. Mahmoud, M. Mokhtar, A. Mousa, H. Ayoub, Y. Elbashar, Raman spectroscopic and advanced signal processing analyses for real-time standoff detection and identification of explosives. Opt. Quant. Electron. **54**(4), 265 (2022). https://doi.org/10.1007/s11082-022-03631-w
37. H. Saeidfirozeh, P. Kubelik, V. Laitl, A. Křivková, J. Vrabel, K. Rammelkamp, S. Schröder, I.B. Gornushkin, E. Képeš, J. Žabka, M. Ferus, P. Pořízka, J. Kaiser, Laser-induced breakdown spectroscopy in space applications: review and prospects. TrAC Trends Anal. Chem. 117991 (2024). https://doi.org/10.1016/j.trac.2024.117991
38. A.R. Vasavada, J.P. Grotzinger, R.E. Arvidson, F.J. Calef, J.A. Crisp, S. Gupta, N. Mangold, J. Hurowitzs, S. Maurice, M.E. Schmidt, R.C. Wiens, R.M.E. Williams, Overview of the mars science laboratory mission: bradbury landing to yellowknife bay and beyond. J. Geophys. Res. Planets **119**(6), 1134–1161 (2014). https://doi.org/10.1002/2014JE004622
39. J.P. Grotzinger, J. Crisp, A.R. Vasavada, R.C. Anderson, C.J. Baker, R. Barry, D.F. Blake, P. Conrad, K.S. Edgett, B. Ferdowski, R. Gellert, J.B. Gilbert, M. Golombek, J. Gómez-Elvira, D.M. Hassler, L. Jandura, M. Litvak, P. Mahaffy, J. Maki, M. Meyer, M.C. Malin, I. Mitrofanov, J.J. Simmonds, D. Vaniman, R.V. Welch, R.C. Wiens, Mars science laboratory mission and science investigation. Space Sci. Rev. **170**, 5–56 (2012). https://doi.org/10.1007/s11214-012-9892-2
40. A.R. Vasavada, Mission overview and scientific contributions from the Mars Science Laboratory Curiosity rover after eight years of surface operations. Space Sci. Rev. **218**, 3 (2022). https://doi.org/10.1007/s11214-022-00882-7
41. R. Sridhar, A. Goswami, K. Lohar, S. Malathi, K. Sriram, Chandrayaan-3 LIBS Sensor: preflight characterization, inflight operations and preliminary observations. IEEE Sensors J. (2024)
42. V. Sundararajan, Overview and technical architecture of India's Chandrayaan2 mission to the Moon. AIAA Aerospace Sci. Meeting 2178 (2018). https://doi.org/10.2514/6.2018-2178
43. R.V.L.N. Sridhar, Adwaita Goswami, K.A. Lohar, Bijoy Raha, M.V.H. Rao, Girish Gowda, T.S.Smaran, A. Senthil, Kumar, S. Bhuvaneswari, S.B. Umesh, K.V.Sriram, B. Prabkaran, T. Rethika, Himanshu Pandey, S. Malathi., Close-range in-situ LIBS (laser-induced breakdown spectroscope) experiment aboard the Chandrayaan-3 Pragyaan rover: operations and initial results. LPI Contributions **3040**, 1135 (2024)
44. N.J. Kanu, E. Gupta, G.C. Verma, An insight into India's Moon mission–Chandrayan-3: the first nation to land on the southernmost polar region of the Moon. Planetary Space Sci. 105864 (2024). https://doi.org/10.1016/j.pss.2024.105864
45. Y. Zou, Y. Zhu, Y. Bai, L. Wang, Y. Jia, W. Shen, Y. Fan, Y. Liu, C. Wang, A. Zhang, G. Yu, J. Dong, R. Shu, Z. Ho, T. Zhang, A. Du, M. Fan, J. Yang, B. Zhou, Y. Wang, Y. Peng, Scientific objectives and payloads of Tianwen-1, China's first Mars exploration mission. Adv. Space Res. **67**(2), 812–823 (2021). https://doi.org/10.1016/j.asr.2020.11.005
46. C. Wang, Y. Jia, C. Xue, Y. Lin, J. Liu, X. Fu, L. Xu, Y. Huang, Y. Zhao, Y. Xu, R. Gao, Y. Wei, Y. Tang, D. Yu, Y. Zou, Scientific objectives and payload configuration of the Chang'E-7 mission. Natl. Sci. Rev. **11**(2), 329 (2024). https://doi.org/10.1093/nsr/nwad329
47. K.A. Farley et al., Mars 2020 mission overview. Space Sci. Rev. **216**, 1–41 (2020). https://doi.org/10.1007/s11214-020-00762-y
48. S.K. Saxena, Uses of optical laser in space dynamic. Int. J. Adv. Res. Arts Sci. Eng. Manag. (IJARASEM) **5**, 2395–7852 (2018)
49. J.R. Sze, A.C. Wei, Compact beam expander based on planar structure to avoid inner focus. Opt. Rev. **23**(5), 842–847 (2016). https://doi.org/10.1007/s10043-016-0251-5
50. J. Dire, Telescopes. *Exploring the Universe.* (Springer, 2024), pp. 1–21. https://doi.org/10.1007/978-3-031-65346-9_1
51. M. Naeem, N.U.A. Fatima, M. Hussain, T. Imran, A.S. Bhatti, Design simulation of Czerny–Turner configuration-based Raman spectrometer using physical optics propagation algorithm. Optics **3**(1), 1–7 (2022). https://doi.org/10.3390/opt3010001

Remote Sensing Technologies in Aerospace: Applications and Opportunities

Carina L. Lopes, Renato Mendes, Leonardo Azevedo, Magda C. Sousa, Américo S. Ribeiro, Afonso Ferreira, Beatriz Biguino, Ana C. Brito, and João M. Dias

Abstract Sensors on board satellites collect detailed and consistent information from the Earth's surface, providing scientists with data that have revolutionised the monitoring and research of estuarine and marine systems. Satellite remote sensing data provide almost continuous information about the Earth's surface, making it possible to monitor the evolution of surface changes and identify and better understand the processes that control them. This chapter reviews the advances in remote sensors used to observe oceans and coastal regions. It presents examples of applications that use data from these sensors to monitor these valuable ecosystems, which are currently under increasing anthropogenic and natural threats. Concrete examples will be presented to demonstrate how data from remotely sensed instruments, such as the Multi-Spectral Instrument (MSI), Thematic Mapper (TM), Enhanced Thematic Mapper Plus (ETM+), Moderate Resolution Imaging Spectroradiometer (MODIS), Ocean and Land Colour Instrument (OLCI), Synthetic Aperture Radar (SAR) and other multi-sensor products, have been pivotal in addressing significant oceanographic and environmental challenges, including:

C. L. Lopes (✉) · M. C. Sousa · A. S. Ribeiro · J. M. Dias
CESAM—Centre for Environmental and Marine Studies, Physics Department, University of Aveiro, Campus de Santiago, Aveiro, Portugal
e-mail: carinalopes@ua.pt

R. Mendes
+ATLANTIC CoLAB, Museu das Comunicações, Lisbon, Portugal

Underwater Systems and Technology Laboratory (LSTS), Associate Laboratory for Energy, Transports and Aerospace (LAETA), Faculdade de Engenharia, Universidade do Porto (FEUP), Porto, Portugal

L. Azevedo
CERENA/DER, Instituto Superior Técnico (IST), Universidade de Lisboa, Lisbon, Portugal

A. Ferreira · B. Biguino · A. C. Brito
Faculdade de Ciências, MARE—Marine and Environmental Sciences Centre/ARNET—Aquatic Research Network, Universidade de Lisboa, Lisboa, Portugal

A. C. Brito
Departamento de Biologia, Faculdade de Ciências, Universidade de Lisboa, Lisboa, Portugal

C. Marques et al. (eds.), *Advanced Optical Sensors for Aerospace Applications*, Progress in Optical Science and Photonics 37,
https://doi.org/10.1007/978-981-95-1626-1_10

- Conservation and protection of estuarine environments: Remotely sensed multi-spectral data will be used to accurately detect long-term changes in salt marsh extent, enabling the identification of mechanisms driving the loss and degradation of salt marsh ecosystems.
- Modelling and mitigation of marine pollution: Earth observation data collected by the MSI sensors on board the Sentinel-2 satellites will be coupled with data from numerical dynamic ocean models and processed using deep learning techniques to automatically identify and model the spatiotemporal evolution of floating debris.
- Improving our understanding of coastal and ocean dynamics: Synergies between satellite images will be explored to improve the analysis of river plumes.
- Identification of climate change effects: Satellite-derived sea surface temperature will be considered to study temperature-related extreme weather events (i.e. Marine Heat Waves and Marine Cold Spells).
- Identification of ecological change: use of ocean colour data to analyse phytoplankton bloom phenology patterns and enhance small pelagic fish recruitment prediction.

Keywords Earth observation · Oceans · Coastal regions · Salt marshes · Floating debris · River plumes · Sea surface temperature · Ocean colour

1 Introduction

1.1 Satellite Remote Sensing Process

Remote sensing techniques enable the collection of information about the physical characteristics of objects, areas or phenomena using electromagnetic radiation sensors installed on board platforms without contact with the objects of investigation. Satellite remote sensing is a specific type of remote sensing that takes advantage of installing sensors on board satellites to obtain periodic and consistent information about vast areas of the Earth's surface, even those that are remote and difficult to access. The satellite remote sensing process can be more readily comprehended by examining Fig. 1, which illustrates the main stages involved in the process, from the energy source (1) to the application (7) [1, 2]:

1. **Energy source**—An energy source must provide electromagnetic radiation to the target or surface of interest. A distinction can be made between active and passive remote sensing depending on the energy source. Active remote sensors use their own energy source (often microwave radiation), while passive remote sensors use ambient energy, usually from the Sun.
2. **Radiation and atmosphere**—Electromagnetic energy interacts with the atmosphere before and after reaching the Earth's surface. Atmospheric constituents scatter, absorb and reflect electromagnetic radiation at specific wavelengths,

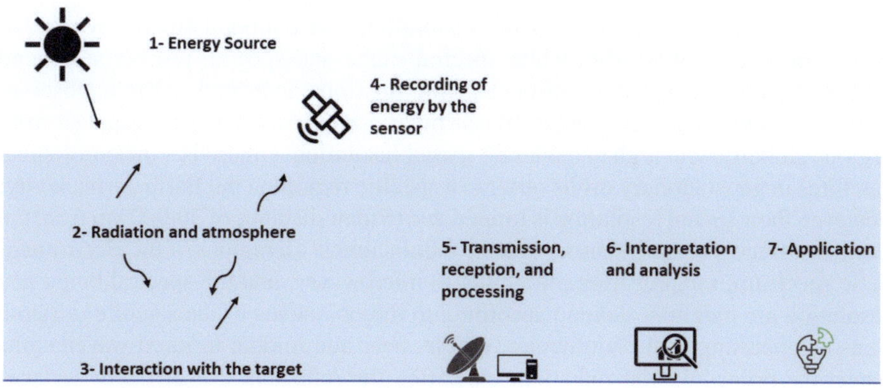

Fig. 1 Stages of satellite remote sensing process

altering the electromagnetic radiation that reaches the Earth's surface and the remote sensor.

3. **Interaction with the target**—After passing the atmosphere the electromagnetic radiation reaches the surface and interacts with the target. This interaction depends on both the characteristics of the target or surface (e.g. type of material, roughness) and on the characteristics of the radiation (e.g. intensity, wavelength, angle of incidence) so that the resulting electromagnetic radiation varies depending on the characteristics of the targets and the incident radiation.
4. **Recording of energy by the sensor**—After interacting with both the surface and atmosphere, the electromagnetic radiation reaches the remote sensor and is recorded. Different sensors (e.g. radiometers, spectrometers, etc.) are specifically designed to efficiently record radiant energy in different regions of the electromagnetic spectrum.
5. **Transmission, reception, and processing**—The records are transmitted to ground stations, processed to correct some radiometric and atmospheric distortions and disseminated to interested parties in standardised formats. Then, data is often subjected to different levels of processing in order to obtain a wide variety of products and variables.
6. **Interpretation and analysis**—Comprise the interpretation and analysis of satellite data in order to extract meaningful information about the object of study (e.g. classification, detection change, etc.). Often, this involves processing satellite products and variables to obtain quantitative and robust indicators of the target of interest.
7. **Application**—The meaningful information extracted from satellite data addresses real-world challenges and deepens our understanding of specific phenomena.

Data from satellite remote sensing technology has unique characteristics that fit specific applications. Spatial coverage and resolution, spectral range and resolution, and temporal frequency are key characteristics of satellite remote sensing data [3, 4]. Spatial coverage, spatial resolution, and temporal frequency depend mostly on the satellite orbit, while spectral range and spectral resolution depend on the characteristics of the sensor. Satellite missions dedicated to Earth Observation (EO) use satellites in low Earth near-polar orbits, providing periodic data from regions globally with high to moderate spatial resolution. Otherwise, meteorological satellites in geostationary orbits observe a specific region on the Earth permanently; however, their spatial resolution is limited due to their distance of 36,000 km from the Earth's surface. Existing sensors operate within almost all regions of the electromagnetic spectrum, ranging from ultraviolet to microwaves, and the spectral range and resolution are carefully defined according to the objectives of the satellite mission. Sensors operating in the ultraviolet, visible, near and middle infrared wavelengths detect the reflected solar radiation, indicating the reflectance of different surfaces [5]. Otherwise, passive sensors operating in the thermal infrared and microwave regions of the electromagnetic spectrum detect the radiation emitted by the Earth and are used to estimate the temperature of emitting objects or surfaces [6]. Most active sensors operate in microwave wavelengths, detecting their own radiation after being reflected by objects and surfaces [7]. Therefore, the spectral characteristics of sensors installed on board a satellite are carefully selected to fit the objectives of the satellite mission. For example, some EO missions use multispectral sensors to detect the visible, near and middle infrared radiation reflected by the Earth's surface in narrow electromagnetic spectrum bands (spectral bands).

1.2 Sensors Used in Ocean and Coastal Applications

Optical Sensors

The first artificial satellite was launched in 1957, marking the beginning of satellite-based remote sensing. One of the first satellite missions' main objectives was to increase the satellites' lifetime. Once this challenge had been met, technological improvements were made to the sensors to optimise the identification of specific characteristics of targets. The launch of Landsat 1 in 1972 represented a significant milestone in developing satellite remote sensing technology. This pioneering satellite was specifically designed for EO, carrying on board the experimental instrument Multispectral Scanner (MSS), which recorded data in four spectral bands (green, red, and two infrared) with a spatial resolution of 60 m [8]. Subsequent Landsat satellites carrying increasingly optimised and advanced sensors (Fig. 2) were launched into space (Landsat 2 to Landsat 9), and the planned launch of Landsat Next by 2030/31 will ensure the continuation of this precious EO record spanning over more than five decades.

Launched in 2016, the Sentinel-2 mission constitutes another significant milestone in EO. It comprises a constellation of two twin satellites, Sentinel-2A and -2B, placed in the same orbit but phased 180°, each equipped with the Multi-spectral Instrument (MSI), sampling 13 spectral bands, 9 in the visible and near-infrared,

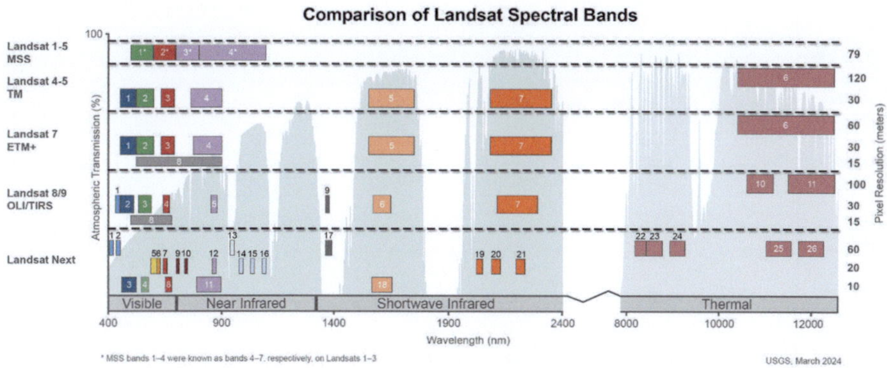

Fig. 2 Spectral bands for the sensors on all Landsat satellites: Multispectral Scanner System (MSS); Thematic Mapper (TM); Enhanced Thematic Mapper Plus (ETM+); Operational Land Imager and the Thermal Infrared Sensor (OLI/TIRS. Image from: https://www.usgs.gov/media/images/spectral-bandpasses-all-landsat-sensors (Public domain permission)

and 3 in the short-wave infrared. This constellation provides data with significantly enhanced capabilities compared to Landsat [9]. Notably, Sentinel-2 offers a higher spatial resolution at 10 m compared to operational Landsat's 30 m, enabling more detailed observations. Furthermore, its revisit time of 5 days, compared to Landsat's 16 days, allows for more frequent monitoring of Earth's surface processes [10]. Optical sensors used for EO measure the radiance at the Top of the Atmosphere (TOA) in specific spectral bands in the visible and near-infrared regions, which result from the interaction of solar radiation with the atmosphere and Earth's surface. These radiances are used in the first stage to estimate the Surface Reflectance (SR) for each wavelength, which measures the fraction of incoming solar radiation reflected from the Earth's surface to the remote sensor [11]. SR derive from the application of atmospheric correction algorithms that estimate and remove the contributions of atmospheric radiance to TOA radiance [11]. A huge variety of products with different applications that can be obtained from SR at specific wavelengths. For example, SR data from Landsat and Sentinel missions have been widely used to map changes in land use [12–15], shoreline position [16–18] and plant phenology [19–21].

Advancements in satellite orbit technology also boosted the development of specialised sensors dedicated to ocean observation. In 1978, the first sensor specially dedicated to monitor the ocean colour was launched into space, and since then, many other satellite missions followed. The most popular instruments, listed in Fig. 3, are:

(1) **CZCS** (Coastal Zone Color Scanner) on board the Nimbus-7 (1978–1985) meteorological satellite, was the first sensor specially dedicated to monitor the ocean colour. Its four visible spectral bands (20 nm bandwidth) provided ocean colour data at a spatial resolution of 825 m [22].
(2) **SeaWiFS** (Sea-viewing Wide Field-of-view Sensor) on board OrbView-2 (1997–2010) had 8 spectral bands in visible and infrared regions that provided ocean colour measurements at 1100 m spatial resolution [23].

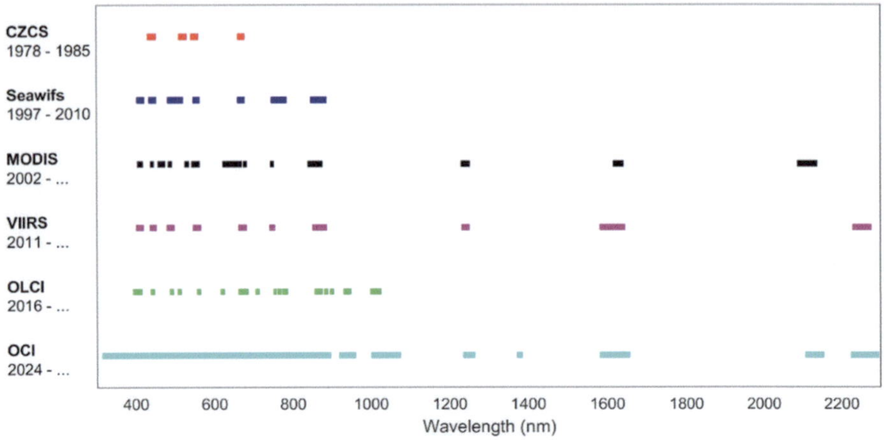

Fig. 3 Comparison of ocean colour sensor specifications between 340 and 2260 nm. Data source: https://oceancolor.gsfc.nasa.gov/resources/docs/rsr_tables/

(3) **MODIS** (Moderate Resolution Imaging Spectroradiometer) on board Terra and Aqua satellites, offer ocean colour data since 2000 and 2002, respectively. It has 16 spectral bands (10–50 nm bandwidth), used for ocean colour products, spanning the visible and infrared, and enables observations of the entire ocean every 1–2 days (combining Terra and Aqua), at spatial resolutions from 250 to 1000 m, depending on the spectral band [24].
(4) **VIIRS** (Visible Infrared Imaging Radiometer Suite) on board Suomi-NPP, NOAA-20 and NOAA-21 satellites, monitor the global ocean since 2012, 2017 and 2022, respectively. It has 10 spectral bands in visible and infrared regions that provide ocean colour measurements at spatial resolutions from 375 to 750 m, depending on the spectral band [25].
(5) **OLCI** (Ocean and Land Colour Instrument) on board Sentinel-3A and Sentinel-3B satellites, offer ocean colour data since 2016 and 2018, respectively, at 300 m spatial resolution. It has 21 spectral bands (3–27 nm bandwidth), used for ocean colour products, spanning the visible and infrared, enabling observations of the entire ocean every 1–2 days (combining Sentinel-3A and 3B) [26].
(6) **OCI** (Ocean Color Instrument), launched in 2024 on board of PACE satellite, is the first hyperspectral radiometer, specially designed for global ocean colour observation [27]. It has 282 spectral bands in the region 340–895 nm (5 nm bandwidth, with 2.5 or 1.25 nm sampling), and 7 bands in the region 940–2260 nm, providing ocean colour products at 1200 m spatial resolution.

In summary, the first ocean colour sensor to provide global data coverage was the CZCS, which was operational from 1978 to 1986 [22]. While CZCS marked the beginning of spaceborne ocean colour observations, it was only in 1997, with the launch of the SeaWiFS, that a continuous, long-term global ocean colour record was established [28, 29]. Since then, coordinated efforts by multiple space agencies (e.g., ESA, NASA) have resulted in a freely available, multi-decadal archive of ocean colour data, which has proven invaluable for studying long-term trends in global primary production [30]. Having ensured a continuous record of data, recent advancements in ocean colour remote sensing have focused on other important aspects, such as improving both the spatial and spectral resolution of observations. The OLCI, aboard the ESA Sentinel-3 constellation, provides daily ocean colour data with a significantly higher spatial resolution (300 m at full resolution) compared to earlier sensors [31]. More recently, the NASA Plankton, Aerosol, Cloud, ocean Ecosystem (PACE) mission, launched in early 2024, introduced the OCI, a hyperspectral sensor capable of unprecedented spectral resolution amongst satellite missions [32]. This advancement enables more detailed characterisation of phytoplankton diversity, further enhancing our ability to monitor and understand oceanic biogeochemical processes from space [32].

Ocean colour sensors measure the TOA radiance in specific spectral bands in the visible and near-infrared regions, which result from the interaction of solar radiation with the atmosphere and seawater (Fig. 4). These radiances are used in the first stage to estimate remote sensing reflectance (Rsr) for each wavelength, which measures the water-leaving radiance normalised by the downwelling solar irradiance. This signal represents the sunlight scattered out of the ocean after interacting with the water and its constituents [33]. Hence, the colour is primarily determined by the optical properties of water and its components. Pure seawater appears blue because it preferentially absorbs red wavelengths, but biotic and abiotic constituents can alter this colour by absorbing and scattering light at different wavelengths. However, it is crucial to account that the ocean colour signal comprises only about 10% of the total radiance detected by a satellite sensor at the top of the atmosphere, making atmospheric correction an essential step in ocean colour retrieval [34].

In the open ocean, phytoplankton—microscopic aquatic organisms that contain multiple pigments and photosynthesize—are the main drivers of ocean colour [33]. Chlorophyll-a, the primary pigment in phytoplankton, absorbs light in the blue and red parts of the spectrum while reflecting green, giving phytoplankton-rich waters a greenish hue. Consequently, high phytoplankton abundance is typically associated with greener waters, contrasting with the deep blue of oligotrophic regions. In addition to chlorophyll, other pigments further contribute to the optical signal, adding complexity to the interpretation of ocean colour data. Coastal waters, however, are optically more complex due to the influence of suspended sediments, riverine inputs, and coloured dissolved organic matter (CDOM) independent of phytoplankton, which can significantly alter ocean colour [33]. As a result, ocean colour remote sensing in coastal regions often requires more careful approaches and advanced processing techniques to distinguish between different optical constituents accurately.

Thermal Sensors

Monitoring of sea surface temperatures by satellite remote sensing began in 1978 with the launch of the radiometer **AVHRR** (Advanced Very High-Resolution Radiometer)

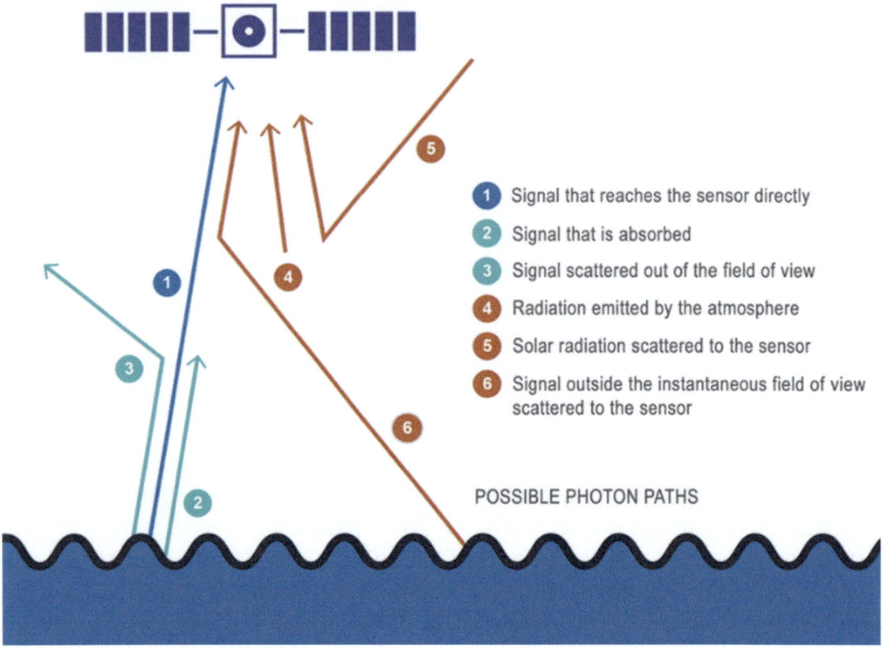

Fig. 4 Possible photon trajectories after interaction of sunlight with the atmosphere and seawater. Adapted from Sutcliffe et al. [35]

on board the polar-orbiting TIROS-N meteorological satellite. This radiometer had 4 spectral bands, 1 in the thermal infrared region (10.30–11.30 um), that recorded the long radiation emitted by the atmosphere and ocean surface. This radiometer evolved and actually has 6 spectral bands, 2 in the thermal infrared region (10.3–11.3 μm; 11.5–12.5 μm). Moreover, it has been installed on the NOAA and EUMETSAT polar-orbiting satellites and has provided continuous daily information on sea surface temperature around the globe since 1978 [36]. Some sensors, such as MODIS and VIIRS, have thermal infrared channels and are also used to estimate sea surface temperature. Another popular sensor is the **SLSTR** (Sea and Land Surface Temperature Radiometer), orbiting the Earth on board Sentinel-3A and 3B satellites. Despite slight differences between thermal infrared sensors (e.g. spectral range and bandwidth), they all use channels within the range of 8 to 13 μm, benefiting from this region's high atmospheric transmittance. Sensors record the thermal infrared energy emitted by the surface and the atmosphere, which can be used to estimate brightness temperatures according to Planck's Function [37]. The SST is determined from the brightness temperatures; however as the atmosphere is not entirely transparent, it is necessary to correct for atmospheric effects [37]. In addition to thermal infrared methods for SST retrieval, alternative techniques use channels around 3.7 μm for night-time observations [38] and passive microwave radiometry for measurements through clouds [39]. However, thermal infrared SST retrieval is the longest and most mature application of ocean remote sensing, providing consistent and continuous data over the global ocean, and fostering our comprehensive knowledge about the climate, oceanic and atmospheric processes.

Microwave Active Sensors

The potential of active remote sensing sensors in observing the global ocean was recognised early on. The Seasat satellite, launched in 1978, was the first to be equipped with active microwave sensors, marking the beginning of ocean observation using active microwave technology. Despite, its short lifetime (~110 days) demonstrated its power in detecting ocean surface topography, ocean waves and surface ocean currents [40]. Active microwave remote sensing involves transmitting microwave radiation from a sensor towards the Earth's surface and measuring the radiation reflected back to the sensor. This reflected signal and the two-way travel time provide information about the surface's properties. There are three types of active microwave sensors used to monitor ocean and coastal surfaces:

Altimeters: Nadir-pointing radars are capable of very precisely determining the distance between the satellite and the reflecting sea surface. This accuracy is achieved by precise orbit determination using systems such as DORIS (Doppler Orbitograpghy and Radiopositioning Integrated by Satellite), SLR (Satellite Laser Ranging) and GNSS (Global Navigation Satellite Systems), together with passive microwave radiometer measurements to correct for atmospheric signal delays [41]. Since 1992 (launch of TOPEX/Poseidon), high-precision global ocean topography data have enabled scientists to study sea level variability and to understand the complex processes that drive mean sea level change [42].

Scatterometers: A scatterometer is a radar instrument that measures backscattered radiation at multiple angles using microwave signals [43]. The analysis of scatterometer signals enables the determination of the ocean's reflectivity, which provides information about the characteristics of the surface (e.g. waves, winds and sea ice) [43]. Due to their longevity, the most popular scatterometers are AMI-SCAT (Active Microwave Instrument—Scatterometer) (1991–2011), SeaWinds (1999–2018) and ASCAT (Advanced Scatterometer) (2006-…) [43].

Synthetic aperture radars (SAR): SAR systems measure the time delay and the amount of the backscattered energy from the transmitted signal and use Doppler/range processing to construct backscatter images [44]. Since the 1990s, several satellite missions have used SAR systems for oceanographic applications, e.g. RADARSAT-1/2 since 1995, ALOS-1/2 since 2006 and Sentinel-1A/B since 2014 [44].

2 Applications to Oceans and Coastal Regions

2.1 Detection of Estuarine Salt Marshes

Salt marshes are vital and productive ecosystems, providing a variety of services ranging from nutrient regeneration and primary production to essential habitat for marine species and coastal protection [45]. Despite this value, salt marshes are experiencing a global decline in both area and condition due to increasing threats, including socio-economic and biogeophysical factors, all exacerbated by climate change [46]. Understanding the drivers of this decline, accurately assessing the conservation status of these habitats, and anticipating future threats are critical for developing effective management solutions [46]. Several studies demonstrate that satellite remotely sensed multispectral data from Landsat archive (TM, ETM+ and OLI sensors) can be used to efficiently monitor the condition of salt marsh communities over the last four decades, enabling the identification of distinct evolutionary trends and, therefore, providing a precise evaluation of vegetation dynamics [47–51]. Moreover, Sentinel-2 data have also proved effective in monitoring salt marshes, with the advantage of providing greater spatial resolution (10 m compared to Landsat's 30 m) and temporal resolution (5 days compared to Landsat's 16 days) [10].

Salt marshes are terrestrial ecosystems found in estuarine systems of temperate zones all over the world. They are found in intertidal flats and are characterized by halophyte plants (salt-tolerant). Salt marsh vegetation has distinct spectral signatures compared to surrounding areas like open water, mudflats, or other vegetation types, and can be detected through appropriate optical sensors. The main spectral features of intertidal flats environments and water have been summarized by Yang et al. [52]:

Salt marsh vegetation: Healthy salt marsh plants exhibit high reflectance in the NIR portion of the electromagnetic spectrum due to their dense leaf structure and internal leaf scattering, and low reflectance in the red and green bands due to chlorophyll absorption for photosynthesis.

Water: Water bodies present very low reflectance over visible, NIR, and SWIR regions, due to the strong absorption capacity of the water.

Mudflat: Mudflats exhibit low reflectance across the visible, NIR, and SWIR bands of the electromagnetic spectrum, to their high moisture content. Although low, their reflectance is higher than that of water bodies.

This unique spectral behaviour enables distinct mudflats from salt marshes, and water bodies, particularly analysing visible, NIR, and SWIR bands. In particular, spectral indices, such as Normalized Difference Water Index (NDWI) and Vegetation Indices, have been widely used for salt marsh identification [50, 53–56]. The (NDWI):

$$NDWI = \frac{Green - SWIR}{Green + SWIR} \tag{1}$$

uses *Green* and *Shortwave Infrared* (SWIR) bands to enhance the contrast between land and water, and is effective in distinguishing inundated and exposed regions [57]. Whereas VI use *Red* and *NIR* bands to highlight vegetated regions, benefiting from the high/low reflectance of NIR/Red radiation over vegetated regions [58].

The Normalized Difference Vegetation Index (NDVI) is the most popular VI. It describes the normalized ratio between the *Red* and *NIR* bands [58]:

$$NDVI = \frac{NIR - Red}{NIR + Red} \tag{2}$$

NDVI is particularly sensitive to vegetation "greenness" and commonly used as an indicator of canopy photosynthesis [59]. When vegetative cover is low and the soil surface is exposed, the reflectance of soil can be similar to the vegetation, smoothing the contrast between classes.

The Modified Soil Adjusted Vegetation Index (MSAVI) was designed by Qi et al. [60] to minimize the soil brightness effect:

$$MSAVI = \frac{(2NIR+1) - \sqrt{(2NIR+1)^2 - 8(NIR - Red)}}{2} \tag{3}$$

The Atmospheric and Soil Vegetation Index (ASVI) is a modification of MSAVI and accounts for atmospheric scattering effects by including the *Blue* band in the formulation [61]:

$$ASVI = \frac{(2NIR+1) - \sqrt{(2NIR+1)^2 - 8(NIR - 2Red + Blue)}}{2} \tag{4}$$

The Global Environmental Monitoring Index (GEMI) proposed by Pinty and Verstraete [62], is a non-linear VI that simultaneously reduces the sensitivity to atmospheric visibility and soil reflectance (except for bright soils) and maximizes the sensitivity to vegetation cover:

$$\begin{aligned} GEMI &= \zeta(1 - 0.25\zeta) - \frac{Red - 0.125}{1 - Red}, \\ \zeta &= \frac{2\left(NIR^2 - Red^2\right) + 1.5NIR + 0.5Red}{NIR + Red + 0.5} \end{aligned} \tag{5}$$

This section demonstrates the application of satellite optical sensor data, from Landsat and Sentinel-2 satellites, for identifying estuarine salt marshes, focusing on two distinct estuarine systems located in the West coast of Portugal (Fig. 5): (1) the Aveiro Lagoon, characterized by its shallow depth, complex web of narrow channels, extensive tidal flats, and degraded salt marshes; and (2) the Tagus Estuary, characterized by a wide and deep bay with salt marshes located upstream on gently sloping, wave-protected banks. This analysis covers the period 1984–2022 and focus on the evolution of São Jacinto head's salt marshes (Fig. 5b) and Pancas salt marsh (Fig. 5c) by analysing spectral indices derived from surface reflectance data.

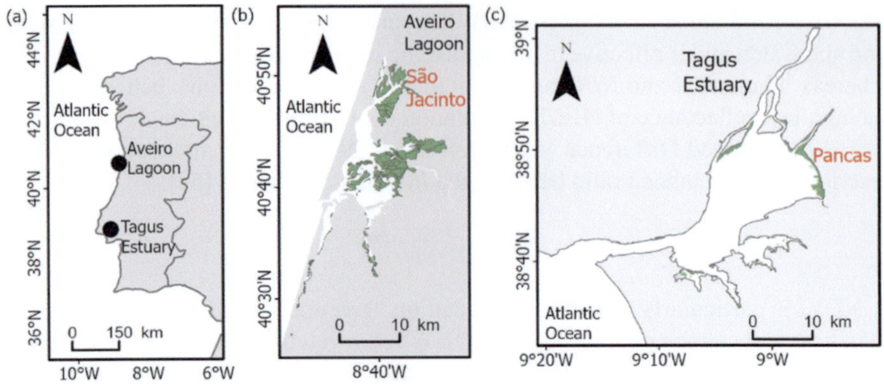

Fig. 5 **a** Map of Portugal depicting the location of the estuarine systems under study; **b** Aveiro Lagoon with the location of São Jacinto head's salt marshes **c** Tagus Estuary with the location of Pancas salt marsh

Aveiro Lagoon

Lopes et al. [47] used data from the Landsat archive to assess long-term changes in the vegetation of the Aveiro Lagoon. A catalogue of 264 images acquired between 1984 and 2017 was compiled and analysed as follows:

(1) NDWI, NDVI, MSAVI, ASVI and GEMI were calculated for each image;
(2) NDWI was used to estimate the flooded area in each image, and the changes in flood patterns.
(3) NDVI, MSAVI, ASVI and GEMI were statistically analysed over tidal flats to characterise the changes in the intertidal vegetation.

Their results showed that the area of intertidal regions increased at the same time as the VI of intertidal regions decreased, indicating a decline in estuarine vegetation. Moreover, the largest flooded area and VI modifications occurred after 1999, suggesting that changes might be associated with dredging activities performed in the main lagoon channels.

This work uses the same dataset and processing as Lopes et al. [47] and analyses in detail the evolution of the marshes at the end of the São Jacinto channel, which were identified by Lopes et al. [48] as one of the most threatened in the lagoon. The Fig. 6 shows the spatial variation of NDVI in the tidal flats surrounding the end of the São Jacinto channel before and after 1999, as well as the changes in estuarine vegetation. The NDVI analysis shows a clear general decline in NDVI values after 1999. The largest changes in vegetation cover were also observed throughout the region, reflected in the shifting green and brown areas on the NDVI maps. Lopes et al. [48] estimated that 24.3% of the total intertidal lagoon area has lost vegetation and that 26.9% has suffered moderate to severe degradation. The salt marshes around São Jacinto head's regions have suffered the most degradation, with an estimated 4.5 km^2 (26%) of the 17.6 km^2 area lost since 1999. In addition, about 7.8 km^2

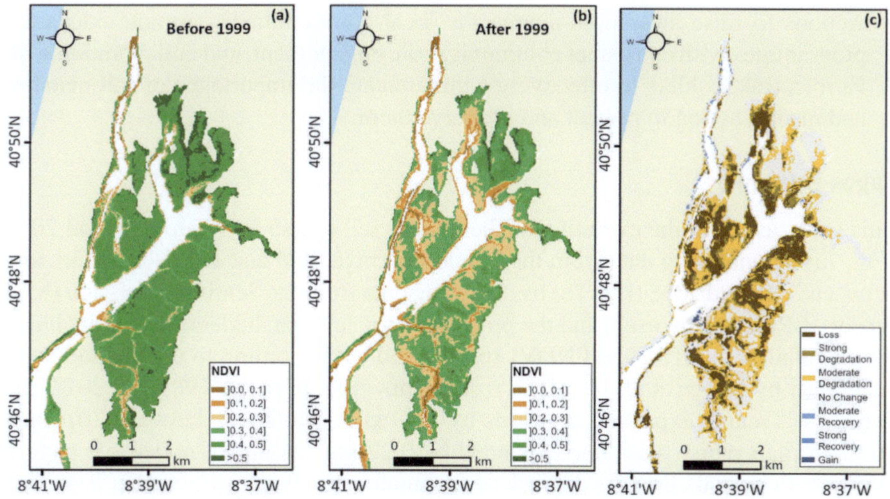

Fig. 6 **a** NDVI average before 1999; **b** NDVI average after 1999; **c** the respective vegetation changes

(44%) of the remaining salt marshes show evidences of moderate to strong degradation. Lopes et al. [48] used a hydrodynamic model to reconstruct the hydrodynamic conditions before and after 1999 in order to identify the mechanisms responsible for the degradation of the estuarine vegetation. They found that the deepening of the lagoon channels, due to dredging activities, increased the tidal effects, particularly increasing the mean currents, the frequency of inundation and the mean sea level. The reported hydrodynamic modifications are concordant with marsh loss and degradation patterns and suggest that these can result from both edge erosion motivated by the intensification of tidal currents and interior drowning triggered by the increase in inundation frequency and mean sea level.

More recently, Castro et al. [63] used Landsat and Sentinel-2 data to assess changes in the shoreline of the fringing salt marshes along the lagoon. They found that the salt marsh shoreline generally retreated at an average rate of 1.2 m/year after 1999, resulting in a reduction in their extent given that the inland boundary was maintained (the salt marshes are bounded by roads). They concluded that salt marsh shoreline retreat poses significant threats to the ecosystem services, namely habitat loss, water quality degradation, reduced protection against storm surges, and decrease in the capacity to sequester blue carbon. They also suggest the implementation of actions to reverse this situation, namely:

- Actions for erosion control and sediment retention—instal wood fences along the shoreline to prevent erosion and promote sediment retention.
- Use dredged sediments to benefit salt marshes—avoid future dredging activities, but if new dredging is deemed necessary, it is imperative that dredged sediments are used for benefiting salt marshes.

- Actions to raise awareness among the local communities—promote education programmes with consistent communication, engagement, and collaboration with various stakeholders to effectively communicate the importance of salt marshes and inspire action to protect and conserve them.

Tagus Estuary

This study assesses the evolution of the Pancas salt marsh between 1984 and 2018 (Fig. 7a) by analysing data from the Landsat archive (TM and ETM+ sensors), and between 2016 and 2018 (Fig. 7b) by analysing data from the Sentinel-2 mission (MSI sensor). SR data were used, and the processing for salt marsh identification included the calculation of NDWI and NDVI spectral indices according to the methodology developed by Lopes et al. [49]. The results show that between 1984 and 2018, the Pancas salt marsh expanded seawards by 1.03 km^2. Moreover, between 2016 and 2022, the salt marsh shoreline expanded by 0.22 km^2. Another, noticeable pattern during both periods was the dieback of the inner salt marsh, estimated at 0.05 km^2 between 1984 and 2018 and 0.1 km^2 between 2016 and 2022.

Based on salt marsh extent derived from satellite imagery and a reference value for the carbon stored in the sediments of the Pancas salt marsh of 9–10 kg C m^{-2}, Freitas et al. [64] determined the amount of blue carbon sequestered and stored in Pancas salt marsh. Their results show that in 2022, Pancas salt marsh would store 66,500 tonnes of blue carbon. Moreover, combining satellite derived salt marsh extent changes with a reference value of 330 gC m^{-2} y^{-1} [65] for the amount of carbon sequestrated in the sediments of the Pancas salt marsh, the same authors estimated the amount of blue carbon sequestered in both periods. Between 1984 and 2018, the

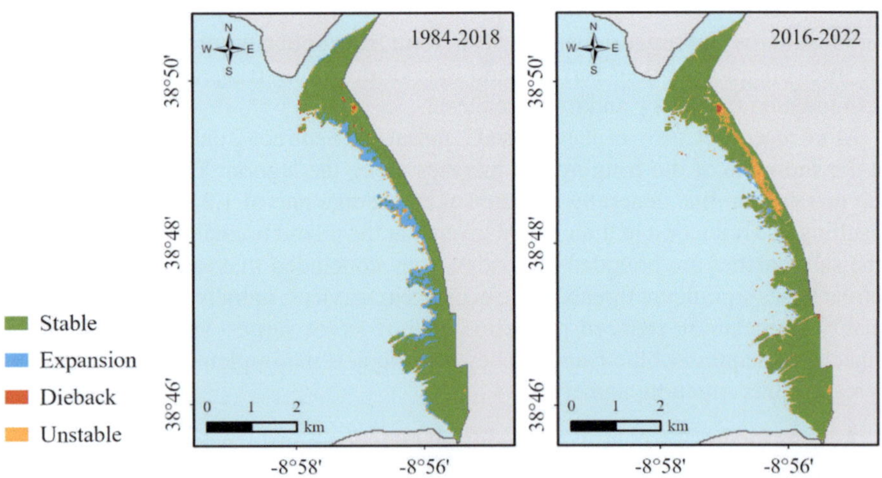

Fig. 7 Changes in the spatial distribution of salt marsh cover from 1984 to2018 and from 2016 to 2022

blue carbon sequestration rate was 339.9 tonnes y^{-1}, while between 2016 and 2022, the rate decreased to 72.6 tonnes y^{-1}.

2.2 *Detection of Plastics in Oceanic Waters*

It is estimated that 80% of all floating debris are marine plastic. These plastics are eventually ingested by marine species or act as a trap, causing severe injuries and deaths to marine life. Marine plastic pollution is a substantial threat to wildlife and human health. The ability to detect floating plastic debris in short periods after it enters the ocean is a mitigating factor, which avoids the degradation of the macroplastics into smaller pieces and the consequential threat to the food chain. The mitigation of marine pollution can benefit from EO data collected by the MSI sensor onboard Sentinel-2A satellites, combined with numerical ocean models and processed using deep learning techniques. This approach aims to identify different types of floating debris from satellite images and explore the potential to distinguish between various types of plastic. Beyond the challenging task of detecting floating plastic debris, it integrates the spatiotemporal modelling of plastic accumulation with high-resolution numerical ocean modelling. Deep learning methods can be used to couple ocean numerical models and marine floating debris automatically classified from Earth observation data. A Lagrangian modelling approach is coupled to this model to mimic the behaviour of particles (i.e., plastics) sourced by the input rivers.

Deep-Classification of Floating Plastic Debris from Satellite Images

Satellite images were processed to correct atmospheric effects, and a mask was applied to separate water masses from land regions. To perform the atmospheric correction of the satellite images, we used the dark spectrum fitting (DSF) algorithm from the atmospheric correction for OLI "lite" (ACOLITE) v.20210802.0 software. This method assumes that the atmosphere is homogeneous and that the scene contains pixels with zero or very close to zero surface reflectance in at least one of the sensor bands (i.e., dark pixels). The spectral signature of the dark pixels, or dark spectrum, is then used to determine the best-fitting combination of the spectral band and aerosol model for the atmospheric correction. With the most appropriate combination selected, the parameters required for the "path-corrected" reflectance computation are chosen from a lookup table. Due to low atmospheric transmittance, Bands 9 (B9) and 10 (B10) are excluded from the output of the corrections.

After processing, features in these images were manually labelled and classified for several types of floating debris. The labelling was performed based on their visual inspection and using published works where these features were previously identified and validated. This task resulted in the generation of a validated library of spectral signatures for different floating debris. A total of 2735 pixels were validated with features comprising water, plastic, wood, seaweed, pumice, sea snot and sea foam. The resulting spectral signatures library is shown in Fig. 8, and curated data

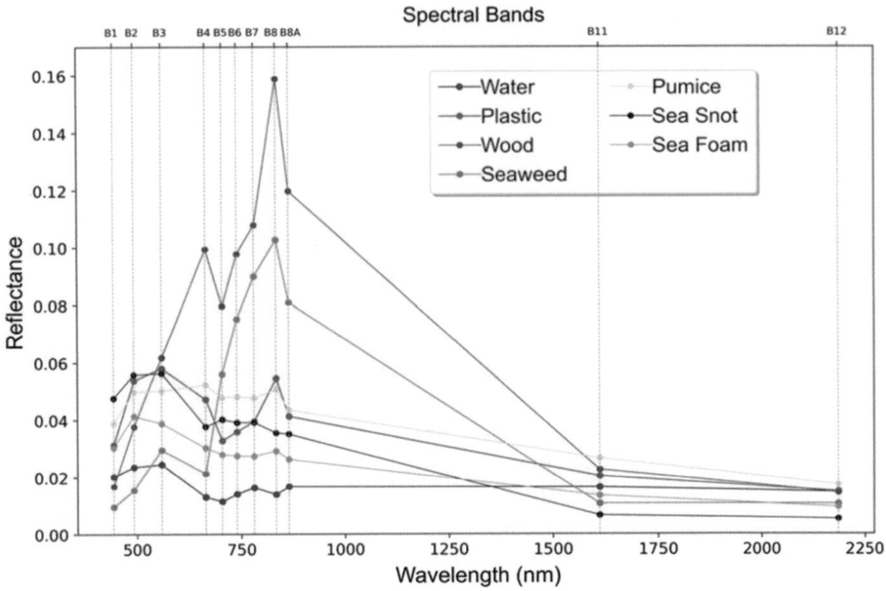

Fig. 8 Spectral signatures derived from the mean reflectance of all data after the atmospheric correction process. Despite the Sentinel-2 satellites' spectral bands having slightly different central wavelengths, this figure uses the same central wavelengths to facilitate interpretation. The atmospheric correction process removed B9 (945 nm) and B10 (1375 nm). Adapted from Duarte and Azevedo [66]

were made publicly available at: https://github.com/miguelmendesduarte/Floating-Marine-Debris-Data [66].

For the automatic classification of the processed images, a supervised deep-learning classification algorithm was applied to identify the features of interest in the processed Sentinel-2 images. The evaluation of the various deep learning classifications was based on the training and test scores obtained during the training of the algorithms (i.e., accuracy of the predicted versus actual classes). Additionally, to include noise within the training data of the XGBoost algorithm, we augmented the original training data set using generative adversarial networks (GANs). The GANs were trained to generate spectra responses that resemble the ones inferred from the data summarized in Fig. 8. Finally, the original XGBoost algorithm was extended to include spatial predictions with uncertainty. This objective was reached using an ensemble method where multiple XGBoost run in parallel. The uncertainty is assessed based on the variability of the predictions after training (Fig. 9).

High-Resolution Numerical Modelling of Floating Debris in Coastal Environments

A high-resolution numerical grid was developed to support the hydrodynamic and Lagrangian models. The grid comprises a single high-resolution domain, covering the northwest Iberian Peninsula and its coastal environments (Fig. 10).

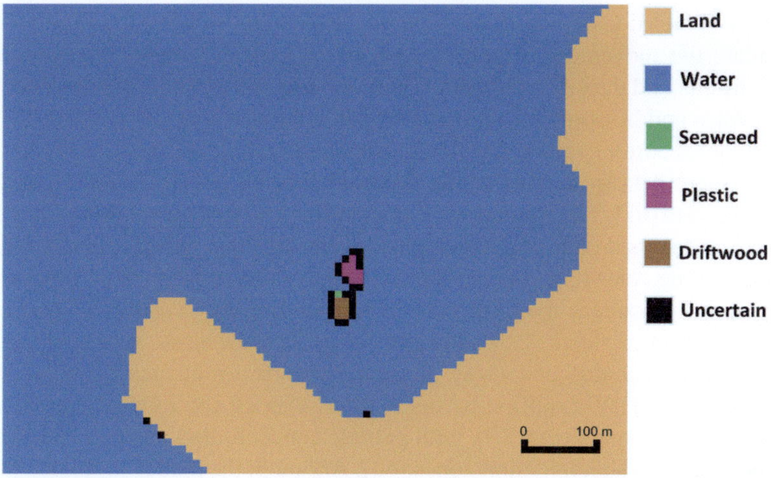

Fig. 9 Ensemble model's predictions. The original data are from a Sentinel-2A image from the 31st of July 2021 in the Gulf of Gera, Greece

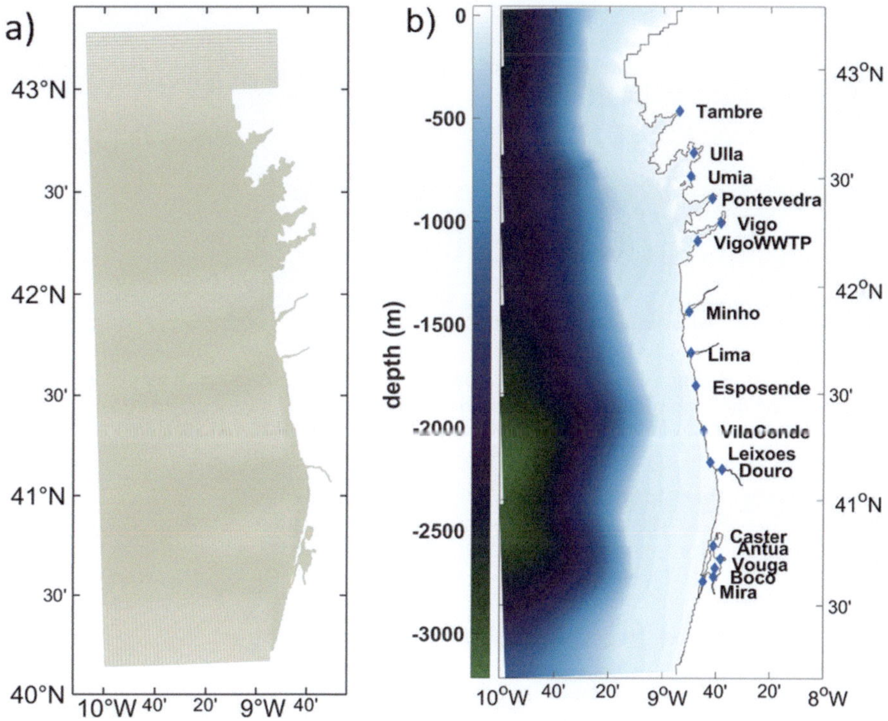

Fig. 10 Study area with the extent of the numerical domain (**a**) and the location of plastic litter sources (**b**)

A three-dimensional hydrodynamic model is coupled with the Lagrangian model to simulate the movement of particles—such as plastics—under the influence of ocean currents and density gradients. These models are essential for understanding the pathways and hotspots for plastic debris in coastal regions. The Delft3D model suite simulates current velocity, water density, and particle dispersion. The model is based on input data from river sources identified in an online database containing over 1000 rivers [67]. The abundance, composition, and trends of marine litter can be identified by analysing one of the indicators of marine litter, which is the part directly deposited on the coast (e.g., beach) that contributes to the transport of marine litter.

The model considers five size categories and varying densities of particles to assess the impact of plastic's size and density on its movement. Sensitivity analysis is conducted to examine how these physical properties influence the behaviour of drifting particles. Half a million particles were released continuously in each river discharge of the domain (Fig. 10), with particle concentration mapped and analysed to predict potential accumulation zones [68]. Although the model simulated particle dispersal across all estuaries within the domain, the validation of the results was limited to these two estuaries (Lima (Fig. 11) and Ria de Aveiro (Fig. 12) estuaries) due to the availability of Sentinel-2A imagery.

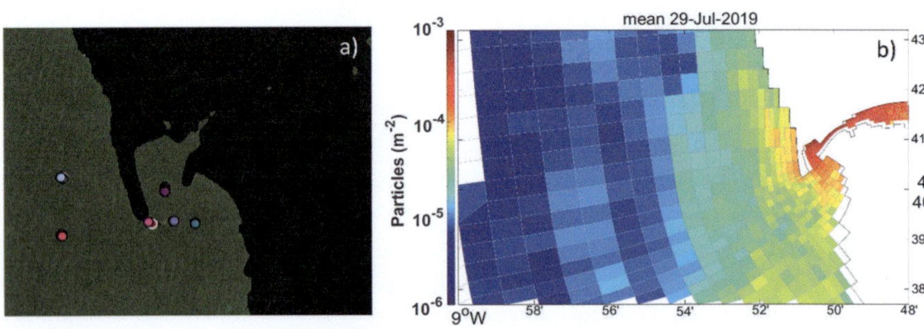

Fig. 11 Pixels containing plastic debris identified from a Sentinel-2A image of the mouth of Lima estuary (**a**) and model particle predictions (**b**) on 29 July 2019

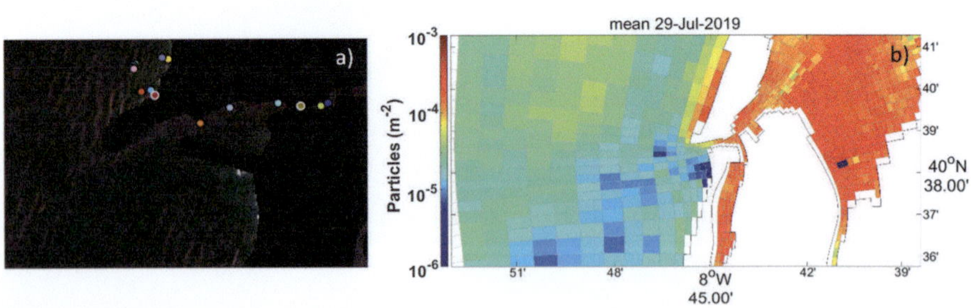

Fig. 12 Pixels containing plastic debris identified from a Sentinel-2A image of the mouth of Aveiro Lagoon (**a**) and model particle predictions (**b**) on 29 July 2019

The model generally shows good agreement with the satellite observations, as higher concentrations of particles are predicted near estuary mouths and coastlines, where plastic debris is more likely to accumulate. Figures 11b and 12b show the particle concentration, obtained by dividing the number of particles by the horizontal area of each model grid cell, averaged over one day for each grid cell, encompassing the date and location of the respective satellite image. This value indicates where the particles are more likely to accumulate and, therefore, be detected by remote sensing data to validate the model.

Overall, the particle concentrations are higher inside the estuaries and rivers and become lower as they reach the open ocean and decrease offshore, with differences reaching one or two orders of magnitude between waters close to the mouths and the offshore waters. The main difference between the Lima Estuary (Fig. 11) and the Ria de Aveiro (Fig. 12) is that, in the Lima Estuary, particle concentrations increase downstream before dispersing at the mouth into the ocean. In contrast, in the Aveiro Lagoon, there is a gradual decrease in particle concentrations along the main channel connecting to the ocean.

Figure 11b shows that near the Lima estuary, the model predicts a higher concentration of particles towards the north, close to the coast and the mouth. This agrees with the satellite observations in Fig. 11a as the pixels where plastic particles were identified are located near the mouth, around the northern breakwater. The same agreement is observed at the Ria de Aveiro (Fig. 12) as the particles detected from the satellite image are located inside the main entrance channel and to the north, outside, close to the northern breakwater, where the model predicts the higher accumulation of particles.

2.3 Detection of River Plumes

River plumes play a crucial role in coastal and ocean dynamics, influencing biogeochemical cycles, sediment transport, and marine ecosystems. These plumes, formed by freshwater discharge into the ocean, carry dissolved and particulate matter, affecting water properties such as salinity, turbidity, and temperature. Understanding their spatial and temporal variability is essential for environmental monitoring, climate studies, and ecosystem management applications [69, 70].

Remote sensing has become a key tool for studying river plumes, offering broad spatial coverage and frequent observations. By leveraging satellite imagery, it is possible to study the extent, dispersion, and evolution of these plumes under different oceanographic and meteorological conditions. With the emergence of more sensors and satellite platforms, the temporal, spatial and radiometric synergies between them allow for a more comprehensive analysis by combining data, for example, from optical, thermal, and radar observations [71–78].

This study focuses on the Douro River plume as a case study to showcase the potential of remote sensing in coastal dynamics analysis (Fig. 13). The Douro River, one of the largest discharging into the Atlantic Ocean, generates a highly dynamic plume influenced by river discharge, tides, and wind forcing [75, 76, 79, 80]. By exploring satellite data synergies, this study aims to present a case study on the detection and characterisation of river plumes while also providing insights into other sub-mesoscale processes associated with these transient coastal phenomena. The following section details the data sources and image processing techniques used in this study.

Typical Remote Sensing Data Sources for Detecting River Plumes

Detecting river plumes using remote sensing relies on identifying key properties distinguishing freshwater discharge from surrounding marine waters. These properties include salinity, turbidity, chlorophyll concentration, temperature, and surface roughness, which can be effectively monitored through satellite-based remote sensing products.

Salinity is traditionally the natural tracer of freshwater plumes. Although remotely sensed surface salinity from the Soil Moisture and Ocean Salinity (SMOS) mission [81] and the Aquarius mission [82] by the European Space Agency (ESA) and National Aeronautics and Space Administration (NASA), respectively, has been used to study large river plumes (e.g., Amazon River plume, Korosov et al. [83]), their spatial and temporal resolutions are insufficient for capturing smaller-scale plumes (mesoscale or smaller). These smaller plumes often exhibit fine-scale structures that evolve rapidly, making them challenging to monitor with current satellite salinity products.

Similarly, sea surface temperature (SST) can be a valuable tracer for river plumes, particularly when water is thermally distinct from the receiving basin. Thermal sensors aboard Sentinel-3 (SLSTR) and MODIS allow for the simultaneous retrieval

Fig. 13 Drone imagery from Douro River Plume. Courtesy of Underwater Systems and Technology Laboratory (LSTS), Faculty of Engineering—University of Porto / INEGI

of SST alongside ocean colour parameters, providing a complementary dataset for plume characterisation (e.g. Congo River Plume, Hopkins et al. [84]). However, SST measurements are susceptible to atmospheric conditions and radiative processes, which can influence surface temperature readings and introduce potential biases when interpreting satellite-derived results.

Due to these limitations, ocean-colour satellite images have been historically used with considerable success to track the behaviour and main scales of riverine plumes because of the optical differences and significant correlation between turbidity and surface salinity [70–72, 76, 85]. The turbidity concept measures water cloudiness [86]. It is usually linked to suspended and dissolved matter, such as clay, silt, finely divided organic matter, plankton and other microscopic organisms, organic acids and dyes, which decreases water transparency [87]. River plumes often present strong suspended sediment signals, affecting water cloudiness, especially in the green and red light wavebands in the visible region of the electromagnetic spectrum [88].

Beyond optical and thermal observations, Synthetic Aperture Radar (SAR) provides an additional means of detecting river plumes. Radar sensors such as Sentinel-1 and TerraSAR-X measure surface roughness variations, which can indicate freshwater intrusion due to differences in wind-wave interaction and current shear at the plume boundary [74, 89]. Unlike optical sensors, SAR can operate independently of cloud cover and sunlight, making it particularly valuable for monitoring plumes in persistently cloudy regions or during nighttime. However, the interpretation of SAR signals for plume detection remains complex, as roughness patterns can also be influenced by wind speed, precipitation, and surface slicks, requiring careful ancillary data analysis.

One of the main challenges in remote sensing of river plumes is the trade-off between spatial and temporal resolution and the phenomena resolution itself. High-spatial-resolution sensors, such as MSI onboard Sentinel-2, provide detailed observations of plume structures (30 m) but have relatively long revisit times (5 days per satellite), limiting their ability to capture short-term dynamics. Conversely, sensors like Sentinel-3 OLCI or MODIS, which offer daily global coverage, have coarser spatial resolutions (300–1000 m) that may not resolve fine-scale plume features.

However, recent advances in satellite mission design help mitigate this issue. For instance, the Sentinel-2 constellation, which includes Sentinel-2A and 2B, effectively reduces the revisit time to 2–3 days, improving temporal coverage without sacrificing spatial detail. Similarly, leveraging data from multiple platforms carrying the same sensor type, such as OLCI onboard Sentinel 3 or MODIS aboard Terra and Aqua, allows for increased observation frequency, enhancing the ability to track rapid changes in plume behaviour. This multi-satellite approach is particularly valuable for monitoring dynamic coastal processes, where both high-resolution spatial mapping and frequent observations are crucial for capturing transient plume dispersal patterns. In addition to spatial and temporal resolution, spectral resolution is critical in river plume detection. Sensors with higher spectral resolution allow for more precise differentiation between dissolved and suspended materials, improving the characterisation of sediment loads, phytoplankton communities, and organic matter within the plume. This differentiation is critical in creating automatic or semi-automatic local or regional tools for detecting river plumes [78]. In this context, the new data from NASA's recent PACE (Plankton, Aerosol, Cloud, ocean Ecosystem) mission will be a significant asset [90]. With hyperspectral capabilities, PACE will provide unprecedented detail in ocean colour measurements, enhancing our ability to distinguish between different water constituents and refine river plume monitoring, particularly in complex coastal environments.

Data Acquisition and Processing

By combining optical, thermal, and radar observations, multi-sensor remote sensing enhances the understanding river plume dynamics. In the river plume detection studies, particularly Douro River Plume, synergies between different missions enable a more detailed assessment of plume dispersion, seasonal variability, and other associated coastal phenomena.

The satellite images used in this study were obtained through the Copernicus DataSpace platform (https://browser.dataspace.copernicus.eu) and processed using SNAP - ESA Sentinel Application Platform v11.0.0 (http://step.esa.int), where image enhancements were applied. Prior to detailed analysis, a preliminary evaluation and cataloguing of Sentinel-2 imagery were conducted using EO Browser (https://apps.sentinel-hub.com/eo-browser/, Sinergise Solutions d.o.o., a Planet Labs company), and for Sentinel 1 and 3 were using OceanDataLab (https://ovl.oceandatalab.com).

Satellite Synergies in the River Plume's Detection. The case of the Douro River Plume

Using the Douro River plume as a case study, we demonstrate how different satellite sensors can be used to detect and characterise river plumes. Data from Sentinel-2 (optical), Sentinel-3 (ocean colour and thermal), and Sentinel-1 (SAR) provide complementary insights, each highlighting distinct plume features. Sentinel-2 captures fine-scale variations in turbidity and suspended matter, Sentinel-3 offers a broader perspective on colour and temperature gradients, while Sentinel-1 detects surface roughness changes linked to freshwater discharge.

The Sentinel-2 image in Fig. 14 captures the Douro River plume extending into the coastal ocean. Its higher turbidity clearly distinguishes it from surrounding marine waters. The plume appears lighter-coloured due to increased concentrations of suspended sediments and dissolved organic matter.

The spectral analysis compares reflectance values inside the plume and adjacent oceanic coastal waters. The most pronounced differences occur between 550 nm (green) and 700 nm (red), where the plume exhibits higher reflectance due to its elevated turbidity and particulate matter. These spectral characteristics are widely used for river plume detection, as they provide a simple yet effective way to distinguish freshwater intrusions from marine environments [71, 72, 75, 76, 78].

The plume front can be discerned thanks to Sentinel-2's high spatial resolution, particularly in the southern region, where a thin white line is visible. This feature results from a convergence zone at the plume boundary, as Fig. 13 demonstrates.

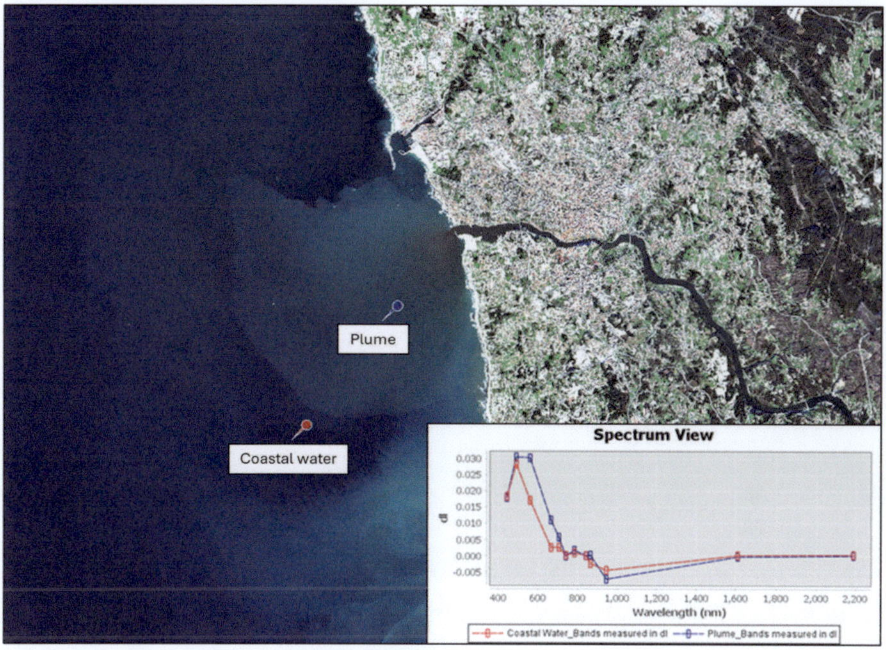

Fig. 14 Sentinel-2B—MSI- true-colour image from January 18, 2025, showing the Douro River plume highlighted by turbidity and colour differences in the water. The panel shows a spectral analysis within and outside the plume (plume and coastal oceanic waters)

By leveraging Sentinel-2's multi-band capabilities, this analysis enhances our ability to monitor plume dispersion, intensity, and seasonal variability, supporting coastal management and environmental monitoring efforts.

Building on the previous Sentinel-2 analysis, which identified the plume front, Sentinel-1 SAR imagery further enhances our ability to detect key plume features through differences in surface roughness.

In the Sentinel 1 image from November 27, 2024, in Fig. 15, the Douro River plume exhibits a distinct concentric shape, a signature of freshwater outflow interacting with coastal dynamics. The contrast between the smoother freshwater plume and the rougher surrounding waters allows for a clear delineation of its extent. Additionally, Sentinel-1's high-resolution radar capabilities enable the identification of vessels, surface slicks, and internal waves. The ability to capture such fine-scale oceanographic processes underscores the importance of SAR data in complementing optical and thermal satellite observations for a more comprehensive analysis of river plumes.

Additionally, using a multi-sensor approach, a time-series analysis of images over consecutive days showcases how the plume's extent and propagation can be monitored with high spatial detail (Fig. 16). This highlights the effectiveness of combining

Fig. 15 Sentinel-1 SAR image from November 27, 2024, depicting the Douro River plume with a distinct concentric structure. The plume front is visible due to differences in the roughness of water, creating a contrasting pattern between the estuarine outflow and the surrounding ocean

different satellite missions to enhance plume detection and improve coastal water monitoring.

Figure 16 illustrates the detection and temporal evolution of the Douro River plume between January 14 and 18, 2025, highlighting the advantages and trade-offs of using different satellite sensors. In the upper part, higher spatial resolution images from Sentinel-1 (SAR) and Sentinel-2 (optical) provide detailed insights into the plume's structure. A zoom-in on the January 16 Sentinel-2 image reveals the plume front with remarkable clarity, showing a thin white line to the south, indicative of a convergence zone. Meanwhile, the Sentinel-1 image from January 14 demonstrates how radar data can effectively characterise the plume, even on cloudy days, ensuring continuous observation regardless of atmospheric conditions.

In contrast, the lower part of the figure presents daily images from Sentinel-3, which, despite their lower spatial resolution, offer an essential advantage: high temporal frequency. This frequency is particularly useful for analysing short-term variations in plume propagation, which in this case are primarily driven by changes in wind intensity and direction. The sequential Sentinel-3 images between January 14 and 18 capture these shifts, showing how the plume disperses differently over time—a dynamic that would be harder to track with less frequent high-resolution imagery.

This comparison underscores the importance of combining data from multiple satellite missions, each offering different spatial, temporal, and spectral resolutions. The synergy between optical, thermal, and radar sensors enhances our ability to detect

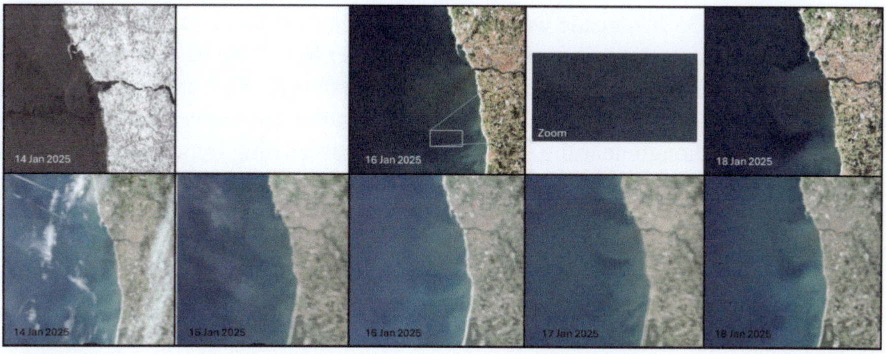

Fig. 16 Temporal evolution of the Douro River plume between January 14 and January 18, 2025, using data from Sentinel-1, Sentinel-2, and Sentinel-3. The top row showcases higher spatial resolution images (Sentinel-1A SAR and Sentinel-2 optical), allowing for detailed observation of the plume front and fine-scale structures. The zoomed-in view of the January 16 Sentinel-2A image highlights the sharp delineation of the plume boundary. The bottom row presents daily observations from Sentinel-3A and B, which provide essential continuity for daily tracking of the plume's coastal propagation despite having a lower spatial resolution

and monitor river plumes with greater accuracy, improving our understanding of their evolution and interaction with coastal dynamics through remote sensing products.

2.4 *Monitoring Sea Surface Temperature*

Sea surface temperature (SST) is one of the most important characteristics of seawater [91]. It is the main determinant of density in most of the ocean and, therefore, influences ocean circulation and nutrient fluxes worldwide [37, 91]. SST variations affect ocean primary production, which bounds the ocean food web and impacts carbon flows throughout the environment [37]. Consequently, it drives species distribution, abundance, diversity, and behaviour and ultimately impacts coastal populations that are dependent on marine resources, whether for food, leisure, or other services [92, 93]. In fact, not only is SST essential to study marine ecosystems, as it is also fundamental to study the global climate. Ocean–atmosphere interactions are the complex interface that regulates Earth's climate, which turns SST into an essential parameter for weather prediction and atmospheric model simulations [94, 95]. In fact, all processes in nature exhibit a temperature dependence [37] and measuring SST variations can help anticipate implications for water quality and ecosystem health [96]. For all these reasons, is it very important to monitor SST variations, and remote sensing techniques greatly improved the ability to do so in a more cost-effective way, that provides a global view of the ocean with a regular and high-resolution coverage.

Satellite-derived SST data is also key in the context of climate change. Global ocean temperatures are increasing, which can greatly affect marine ecosystems [97].

An increase of 0.035 °C/decade was observed between 1854 and 2017 for the surface of the global ocean, and this increase has been accelerating [98]. Regionally, changes in the magnitude of this increase are observed. For example, SST has increased since 1982 along the Western Iberian Coast, with local variations between 0.10 and 0.25 °C per decade, although near the shore, significant decreasing trends were observed apparently related to the seasonal upwelling that characterises the region [99]. The Mediterranean Sea is one of the areas on the planet most susceptible to increases in SST. An increase of 0.040 ± 0.001 °C/yr was observed by García-Monteiro et al. [100] between 2003 and 2019, showing warming trends from 0.02 °C/yr for the Alboran Sea to 0.07 °C/yr for the Adriatic Sea. Allied to the general long-term increase of the SST, extreme temperature-related events, which include storms, floods and marine heat waves or marine cold spells, are becoming more frequent and intense [101].

Marine heat waves and Marine Cold Spells

Marine heat waves (MHW) are periods of prolonged anomalously warm water at a particular location that can last from days to months and extend up to thousands of kilometres [102, 103]. These can be caused by local oceanic and atmospheric processes (*e.g.*, air-sea flux and horizontal advection) and could be influenced by large-scale climate variability (*e.g.*, teleconnections) [103]. According to Oliver et al. [103], temperatures are considered anomalously warm if they are above a specific value, *i.e.* a threshold, that can be fixed or variable over time or space. Therefore, there are several methods to quantitatively define MHWs but a commonly adopted approach that allows a coherent and comparable framework for the study of these type of events is the definition proposed by Hobday et al. [101], which states that a MHW is a period of at least five consecutive days with water temperatures above the 90th percentile threshold of the long-term mean seasonal climatology. In turn, marine cold spells (MCS) can be defined as the inverse of the MHWs, *i.e.* periods of anomalous cold-water temperature below the 10th percentile threshold of the long-term mean seasonal climatology for more than five continuous days, often developed due to wind-induced ocean processes [104]. Both MHWs and MCSs can shape marine ecosystem structure and functioning, pushing marine organisms and ecosystems to the limits of their resilience [102]. Coral mortality, shifts in the phenology, distribution, and population of molluscs and crustaceans and consequent economic implications (*e.g.*, reduction of commercial catch), local extinction of species, altered patterns of phytoplankton production and mass mortalities of birds and mammals are just some of the impacts observed with MHWs and MCSs worldwide [104, 105]. Thus, measuring the severity of these phenomena and understanding why different events can have wide-ranging biological, economic or political impacts, is also essential [105]. Following this line, Hobday et al. [105] proposed four categories for MHWs by framing the maximum temperature observed in an event in multiples of the local difference between the climatological mean and the 90th percentile. Therefore, MHWs can be moderate (Category I), strong (Category II), severe (Category III) and extreme (Category IV), as their intensity is framed in the 1-2x, 2-3x, 3-4x and > 4x of the difference between the climatological mean and the 90th percentile,

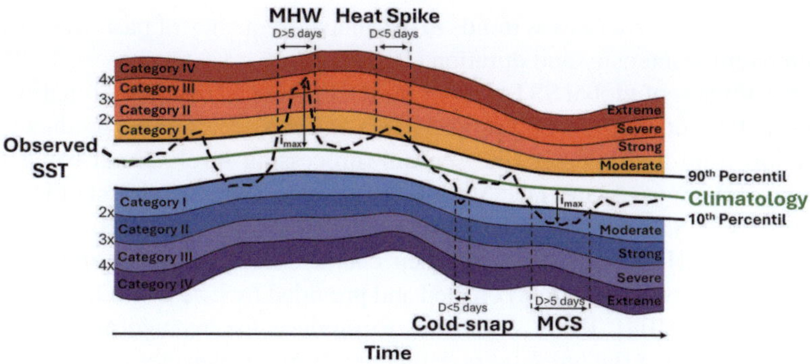

Fig. 17 Schematic of the metrics used to identify and categorize Marine Heat Waves (MHW) and Marine Cold Spells (MCS). This figure is based and partially adapted from Hobday et al. [101, 105] and Schlegel et al. [104]. D refers to the duration of the event and i_{max} corresponds to the maximum temperature above the climatological mean observed during an event

respectively. The same approach was adopted for MCSs [104]. Figure 17 illustrates how MHWs and MCSs are identified and categorized, following the definitions above mentioned.

Current studies and future projections suggest that MHWs have and will become more frequent, longer lasting, and more intense, as a consequence of long-term ocean warming [97, 102, 103, 106]. As for MCSs, studies have indicated a global decrease in their duration and intensity, except in the Southern Ocean where an increase has been observed for the past 15 years [104]. Still, we have limited knowledge on past occurrences and the future progression of this type of events, especially in coastal and transitional regions (*e.g.* estuaries, lagoons). This is aggravated by the fact that changes and characteristics of reported MHWs and MCSs are known to differ regionally [102, 107]. SST is now one of the variables with the longest record of data, given that it was one of the first variables to be monitored using remote sensing techniques. High resolution observations, with daily records covering entire coastal regions can be an extremely valuable and unique tool for climatological assessments, such as the study of MHWs and MCSs.

Case-Study: MHWs and MCSs Along the Western Iberian Coast

The main aim of this study was to identify features and patterns of past events of MHWs/MCSs along strategic regions of the Western Iberian Coast (WIC). The study area included the coastal region of Galicia and Algarve (Fig. 18), located on opposite shores of the WIC, and is framed in the context of a regional study (see Biguino et al., in review). Both regions are well known tourist centres, where strong anthropogenic pressure is observed and, at the same time, maritime areas with high ecological and socio-economic value are located. To achieve the main aim, three objectives were outlined: *(i)* identify and categorize past events observed in four points of the study regions (named from A to D, as in Fig. 18); *(ii)* evaluate the average patterns of MHWs/MCSs regarding the average intensity and duration of the events considering the last 41 years; and *(iii)* assess trends in the annual frequency of past MHWs/MCSs and their average intensity and duration.

Daily estimates of global SSTs from a product produced by ESA Climate Change Initiative (CCI) and the SST CCI project [38] were considered. This product gathers information from multiple satellite sensors (processing level 4) and has a spatial resolution of 0.05° (≈ 5 km). Data from 1st of January 1982 to the 31st of December 2022 were used. SST data was previously validated for the WIC [99].

To identify MHWs and MCSs and their characteristics, a python module called *marineHeatWaves* (version 0.28) created and provided by Eric C. J. Oliver was used (detect function) [108]. Event's statistics to produce the average and trend maps were calculated using the *blockAverage* function from the *marineHeatWaves* python module. Only significant trends were coloured in the trend maps (significance level of 0.05). It should be noted that higher intensities correspond to lower values (negative values) in MCSs.

(i) *Identification and categorization of past events*

The first objective was to identify and categorize past MHWs and MCSs events over four coastal locations of Galicia and Algarve. Table 1 shows the total number

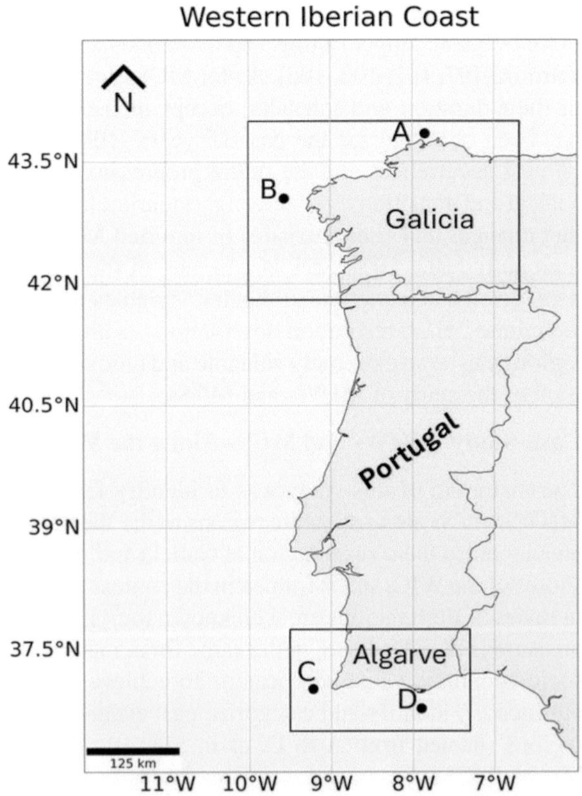

Fig. 18 Western Iberian Coast (WIC) and the study site regions of Galicia and Algarve

of MHWs and MCSs identified between 1982 and 2022 (41 years) as well as the number of events observed according to the four intensity categories.

With the exception of region A, there were more MCSs than MHWs. Although relatively similar, it was near Cape São Vincente (C) that the highest number of MHWs and MCSs was observed. This region is particularly favourable to the occurrence of both types of events. No extreme events were identified for MHWs or MCSs, but a Severe MHW (Category III) was observed in regions B, C and D. The most intense MHW event was observed near Cape São Vicente. It occurred between 13th April and 12th June 2011 (Spring, 61 days) and reached a maximum intensity of 4.88 °C, being classified as Severe. As for MCSs, the most intense event occurred between the 3rd and the 22nd of September 2007 (late Summer), lasted 20 days and had − 3.96 °C of maximum intensity (Strong).

(ii) *Evaluation of average patterns*

The second objective consisted in evaluating MHWs and MCSs average features to assess spatial patterns. Figures 19 and 20 show the total number of MHWs (A) and MCS (D) observed throughout the period of study, their average mean intensity (B and E) and duration (C and F), for the region of Galicia and Algarve, respectively.

A higher incidence of MCSs, compared to MHWs, was observed near the continental margin of both regions. Overall, the Algarve is more favourable to the occurrence of MHW than the coast of Galicia, with a particularly higher incidence observed near Cape São Vicente, Ria Formosa and the Guadiana Estuary. The same spatial patterns were observed for MCSs, although the region off Cape São Vicente is more prone to these events. In Galicia, the regions of Rias Baixas and the Minho Estuary stand out for MCSs. The regions with the highest incidence of MHWs or MCSs are not always the regions with the most intense events. This is observed in Rias Baixas and near Cape São Vicente. The Algarve coastal region has higher intensities MHW/MCS near its southern margin and increases/decreases southwards. As for the duration, regions with more events tend to correspond to regions where events lasted less time, and vice versa. This is observed for both MHWs and MCSs and is particularly evident in the Algarve.

(iii) *Trend assessment*

Table 1 Number of Marine Heat Waves (MHWs) and Marine Cold Spells (MCSs) identified in total and in each category of intensity, between1982 and 2022 in four locations off the coast of Galicia (A and B) and Algarve (C and D)

Coastal Points	Number of MHWs					Number of MCSs				
	I	II	III	IV	Total	I	II	III	IV	Total
A—Ortigueira Coast	72	12	0	0	85	63	20	0	0	83
B—Finisterre Coast	67	14	1	0	82	72	13	0	0	85
C—Cape São Vicente	82	10	1	0	**93**	91	8	0	0	**99**
D—Ria Formosa Coast	73	12	1	0	86	82	12	0	0	94

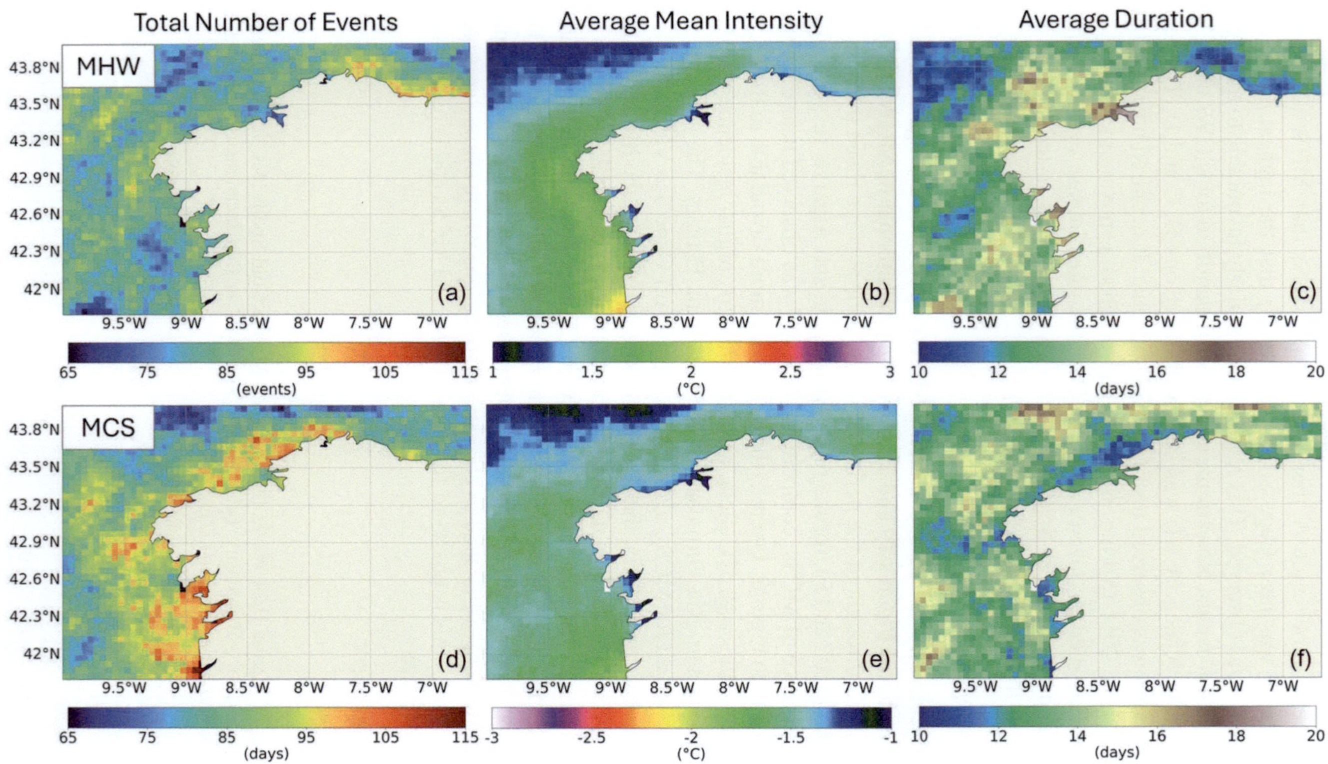

Fig. 19 Total number of MHWs (**a**) and MCSs (**d**) observed throughout the period of study and their average mean intensity (**b** and **e**) and duration (**c** and **f**), for the region of Galicia

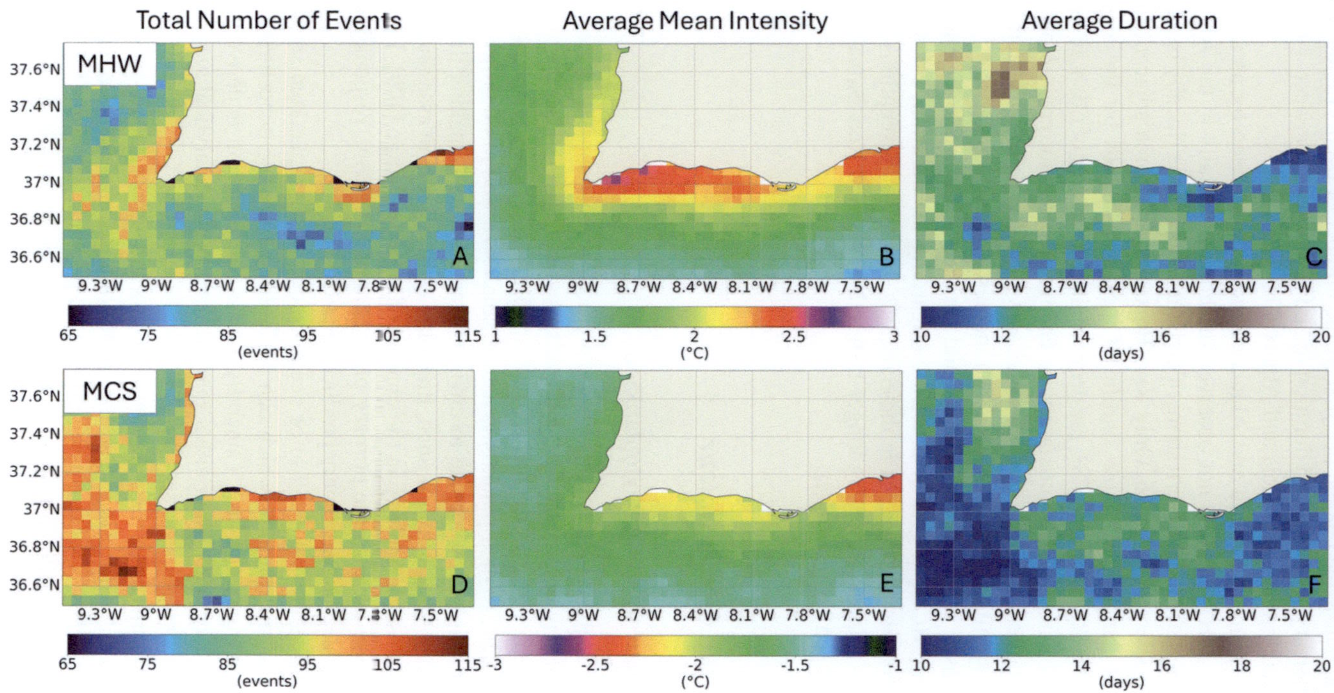

Fig. 20 Total number of MHWs (**a**) and MCSs (**d**) observed throughout the period of study and their average mean intensity (**b** and **e**) and duration (**c** and **f**), for the region of Algarve

The third objective was to conduct a trend analysis of MHWs and MCSs events. Figures 21 and 22 show the change per year of the number of events (a and d) and of their mean intensity (b and e) and duration (c and f), observed between 1982 and 2022 in the coastal regions of Galicia and Algarve, respectively.

The coastal regions of Galicia and Algarve have been affected by an increase in the number of MHWs that is not observed offshore, especially in the Algarve. Particularly relevant are the northern shore of Galicia, Rias Baixas (locally isolated), and the southern shore of Algarve. Only Rias Baixas and the area adjacent to the southern coast of the Algarve show significant increasing trends for the average intensity of MHWs. Regarding MCSs, the decrease in the number of offshore events is evident, which translates into non-significant changes close to the coast. However, the first few kilometres of the Algarve coastline, as well as some very coastal regions of Galicia, show significant trends towards an increase in the occurrence of MCSs. Likewise, the coastal region of NW Galicia and the area of Cape São Vicente, has been experiencing MCSs of greater intensity. Regarding the duration, events have not undergone any relevant changes in these regions. Interesting exceptions are the offshore region of Algarve, showing longer-lasting MHWs, and Rias Baixas where MCSs have also been lasting longer.

2.5 *Monitoring Ocean Colour*

Using ocean Colour to Study Phytoplankton

One of the most frequent applications of ocean colour remote sensing is the study and monitoring of phytoplankton. These organisms are ubiquitous, occurring even in the most remote oceanic regions where in-situ observations are logistically challenging (Fig. 23). Due to their planktonic nature, small size, and rapid growth rates, phytoplankton are highly responsive to environmental changes, making them key indicators of oceanic variability. As such, they play a crucial role in assessing shifts in marine ecosystems, particularly in the context of climate change [109]. Furthermore, phytoplankton are also the primary producers of the ocean, responsible for over half of the planet's total primary production [110, 111].

The most fundamental application of ocean colour in phytoplankton studies is estimating biomass distribution, which can be inferred from the concentration of the photosynthetic pigment Chl-*a*. Generally, Chl-*a* concentration is derived from remote sensing reflectances using either empirical and semi-analytical algorithms [112].

Empirical algorithms relate measurements of Chl-a and remote sensing reflectance using a mathematical expression. One the most common and accurate empirical algorithms, employed by NASA in its satellite Chl-*a* products, estimates Chl-*a* from the ratio of the remote sensing reflectances between the violet-blue and green region of the light spectrum [29]:

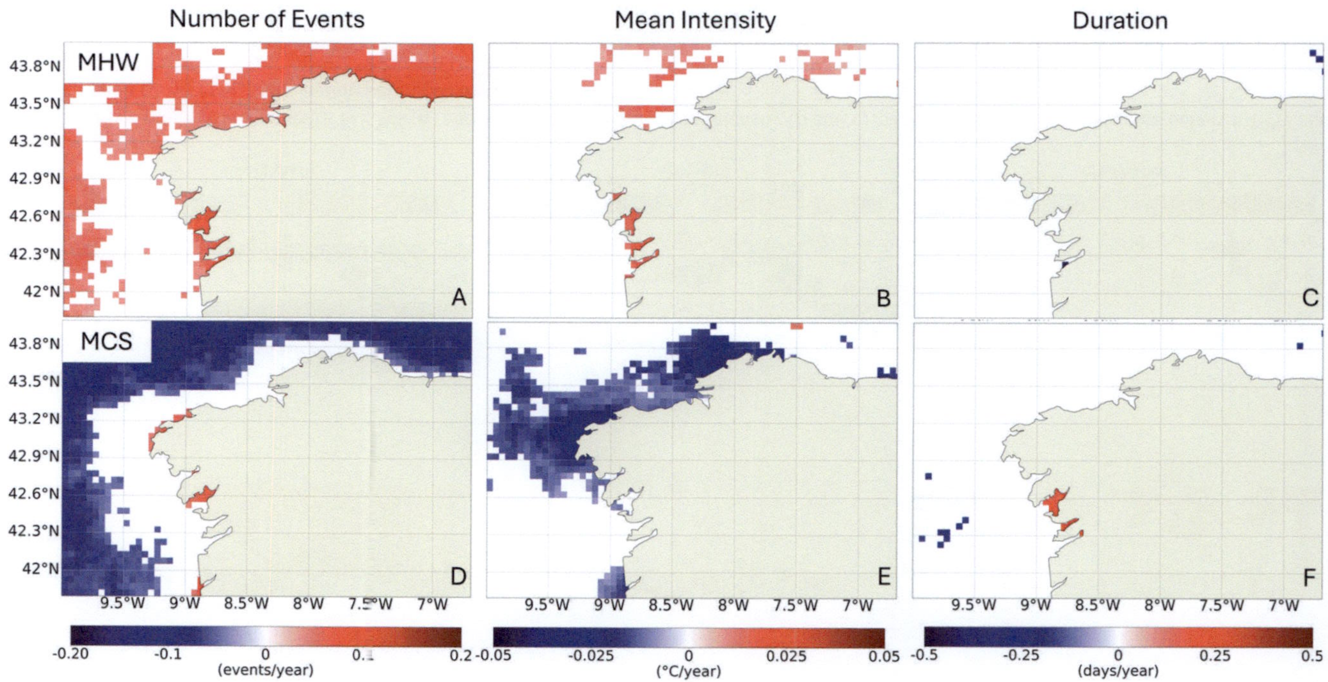

Fig. 21 Change per year of the number of MHWs (**a**) and MCSs (**d**), of their mean intensity (**b** and **e**) and duration (**c** and **f**), observed between 1982 and 2022 in the coastal region of Galicia. Only significant trends are coloured (p-value < 0.05)

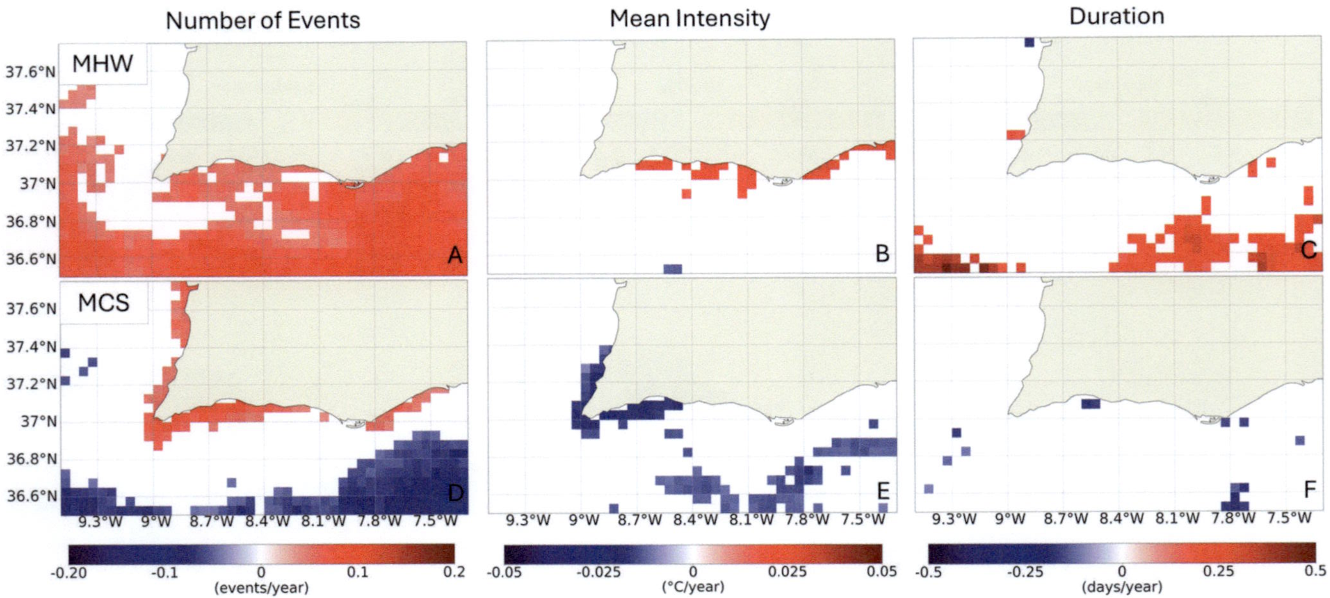

Fig. 22 Change per year of the number of MHWs (**a**) and MCSs (**d**), of their mean intensity (**b** and **e**) and duration (**c** and **f**), observed between 1982 and 2022 in the coastal region of Algarve. Only significant trends are coloured (p-value < 0.05)

Fig. 23 Mean phytoplankton biomass (using chlorophyll-a concentration as a proxy; mg/m^3) in the Atlantic Ocean during 2002–2025 (MODIS Aqua)

$$log_{10}(Chl\alpha) = a_0 + a_1X + a_2X^2 + a_3X^3 + a_4X^4 \quad (6)$$

$$X = log_{10}\left(\frac{\max[Rrs_{443}, Rrs_{490}, Rrs_{510}]}{Rrs_{555}}\right), \quad (7)$$

where, a0 to a4 are polynomial fit coefficients calculated using in situ Chl *a* data [29]. Hence, using this algorithm, lower ratios (greenish waters) correspond to higher concentrations of Chl-*a*.

Semi-analytical algorithms, on the other hand, estimate Chl-*a* through a combination of empirically acquired coefficients and radiative transfer equations (e.g. Lee and Carder [113]). Using these algorithms, Chl-*a* is typically derived from the estimated spectral absorption of phytoplankton (a_{ph}). These algorithms often provide estimates of other aquatic constituents, including the spectral backscattering by particles (b_{bp}) and the combined absorption of by non-algal particles and coloured dissolved organic material (a_{dg}). Hence, they are able to deconstruct the R_{rs} signal into multiple components, which can often be complex and require several assumptions [114, 115].

Accurate observations of Chl-*a* help researchers to map global primary productivity and monitor its variability over time. More advanced applications use ocean colour data to study additional phytoplankton properties, including pigment diversity, size structure and community composition (e.g. Hirata et al. [116]). One key aspect of phytoplankton dynamics that can be studied using ocean colour is bloom phenology—the timing, duration, and intensity of phytoplankton blooms [117]. Similar to terrestrial plants, phytoplankton exhibit seasonal growth cycles that depend on environmental factors such as light availability, nutrient availability, and temperature.

The first step in studying bloom phenology is bloom detection, typically achieved through the application of biomass-based thresholds to satellite Chl-*a* data [118, 119]. Moreover, since phytoplankton biomass can change rapidly from day-to-day, it is crucial that the Chl-*a* data used contains the least amount of data gaps possible. Hence, phenology studies often use multi-sensor products (e.g. ESA OC-CCI), gap-free L4 products (e.g. Globcolour), or custom interpolation algorithms to fill the existent gaps (e.g. Racault et al. [118]). In offshore waters, cumulative biomass thresholds are often used, e.g., the bloom onset is identified when cumulative Chl-*a* reaches 15% of the annual total [119]. However, in more complex regions multiple blooms can be observed, thus alternative thresholds based on the annual median Chl-*a* (e.g., ranging from 5 to 15%) may be more suitable [117, 119].

Once blooms are identified for a given region or pixel, key phenological metrics such as bloom onset, end, peak, magnitude, and duration can be extracted (Fig. 24). In temperate regions, such as in the North Atlantic, a characteristic spring bloom occurs following increased solar radiation and the stratification of the upper ocean after winter mixing [120]. While short-lived, the North Atlantic spring bloom is responsible for nearly half of this basin's total annual biologically-exported carbon (0.55–1.94 GtC y^{-1}; Sanders et al. [121]). In coastal areas, mesoscale processes such as upwelling and nutrient inputs from riverine discharge or atmospheric dust deposition can further shape bloom dynamics [118, 122].

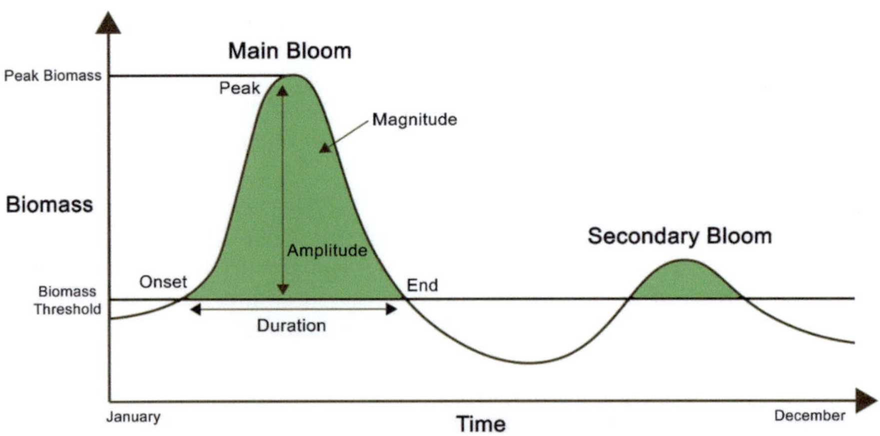

Fig. 24 Example of phytoplankton bloom phenology metrics that can be acquired by applying a biomass threshold method to satellite derived chlorophyll-a (adapted from Ferreira et al. [117])

Beyond its ecological significance, monitoring phytoplankton blooms via ocean colour remote sensing has multiple practical applications. It can be used to support fisheries and aquaculture by monitoring the environment and the productivity of a region of interest, detecting changes in primary production and contributing to cost-effective strategies for fisheries management [123]. Moreover, it can be used to monitor harmful algal blooms (HABs), caused by the rapid proliferation of toxin-producing phytoplankton species and can lead to mass mortality events in fish farms, shellfish poisoning, and other economic and environmental damages [124].

Because phytoplankton serve as the base of the marine food web, changes in phytoplankton bloom phenology can have cascading effects on higher trophic levels. Hence, ocean colour data have been used to analyse and predict changes in recruitment and abundance of upper trophic levels, including zooplankton [125], fish [126], whales [127], and even seabirds [128].

A clear example of the link between phytoplankton phenology and fish recruitment, visible from space, is found in temperate regions. Here, the spring phytoplankton bloom is typically followed by a zooplankton bloom, which in turn supports fish larvae and juvenile fish. Many fish species spawn in synchrony with this cycle to ensure sufficient food availability for their offspring [129]. However, shifts in bloom timing due to climate change or other environmental perturbations can disrupt this synchronization, leading to reduced fish recruitment and, consequently, declines in fisheries yields ("Match/Mismatch Hypothesis"; Cushing [130]; Fig. 25). Using phytoplankton bloom phenology metrics, derived from ocean colour data, these phenomena have already been observed in several commercially important species including haddock (*Melanogrammus aeglefinus;* Platt et al. [123]), herring (*Clupea harengus;* Schweigert et al. [131]), and salmon (*Oncorhynchus gorbuscha*; Malick et al. [132]).

Case Study: Phytoplankton and Small Pelagic Fishes in Western Iberia

The Western Iberian Coast (NE Atlantic) is a dynamic coastal region, encompassing the western coasts of Portugal and Spain, that serves as a transition region between temperate and subtropical climates and is located within the northern sector of the Canary Current Upwelling System [133]. The coastal upwelling off Western Iberia

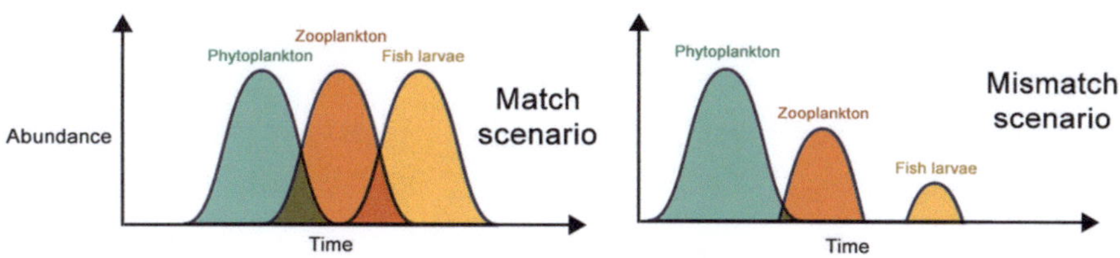

Fig. 25 Diagram of the "Match/Mismatch Hypothesis", presenting a "Match" scenario where there is an adequate overlap between the phytoplankton and zooplankton blooms and fish spawning and, alternatively, a "Mismatch" scenario, where an earlier onset of the phytoplankton bloom leads to a cascading negative effect, leading to a decrease in the survivability of fish larvae and, consequently, of fish recruitment. Our own figure based on Cushing et al. [130]

supports productive fisheries of small pelagic fishes (SPF), such as the European sardine (*Sardina pilchardus*), chub mackerel (*Scomber colias*), and horse mackerel (*Trachurus trachurus*). Among these, the European sardine is typically the most abundant and holds substantial socioeconomic importance in the region. However, since the early 2000s, sardine abundance has sharply declined due to consecutive years of low recruitment [134] (Fig. 26a), affecting the ecosystem and commercial fisheries [135].

Despite often being disregarded in fisheries studies, phytoplankton can have a key role in the life cycle of small pelagic fishes, acting both as a direct and indirect food source [136]. In the case of the planktivorous European sardine, sardine larvae feed almost exclusively on zooplankton (dependent on phytoplankton), while the diet of juvenile and adult individuals include both phyto- and zooplankton [136]. In Western Iberia, the contribution of phytoplankton to the sardine's diet is especially important during the spring and summer months, coinciding with the phytoplankton spring bloom and more intense coastal upwelling [137].

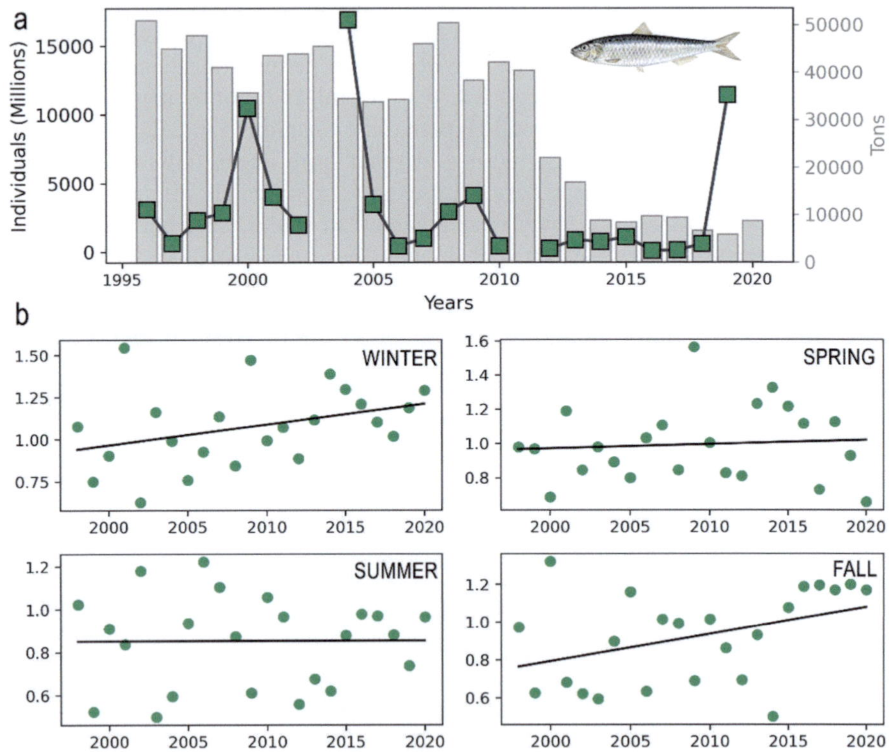

Fig. 26 Recent changes in sardine recruitment and landings in NW Portugal (Western Iberia) (1996–2020) (**a**) and in the average chlorophyll-a concentration for each season of the year (1998–2020) (**b**). Partially adapted from Ferreira et al. [138]

This case study focuses on how satellite estimations of phytoplankton bloom phenology can be used to study changes in the recruitment of a small pelagic fish, namely, of the European sardine. In the past, satellite-derived Chl-*a* has already been used as a proxy of sardine feeding intensity, helping distinguish between good and bad recruitment years [137, 139]. Nevertheless, existent satellite data indicates that Chl-*a* alone cannot explain the recent decline in sardine recruitment since no decline in Chl-*a* has appeared to occur during the same period in Western Iberia (Fig. 26b). Phytoplankton bloom phenology, however, has only very recently been related to sardine recruitment [138], as described below.

The first step in this case study was to characterize phytoplankton bloom phenology off Western Iberia by identifying areas with similar bloom phenology patterns (i.e., phenoregions). To do this, the average Chl-*a* seasonal cycle between 1998 and 2018 was first calculated for each pixel along the Western Iberian Coast using data from the OC-CCI v5.0 product. Subsequently, climatological bloom phenology metrics were estimated for each pixel and used to run a hierarchical clustering analysis which grouped phenoregions. As a result, five distinct phenoregions were identified along the Western Iberian Coast, each characterized by unique seasonal Chl-*a* cycles [117].

The results showed that offshore waters exhibit low productivity and a well-defined spring bloom (Fig. 27), with latitudinal gradients influencing bloom timing—southern offshore waters experience blooms nearly a month earlier than northern counterparts. Coastal regions, however, influenced by upwelling and riverine discharges, display higher productivity, frequent multi-annual blooms, and significant interannual variability. After establishing the complexity of phytoplankton bloom phenology off Western Iberia, it was possible to observe that bloom phenology off NW Portugal, a region within Western Iberia often considered as a hotspot for European sardine recruitment, is highly influenced by coastal upwelling (Fig. 27).

When comparing sardine recruitment with the newly-estimated phytoplankton bloom phenology metrics, the findings indicated that years with high sardine recruitment appear to coincide with periods of relatively high bloom magnitude, suggesting that sardine recruitment may benefit from higher phytoplankton (and consequently zooplankton) availability during their spawning season (Fig. 28). Given the sardine's spawning season (October–April), a larger spring bloom likely supports the growth of larvae and juveniles during critical early life stages. These results align with the previous studies linking sardine recruitment to Chl-*a* concentrations off NW Portugal [137, 139].

Interestingly, no clear relationship was found between bloom timing and sardine recruitment, indicating that sardines may be less susceptible to phenological mismatches than other temperate species. Instead, rising sea temperatures emerged as the primary factor associated with recent recruitment declines [138]. This aligns with Asch et al. [129], who suggested that 'geographic spawners' confined to specific habitats are more vulnerable to match-mismatch dynamics compared to highly dynamic and motile species like the European sardine, which are more dependent on broader environmental conditions.

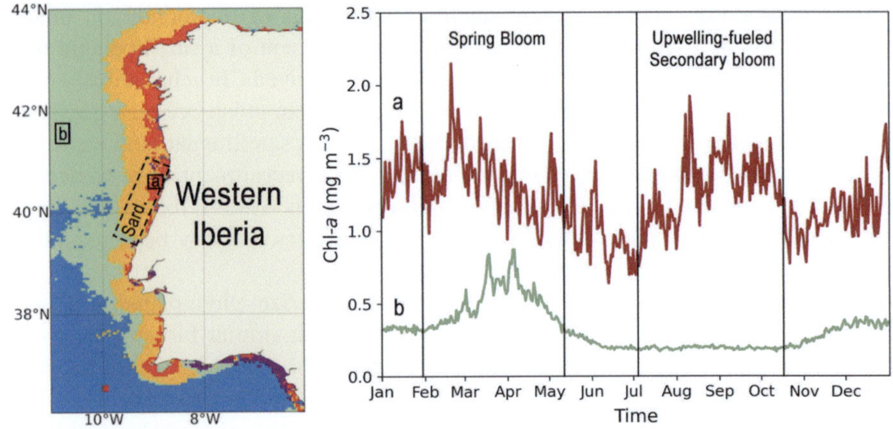

Fig. 27 Phytoplankton bloom phenology in two key phenoregions identified off Western Iberia: a coastal area highly influenced by coastal upwelling (**a**; red), and an offshore area characterized by a typical temperate spring bloom (**b**; green). The dashed rectangle in the map signals the sardine recruitment hotspot area off NW Portugal. Adapted from Ferreira et al. [117]

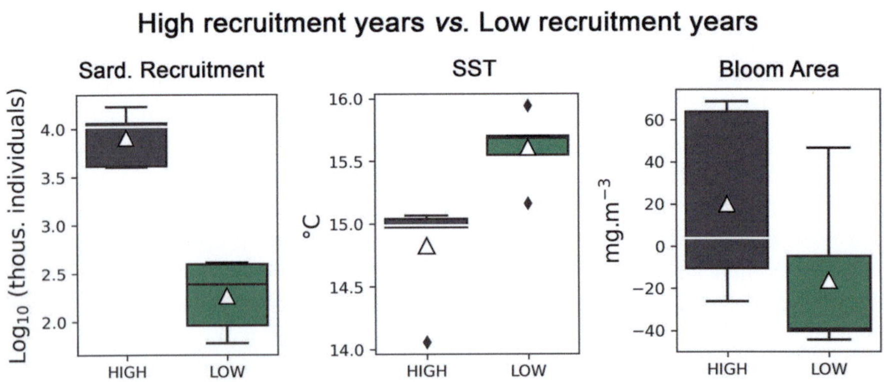

Fig. 28 Comparison of the sea surface temperature (SST; °C), and bloom area or magnitude (mg m^{-3}) between the five years with best sardine recruitment and the years with lowest sardine recruitment. Adapted from Ferreira et al. [138]

3 Conclusions

The advent of satellites has revolutionized Earth observation, providing scientists with consistent and regular data from across the globe, including regions inaccessible or challenging to monitor using traditional *in-situ* methods. Moreover, advancements in computing power and data analysis techniques have enabled this vast dataset's effective processing and interpretation, empowering researchers to address a wide range of critical environmental and scientific challenges. This chapter analysed the advances in remote sensing technologies employed for the ocean and coastal observation, highlighting how data acquired from various sensors on board satellites have been pivotal in addressing significant oceanographic and coastal challenges.

The first application demonstrated the effectiveness of using multispectral data from TM, ETM+, and MSI sensors onboard Landsat and Sentinel-2 satellites missions to efficiently monitor the evolution of estuarine salt marshes. In particular, they have made it possible: (1) to assess the degradation of salt marshes and the mechanisms of degradation, promoting the development of actions to reverse the decline and promote their recovery, thus contributing to the protection of these valuable ecosystems; (2) to detect progradation within salt marshes, and related evolution of the carbon sequestration rate and the total carbon stored, demonstrating the role of marshes in mitigating climate change.

The comprehensive approach of the second application demonstrates the potential of integrating remote sensing data, numerical modelling, and deep learning techniques to effectively detect and monitor floating plastic debris in oceanic waters, thereby contributing to mitigating marine pollution.

The third application briefly reviewed river plume detection using remote sensing, using the Douro River plume as a case study. By integrating data from multiple satellite sensors, we demonstrated how optical, thermal, and radar observations complement each other in identifying plume fronts and tracking their evolution. The synergy between different missions enhances monitoring capabilities, allowing for a more comprehensive understanding of plume dynamics. Future developments could explore AI/ML techniques to improve proxy identification between spectral bands and new satellite sensors (e.g., hyperspectral missions such as PACE), optimise regional detection strategies, and enable more precise tracking of river plumes over time.

The fourth application shows that at the WIC scale, it was possible to highlight the main patterns of MHWs and MCSs:

(i) Cape São Vincente registered the highest number of both MHWs (93) and MCSs (99). No extreme events were identified for MHWs or MCSs.
(ii) Algarve is more favourable to the occurrence of MHW than the coast of Galicia. Cape São Vicente is especially prone to MCSs. The regions with more events had events lasted less time, and vice versa.
(iii) The number of MHWs has increased along Galicia and Algarve and the events have become more intense locally. MCSs have been increasing near shore and increasing offshore. The NW of Galicia and Cape São Vicente have been experiencing more intense MCSs.

The use of satellite-derived SSTs allowed to obtain a consistent climatology based on more than 30 years of daily data (with no gaps), and covering the entire region of interest. This kind of assessment contributes to the development and implementation of appropriate regional mitigation and adaptation plans needed in the context of climate change and can help understand the response of several ecosystem components to extreme temperature-related events.

The fifth application underscores the value of using ocean colour remote sensing for predicting small pelagic fish recruitment, especially when integrated with other remotely sensed environmental variables (e.g. SST). While no clear evidence of match-mismatch dynamics was seen in this study, the observed link between low recruitment and reduced bloom magnitude highlights the complex connections between primary producers and higher trophic levels. Overall, phytoplankton bloom phenology offers valuable insights into primary production and ecosystem health, especially when using ocean colour data from space. It also opens new possibilities for understanding ecosystem responses to climate variability, improving biogeochemical models, and supporting fisheries management and conservation efforts. Ultimately, satellite-based phenological studies enhance our ability to predict future changes to marine ecosystems and their ecological and socio-economic impacts.

References

1. J.L. Awange, J.B. Kyalo Kiema, Fundamentals of remote sensing, in *Environmental Science and Engineering* (2013). https://doi.org/10.1007/978-3-642-34085-7_7
2. CCRS, Fundamentals of Remote Sensing. A Canada Centre for Remote Sensing Tutorial, no. November. 1999
3. J.A. Richards, X. Jia, Sources and characteristics of remote sensing image data. Remote. Sens. Digit. Image Anal. (1999). https://doi.org/10.1007/978-3-662-03978-6_1
4. M. Barnsley, Digital remotely-sensed data and their characteristics. Geogr. Inf. Syst. **1**(1) (1999)
5. C. Elachi, J. Zyl, Solid surfaces sensing in the visible and near infrared, in *Introduction to the Physics and Techniques of Remote Sensing* (2021). https://doi.org/10.1002/9781119523048.ch3
6. C. Elachi, J. van Zyl, Solid-surface sensing: thermal infrared, in *Introduction to the Physics and Techniques of Remote Sensing* (2006). https://doi.org/10.1002/0471783390.ch4
7. C. Elachi, J. van Zyl, Solid-surface sensing: microwave and radio frequencies, in *Introduction to the Physics and Techniques of Remote Sensing* (2006). https://doi.org/10.1002/0471783390.ch6
8. M.A. Wulder et al., Fifty years of Landsat science and impacts (2022). https://doi.org/10.1016/j.rse.2022.113195
9. S.L. Ustin, E.M. Middleton, Current and near-term advances in Earth observation for ecological applications (2021). https://doi.org/10.1186/s13717-020-00255-4
10. H. Astola, T. Häme, L. Sirro, M. Molinier, J. Kilpi, Comparison of Sentinel-2 and Landsat 8 imagery for forest variable prediction in boreal region. Remote Sens. Environ. **223** (2019). https://doi.org/10.1016/j.rse.2019.01.019
11. D.P. Roy et al., Characterization of Landsat-7 to Landsat-8 reflective wavelength and normalized difference vegetation index continuity. Remote Sens. Environ. (2016). https://doi.org/10.1016/j.rse.2015.12.024
12. V. Nasiri, A. Deljouei, F. Moradi, S.M.M. Sadeghi, S.A. Borz, Land use and land cover mapping using Sentinel-2, Landsat-8 satellite images, and google earth engine: a comparison of two composition methods. Remote Sens. **14**(9) (2022). https://doi.org/10.3390/rs14091977
13. G. Forkuor, K. Dimobe, I. Serme, J.E. Tondoh, Landsat-8 versus Sentinel-2: examining the added value of sentinel-2's red-edge bands to land-use and land-cover mapping in Burkina Faso. GIScience Remote Sens. **55**(3) (2018). https://doi.org/10.1080/15481603.2017.1370169

14. Z. Zhu, S. Qiu, S. Ye, Remote sensing of land change: a multifaceted perspective (2022). https://doi.org/10.1016/j.rse.2022.113266
15. S. Jombo, S. Adelabu, Evaluating Landsat-8, Landsat-9 and Sentinel-2 imageries in land use and land cover (LULC) classification in a heterogeneous urban area. GeoJournal **88** (2023). https://doi.org/10.1007/s10708-023-10982-8
16. K. Darwish, S. Smith, A comparison of Landsat-8 OLI, Sentinel-2 MSI and planetscope satellite imagery for assessing coastline change in El-Alamein, Egypt †. Eng. Proc. **10**(1) (2021). https://doi.org/10.3390/ecsa-8-11258
17. J.E. Pardo-Pascual et al., Assessment of satellite-derived shorelines automatically extracted from Sentinel-2 imagery using SAET. Coastal Eng. **188** (2024). https://doi.org/10.1016/j.coastaleng.2023.104426
18. K. Vos et al., Benchmarking satellite-derived shoreline mapping algorithms. Commun. Earth Environ. **4**(1) (2023). https://doi.org/10.1038/s43247-023-01001-2
19. G. Misra, F. Cawkwell, A. Wingler, Status of phenological research using sentinel-2 data: A review. Remote Sens. **12**(17) (2020). https://doi.org/10.3390/RS12172760
20. J.A. Caparros-Santiago, L.C. Quesada-Ruiz, V. Rodriguez-Galiano, Can land surface phenology from Sentinel-2 time-series be used as an indicator of Macaronesian ecosystem dynamics? Ecol. Inf. **77** (2023). https://doi.org/10.1016/j.ecoinf.2023.102239
21. K. Kowalski, C. Senf, P. Hostert, D. Pflugmacher, Characterizing spring phenology of temperate broadleaf forests using Landsat and Sentinel-2 time series. Int. J. Appl. Earth Obs. Geoinformation **92** (2020). https://doi.org/10.1016/j.jag.2020.102172
22. W.A. Hovis et al., Nimbus-7 coastal zone color scanner: System description and initial imagery. Science **210**(4465) (1980). https://doi.org/10.1126/science.210.4465.60
23. G.C. Feldman, Sea-viewing wide field-of-view sensor (Sea Wi FS). Van Nostrand's Scientific Encyclopedia (2007). https://doi.org/10.1002/9780471743989.vse10144
24. V.V. Salomonson, W.L. Barnes, P.W. Maymon, H.E. Montgomery, H. Ostrow, MODIS: advanced facility instrument for studies of the earth as a system. IEEE Trans. Geosci. Remote. Sens. **27**(2) (1989). https://doi.org/10.1109/36.20292
25. C. Cao, F.J. De Luccia, X. Xiong, R. Wolfe, F. Weng, Early on-orbit performance of the visible infrared imaging radiometer suite onboard the suomi national polar-orbiting partnership (S-NPP) satellite. IEEE Trans. Geosci. Remote. Sens. **52**(2) (2014). https://doi.org/10.1109/TGRS.2013.2247768
26. C. Donlon et al., The global monitoring for environment and security (GMES) Sentinel-3 mission. Remote. Sens. Environ. **120** (2012). https://doi.org/10.1016/j.rse.2011.07.024
27. G. Meister et al., The Ocean Color Instrument (OCI) on the Plankton, Aerosol, Cloud, ocean Ecosystem (PACE) Mission: system design and prelaunch radiometric performance. IEEE Trans. Geosci. Remote Sens. **62**, 1–18 (2024). https://doi.org/10.1109/TGRS.2024.3383812
28. S. Djavidnia, F. Mélin, N. Hoepffner, Comparison of global ocean colour data records. Ocean Sci. **6**(1) (2010). https://doi.org/10.5194/os-6-61-2010
29. J.E. O'Reilly et al., Ocean color chlorophyll algorithms for SeaWiFS. J. Geophys. Res. Ocean. **103**(C11) (1998). https://doi.org/10.1029/98JC02160
30. S.B. Groom et al., Satellite ocean colour: current status and future perspective (2019). https://doi.org/10.3389/fmars.2019.00485
31. J. Nieke et al., Ocean and land color imager on sentinel-3. Opt. Payloads Space Mission. (2015). https://doi.org/10.1002/9781118945179.ch10
32. I. Cetinić et al., Phytoplankton composition from sPACE: requirements, opportunities, and challenges (2024). https://doi.org/10.1016/j.rse.2023.113964
33. IOCCG, IOCCG Report Number 03: Remote sensing of ocean colour in coastal, and other optically-complex. Waters **3** (2000)
34. IOCCG, Atmospheric correction for remotely-sensed ocean-colour products. IOCCG Report Number **10**(10) (2010)
35. A. Sutcliffe, A.C. Brito, C. Sá, F. Sousa, D. Boutov, V. Brotas, Observação da Terra: uso de imagens de temperatura da superfície do mar e cor do oceano para a monitorização de águas costeiras e oceânicas. Lisboa (2016)

36. S. Kalluri, C. Cao, A. Heidinger, A. Ignatov, J. Key, T. Smith, The advanced very high resolution radiometer contributing to earth observations for over 40 years (2021). https://doi.org/10.1175/BAMS-D-20-0088.1
37. P.J. Minnett et al., Half a century of satellite remote sensing of sea-surface temperature. Remote. Sens. Environ. **233** (2019). https://doi.org/10.1016/j.rse.2019.111366
38. C.J. Merchant et al., Satellite-based time-series of sea-surface temperature since 1981 for climate applications. Sci. Data **6**(1) (2019). https://doi.org/10.1038/s41597-019-0236-x
39. E. Alerskans, J.L. Høyer, C.L. Gentemann, L.T. Pedersen, P. Nielsen-Englyst, C. Donlon, Construction of a climate data record of sea surface temperature from passive microwave measurements. Remote Sens. Environ. **236** (2020). https://doi.org/10.1016/j.rse.2019.111485
40. M. National Academies of Sciences, Engineering, A strategy for active remote sensing amid increased demand for radio spectrum (2015). https://doi.org/10.17226/21729
41. P. Schreiner, R. König, K.H. Neumayer, A. Reinhold, On precise orbit determination based on DORIS, GPS and SLR using Sentinel-3A/B and -6A and subsequent reference frame determination based on DORIS-only. Adv. Space Res. **72**(1) (2023). https://doi.org/10.1016/j.asr.2023.04.002
42. J.F. Legeais et al., Copernicus sea level space observations: a basis for assessing mitigation and developing adaptation strategies to sea level rise. Front. Marine Sci. **8** (2021). https://doi.org/10.3389/fmars.2021.704721
43. S. Singh, R.K. Tiwari, V. Sood, R. Kaur, S. Prashar, The legacy of scatterometers: review of applications and perspective (2022). https://doi.org/10.1109/MGRS.2022.3145500
44. R.M. Asiyabi, A. Ghorbanian, S.N. Tameh, M. Amani, S. Jin, A. Mohammadzadeh, Synthetic Aperture Radar (SAR) for ocean: a review. IEEE J. Sel. Top. Appl. Earth Obs. Remote. Sens. **16** (2023). https://doi.org/10.1109/JSTARS.2023.3310363
45. L.A. Boorman, Salt marshes—present functioning and future change. Mangrove Salt Marshes **3**(4), 227–241 (1999). https://doi.org/10.1023/A:1009998812838
46. G. Mariotti, J. Carr, Dual role of salt marsh retreat: long-term loss and short-term resilience. Water Resour. Res. **50**(4), 2963–2974 (2014). https://doi.org/10.1002/2013WR014676
47. C.L. Lopes, R. Mendes, I. Caçador, J.M. Dias, Evaluation of long-term estuarine vegetation changes through Landsat imagery. Sci. Total. Environ. **653**, 512–522 (2019). https://doi.org/10.1016/j.scitotenv.2018.10.381
48. C.L. Lopes, R. Mendes, I. Caçador, J.M. Dias, Assessing salt marsh loss and degradation by combining long-term LANDSAT imagery and numerical modelling. Land Degrad. Dev. **32**(16), 4534–4545 (2021). https://doi.org/10.1002/ldr.4050
49. C.L. Lopes, R. Mendes, I. Caçador, J.M. Dias, Assessing salt marsh extent and condition changes with 35 years of Landsat imagery: Tagus Estuary case study. Remote Sens. Environ. **247** (2020). https://doi.org/10.1016/j.rse.2020.111939
50. G.J. Miller, J.T. Morris, C. Wang, Mapping salt marsh dieback and condition in South Carolina's North Inlet-Winyah Bay National Estuarine Research Reserve using remote sensing. AIMS Environ. Sci. **4**(5), 677–689 (2017). https://doi.org/10.3934/environsci.2017.5.677
51. Y. Mo, M.S. Kearney, J.C.A. Riter, F. Zhao, D.R. Tilley, Assessing biomass of diverse coastal marsh ecosystems using statistical and machine learning models. Int. J. Appl. Earth Obs. Geoinf. **68**, 189–201 (2018). https://doi.org/10.1016/J.JAG.2017.12.003
52. G. Yang et al., MFI: a mudflat index based on hyperspectral satellite images for mapping coastal mudflats. Int. J. Appl. Earth Obs. Geoinformation **133**, 104140 (2024)
53. T. Kuleli, A. Guneroglu, F. Karsli, M. Dihkan, Automatic detection of shoreline change on coastal Ramsar wetlands of Turkey. Ocean Eng. **38**(10), 1141–1149 (2011). https://doi.org/10.1016/j.oceaneng.2011.05.006
54. M.L. Laengner, K. Siteur, D. van der Wal, Trends in the seaward extent of saltmarshes across Europe from long-term satellite data. Remote Sensing **11**(14), 1653 (2019). https://doi.org/10.3390/rs11141653
55. N. Murray, S. Phinn, R. Clemens, C. Roelfsema, R. Fuller, Continental scale mapping of tidal flats across east asia using the landsat archive. Remote Sens. **4**(11), 3417–3426 (2012). https://doi.org/10.3390/rs4113417

56. C. Sun, S. Fagherazzi, Y. Liu, Classification mapping of salt marsh vegetation by flexible monthly NDVI time-series using Landsat imagery. Estuar. Coast. Shelf Sci. **213**, 61–80 (2018). https://doi.org/10.1016/j.ecss.2018.08.007
57. H. Xu, Modification of normalised difference water index (NDWI) to enhance open water features in remotely sensed imagery. Int. J. Remote Sens. **27**(14), 3025–3033 (2006). https://doi.org/10.1080/01431160600589179
58. J.W. Rouse, R.H. Haas, J.A. Schell, D.W. Deering, Monitoring vegetation systems in the Great Okains with ERTS, in *Third Earth Resources Technology Satellite-1 Symposium*, vol. 1. (1973), pp. 325–333, 10/citeulike-article-id:12009708
59. J.A. Gamon et al., Relationships between NDVI, canopy structure, and photosynthesis in three Californian vegetation types. Ecol. Appl. **5**(1), 28–41 (1995). https://doi.org/10.2307/1942049
60. J. Qi, A. Chehbouni, A.R. Huete, Y.H. Kerr, S. Sorooshian, A modified soil adjusted vegetation index. Remote Sens. Environ. **48**(2), 119–126 (1994). https://doi.org/10.1016/0034-4257(94)90134-1
61. J. Qi, Y. Kerr, A. Chehbouni, External factor consideration in vegetation index development, in *Proceedings of 6th International Symposium on Physical Measurements and Signatures in Remote Sensing*. (1994), pp. 723–730
62. B. Pinty, M.M. Verstraete, GEMI: a non-linear index to monitor global vegetation from satellites. Vegetatio **101**(1), 15–20 (1992). https://doi.org/10.1007/BF00031911
63. I.J. Castro, J.M. Dias, C.L. Lopes, Assessing shoreline changes in fringing salt marshes from satellite remote sensing data. Remote Sens. **15**(18) (2023). https://doi.org/10.3390/rs15184475
64. A. Freitas, J.M. Dias, C.L. Lopes, Application of remote sensing methods for monitoring extent, condition and blue carbon storage in salt marshes. Remote Sens. Appl. Soc. Environ. **35**, 101226 (2024). https://doi.org/10.1016/j.rsase.2024.101226
65. A.I. Sousa, A.I. Lillebø, M.A. Pardal, I. Caçador, Productivity and nutrient cycling in salt marshes: Contribution to ecosystem health. Estuarine, Coastal Shelf Sci. **87**(4) (2010). https://doi.org/10.1016/j.ecss.2010.03.007
66. M.M. Duarte, L. Azevedo, Automatic detection and identification of floating marine debris using multispectral satellite imagery. IEEE Trans. Geosci. Remote. Sens. **61** (2023). https://doi.org/10.1109/TGRS.2023.3283607
67. L.J.J. Meijer, T. van Emmerik, R. van der Ent, C. Schmidt, L. Lebreton, More than 1000 rivers account for 80% of global riverine plastic emissions into the ocean. Sci. Adv. **7**(18) (2021). https://doi.org/10.1126/sciadv.aaz5803
68. M.C. Sousa et al., Modelling the distribution of microplastics released by wastewater treatment plants in Ria de Vigo (NW Iberian Peninsula). Mar. Pollut. Bull. **166**, 112227 (2021). https://doi.org/10.1016/j.marpolbul.2021.112227
69. K.R. Dyer, *Estuaries: A Physical Introduction, Estuarine and Coastal Marine Science*. Wiley (1973). https://doi.org/10.1016/0302-3524(73)90042-X
70. B. Dzwonkowski, X.H. Yan, Tracking of a Chesapeake Bay estuarine outflow plume with satellite-based ocean color data. Cont. Shelf Res. **25**(16), 1942–1958 (2005)
71. S.L. Palacios, T.D. Peterson, R.M. Kudela, Development of synthetic salinity from remote sensing for the Columbia River plume. J. Geophys. Res. Ocean. **114**(4) (2009). https://doi.org/10.1029/2008JC004895
72. G.S. Saldías, M. Sobarzo, J. Largier, C. Moffat, R. Letelier, Seasonal variability of turbid river plumes off central Chile based on high-resolution MODIS imagery. Remote Sens. Environ. **123** (2012). https://doi.org/10.1016/j.rse.2012.03.010
73. C. Petus, V. Marieu, S. Novoa, G. Chust, N. Bruneau, J. M. Froidefond, Monitoring spatio-temporal variability of the Adour River turbid plume (Bay of Biscay, France) with MODIS 250-m imagery. Continental Shelf Res. **74** (2014). https://doi.org/10.1016/j.csr.2013.11.011
74. R. Mendes et al., On the generation of internal waves by river plumes in subcritical initial conditions. Sci. Rep. **11**(1) (2021). https://doi.org/10.1038/s41598-021-81464-5

75. R. Mendes, G.S. Saldías, M. deCastro, M. Gómez-Gesteira, N. Vaz, J.M. Dias, Seasonal and interannual variability of the Douro turbid river plume, northwestern Iberian Peninsula. Remote Sens. Environ. **194** (2017). https://doi.org/10.1016/j.rse.2017.04.001
76. R. Mendes et al., Observation of a turbid plume using MODIS imagery: the case of Douro estuary (Portugal). Remote Sens. Environ. **154**(1) (2014). https://doi.org/10.1016/j.rse.2014.08.003
77. F. Toublanc, N.K. Ayoub, P. Marsaleix, On the role of wind and tides in shaping the Gironde River plume (Bay of Biscay). Continental Shelf Res. **253** (2023). https://doi.org/10.1016/j.csr.2022.104891
78. J. Tavora, G.A. Gonçalves, E.H. Fernandes, M.S. Salama, D. van der Wal, Detecting turbid plumes from satellite remote sensing: State-of-art thresholds and the novel PLUMES algorithm. Front. Marine Sci. **10** (2023). https://doi.org/10.3389/fmars.2023.1215327
79. A.C. Teodoro, H. Goncalves, F. Veloso-Gomes, J.A. Gonçalves, Modeling of the douro river plume size, obtained through image segmentation of MERIS data. IEEE Geosci. Remote Sens. Lett. **6**(1) (2009). https://doi.org/10.1109/LGRS.2008.2008446
80. R. Mendes, M.C. Sousa, M. deCastro, M. Gómez-Gesteira, J.M. Dias, New insights into the Western Iberian Buoyant Plume: iInteraction between the Douro and Minho River plumes under winter conditions. Prog. Ocean. (2016). https://doi.org/10.1016/j.pocean.2015.11.006
81. J. Font et al., SMOS first data analysis for sea surface salinity determination. Int. J. Remote. Sens. **34**(9–10) (2013). https://doi.org/10.1080/01431161.2012.716541
82. D.M. Le Vine, G.S.E. Lagerloef, F.R. Colomb, S.H. Yueh, F.A. Pellerano, Aquarius: an instrument to monitor sea surface salinity from space. IEEE Trans. Geosci. Remote Sens. (2007). https://doi.org/10.1109/TGRS.2007.898092
83. A. Korosov, F. Counillon, J.A. Johannessen, Monitoring the spreading of the Amazon freshwater plume by MODIS, SMOS, Aquarius, and TOPAZ. J. Geophys. Res. Ocean. **120**(1) (2015). https://doi.org/10.1002/2014JC010155
84. J. Hopkins et al., Detection and variability of the Congo River plume from satellite derived sea surface temperature, salinity, ocean colour and sea level. Remote Sens. Environ. **139** (2013). https://doi.org/10.1016/j.rse.2013.08.015
85. R.V. Martins, L.D. Lacerda, S. Mounier, H.H.M. Paraquetti, W.S. Marques, Caracterização hidroquímica, distribuição e especiação de mercúrio nos estuários dos rios Ceará e Pacoti, região metropolitana de Fortaleza, Ceará, Brasil. Geochimica Brasiliensis **16**(1), 37 (2002). https://doi.org/10.21715/GB.V16I1.497
86. R.J. Davies-Colley, D.G. Smith, Turbidity, suspended sediment, and water clarity: a review (2001). https://doi.org/10.1111/j.1752-1688.2001.tb03624.x
87. C. Petus, G. Chust, F. Gohin, D. Doxaran, J.M. Froidefond, Y. Sagarminaga, Estimating turbidity and total suspended matter in the Adour River plume (South Bay of Biscay) using MODIS 250-m imagery. Continental Shelf Res. **30**(5) (2010). https://doi.org/10.1016/j.csr.2009.12.007
88. H. Loisel, V. Vantrepotte, C. Jamet, D. Ngoc Dat, Challenges and new advances in ocean color remote sensing of coastal waters. Top. Ocean. (2013). https://doi.org/10.5772/56414
89. J.D. Nash, J.N. Moum, River plumes as a source of large-amplitude internal waves in the coastal ocean. Nature **437**(7057) (2005). https://doi.org/10.1038/nature03936
90. H.M. Dierssen et al., Synergies between NASA's hyperspectral aquatic missions PACE, GLIMR, and SBG: opportunities for new science and applications. J. Geophys. Res. Biogeosciences **128**(10) (2023). https://doi.org/10.1029/2023JG007574
91. L.D. Talley, G.L. Pickard, W.J. Emery, J.H. Swift, *Descriptive Physical Oceanography: An Introduction*, 6th edn. (2011). https://doi.org/10.1016/C2009-0-24322-4
92. EPA, *Climate Change Indicators: Sea Surface Temperature*. Environmental Protection Agency.
93. R.A. Hastings, L.A. Rutterford, J.J. Freer, R.A. Collins, S.D. Simpson, M.J. Genner, Climate change drives poleward increases and equatorward declines in marine species. Curr. Biol. **30**(8) (2020). https://doi.org/10.1016/j.cub.2020.02.043
94. NOAA, Why do scientists measure sea surface temperature? Natl. Ocean. Serv.

95. D. Zardi, Atmosphere and ocean interactions. Rendiconti Lincei. Scienze Fisiche e Naturali **35**(2), 311–325 (2024). https://doi.org/10.1007/s12210-024-01243-y
96. EPA, Factsheet on water quality parameters—temperature, Environmental Protection Agency, no. C, 2021
97. IPCC, "IPCC, 2019: IPCC Special Report on the Ocean and Cryosphere in a Changing Climate," The Ocean and Cryosphere in a Changing Climate, no. October, 2019
98. Z. Wu, C. Jiang, M. Conde, J. Chen, B. Deng, The long-term spatiotemporal variability of sea surface temperature in the northwest Pacific and China offshore. Ocean Sci. **16**(1) (2020). https://doi.org/10.5194/os-16-83-2020
99. B. Biguino et al., 40 years of changes in sea surface temperature along the Western Iberian Coast. Sci. Total. Environ. **888** (2023). https://doi.org/10.1016/j.scitotenv.2023.164193
100. S. García-Monteiro, J.A. Sobrino, Y. Julien, G. Sòria, D. Skokovic, Surface temperature trends in the Mediterranean Sea from MODIS data during years 2003–2019. Reg.Nal Stud. Mar. Sci. **49** (2022). https://doi.org/10.1016/j.rsma.2021.102086
101. A.J. Hobday et al., A hierarchical approach to defining marine heatwaves. Prog. Ocean. **141** (2016). https://doi.org/10.1016/j.pocean.2015.12.014
102. T.L. Frölicher, E.M. Fischer, N. Gruber, Marine heatwaves under global warming. Nature **560**(7718) (2018). https://doi.org/10.1038/s41586-018-0383-9
103. E.C.J. Oliver et al., Marine heatwaves. Annu. Rev. Mar. Sci. **13**, 313–342 (2021). https://doi.org/10.1146/annurev-marine-032720-095144
104. R.W. Schlegel, S. Darmaraki, J.A. Benthuysen, K. Filbee-Dexter, E.C.J. Oliver, Marine cold-spells. Prog. Ocean. **198** (2021). https://doi.org/10.1016/j.pocean.2021.102684
105. A.J. Hobday et al., Categorizing and naming marine heatwaves. Oceanography **31**(2 Special Issue) (2018). https://doi.org/10.5670/oceanog.2018.205
106. IPCC, Summary for policymakers in climate change 2022: mitigation of climate change, in *Contribution of Working Group III to the Sixth Assessment Report of the IPCC* (2023)
107. A. Simon, C. Poppeschi, S. Plecha, G. Charria, A. Russo, Coastal and regional marine heatwaves and cold spells in the northeastern Atlantic. Ocean Sci. **19**(5) (2023). https://doi.org/10.5194/os-19-1339-2023
108. E.C.J. Oliver, Marine heatwaves detection code (2018)
109. J.E. Cloern, T.S. Schraga, E. Nejad, T. Eddy, Phytoplankton as indicators of global warming? Limnol. Ocean. Lett. **9**(3) (2024). https://doi.org/10.1002/lol2.10354
110. P.G. Falkowski, The role of phytoplankton photosynthesis in global biogeochemical cycles (1994). https://doi.org/10.1007/BF00014586
111. M.J. Behrenfeld et al., Biospheric primary production during an ENSO transition. Science **291**(5513) (2001). https://doi.org/10.1126/science.1055071
112. IOCCG, Remote Sensing of inherent optical properties : fundamentals, tests of algorithms, and applications, Rep. Int. Ocean. Colour Coord. Group **5**(5) (2006)
113. Z. Lee, K.L. Carder, Absorption spectrum of phytoplankton pigments derived from hyperspectral remote-sensing reflectance. Remote Sens. Environ. **89**(3) (2004). https://doi.org/10.1016/j.rse.2003.10.013
114. Z. Lee, K.L. Carder, R.A. Arnone, Deriving inherent optical properties from water color: a multiband quasi-analytical algorithm for optically deep waters. Appl. Opt. **41**(27) (2002). https://doi.org/10.1364/ao.41.005755
115. A.A. Gitelson et al., A simple semi-analytical model for remote estimation of chlorophyll-a in turbid waters: Validation. Remote Sens. Environ. **112**(9) (2008). https://doi.org/10.1016/j.rse.2008.04.015
116. T. Hirata, J. Aiken, N. Hardman-Mountford, T.J. Smyth, R.G. Barlow, An absorption model to determine phytoplankton size classes from satellite ocean colour. Remote Sens. Environ. **112**(6) (2008) https://doi.org/10.1016/j.rse.2008.03.011
117. A. Ferreira, V. Brotas, C. Palma, C. Borges, A.C. Brito, Assessing phytoplankton bloom phenology in upwelling-influenced regions using ocean color remote sensing. Remote Sens. **13**(4) (2021). https://doi.org/10.3390/rs13040675

118. M.F. Racault, C. Le Quéré, E. Buitenhuis, S. Sathyendranath, T. Platt, Phytoplankton phenology in the global ocean. Ecol. Indicators **14**(1) (2012). https://doi.org/10.1016/j.ecolind.2011.07.010
119. S R. Brody, M.S. Lozier, J.P. Dunne, A comparison of methods to determine phytoplankton bloom initiation. J. Geophys. Res. Ocean. **118**(5) (2013). https://doi.org/10.1002/jgrc.20167
120. S.M. Chiswell, P.H.R. Calil, P.W. Boyd, Spring blooms and annual cycles of phytoplankton: a unified perspective. J. Plankton Res. **37**(3) (2014). https://doi.org/10.1093/plankt/fbv021
121. R. Sanders et al., The biological carbon pump in the north Atlantic. Prog. Ocean. **129**(PB) (2014). https://doi.org/10.1016/j.pocean.2014.05.005
122. A. Corredor-Acosta, C.E. Morales, S. Hormazabal, I. Andrade, M.A. Correa-Ramirez, Phytoplankton phenology in the coastal upwelling region off central-southern Chile (35°S-38°S): Time-space variability, coupling to environmental factors, and sources of uncertainty in the estimates. J. Geophys. Res.: Ocean. **120**(2) (2015). https://doi.org/10.1002/2014JC010330
123. T. Platt, C. Fuentes-Yaco, K.T. Frank, Spring algal bloom and larval fish survival. Nature 423 (2003)
124. W.A. Hovis, The Nimbus-7 Coastal Zone Color Scanner (CZCS) Program, in *Oceanography from Space*, ed by J.F.R. Gower (Springer US, Boston, MA, 1981), pp. 213–225. https://doi.org/10.1007/978-1-4613-3315-9_27
125. K.H.P. Strömberg, T.J. Smyth, J.I. Allen, S. Pitois, T.D.O'Brien, Estimation of global zooplankton biomass from satellite ocean colour. J. Mar. Syst. **78**(1) (2009). https://doi.org/10.1016/j.jmarsys.2009.02.004
126. M.K. Trzcinski, E. Devred, T. Platt, S. Sathyendranath, Variation in ocean colour may help predict cod and haddock recruitment. Mar. Ecol. Prog. Ser. **491** (2013). https://doi.org/10.3354/meps10451
127. J.C. Burtenshaw et al., Acoustic and satellite remote sensing of blue whale seasonality and habitat in the Northeast Pacific. Deep.-Sea Res. Part II: Top. Stud. Ocean. **51**(10–11) SPEC. ISS. (2004). https://doi.org/10.1016/j.dsr2.2004.06.020
128. R.M. Suryan, J.A. Santora, W.J. Sydeman, New approach for using remotely sensed chlorophyll a to identify seabird hotspots. Mar. Ecol. Prog. Ser. **451** (2012). https://doi.org/10.3354/meps09597
129. R.G. Asch, C.A. Stock, J.L. Sarmiento, Climate change impacts on mismatches between phytoplankton blooms and fish spawning phenology. Glob. Chang. Biol. **25**(8) (2019). https://doi.org/10.1111/gcb.14650
130. D.H. Cushing, Plankton production and year-class strength in fish populations: an update of the match/mismatch hypothesis. Adv. Mar. Biol. **26**(C) (1990). https://doi.org/10.1016/S0065-2881(08)60202-3
131. J.F. Schweigert, M. Thompson, C. Fort, D.E. Hay, T.W. Therriault, L.N. Brown, Factors linking pacific herring (clupea pallasi) productivity and the spring plankton bloom in the strait of Georgia, British Columbia, Canada. Prog. Ocean. **115** (2013). https://doi.org/10.1016/j.pocean.2013.05.017
132. M.J. Malick, S.P. Cox, F.J. Mueter, R.M. Peterman, Linking phytoplankton phenology to salmon productivity along a north-south gradient in the Northeast Pacific Ocean. Can. J. Fish. Aquat. Sci. **72**(5) (2015). https://doi.org/10.1139/cjfas-2014-0298
133. J. Arístegui et al., Sub-regional ecosystem variability in the Canary Current upwelling. Prog. Oceanogr. **83**(1–4), 33–48 (2009). https://doi.org/10.1016/j.pocean.2009.07.031
134. D. Szalaj, A. Silva, P. Ré, H. Cabral, Detecting regime shifts in the Portuguese continental shelf ecosystem within the last three decades. Front. Marine Sci. **8** (2021). https://doi.org/10.3389/fmars.2021.629130
135. T. Veiga-Malta et al., First representation of the trophic structure and functioning of the Portuguese continental shelf ecosystem: Insights into the role of sardine. Mar. Ecol. Prog. Ser. **617–618** (2019). https://doi.org/10.3354/meps12724
136. S. Garrido, C.D. van der Lingen, *Feeding Biology and Ecology*. (2014). https://doi.org/10.1201/b16682

137. S. Garrido, R. Ben-Hamadou, P.B. Oliveira, M.E. Cunha, M.A. Chícharo, C.D. Van Der Lingen, Diet and feeding intensity of sardine *Sardina pilchardus*: correlation with satellite-derived chlorophyll data. Mar. Ecol. Prog. Ser. **354** (2008). https://doi.org/10.3354/meps07201
138. A. Ferreira, S. Garrido, J.L. Costa, A. Teles-Machado, V. Brotas, A.C. Brito, What drives the recruitment of European sardine in Atlanto-Iberian waters (SW Europe)? Insights from a 22-year analysis. Sci. Total. Environ. **881** (2023). https://doi.org/10.1016/j.scitotenv.2023.163421
139. S. Garrido et al., Temperature and food-mediated variability of European Atlantic sardine recruitment. Prog. Ocean. **159** (2017). https://doi.org/10.1016/j.pocean.2017.10.006

Optical Sensor Innovations for Space Sustainability: Managing Space Debris in Aerospace Applications

S. J. Rudresha, Usha Raghunathagouda Patil, and S. K. Vinay

Abstract This chapter addresses a critical point of intersection between optical sensor technology and space sustainability, particularly with the management of space debris in aerospace applications. As the number of satellites and other spacecraft proliferates in low Earth orbit (LEO), engineers are increasingly confronted by the heightened risk provided by space debris. High-resolution optical sensors hold promising methodologies to support the detection, monitoring, and mitigation of this fast-growing threat. This chapter demonstrates how advanced concepts in optical sensing technology resolution imaging and spectral analysis can be applied to debris detection and real-time tracking. Those capabilities help ensure safe navigation of operational satellites and improve collision avoidance systems. To set the stage for such discussion, the chapter first describes basic physics governing the operation of optical sensors which may be exploited for beneficial debris management applications. The chapter then develops recent advances in optical sensing technology, especially in materials science that allow for enhanced sensor designs and improvements in data processing algorithms that impact sensor performance and reliability. A few case studies follow this investigation and illustrate how optical sensors are implemented effectively to monitor large objects as well as smaller fragments of space debris. The case studies then present very practical examples of the optical sensor contribution toward debris management and reflect various operational contexts where optical sensors prove their utility. Then, it's a mixed scenario of problems with currently used optical sensor systems related to environmental constraints, limitations of resolution, and integrating data into usable products. These challenges require constant innovation in optical sensing technologies. In the discussion on future trends, the trend will focus on the inclusion of AI and ML techniques in debris tracking systems. With AI and ML, researchers and engineers will be able to enhance predictive analytics to make better decisions regarding debris avoidance management. This forward-looking perspective forms a crucial enabler for smarter and more

S. J. Rudresha · U. R. Patil (✉)
Department of EEE, PESITM, Sivamogga, Karnataka, India
e-mail: usharp128@gmail.com

S. K. Vinay
Department of ISE, PESITM, Sivamogga, Karnataka, India

C. Marques et al. (eds.), *Advanced Optical Sensors for Aerospace Applications*, Progress in Optical Science and Photonics 37,
https://doi.org/10.1007/978-981-95-1626-1_11

resilient aerospace operations to be prepared for enhanced safety and sustainability in space activities. The current chapter therefore aims to provide a comprehensive source for engineers, researchers, and industry professionals interested in the potential application of optical sensor technology in the context of effective space debris management. The critical focus on sustainable practices in the aerospace industry this chapter will ensure that long-term space exploration and utilization are feasible and sustainable only with the assistance of optical sensors. It combines theoretical understanding with case studies from real-world examples to ensure better insights into the contribution of advanced optical sensors in the management of space debris toward making aerospace operations safer and more sustainable.

Keywords Quantum entanglement · Ultra-precise tracking · Space debris monitoring · Optical ground-based tracking · Quantum sensors · Entangled photons · Satellite tracking · Laser ranging · Quantum metrology · Quantum information processing · Photonics in aerospace · Quantum-assisted navigation · Space situational awareness (SSA) · High-precision measurements · Quantum optics for space applications

1 Introduction

Satellite and space exploration technologies have reshaped contemporary existence, driving international communications, navigation, and space research. Increased satellites, however, have fuelled fears of rising space debris in space. Increased satellites are no longer operational or fully functional and remain in orbits (spent rocket stages). They risk hazardous collisions among them and/or operating spacecraft, seriously threatening normal operating spacecraft performance. Optical sensors have thus become a vital solution to the detection and observation of objects, maintaining space situational awareness (SSA) as well as collision avoidance [1]. From laser trackers to ground space-based telescopes, they provide vital inputs, even braving obstacles such as vision restriction, and illumination levels, and identifying small-sized debris [2].

To address these issues, scientists have come up with new optical sensing technologies such as the Wide Eye Debris (WED) telescope for better debris detection and tracking [3]. Through better monitoring, such technologies enable improved prediction of possible collisions and inform space traffic management policy. In addition, image and sensing material technology advancements, signal processing software, and optical imaging technology improve the precision and effectiveness of the optical tracking system to provide maximum protection of strategic space infrastructure (Fig. 1).

With space-based services entering the mainstream of daily life, there is a critical demand for intelligent debris management systems. The integration of artificial

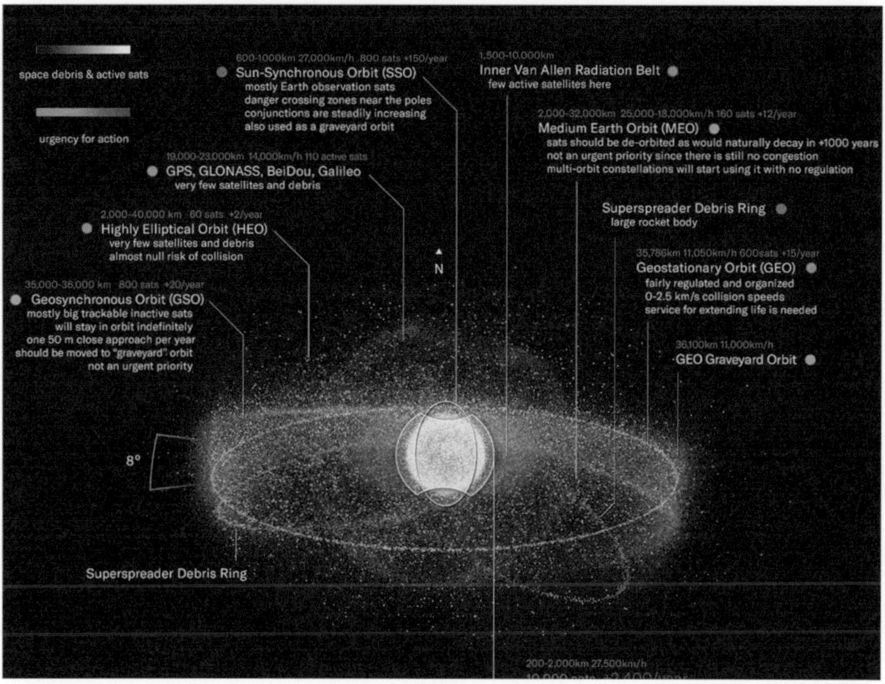

Fig. 1 Infographic showing the space debris situation in different kinds of orbits around Earth. (Image by Pablo Carlos Budassi, licensed under CC BY 4.0. *Source* Wikipedia—Space Debris)

intelligence (AI) and machine learning (ML) with optical sensing systems will revolutionize the process of monitoring and analyzing debris, enabling improved forecasting and automated decision-making [4]. This chapter investigates how the role of space debris mitigation with optical sensors is evolving with the help of advanced technology, practical implementations, and ongoing developments in work to create a cleaner, greener space.

2 Fundamentals of Optical Sensors for Space Debris Management

2.1 *Principles of Optical Sensing*

Optical sensing technology makes use of light to sense a number of parameters with high accuracy and provides some of the key benefits of small size, immunity to electromagnetic interference, and ruggedness in extreme environments. The sensors work on the principles of reflection, absorption, fluorescence, and interferometry and are thus extremely versatile for a range of applications [129]. Optical fiber sensors, in particular, are also extremely popular as they are compact and can deliver extensive networks for monitoring. They have tremendous applications in industrial strain measurement, biomedical diagnosis, and environmental measurements, where the need for acquiring data in real-time as well as high sensitivity is required [130]. Commercialization of the technologies has enabled the development of oilfield monitoring and navigation fiber optic gyroscopes and oilfield monitoring optical sensors, exemplifying their usage [130]. Use of optical sensors in wireless sensor networks is also revolutionizing remote and large-scale monitoring systems.

Optical fiber sensors are based on the principle of utilization of light propagation, interference, and polarization to enable accurate measurement in a wide range of applications. Optical fiber sensors can withstand electromagnetic interferences and are therefore even more valuable when utilized in such applications where electronic sensors in general cannot perform at their best. Optical sensing also extended beyond the molecular scale, making use of sophisticated technologies such as surface plasmon resonance (SPR) and spectroscopy to detect extremely subtle chemical and biochemical variations with great precision. It has greatly enhanced their application in medical diagnosis, environmental monitoring, and process control. While problems like environmental sensitivity and integration remain, ongoing developments in materials and data processing continue to enhance the precision and effectiveness of optical sensing technology, further solidifying its place in contemporary scientific and engineering fields.

Optical sensors are based on simple principles of light interaction like reflection, absorption, and scattering. The primary types of optical sensors used in the detection of space debris are the following sensors:

Electro-optical (EO) Sensors

Electro-optical (EO) sensors are very versatile sensors employed to measure electric fields and monitor objects in marine environments. Although originally designed to optimize EO tensor coefficients, their performance can be optimized further by tuning along low static birefringence crystal directions [5]. These sensors have also progressed significantly, with onboard EO systems now capable of measuring intense electromagnetic pulses, meeting strict military standards for transient field testing [6]. However, temperature stability remains a critical challenge, as temporary variations occur as a result of thermo-optic effects and thermal expansion of the crystal resonator [7]. Detection and tracking of objects in these environments are also disrupted by movement of the background, camera shake, and low contrast. To counter such limitations, multi-sensor fusion techniques and top-of-the-line commercial systems have been adopted [8].

EO sensors are also important in sensing inorganic anions and measuring radiofrequency-specific absorption rates. Their operation is based on EO materials such as conjugated chromophores [9], lithium niobate crystals [10], and EO polymers [11]. Most significantly, EO polymer-based sensors surpass inorganic crystals through more EO coefficients, reduced dielectric constants, and quicker response

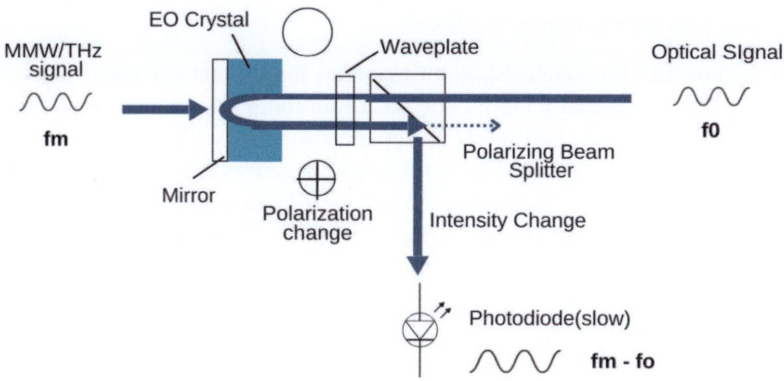

Fig. 2 Block diagram of electro-optic sensor as photonic MMW down-converter [13]

[11]. An improvement in sensitivity has been reached by exploiting the dielectric relaxation effects [10] and through integrating resonant elements like microring resonators [11]. Recent advancements consist of an EO sensor array tailored for measurements of specific absorption rates [12] and an all-dielectric sensor employing a polymer microresonator inductively coupled with a side-polished optical fiber that demonstrates ultra-high sensitivity of 100 mV/m over the frequency range up to 550 MHz [11]. All these developments keep widening the applicability of EO sensors in scientific, military, and industrial areas (Fig. 2).

Infrared (IR) Sensors

Infrared (IR) sensors find application in a large number of instances with technological developments making them more sensitive and accurate. Near-infrared sensors, which are compatible with Full-CMOS, have facilitated better optoelectronic utilization, while low-cost reflective IR sensors facilitate improved motion and positioning [14, 15]. Nanomaterials have made smart IR thermal imaging sensors available, which are unable to respond, providing a boost in their applications for the military, medical, and environmental monitoring areas [16, 17]. Despite the limitations, studies are still focused on enhancing material and nanostructure optimization for improved efficiency in IR sensors [16]. IR sensors enhance car safety, human activity recognition, and fire detection.

Integrated IR sensors and radar offer increased situational awareness inside the vehicle, and pyroelectric IR sensors improve motion tracking for automation and security [18, 19]. Biomimetic IR sensors, modeled on insect photo mechanic receptors, provide effective fire detection [20]. Latest advances in van der Waals heterostructure IR sensors allow high sensitivity, ultrafast photoresponse, and wide-band detection for next-generation applications [21]. These advances keep pushing the range of IR sensing technology into ever-more industries.

Light Detection and Ranging (LIDAR)

LIDAR technology has emerged as an essential instrument for uses varying from forest stock and biodiversity monitoring to autonomous car perception. It facilitates accurate terrain modeling and forest attribute estimation, with LIDAR-derived variables exhibiting robust relationships with species richness in a variety of ecosystems [22, 23]. New technologies, including optical modulation methods, new laser sources such as photonic crystal surface-emitting lasers (PCSELs), and more efficient scanning systems, are improving their functionality [24]. New techniques such as pulse coding with direct-sequence optical code division multiple access (DS-OCDMA) also enable continuous distance measurement with less idle time and higher efficiency [25].

LIDAR's capability to record high-resolution 3D data has revolutionized forest structure analysis and biomass estimation, giving direct information on canopy height, vertical structure, and habitat mapping [26, 27]. It is an important tool in conservation work by providing detailed habitat information for wildlife research [21]. Its precision can be affected by weather. To reverse this, researchers have explored sensor fusion and 3D-to-2D point cloud transformation via interpolation techniques like inverse distance weighted interpolation to enable data usability and integration with other sensing modalities [28]. Such technologies continue to improve LIDAR's applications, making them better, at increasing environmental monitoring and autonomous system performance (Table 11.1).

Table 11.1 Comparison of optical sensor types for space debris detection [29–31]

Sensor type	Detection method	Strengths	Limitations
Electro-optical (EO)	Passive imaging through reflected sunlight	• High spatial resolution imaging ability • Effective against high-albedo objects	• Needs sufficient sunlight; performance in Earth's shadow is poor • Performed by atmospheric conditions
Infrared (IR)	Detection of thermal radiation	• Can sense objects in the Earth's shadow • Good against low-albedo objects with thermal emission	• Typically has poorer spatial resolution than EO sensors • Its performance can be degraded by thermal background noise
LiDAR	Active laser-ranging	• Offers accurate ranging and 3D mapping of objects • Independent of ambient light conditions	• High power consumption because of active laser emission • Limited by the range and field of view limitations

2.2 *Optical Sensor Technologies Used in Space Applications*

Optical sensor technologies are pivotal in space applications, including the support of spacecraft engineering, microbial monitoring, sea surveillance, and data management. Fiber optic sensors have been explored extensively for control and monitoring in spacecraft across a range of different environments [32]. Miniature optical sensors enable online, non-penetrating monitoring of bacterial growth in microgravity by monitoring turbidity from blue LED light sources [33]. Spaceborne optical sensors, satellite AIS, and synthetic aperture radar provide augmentation of global maritime surveillance through the complementation of land systems [34]. Compressed sensing methods in optical imagery also provide solutions for effective compression, encryption, and onboard analysis of images, maximizing spaceborne optical systems and data management [35].

Charge-Coupled Devices (CCD) and Complementary Metal–Oxide–Semiconductor (CMOS) Sensors

Charge-Coupled Devices (CCDs) and Complementary Metal–Oxide–Semiconductor (CMOS) sensors have revolutionized high-resolution imaging. Although CCDs have been the norm for a long time, CMOS sensors are becoming popular because they are cheaper, consume less power, and operate at high speeds [36]. CMOS technology is especially useful for high-resolution video applications, with pixel rates exceeding gigapixels per second [37]. sCMOS scientific CMOS cameras now compete with electron-multiplying CCDs in superresolution imaging [38], whereas CCD sensors incorporating compressed sensing methods boost efficiency in streaming IoT applications, particularly in low-light environments [39].

The transition away from CCDs to CMOS sensors is ongoing as developments enhance power efficiency and integration [40, 41]. CCDs continue to be essential for astronomy and scientific imaging because of their high sensitivity [42], although contemporary CMOS sensors equal or near CCD performance, with new developments such as 3D stacking and photon-counting [41]. For high-speed imaging, sCMOS cameras are more general-purpose, and EM-CCD cameras excel at live-cell imaging under low light [43]. Application-wise, CCD or CMOS would depend on the signal-to-noise ratio, resolution, and speed of the imaging (Figs. 3 and 4).

Multi-spectral and Hyper-Spectral Imaging

Hyperspectral and multispectral imaging (HSI/MSI) technologies are revolutionizing medical diagnostics, especially in gastroenterology and pathology, by recording a wider spectrum of wavelengths than traditional RGB imaging [45, 46]. In gastroenterology, they assist in the detection of precancerous lesions and inflammatory diseases, enhancing early diagnosis. In pathology, HSI/MSI is used to perform accurate digital staining, lowering variability from laboratory to laboratory. In wound healing as well, the systems provide tissue oxygen maps for predicting hazard in wounds and for tracking healing, most helpful to diabetic and chronic wound patients

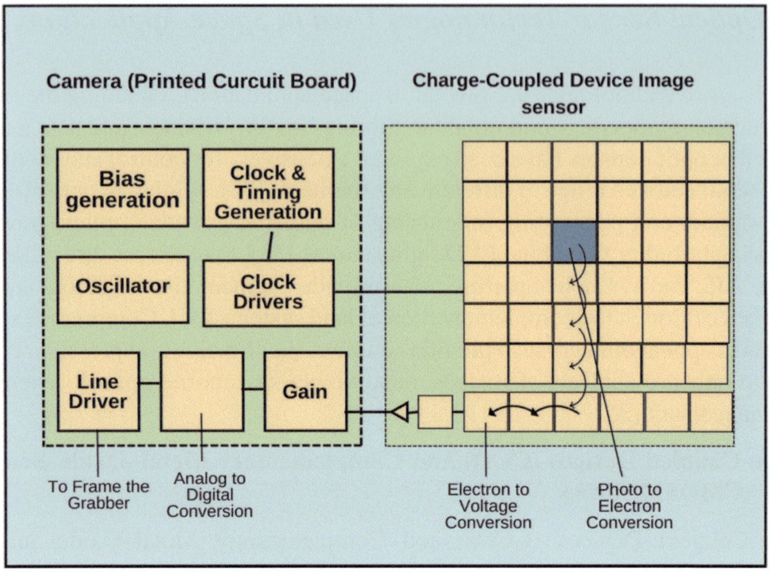

Fig. 3 Diagram of charge-coupled devices (CCD) [44]

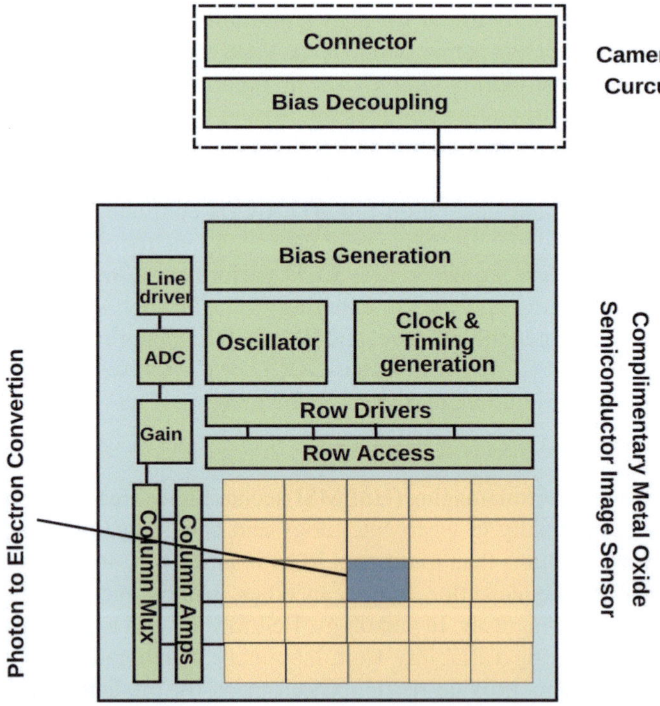

Fig. 4 Diagram of CMOS [44]

[47]. By employing HSI/MSI as an integral component in treatment procedures, physicians are able to maximize diagnostic accuracy and maximize patient outcome.

Outside of medicine, HSI/MSI is transforming agriculture, environmental monitoring, and industry. Machine learning-enabled handheld devices and phone-integrated sensors make it possible for in-situ real-time detection of stress in plants, enabling enhanced crop health and yield [48]. In the field of environmental science, the technologies help monitor levels of contamination and water purity, while industrially, they ensure food safety and precision in material inspection. While promising, studies will need to be conducted to enhance further affordability and integration to allow more extensive application across fields [46, 47] (Fig. 5).

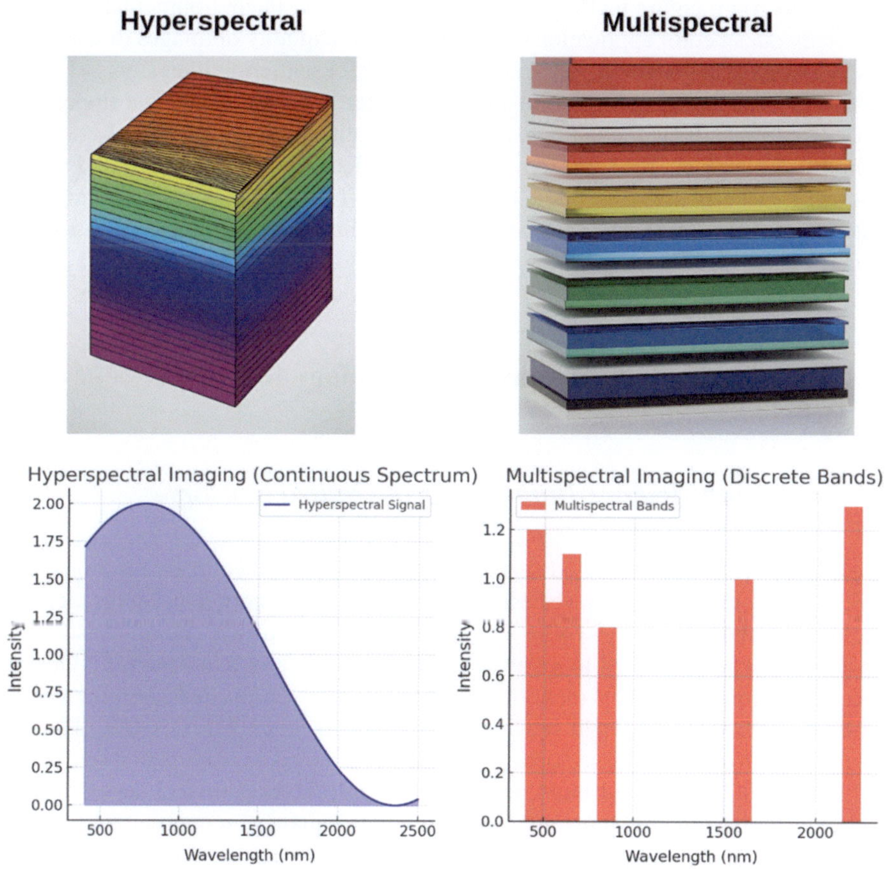

Fig. 5 Comparison of the image stacks in multispectral imaging, in which there are images taken in several different spectra, and hyperspectral imaging, in which there are images taken in many different spectra [49]

Polarimetric Optical Sensors

Polarimetric optical sensors facilitate object characterization by studying surface reflection properties using a light polarization state [50]. The sensors have been important in aerosol monitoring, target detection, and surface characterization. At millimeter-wave frequencies, they identify road debris and faults, with target size, orientation, and polarization state influencing their performance [51]. New developments consist of sensitivity increase through nonlinear phase modulation through the Kerr effect and outperforming classical polarimetric sensors [52]. Nonetheless, proximal remote sensing remains challenging because penetration and scattering effects hinder the reliability of hyperspectral reflectance data through varying wavelengths [53].

Polarimetric imaging methods have also been helpful in environmental research, especially ocean color applications. They enhance the quality of images in scattering media by demixing target signals and estimating backscattering intensity [54]. They are useful in gathering information about air–water interfaces and wave slopes [55] and complement lidar technologies in ocean depth, particle shape, and chemical composition analysis [56]. Moreover, mobile reflectometry methods based on in-built camera-flash modules or LCDs enable free-form handheld measurement of material mesostructure and surface reflectance, further expanding applications in surface analysis [57].

3 Advances in Optical Sensing Technology for Space Debris Monitoring

3.1 Materials Science Innovations in Optical Sensors

There has been considerable advancement in material science in recent times, and optical sensor technologies have been greatly improved. Nanoporous anodic alumina photonic crystals (NAA-PCs) offer a broad platform for sensitive chemo-sensors and biosensors and can be used to monitor molecular binding in real-time [58]. Optical microresonators have been identified as good sensors and transducers for use in applications like electric field sensing and single-molecule detection [59]. High-Q photonic microstructure-based optical sensors have shown unprecedented sensitivity in label-free chemical and biological analysis [60]. Hybrid optical sensors using molecularly imprinted polymers (MIPs) also show enhanced recognition capability and optical signal amplitude [61]. This leads to superior sensitivity, selectivity, and versatility in applications such as environmental monitoring, medical diagnostics, and biotechnology.

Improved sensitivity and durability in sensors have been the results of materials science breakthroughs in recent years, especially for space conditions. Nanostructured coatings reduce degradation by radiation, and meta-optics and plasmonic sensors optimize the detection of light and signal processing. Radiation-hardened semiconductor material enhances longevity and performance in harsh radiation environments. Electrochemical nanostructuring of GaN epilayers has been used to boost radiation hardness by more than an order of magnitude [62]. Nanomaterials like quantum dots, nanowires, and plasmonic structures in infrared photodetectors increase sensitivity and light-matter interaction [63]. Photoconductive antennas enhanced by nanostructures enhance terahertz wave generation and detection [64]. Ion irradiation of III-V semiconductor surfaces allows self-organized nanostructures with plasmonic characteristics for optoelectronics [65].

Radiation-hardened graphene field-effect transistors (GFETs) show negligible performance degradation following gamma radiation exposure [66]. Bulk nanolayered Cu-Nb composites with engineered interfaces demonstrate enhanced irradiation tolerance and thermal stability [67]. Radiation-induced disorder in GaAs impacts bright exciton lifetimes by enhancing band tailing, affecting radiative and nonradiative recombination efficiency [68]. These developments help in the creation of radiation-tolerant materials for space, military, and other high-radiation applications.

3.2 Enhanced Data Processing Algorithms

Advances in algorithms for data processing over the past few years have dramatically enhanced precision, efficiency, and reliability across most scientific and industrial applications. In dispersive interferometry, the truncated-spectrum and high-order-angle algorithms have reduced phase distortion to the minimum and errors to a limit of 1.3 μm, thus making optical metrology very precise [69]. Continuous glucose monitoring has been aided by advanced algorithms minimizing sensor error rate and duration but preserving accuracy to enhance real-time diabetes care [70]. Likewise, in terahertz pulsed imaging, methods of baseline subtraction and impulse response extraction have aided spectroscopic and time-domain analysis to enhance contrast in imaging for biomedical and pharmaceutical use [71].

Within the field of chromatographic data analysis, self-improving algorithms for peak detection currently allow real-time chemical compound analysis regardless of the conditions of the ambient environment, resulting in high-throughput quality control [72]. Machine-learning algorithms are extending signal processing pattern recognition and anomaly detection further, opening new fields like hyperspectral imagery, biomedical diagnostic tools, and environmental monitoring. These advances together improve the speed, accuracy, and automation of data analysis, propelling progress in various scientific and technological fields (Fig. 6).

New technologies in real-time image processing, computational photography, and object recognition using deep learning have greatly enhanced debris detection and identification in different environments. Deep learning models and high-resolution imagery from UAVs have greatly increased marine debris detection with 81.4% accuracy on the TrashCan dataset using diffusion model-based methods [73]. UAVs incorporating the YOLO v2 algorithm have yielded excellent precision and recall

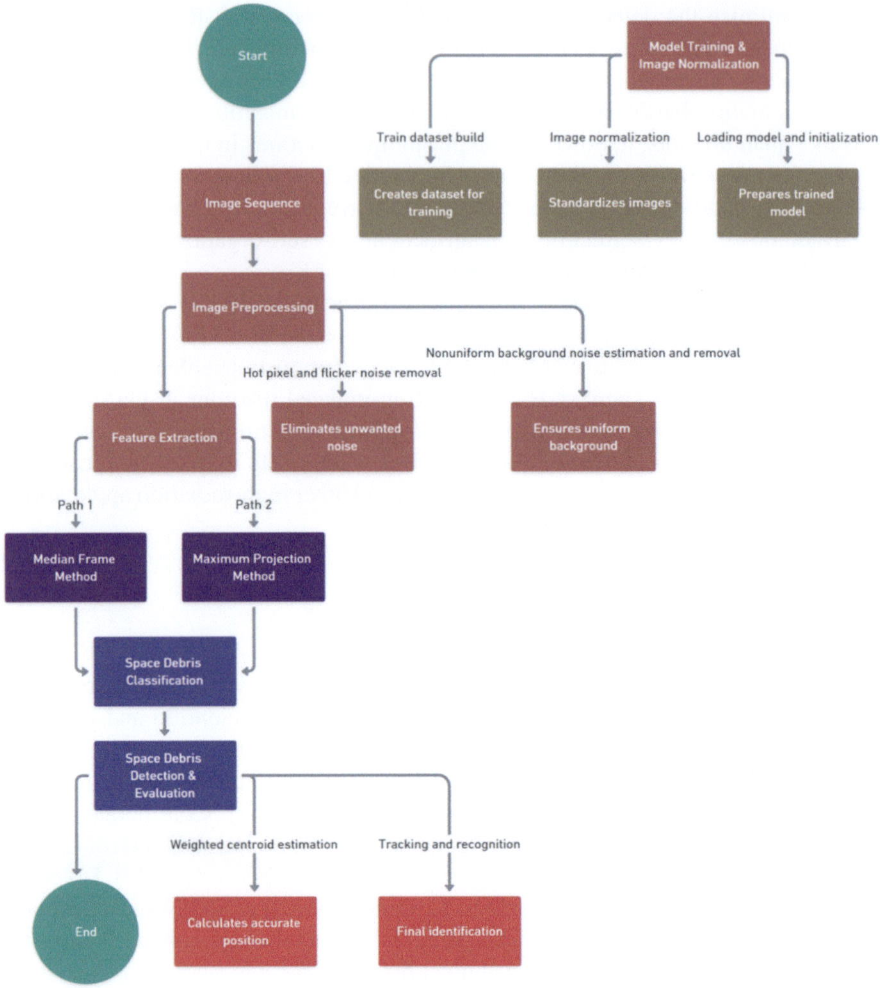

Fig. 6 The flowchart of space debris detection

rates of bottle debris identification, with peak performance at 0.5 cm/pixel resolution [74]. Besides, YOLO-inferred image processing techniques have improved small UAV detection to 96.7% mAP at an IoU threshold of 50% [75]. The technology is assisting in the development of better environmental monitoring, particularly marine conservation and rescue operations.

Aside from maritime use, machine learning and image processing methods have been applied in railway maintenance, disaster relief, and deep-sea object recognition. Convolutional neural networks (CNNs) are highly accurate for the detection of railway fasteners [76] and deep-sea trash [77]. A hybrid system of crowdsourcing and machine learning has been proposed for rapid processing of aerial imagery, extendable to wildlife conservation and disaster response [78]. For coastal regions, very high-resolution satellite imagery processed using machine learning algorithms, specifically Shannon's entropy, has been shown to map sea trash [79]. These investigations demonstrate the growing importance of AI-based imaging technologies in solving environmental and safety concerns in a variety of industries.

4 Optical Sensor Implementation for Space Debris Monitoring

Technological developments in optical sensors have enhanced the detection of space debris. [80] suggested a star sensor-based approach to the detection of small, faint debris, whereas NASA's Space Debris Sensor scans micro debris on the ISS [81]. Yang et al. [82] designed a wide-field optical telescope array for observing geosynchronous objects. For enhancing accuracy, [83] provided an image restoration algorithm through point-spread function modeling. They support tracking and characterizing space debris, enhancing space situational awareness and asset security.

4.1 Optical Ground-Based Tracking Systems

Technological advancements in optical ground-based tracking systems have greatly improved space surveillance and situational awareness. Wide-field telescopes, like the Zhulong network, have angular data errors of less than 3 arcseconds, which allows for high-precision orbit determination and prediction [84]. Such accuracy is important for the monitoring of operational satellites and space debris to decrease the chances of collisions. Changchun Observatory 280 mm telescope array enhances geosynchronous debris monitoring via image differencing and trajectory association algorithms, facilitating observation of distant objects in crowded orbital areas [82]. Moreover, network-based methods, for example, the OSIRIS-REx tracking experiment, validate the capability of wide-field optical telescope arrays in tracking spacecraft between various observing stations, thus allowing real-time monitoring of space bodies [85]. Improvements in telescope technology have stretched the scope of observations of celestial and space bodies.

Liquid-mirror telescopes, with rotating mercury forming a parabolic reflector, provide wide-field infrared imaging well adapted to large-area sky surveys and transient object searching [86]. Likewise, freeform mirror-based systems provide search and track combined, with compact payload and low mass but high image quality [87]. Falcon Telescope Network, which is a global network of small-aperture optical telescopes, monitors artificial satellites and celestial bodies simultaneously and in real-time at multiple sites worldwide, enhancing multi-angle tracking accuracy [88]. Adaptive optics technology has also transformed ground-based astronomy through real-time distortion compensation by the atmosphere, allowing telescopes to be diffraction-limited in resolution, something frequently beyond the capability of space-based telescopes [89]. All these technologies add to increased space situational awareness, planetary defense, and deep-space exploration.

4.2 Space-Based Optical Sensors

Recent developments in space-based optical remote sensing have dramatically improved Earth observation capacity, with CubeSat platforms at the forefront of sensor miniaturization and affordable imaging solutions [90]. These technologies enable real-time monitoring uses like disaster response, climate analysis, and resource management. The European Space Agency has also increased the application of fiber optic sensors on spacecraft to enhance thermal control, structural health monitoring, and radiation measurement [32]. Increased spatial, temporal, and spectral resolutions in missions such as Landsat 8 and Sentinel 2 have also enhanced ecosystem monitoring, facilitating more accurate vegetation health monitoring and land cover mapping [91].

Aside from Earth observation, space-based sensing technologies also play a pivotal function in tracking and preventing space debris. The NASA Space Debris Sensor (SDS) onboard the International Space Station performs an important role in analyzing microdebris with diameters smaller than one millimeter, complementing ground radar sensing of larger debris [81]. To further mitigate debris threats, spaceborne single-photon Lidar (SSPL) offers accurate debris localization via non-parametric Bayesian inference and extended Kalman filtering [92]. Furthermore, the CubeMAP mission will improve atmospheric research by employing nanosatellites with high-end thermal infrared sounders to investigate the upper troposphere and stratosphere composition [93].

Sustainability and innovation in space technology continue to be a priority, as evidenced by projects such as Eco-LeanSat, encouraging sustainable satellite production and reuse [94]. Space-based Lidar has been further advanced to provide accurate measurements of atmospheric winds, ozone concentration, and carbon dioxide concentration, all important for climate monitoring and predictive analysis [95]. However, funding continues to be the major impediment to large-scale implementation, requiring tighter coordination between space agencies, academia, and industry champions. As optical sensing technology advances, missions in the future will probably accentuate improved automation, advanced debris surveillance, and eco-friendly satellite designs, making space exploration and environmental monitoring long-term sustainable.

4.3 Multi-sensor Fusion for Enhanced Tracking

Multi-sensor integration of radar, optical, infrared, GNSS, and IMU technologies has greatly improved debris tracking and collision avoidance in autonomous maritime navigation [96]. AI-based fusion methods enable real-time vessel classification and situational awareness, promoting safety in intricate marine environments. Machine learning, and specifically deep learning, has proved to be effective in multi-sensor data processing for maritime applications [96, 97]. The MIIDAPS-AI algorithm processes microwave and infrared sensor data effectively for environmental parameter analysis, which aids in enhanced maritime operations [97]. Sensor fusion and tracking algorithms have also been effectively used in unmanned surface vehicle (USV) navigation and collision avoidance systems to comply with COLREGs. For monitoring marine debris, an integrated observing system (IMDOS) has been suggested, which incorporates remote sensing, in situ measurements, and predictive modeling to offer large-scale, long-term monitoring of marine pollution [98].

Sensor fusion technologies have also revolutionized precision agriculture by combining UAVs, AI, and remote sensing to track crops in real time and allocate resources [99]. Optical, thermal, multispectral, and hyperspectral sensor data fusion methods enhance crop parameter monitoring on a site-specific basis, enhancing yield estimation and disease identification [100]. Artificial intelligence techniques are transforming remote sensing and numerical weather forecasting by enhancing data processing, quality verification, and prediction accuracy [101]. Vision-based collision warning systems with UAVs have been used in precision agriculture to identify objects at 400–900 m and give 8–10 s of warning [102].

Even with all of these developments, there are nevertheless some challenges such as high running costs, UAV battery life constraints, and a requirement for specialist staff [99]. Future research involves enhancing sensor quality, reducing complexity in AI models to reduce the power needed to run them, and adding autonomous systems to enable improved real-world functionality. Emerging technologies in efficient AI, autonomous sensor networks, and self-powered UAV platforms will continue to broaden the application of multi-sensor fusion across many domains, ranging from naval defense to precision agriculture.

5 Challenges in Optical Sensor-Based Space Debris Management

5.1 Environmental Constraints

Atmospheric interference strongly affects ground-based optical tracking and space-based sensor performance. In ground-based interferometry, group delay fringe tracking locks optical path differences in low light but is prone to atmospheric dispersion and temporal phase variations, which influence measurement accuracy [103].

To minimize these problems, optimized fringe sensor designs based on co-axial pair-wise combination schemes have been created to enhance sensitivity and immunity under changing atmospheric conditions [104]. Further, transportable optical ground terminals with fine-pointing systems present a viable means of free-space optical communication and can be repositioned in areas with less atmospheric disturbance to ensure more reliable communication.

Environmental influences in the form of high-energy particles, eclipses, and thermal fluctuations in low Earth orbit present performance challenges to space-based sensors, calling for sophisticated data correction methodologies [105]. The corrections assist in the maintenance of sensor precision and dependability in changing space conditions, aiding remote sensing, astronomy, and optical communication using satellites. Resolution of atmospheric as well as space-based interference via technology helps improve tracking, measurement, and data transmission functionalities in numerous scientific and operational areas.

5.2 *Resolution and Detection Limitations*

Optical sensors are critically tested in detecting and monitoring sub-centimeter space junk due to the limitations in resolution, loss of signal over long distance, and the natural difficulty in imaging low-light objects against the void of space. Super-resolution techniques have been explored to promote imaging ability in the observation at a long distance, extending the spatial resolution above the diffraction limit of the optical system [106]. But minute, weak debris is still a problem to detect with low reflectivity upon illumination and inadequate pixel utilization in imaging detectors, affecting visibility and tracking precision [80]. Optical tracking of nanoscale particles using light requires a balance between several system components like the light source, sensor sensitivity, and optical arrangements for the task of achieving maximum detectability under varying observation conditions [107]. Even as machine learning and adaptive optics image processing methodologies have proved to enhance debris detection, they are still overshadowed by atmospheric distortion limits in the case of earth-based platforms and the expense of placing high-quality optics sensors in space.

Despite these advances, present space-based surveillance platforms cannot precisely locate and reliably follow small fragments of debris, detracting from space situational awareness as well as satellite operation. High-precision astrometric methods and high-level trajectory estimation algorithms are in the planning stage to enhance tracking precision for such evasive objects [80]. Increasing volumes of space debris, particularly in low Earth orbit (LEO), have made long-term observation and real-time tracking even more vital. Optical sensors combined with radar and laser ranging are under consideration for detection capability improvement by applying multi-sensor fusion in an attempt to overcome the frailties of a single system. These hybrid methods try to give a better picture of debris distribution, motion behavior, and possible collision hazards to enable active debris removal (ADR) policy and long-term space traffic management.

To respond to the growing danger of sub-centimeter debris, NASA has designed the Space Debris Sensor (SDS), a high-tech device intended to study micro debris smaller than a millimeter in diameter. The International Space Station (ISS) sensor collects data on the velocity, direction, and dimensions of these tiny pieces of debris, which are hugely common and pose a significant threat to operational spacecraft [81]. The findings of SDS and similar missions assist in improving space debris dynamics predictive models and mitigation strategies to safeguard space assets. Future improvements in optical tracking through artificial intelligence-based data processing and emerging imaging technologies will have key roles to play in safeguarding space safety, providing necessary satellite infrastructure, and making space exploration sustainable for future generations.

5.3 Integration and Data Usability Issues

The latest developments in space-borne optical remote sensing have significantly enhanced data resolution and accessibility for the monitoring of terrestrial ecosystems, facilitating accurate analysis of vegetation, land use change, and climate trends [91]. High-resolution missions such as Landsat 8 and Sentinel-2 provide improved spectral and temporal resolution, but interoperability challenges in multi-source sensor data fusion continue to exist because of the intricacies involved in managing multiple sensors [108]. To tackle this, standardized building blocks for data space and open-access frameworks are being created to enable data exchange [109]. This enhances cross-platform interoperability, making remote sensing applications more useful.

Data integration advancements are also revolutionizing decision-support systems. In space situational awareness, cognitive systems engineering has maximized sensor network tasking, enhancing information fusion as well as space object tracking [110]. Solving statistical issues in merging multi-resolution spatial data remains an area of interest in geography and ecology [111]. Semantic data integration techniques are on the rise to combine heterogeneous datasets for better decision-making [112]. These technologies allow more efficient data-driven information in environmental science, urban planning, and disaster management. In biosensor technology, advances in molecular recognition, microfluidics, and fabrication methods are driving the development of small, low-cost optical biosensors for public health and environmental monitoring [113]. Nanophotonics and lab-on-a-chip technologies improve detection precision, while cross-industry collaborations encourage real-world implementation [114]. Albeit advances, problems of data quality, affordability, and continuity in acquisition persist, which highlight the necessity for continuing research and development of optical remote sensing and biosensing technologies.

6 Future Trends in Optical Sensor-Based Space Debris Mitigation

As the space traffic grows, mitigating space junk has emerged as a major issue for sustainable space activities. Optical sensors are critical in detecting, tracking, and characterizing debris in different regimes of orbit. Future developments will concentrate on increasing sensor resolution, adaptive optics, and AI-based data processing to better identify debris. New technologies, including space-based optical networks, laser ranging, and AI-based tracking systems, will enhance real-time observation and collision avoidance. Moreover, advancements in optical communication and debris removal technologies will further augment space sustainability initiatives.

6.1 AI and Machine Learning in Optical Debris Tracking

Artificial intelligence (AI) and machine learning (ML) are revolutionizing scientific research, especially in astronomy, optics, and environmental observation. In astronomy, ML boosts exoplanet detection, prediction of solar activity, and analysis of gravitational waves, enhancing the accuracy of detection and data processing [115]. Artificial intelligence (AI)-derived models of optical communications, such as convolutional neural networks (CNNs) and recurrent neural networks (RNNs), enhance image detection, noise removal, and data transmission effectiveness [116]. AI is also revolutionizing satellite remote sensing and numerical weather forecasting by refining data exploitation and prediction accuracy [101]. These innovations streamline decision-making in scientific and industrial usage, increasing efficiency and precision.

ML is also important in nanoscale imaging and spectroscopy, enhancing data collection and analysis in optical scanning probe nanoscopy [117]. The Zwicky Transient Facility utilizes ML in transient detection and asteroid monitoring when observing the sky, enhancing observation and classification efficiency [118]. Adaptive optics are facilitated by AI-optimized wavefront correction, resulting in sharper, real-time astronomical images [119]. The advent of all-optical deep learning, e.g., 3D-printed diffractive deep neural networks, allows ultra-high-speed image recognition and signal processing at light speed [120]. AI also extends to environmental surveillance, and ML algorithms interpret high-resolution satellite images to detect and chart marine coastal trash, providing solutions to reduce litter in remote areas where data may be limited [79]. AI and ML are also revolutionizing space exploration, optical technology, and sustainability, with cutting-edge solutions in many industries.

6.2 Quantum Optical Sensing

Quantum entanglement improves optical sensing and high-precision measurements by noise-immunity and high-resolution imaging. Entangled photon pairs improve LiDAR technology by separating signals from noise, providing safe depth imaging [121]. Path-entangled NOON states in gyroscopes overcome classical shot-noise limits, improving phase sensitivity [122]. Quantum radar uses entanglement to detect distant targets, but its concepts are difficult to deal with for classical radar engineers [123]. Breakthroughs in quantum manipulation, especially in cold atoms and trapped ions, provide multi-particle entanglement, opening the door to better quantum sensors [124].

Quantum-aided sensing methods extend optical phase measurement beyond the classical limit. Phase-squeezed quantum states surpass coherent-state limits for monitoring fluctuating optical phases [125]. Nanophotonic and nanoplasmonic quantum sensing combines non-classical light with chip-based technologies, with the potential for single-molecule detection under low uncertainty [126]. A method for magnifying a quantum phase with cold rubidium atoms has been shown to perform squeezed-state metrology 8 dB below the standard quantum limit, even in noisy environments [127]. Entanglement-based methods are constantly pushing the boundaries of high-precision optical measurements, as evidenced by four-photon interference experiments beyond the standard quantum limits [128]. These developments push quantum metrology applications in measuring distance, position monitoring, and acceleration detection into new realms for scientific and industrial technologies.

6.3 Autonomous Optical Sensor Networks

Current research indicates sensor fusion, AI-driven decision-making, and autonomous space and surface mission collision avoidance. Autonomous ships use radar, LiDAR, and cameras to identify objects and navigate, and human operators can intervene via AI transparency. Machine learning supports vessel safety classification and anomaly detection [96]. In satellite constellations, economic competition jeopardizes the inefficient occupation of orbits, and thus there is a need for regulatory policies.

Innovative approaches enhance connectivity and security in autonomous systems. Non-terrestrial networks (NTNs) based on satellites and airships extend IIoT radio coverage, with AI-enhancing performance. UAVs take advantage of deep reinforcement learning for energy-efficient collision avoidance (Ouahouah et al. 2022), and mega-constellation satellite architectures maximize connectivity through optical inter-satellite links. Also, an insect-inspired spike-based system with mems transistors provides energy-efficient night-time collision detection. These developments enhance the reliability and efficiency of autonomous systems across domains.

7 Conclusion

The growing menace of space debris threatens the sustainability of space operations, making advanced optical sensors indispensable for detection, tracking, and mitigation. Advanced optical sensors combined with AI-driven analytics and advanced material sciences provide high-resolution imaging and live monitoring for space situational awareness. Breakthroughs in adaptive optics, radiation-hardened materials, and deep learning algorithms have significantly improved accuracy in tracking, and sophisticated data fusion enables enhanced cooperation among global space agencies and organizations.

Even with these advancements, atmospheric interference, resolution limitations, and interoperability continue to be issues. Ground sensors are plagued by environmental limitations, while space systems contend with radiation exposure and power constraints. Sub-centimeter debris detection is especially difficult, and it represents a high-risk hazard due to high-impact velocities. Meeting these challenges involves continuing research, standardized data exchange frameworks, and ongoing international collaboration to enhance sensor efficiency and integration.

In the years to come, advancements in AI, quantum optics, and autonomous sensor networks will keep transforming space debris management. Automation enabled by AI will further enhance predictive analytics, while sensors that have been upgraded with quantum may improve detection at smaller scales. Autonomous sensor constellations will enable real-time, adaptive monitoring, improving mitigation strategies. This chapter highlights the importance of ongoing innovation and interdisciplinary collaboration to ensure the long-term safety and sustainability of space operations.

References

1. J.L. Santos, Optical sensors for industry 4.0. IEEE J. Selected Top. Quantum Electron. **27**(6), 1–11 (2021). https://doi.org/10.1109/JSTQE.2021.3078126
2. N.S. Shrikrishna, R. Sharma, J. Sahoo, A. Kaushik, S. Gandhi, Navigating the landscape of optical biosensors. Chem. Eng. J. **490**, 151661–151661 (2024). https://doi.org/10.1016/j.cej.2024.151661
3. L. Cibin, J. Martin, T. Becker, Wide eye debris telescope: a novel approach to space debris surveillance, in *Proceedings of the International Astronautical Congress 2012*, pp 1–10
4. R. Howard, P. Richards, Optical sensor systems for space situational awareness. Aerospace Res. J. **8**(2), 120–134 (2007)
5. A. Garzarella, C.M. Sorace, B.T. O'Connor, Optimization of electro-optic sensors through low static birefringence crystal directions. J. Appl. Phys. **105**(3), 033102 (2009). https://doi.org/10.1063/1.3072612
6. H. Lee, J. Kim, S. Park, Integrated electro-optic sensors for high-intensity electromagnetic pulse measurements in military applications. IEEE Sens. J. **23**(7), 9503–9512 (2023). https://doi.org/10.1109/JSEN.2023.3245678
7. Y. Xiao, W. Liu, H. Chen, Temperature stability challenges in electro-optic sensors: the impact of thermo-optic effects and thermal expansion. J. Lightwave Technol. **42**(1), 112–120 (2024). https://doi.org/10.1109/JLT.2023.3278901

8. D.K. Prasad, S. Rajasekaran, L. Rachmawati, Video-based maritime object detection and tracking: challenges, trends, and future directions. IEEE Trans. Intell. Transp. Syst. **17**(5), 1275–1292 (2016). https://doi.org/10.1109/TITS.2015.2506579
9. P. Anzenbacher, F. Li, M.A. Palacios, W. Wang, Conjugated polymers for optical sensing of anions. Chem. Soc. Rev. **34**(6), 488–500 (2005). https://doi.org/10.1039/B417584F
10. A. Garzarella, C.M. Sorace, B.T. O'Connor, Enhancing the sensitivity of electro-optic sensors through dielectric relaxation effects. Opt. Eng. **46**(12), 125003 (2007). https://doi.org/10.1117/1.2826361
11. S. Sun, H. Zhang, Y. Zhao, Electro-optic polymer-based sensors for high-frequency electromagnetic field detection. Opt. Lett. **32**(18), 2684–2686 (2007). https://doi.org/10.1364/OL.32.002684
12. S. Euphrasie, A. Chousseaud, E. Poizot, A matrix of electro-optic probes dedicated to specific absorption rate measurement. IEEE Trans. Microw. Theory Tech. **61**(5), 2092–2101 (2013). https://doi.org/10.1109/TMTT.2013.2255882
13. T. Nagatsuma, Y. Kado, Microwave photonic devices and their applications to communications and measurements. PIERS Online **4**(3), 376–380 (2008). https://doi.org/10.2529/piers070906210010
14. Z.-H. Chen, P.-H. Huang, C.-P. Wang, Y.-D. Chih, C.-J. Lin, Y.-C. King, Embedded near-infrared sensor with tunable sensitivity for nanoscale CMOS technologies. IEEE Sens. J. **19**(3), 933–939 (2019). https://doi.org/10.1109/jsen.2018.2880414
15. Y. Shan, J.E. Speich, K.K. Leang, Low-cost IR reflective sensors for submicrolevel position measurement and control. IEEE/ASME Trans. Mechatron. **13**(6), 700–709 (2008). https://doi.org/10.1109/tmech.2008.2005407
16. X. Yu, Y. Ji, X. Shen, X. Le, Progress in advanced infrared optoelectronic sensors. Nanomaterials **14**(10), 845–845 (2024). https://doi.org/10.3390/nano14100845
17. S. Fahad, S. Li, Y. Zhai, C. Zhao, Z. Pikramenou, M. Wang, Luminescence-based infrared thermal sensors: Comprehens. Insights. Small **20**(3) (2023). https://doi.org/10.1002/smll.202304237
18. M.E. Russell, C.A. Drubin, A.S. Marinilli, W.G. Woodington, M.J. Del Checcolo, Integrated Automotive Sensors. IEEE Trans. Microw. Theo. Tech. **50**(3), 674–677 (2002) https://doi.org/10.1109/22.989952
19. Q. Guan, C. Li, L. Qin, G. Wang, Daily activity recognition using pyroelectric infrared sensors and reference structures. IEEE Sens. J. **19**(5), 1645–1652 (2019). https://doi.org/10.1109/jsen.2018.2885168
20. H. Schmitz, H. Soltner, H. Bousack, Biomimetic infrared sensors based on Photomechanic infrared receptors in Pyrophilous ("Fire-Loving") insects. IEEE Sens. J. **12**(2), 281–288 (2012). https://doi.org/10.1109/jsen.2010.2076324
21. K.T. Vierling, L.A. Vierling, W.A. Gould, S. Martinuzzi, R.M. Clawges, Lidar: shedding new light on habitat characterization and modeling. Front. Ecol. Environ. **6**(2), 90–98 (2008). https://doi.org/10.1890/070001
22. S.E. Reutebuch, H.-E. Andersen, R.J. McGaughey, Light detection and ranging (LIDAR): an emerging tool for multiple resource inventory. J. Forest. **103**(6), 286–292 (2005). https://doi.org/10.1093/jof/103.6.286
23. J.E. Moeslund, A. Zlinszky, R. Ejrnæs, A.K. Brunbjerg, P.K. Bøcher, J. Svenning, S. Normand, Light detection and ranging explains diversity of plants, fungi, lichens, and bryophytes across multiple habitats and large geographic extent. Ecol. Appl. **29** (5). https://doi.org/10.1002/eap.1907
24. J. Chen, W. Miao, Y. Hong, H. Kuo, Recent advances in light detection and ranging: optical modulation solutions and novel nanotechnologies. Adv. Quantum Technol. (2023). https://doi.org/10.1002/qute.202300157
25. G. Kim, Y. Park, LIDAR pulse coding for high-resolution range imaging at improved refresh rate. Opt. Express **24**(21), 23810 (2016). https://doi.org/10.1364/oe.24.023810
26. K. Lim, P. Treitz, M. Wulder, B. St-Onge, M. Flood, Lidar remote sensing of forest structure. Prog. Phys. Geography: Earth and Environ. **27**(1), 88–106 (2003). https://doi.org/10.1191/0309133303pp360ra

27. M.A. Lefsky, W.B. Cohen, D.J. Harding, G.G. Parker, S.A. Acker, S.T. Gower, Lidar remote sensing of above-ground biomass in three biomes. Glob. Ecol. Biogeogr. **11**(5), 393–399 (2002). https://doi.org/10.1046/j.1466-822x.2002.00303.x
28. I. Ashraf, S. Hur, Y. Park, An investigation of interpolation techniques to generate 2D intensity image from LIDAR data. IEEE Access **5**, 8250–8260 (2017). https://doi.org/10.1109/access.2017.2699686
29. F. Vitiello, Performance analysis of space surveillance using space based optical sensors. Aerotecnica Missili & Spazio **99**(4), 263–273 (2020). https://doi.org/10.1007/s42496-020-00063-1
30. G.M. Koretsky, J.F. Nicoll, M.S. Taylor, A tutorial on electro-optical/infrared (EO/IR) theory and systems (2013)
31. I. Lopez-Calle, Space debris in-orbit detection with commercial automotive LiDAR sensors. Sensors **24**(22), 7293–7293 (2024). https://doi.org/10.3390/s24227293
32. I. McKenzie, S.K. Ibrahim, E. Haddad, S. Abad, A. Hurni, L.K. Cheng, Fiber optic sensing in spacecraft engineering: an historical perspective from the European space agency. Front. Phys. **9** (2021). https://doi.org/10.3389/fphy.2021.719441
33. V. Benthem, V. Den, J. Krooneman, Compact optical sensor for real-time monitoring of bacterial growth for space applications. Ann. New York Acad. Sci. **974**(1), 541–555 (2002). https://doi.org/10.1111/j.1749-6632.2002.tb05929.x
34. G. Soldi, D. Gaglione, N. Forti, A.D. Simone, F.C. Daffina, G. Bottini, D. Quattrociocchi, L.M. Millefiori, P. Braca, S. Carniel, P. Willett, A. Iodice, D. Riccio, A. Farina, Space-based global maritime surveillance. Part I: Satellite Technologies. IEEE Aerospace Electron. Syst. Magaz. **36**(9), 8–28 (2021). https://doi.org/10.1109/maes.2021.3070862
35. G. Coluccia, C. Lastri, D. Guzzi, E. Magli, V. Nardino, L. Palombi, I. Pippi, V. Raimondi, C. Ravazzi, F. Garoi, D. Coltuc, R. Vitulli, A. Zuccaro Marchi, Optical compressive imaging technologies for space big data. IEEE Trans. Big Data **6**(3), 430–442 (2020) https://doi.org/10.1109/tbdata.2019.2907135
36. J. Janesick, G. Putnam, Developments and applications of high-performance ccd and cmos imagingarrays. Annu. Rev. Nucl. Part. Sci. **53**(1), 263–300 (2003). https://doi.org/10.1146/annurev.nucl.53.041002.110431
37. I. Takayanagi, J. Nakamura, High-resolution CMOS video image sensors. Proc. IEEE 2013, **101**(1), 61–73. https://doi.org/10.1109/jproc.2011.2178569
38. T. Quan, S. Zeng, Z.-L. Huang, Localization capability and limitation of electron-multiplying charge-coupled, scientific complementary metal-oxide semiconductor, and charge-coupled devices for superresolution imaging. J. Biomed. Opt. **15**(6), 066005 (2010). https://doi.org/10.1117/1.3505017
39. R. Gambheer, M.S. Bhat, CCD sensor based cameras for sustainable streaming IoT applications with compressed sensing. IEEE Access **11**, 67882–67892 (2023). https://doi.org/10.1109/access.2023.3291396
40. E.R. Fossum, N. Teranishi, A.J.P. Theuwissen, J. Hynecek, Foreword special issue on solid-state image sensors. IEEE Trans. Electron Dev. **50**(1), 1–3 (2003) https://doi.org/10.1109/ted.2002.807524
41. E.R. Fossum, N. Teranishi, A.J.P. Theuwissen, Digital image sensor evolution and new frontiers. Ann. Rev. Vis. Sci. **10**(1), 171–198 (2024). https://doi.org/10.1146/annurev-vision-101322-105538
42. M. Lesser, A summary of charge-coupled devices for astronomy. Publ. Astron. Soc. Pac. **127**(957), 1097–1104 (2015). https://doi.org/10.1086/684054
43. H.T. Beier, B.L. Ibey, Experimental comparison of the high-speed imaging performance of an EM-CCD and SCMOS camera in a dynamic live-cell imaging test case. PLoS ONE **9**(1), e84614 (2014). https://doi.org/10.1371/journal.pone.0084614
44. CCD and CMOS Technology. www.tedpella.com. https://www.tedpella.com/cameras_html/ccd_cmos.aspx (Accessed 29 11 2022).
45. S. Ortega, H. Fabelo, D.K. Iakovidis, A. Koulaouzidis, G.M. Calicoo, Use of hyperspectral/multispectral imaging in gastroenterology. Shedding some–different–light into the dark. J. Clin. Med. **8**(1) (2019). https://doi.org/10.3390/jcm8010036

46. S. Ortega, M. Halicek, H. Fabelo, G.M. CCalico, B. Fei, Hyperspectral and multispectral imaging in digital and computational pathology: a systematic review [Invited]. Biomed. Opt. Express **11**(6), 3195–3233 (2020) https://doi.org/10.1364/BOE.386338
47. G. Saiko, P. Lombardi, Y. Au, D. Queen, D. Armstrong, K. Harding, Hyperspectral imaging in wound care: a systematic review. Int. Wound J. (2020). https://doi.org/10.1111/iwj.13474
48. A.V. Zubler, J.-Y. Yoon, Proximal methods for plant stress detection using optical sensors and machine learning. Biosensors **10**(12), 193 (2020). https://doi.org/10.3390/bios10120193
49. Hyperspectral and Multispectral Imaging | Edmund Optics. www.edmundoptics.com. https://www.edmundoptics.com/knowledge-center/application-notes/imaging/hyperspectral-and-multispectral-imaging/
50. W. Chen, L. Yan, V. Chandrasekar, Optical polarization remote sensing. Int. J. Remote Sens. **41**(13), 4849–4852 (2020). https://doi.org/10.1080/01431161.2020.1743529
51. K. Sarabandi, E.S. Li, Polarimetric characterization of debris and faults in the highway environment at millimeter-wave frequencies. IEEE Trans. Antennas Propagation **48**(11), 1756–1768 (2000) https://doi.org/10.1109/8.900234
52. O. Krarup, C. Baker, L. Chen, X. Bao, Sensitivity enhancement of fiber optical polarimetric sensors using self-induced nonlinear phase modulation via the Kerr effect. Optics Express **30**(9), 13985–13985 (2022). https://doi.org/10.1364/oe.454492
53. C. Nansen, Penetration and scattering—two optical phenomena to consider when applying proximal remote sensing technologies to object classifications. PLoS ONE **13**(10), e0204579 (2018). https://doi.org/10.1371/journal.pone.0204579
54. X. Li, Y. Han, H. Wang, T. Liu, S.-C. Chen, H. Hu, Polarimetric imaging through scattering media: a review. Front. Phys. **10** (2022). https://doi.org/10.3389/fphy.2022.815296
55. M. Malinowski, A. Gilerson, E. Herrera-Estrella, R. Foster, J. Agagliate, M. Ondrusek, Surface roughness and wave slope statistics from the multi-spectral polarimetric imaging of the ocean surface. Opt. Express **32**(13), 22110 (2024). https://doi.org/10.1364/oe.521650
56. C. Jamet, Amir, Z. Ahmad, F. Angelini, M. Babin, M.J. Behrenfeld, E. Boss, B. Cairns, J.H. Churnside, J. Chowdhary, A. Davis, D. Dionisi, L. Duforêt-Gaurier, B.J. Franz, R. Frouin, M. Gao, D. Gray, O. Hasekamp, X. He, C.A. Hostetler going beyond standard ocean color observations: lidar and polarimetry. Front. Marine Sci. **6** (2019) https://doi.org/10.3389/fmars.2019.00251
57. J. Riviere, P. Peers, A. Ghosh, Mobile surface reflectometry. Comput. Graph. Forum **35**(1), 191–202 (2015). https://doi.org/10.1111/cgf.12719
58. C.S. Law, S.Y. Lim, A.D. Abell, N.H. Voelcker, A. Santos, Nanoporous anodic alumina photonic crystals for optical chemo- and biosensing: fundamentals, advances, and perspectives. Nanomaterials **8**(10), 788 (2018). https://doi.org/10.3390/nano8100788
59. K.D. Heylman, K.A. Knapper, E.H. Horak, M.T. Rea, S.K. Vanga, R.H. Goldsmith, Optical microresonators for sensing and transduction: a materials perspective. Adv. Mater. **29**(30), 1700037–1700037 (2017). https://doi.org/10.1002/adma.201700037
60. M.S. Luchansky, R.C. Bailey, High-Q optical sensors for chemical and biological analysis. Anal. Chem. **84**(2), 793–821 (2011). https://doi.org/10.1021/ac2029024
61. A. Rico-Yuste, S. Carrasco, Molecularly Imprinted Polymer-based hybrid materials for the development of optical sensors. Polymers **11**(7), 1173 (2019). https://doi.org/10.3390/polym11071173
62. V.V. Ursaki, I.M. Tiginyanu, O. Volciuc, V. Popa, V.A. Skuratov, H. Morkoç, Nanostructuring induced enhancement of radiation hardness in GaN Epilayers. Appl. Phys. Lett. **90**(16) (2007). https://doi.org/10.1063/1.2723076
63. C.L. Tan, H. Mohseni, Emerging technologies for high-performance infrared detectors. Nanophotonics **7**(1), 169–197 (2018). https://doi.org/10.1515/nanoph-2017-0061
64. N.T. Yardimci, M. Jarrahi, Nanostructure-enhanced photoconductive terahertz emission and detection. Small **14**(44), 1802437 (2018). https://doi.org/10.1002/smll.201802437
65. M. Kang, R.S. Goldman, Ion irradiation of III–v semiconductor surfaces: from self-assembled nanostructures to plasmonic crystals. Appl. Phys. Rev. **6**(4), 041307 (2019). https://doi.org/10.1063/1.5079908

66. K. Alexandrou, A. Masurkar, H. Edrees, J.F. Wishart, Y. Hao, N. Petrone, J. Hone, I. Kymissis, Improving the radiation hardness of graphene field effect transistors **109**(15), 153108–153108 (2016). https://doi.org/10.1063/1.4963782
67. W.-Z. Han, M.J. Demkowicz, N.A. Mara, E. Fu, S. Sinha, A.D. Rollett, Y. Wang, J.S. Carpenter, I.J. Beyerlein, A. Misra, Des. Radiation Tolerant Mater. via Interface Eng. **25**(48), 6975–6979 (2013). https://doi.org/10.1002/adma.201303400
68. C.N. Singh, X.-Y. Liu, B.P. Uberuaga, S.J. Tobin, Reduction of bright exciton lifetimes by radiation-induced disorder. Phys. Rev. Mater. **5**(7) (2021). https://doi.org/10.1103/physrevmaterials.5.073802
69. T. Liu, H. Matsukuma, A. Suzuki, R. Sato, W. Gao, Enhanced data-processing algorithms for dispersive interferometry using a femtosecond laser. Sensors **24**(2), 370–370. https://doi.org/10.3390/s24020370
70. R.A. Harvey Towers, X. Zhang, R. Yousefi, G. Esmaili, L. Wang, A. Garcia, S.E. Beck, A modified CGM algorithm enhances data availability while retaining ICGM performance. J. Diabetes Sci. Technol. **16**(5), 1224–1227 (2021). https://doi.org/10.1177/19322968211007521
71. E.P.J. Parrott, S.M.Y. Sy, T. Blu, V.P. Wallace, E. Pickwell-MacPherson, Terahertz pulsed imaging in vivo: measurements and processing methods. J. Biomed. Opt. **16**(10), 106010 (2011). https://doi.org/10.1117/1.3642002
72. F.E. Lytle, R.K. Julian, Automatic processing of chromatograms in a high-throughput environment. Clin. Chem. **62**(1), 144–153 (2015). https://doi.org/10.1373/clinchem.2015.238816
73. P. Pachaiyappan, G. Chidambaram, A. Jahid, M.H. Alsharif, Enhancing underwater object detection and classification using advanced imaging techniques: a novel approach with diffusion models. Sustainability **16**(17), 7488–7488 (2024). https://doi.org/10.3390/su16177488
74. T.L.C. Tran, Z.-C. Huang, K.-H. Tseng, P.-H. Chou, Detection of bottle marine debris using unmanned aerial vehicles and machine learning techniques. Drones **6**(12), 401 (2022). https://doi.org/10.3390/drones6120401
75. V. Dewangan, A. Saxena, R. Thakur, S. Tripathi, Application of image processing techniques for UAV detection using deep learning and distance-wise analysis. Drones **7**(3), 174 (2023). https://doi.org/10.3390/drones7030174
76. P. Chandran, J. Asber, F. Thiery, J. Odelius, M. Rantatalo, An investigation of railway fastener detection using image processing and augmented deep learning. Sustainability **13**(21), 12051 (2021). https://doi.org/10.3390/su132112051
77. B. Xue, B. Huang, G. Chen, H. Li, W. Wei, Deep-sea debris identification using deep convolutional neural networks. IEEE J. Selected Top. Appl. Earth Observat. Remote Sens. **14**, 8909–8921 (2021). https://doi.org/10.1109/jstars.2021.3107853
78. F. Ofli, P. Meier, M. Imran, C. Castillo, D. Tuia, N. Rey, J. Briant, P. Millet, F. Reinhard, M. Parkan, S. Joost, Combining human computing and machine learning to make sense of big (Aerial) data for disaster response. Big Data **4**(1), 47–59 (2016). https://doi.org/10.1089/big.2014.0064
79. K. Sasaki, T. Sekine, L.-J. Burtz, W.J. Emery, Coastal marine debris detection and density mapping with very high-resolution satellite imagery. IEEE J. Selected Top. Appl. Earth Observat. Remote Sens. **15**, 6391–6401 (2022). https://doi.org/10.1109/jstars.2022.3193993
80. M. Liu, Research on star sensor-based space debris detection and positioning technology. Publ. Astron. Soc. Pac. **135**(1046), 047001 (2023). https://doi.org/10.1088/1538-3873/acc7c9
81. I. Loomis, NASA sensor to study space junk the size of dust. Science **358**(6368), 1234–1234 (2017). https://doi.org/10.1126/science.358.6368.1234
82. W. Yang, Z. Li, M. Liu, D. Liu, J. Sun, Z. Kang, H. Yu, Detection approach for geo space objects with a wide-field optical telescope array. Opt. Express **31**(12), 18717–18717 (2023). https://doi.org/10.1364/oe.483737
83. R. Sun, P. Jia, Image restoration based on statistic psf modeling for improving the astrometry of space debris. Publ. Astron. Soc. Pac. **129**(974), 044502–044502 (2017). https://doi.org/10.1088/1538-3873/aa611a

84. X. Lei, Z. Lao, L. Liu, J. Chen, L. Wang, S. Jiang, M. Li, Telescopic network of Zhulong for orbit determination and prediction of space objects. Remote Sens. **16**(13), 2282–2282 (2024). https://doi.org/10.3390/rs16132282
85. T. Jansen-Sturgeon, Benjamin; G.J. Madsen, P.A. Bland, E.K. Sansom, A.R. Hadrien, Deville-poix; R.M. Howie, M. Cupák, M.C. Towner, M.A. Cox, N.D. Nevill, P. Hoskins, G.P. Bonning, J. Calcino, J.T. Clark, B.M. Henson, A. Langendam, S.J. Matthews, T.P. McClafferty, J.T. Mitchell, Recreating the OSIRIS-REx slingshot manoeuvre from a network of ground-based sensors. Publ. Astronom. Soc. Australia **37** (2020). https://doi.org/10.1017/pasa.2020.36
86. P. Hickson, Wide-field tracking with zenith-pointing telescopes. Mon. Not. R. Astron. Soc. **330**(3), 540–546 (2002). https://doi.org/10.1046/j.1365-8711.2002.05098.x
87. M. Li, X. Xiao, Q. Yu, S. Sun, Design of integrated searching and tracking optical systems based on freeform mirrors. Opt. Express **30**(1), 286–286 (2021). https://doi.org/10.1364/oe.446012
88. F.K. Chun, R.D. Tippets, D.M. Strong, D.J. Della-Rose, D.E. Polsgrove, K.C. Gresham, J.A. Reid, C.P. Christy, M. Korbitz, J. Gray, S. Gartin, D. Coles, R.K. Haaland, R. Walker, J. Workman, J. Mansur, V. Mansur, T. Hancock, J.D. Erdley, T.S. Taylor, A new global array of optical telescopes: the falcon telescope network. Publ. Astron. Soc. Pac. **130**(991), 095003 (2018). https://doi.org/10.1088/1538-3873/aad03f
89. R. Davies, M. Kasper, Adaptive optics for astronomy. Ann. Rev. Astron. Astrophys. **50**(1), 305–351 (2012). https://doi.org/10.1146/annurev-astro-081811-125447
90. G. Stephens, A. Freeman, E. Richard, P. Pilewskie, P. Larkin, C. Chew, S. Tanelli, S. Brown, D. Posselt, E. Peral, The emerging technological revolution in earth observations. Bull. Am. Meteor. Soc. **101**(3), E274–E285 (2020). https://doi.org/10.1175/bams-d-19-0146.1
91. J. Dash, B.O. Ogutu, Recent advances in space-borne optical remote sensing systems for monitoring global terrestrial ecosystems. Progress Phys. Geography: Earth Environ. **40**(2), 322–351 (2016). https://doi.org/10.1177/0309133316639403
92. Y. Tian, X. Hu, S. Chen, Y. Zhao, X. Zhang, D. Wang, W. Xu, M. Xie, W. Hao, X. Su, High accuracy ranging for space debris with Spaceborne single photon lidar. Opt. Express **32**(7), 12318–12318 (2024). https://doi.org/10.1364/oe.519002
93. D. Weidmann, K. Antonini, D.M. Pinto, B.K. Brodersen, G. Patel, M. Hegglin, C. Sioris, W. Bell, K. Miyazaki, L.K. Alminde, A. Gabriele, M. Pastena, A. Hoffmann, Cubesats for monitoring atmospheric processes (CubeMAP): a constellation mission to study the middle atmosphere. CentAUR (University of Reading), 26–26 (2020). https://doi.org/10.1117/12.2573727
94. J.N. Buitrago-Leiva, A. Camps, A. Moncada Niño, Considerations for Eco-LeanSat satellite manufacturing and recycling. Sustainability **16**(12), 4933 (2024). https://doi.org/10.3390/su16124933
95. U. Singh, G. Tzeremes, Second international workshop on space-based lidar remote sensing techniques and emerging technologies [Technical Committees]. IEEE Geosc. Remote Sens. Magaz. **7**(3), 129–132 (2019). https://doi.org/10.1109/mgrs.2019.2928930
96. S. Thombre, Z. Zhao, H. Ramm-Schmidt, J.M. Vallet, T. Malkamäki, S. Nikolskiy, T. Hammarberg, H. Nuortie, A. Bhuiyan, S. Särkkä, V.V. Lehtola, Sensors and AI techniques for situational awareness in autonomous ships: a review **23**(1), 64–83. https://doi.org/10.1109/tits.2020.3023957
97. E.S. Maddy, S.A. Boukabara, MIIDAPS-AI: an explainable machine-learning algorithm for infrared and microwave remote sensing and data assimilation preprocessing—application to LEO and GEO sensors. IEEE J. Selected Top. Appl. Earth Observat. Remote Sens. **14**, 8566–8576 (2021). https://doi.org/10.1109/jstars.2021.3104389
98. N. Maximenko, P. Corradi, K.L. Law, E. Van Sebille, S.P. Garba, R.S. Lampitt, F. Galgani, V. Martinez-Vicente, L. Goddijn-Murphy, J.M. Veiga, R.C. Thompson, C. Maes, D. Moller, C.R. Löscher, A.M. Addamo, M.R. Lamson, L.R. Centurioni, N.R. Posth, R. Lumpkin, M. Vinci, Toward the integrated marine debris observing system. Front. Marine Sci. **6** (2019) https://doi.org/10.3389/fmars.2019.00447

99. J. Agrawal, M.Y. Arafat, Transforming Farming: A Review of AI-Powered UAV Technologies in Precision Agriculture. Drones **8**(11), 664–664 (2024). https://doi.org/10.3390/drones8110664
100. U. Ahmad, A. Nasirahmadi, O. Hensel, S. Marino, Technology and data fusion methods to enhance site-specific crop monitoring. Agronomy **12**(3), 555 (2022). https://doi.org/10.3390/agronomy12030555
101. S.-A. Boukabara, V. Krasnopolsky, J.Q. Stewart, E.S. Maddy, N. Shahroudi, R. N. Leveraging modern artificial intelligence for remote sensing and NWP: benefits and challenges. Bullet. Am. Meteorol. Soc. **100**(12), ES473–ES491 (2019). https://doi.org/10.1175/bams-d-18-0324.1
102. J. Lai, L. Mejias, J.J. Ford, Airborne vision-based collision-detection system. J. Field Robot. **28**(2), 137–157 (2011). https://doi.org/10.1002/rob.20359
103. A.G. Basden, D.F. Buscher, Improvements for group delay fringe tracking. Mon. Not. R. Astron. Soc. **357**(2), 656–668 (2005). https://doi.org/10.1111/j.1365-2966.2005.08681.x
104. N. Blind, J.-B. Olivier Absil, Le Bouquin, J.-P. Berger, A. Chelli optimized fringe sensors for the VLTI next generation instruments. Astronomy and Astrophysics **530**, A121–A121 (2011). https://doi.org/10.1051/0004-6361/201016222
105. M. Auvergne, P. Bodin, L. Boisnard, J.-T. Buey, S. Chaintreuil, G. Epstein, M. Jouret, T. Lam-Trong, P. Levacher, A. Magnan, R. Perez, P. Plasson, J. Plesseria, G. Peter, M. Steller, D. Tiphène, A. Baglin, P. Agogué, T. Appourchaux, D. Barbet, The CoRoT satellite in flight: description and performance. Astron. Astrophys. **506**(1), 411–424 (2009). https://doi.org/10.1051/0004-6361/200810860
106. Z.-P. Li, X. Huang, P.-Y. Jiang, Y. Hong, C. Yu, Y. Cao, J. Zhang, F. Xu, Pan, J.-W. Super-resolution single-photon imaging at 8.2 kilometers. Optics Express **28**(3), 4076–4076 (2020). https://doi.org/10.1364/oe.383456
107. P.P. Mathai, J.A. Liddle, S.M. Stavis, Optical tracking of nanoscale particles in microscale environments. Appl. Phys. Rev. **3**(1), 011105 (2016). https://doi.org/10.1063/1.4941675
108. D. Snowden, V.M. Tsontos, N.O. Handegard, M. Zarate, K. O'Brien, K.S. Casey, N. Smith, H. Sagen, K. Bailey, M.N. Lewis, S.C. Arms, Data interoperability between elements of the global ocean observing system. Front. Mar. Sci. **6** (2019). https://doi.org/10.3389/fmars.2019.00442
109. F. Noardo, R. Atkinson, L. Bastin, J. Maso, I. Simonis, A. Villar, M.-F. Voidrot, P. Zaborowski, Standards for data space building blocks. Remote Sens. **16**(20), 3824 (2024). https://doi.org/10.3390/rs16203824
110. A.D. Jaunzemis, K.M. Feigh, M.J. Holzinger, D. Minotra, M.W. Chan, Cognitive systems engineering applied to decision support in space situational awareness. J. Cognit. Eng. Dec. Mak. **14**(1), 3–33 (2019). https://doi.org/10.1177/1555343419872050
111. C.A. Gotway, L.J. Young, Combining incompatible spatial data. J. Am. Stat. Assoc. **97**(458), 632–648 (2002). https://doi.org/10.1198/016214502760047140
112. M. Masmoudi, S. Ben, M. Hedi Karray, B. Archimede, H. Baazaoui Zghal, Semantic data integration and querying: a survey and challenges. ACM Comput. Surveys (2024). https://doi.org/10.1145/3653317
113. F.S. Ligler, Perspective on optical biosensors and integrated sensor systems. Anal. Chem. **81**(2), 519–526 (2009). https://doi.org/10.1021/ac8016289
114. I. Clay, C. Angelopoulos, A.L. Bailey, A. Blocker, S. Carini, R. Carvajal, D. Drummond, K.F. McManus, I. Oakley-Girvan, K.B. Patel, P. Szepietowski, J.C. Goldsack, Sensor data integration: a new cross-industry collaboration to articulate value, define needs, and advance a framework for best practices. J. Med. Internet Res. **23**(11), e34493 (2021). https://doi.org/10.2196/34493
115. C.J. Fluke, C. Jacobs, Surveying the reach and maturity of machine learning and artificial intelligence in astronomy. WIREs Data Mining Knowl. Discovery **10**(2) (2019). https://doi.org/10.1002/widm.1349
116. D. Wang, M. Zhang, Artificial intelligence in optical communications: from machine learning to deep learning. Front. Commun. Netw. **2** (2021). https://doi.org/10.3389/frcmn.2021.656786

117. X. Chen, S. Xu, S. Shabani, Y. Zhao, M. Fu, A.J. Millis, M.M. Fogler, A.N. Pasupathy, M. Liu, D.N. Basov, Machine learning for optical scanning probe Nanoscopy. Adv. Mater. **35**(34) (2022). https://doi.org/10.1002/adma.202109171
118. A. Mahabal, U. Rebbapragada, R. Walters, F.J. Masci, N. Blagorodnova, J. van Roestel, Q.-Z. Ye, R. Biswas, K. Burdge, C.-K. Chang, D.A. Duev, V.Z. Golkhou, A.A. Miller, J. Nordin, C. Ward, S. Adams, Bellm, E.C., D.B. Bue, C. Cannella, Machine learning for the Zwicky transient facility. Publ. Astronom. Soc. Pacific **131**(997), 038002 (2019). https://doi.org/10.1088/1538-3873/aaf3fa
119. Y. Guo, L. Zhong, L. Min, J. Wang, Y. Wu, K. Chen, K. Wei, C. Rao, Adaptive optics based on machine learning: a review. Opto-Electron. Adv. **0**(0), 200082–200082 (2020). https://doi.org/10.29026/oea.2022.200082
120. X. Lin, Y. Rivenson, N.T. Yardimci, M. Veli, Y. Luo, M. Jarrahi, A. Ozcan, All-optical machine learning using diffractive deep neural networks. Science **361**(6406), 1004–1008 (2018). https://doi.org/10.1126/science.aat8084
121. J. Zhao, A. Lyons, Ulku, A.C., H. Defienne, D. Faccio, E. Charbon, Light detection and ranging with entangled photons. Opt. Exp. **30**(3), 3675–3675. https://doi.org/10.1364/oe.435898
122. M. Fink, F. Steinlechner, J. Handsteiner, J.P. Dowling, T. Scheidl, R. Ursin, Entanglement-enhanced optical gyroscope. New J. Phys. **21**(5), 053010 (2019). https://doi.org/10.1088/1367-2630/ab1bb2
123. Sorelli, G., N. Treps, Grosshans, F., Boust, F., Detecting a target with quantum entanglement. IEEE Aerospace Electron. Syst. Magaz. **37**(5), 68–90 (2022). https://doi.org/10.1109/maes.2021.3116323
124. J. Huang, M. Zhuang, C. Lee, Entanglement-enhanced quantum metrology: from standard quantum limit to Heisenberg limit. Appl. Phys. Rev. 11(3) (2024). https://doi.org/10.1063/5.0204102
125. H. Yonezawa, D. Nakane; T.A. Wheatley, K. Iwasawa, S. Takeda, H. Arao, K. Ohki. K. Tsumura, D.W. Berry, T.C. Ralph, Wiseman, H.M., E.H. Huntington, A. Furusawa. Quantum-enhanced optical-phase tracking. **337** (6101), 1514–1517 (2012). https://doi.org/10.1126/science.1225258
126. J. Xavier, D. Yu, C. Jones, E. Zossimova, F. Vollmer, Quantum nanophotonic and nanoplasmonic sensing: towards quantum optical bioscience laboratories on chip. Nanophotonics **10**(5), 1387–1435 (2021). https://doi.org/10.1515/nanoph-2020-0593
127. O. Hosten, R. Krishnakumar, N.J. Engelsen, M.A. Kasevich, Quantum phase magnification. Science **352**(6293), 1552–1555 (2016). https://doi.org/10.1126/science.aaf3397
128. T. Nagata, R. Okamoto, J.L. O'Brien, K. Sasaki, S. Takeuchi, Beating the standard quantum limit with four-entangled photons. Science **316**(5825), 726–729 (2007). https://doi.org/10.1126/science.1138007
129. B. E. Paton, Optical sensing principles and optic sensor buses, in *Proceedings of SPIE—Optical Sensing & Detection II, SPIE*, vol. 954. (1989). https://doi.org/10.1117/12.947640. SPIE Digital Library
130. J. M. López-Higuera, *Handbook of Optical Fibre Sensing Technology*. (Wiley, Hoboken, NJ, 2014)

Towards Psychophysiological Monitoring of Aircraft Pilot Crew

Ana Sofia Antunes Calado, Tiago Filipe Rodrigues Fernandes, Hugo Humberto Plácido da Silva, and Paulo Sérgio de Brito André

Abstract Physiological sensing within aircraft cabins offers the potential to enhance safety, comfort and overall flight performance for both crew and passengers. This chapter explores the application of physiological sensing technologies to aircraft environments; in particular, we discuss the integration of these sensors within aircraft cabins and on pilot crews, focusing on psychophysiological monitoring—an essential aspect of understanding and mitigating in-flight fatigue, stress, and other critical cognitive load factors affecting pilot performance. Key topics include non-invasive sensors capable of monitoring heart rate, skin conductance, and oxygenated hemoglobin levels without hindering cabin or crew operations. We examine the role of advanced materials and highlight their capability for seamless integration with current cabin and cockpit designs on the pilot, paving the way for the broader utilization of optical sensing techniques. Emphasis is placed on the technological challenges in implementing these systems, particularly within the confined, pressurized, and controlled environment of aircraft interiors. *Our focus is* on psychophysiological responses, and on how these sensors can be leveraged to monitor cognitive and emotional states.

A. S. A. Calado · T. F. R. Fernandes · H. H. P. da Silva · P. S. de Brito André (✉)
Instituto de Telecomunicações e Instituto Superior Técnico, Universidade de Lisboa, Lisboa, Portugal
e-mail: paulo.andre@tecnico.ulisboa.pt

A. S. A. Calado
e-mail: ana.calado@tecnico.ulisboa.pt

T. F. R. Fernandes
e-mail: tiago.filipe.rodrigues.fernandes@tecnico.ulisboa.pt

H. H. P. da Silva
e-mail: hsilva@lx.it.pt

C. Marques et al. (eds.), *Advanced Optical Sensors for Aerospace Applications*, Progress in Optical Science and Photonics 37,
https://doi.org/10.1007/978-981-95-1626-1_12

1 Introduction

Nowadays, air travel is a common means of transportation, playing a key role in accelerating globalization and enhancing connectivity. It has a significant impact by enabling the flow of people, goods and services, transforming how individuals live, connect, and experience the world on a global scale. Over the past few decades, technological advancements have not only improved the design and functionality of an aircraft, but have significantly enhanced safety and efficiency, establishing air travel as one of the most reliable modes of transportation. Today, flying is no longer a luxury but an integral part of everyday life [1].

Beyond sheer passenger volume, aviation plays a critical role in the global economy, facilitating international trade, tourism and cultural exchange. The industry contributes with 3.5 trillion USD to the world's Gross Domestic Product (GDP). If aviation were a country, it would rank 17th in terms of GDP, comparable to nations like the Netherlands. But it does not stop there, the sector is projected to keep growing; by 2038 it is expected to support 143 million jobs and generate 6.3 trillion USD for the global economy [2].

According to Statista, the global number of flights performed by the airline industry has been steadily increasing since the early 2000s. In 2004 approximately 23.8 million flights were recorded, with a projected forecast of 38.7 million by the end of 2024. In 2023, approximately 4500 million passengers boarded global airline flights, highlighting how air travel has become a fundamental part of daily life [2]. With the growing prevalence of air travel, ensuring passenger and crew safety has become a critical concern and a pressing problem. The continued evolution of the aviation industry relies heavily on maintaining high standards of safety. Minor flight incidents can have fatal consequences, emphasizing the need for continuous improvements in aviation safety protocols and technology [3].

Many factors can contribute to aviation accidents. According to a NASA report, the main causes of plane crashes include pilot error, other human errors, weather, mechanical failure, sabotage, and other causes. According to this report, pilot error is listed as the primary cause of 75.5% of general aviation accidents [4]. Consequently, pilot well-being is crucial to flight safety; yet the demands of the job—including irregular hours, disrupted sleep, high stress, fatigue, and prolonged sedentary periods at high altitudes—elevate the risk of both physical and mental health issues. In aviation, the combination of complex cockpit environments, long operational hours, and the monotony of automation can lead to cognitive overload, fatigue, and impaired vigilance. These factors compromise situational awareness, increasing the likelihood of errors that threaten flight safety and may result in catastrophic accidents [5].

This work explores physiological monitoring techniques for assessing pilots' health during flights, aiming to better understand how mental and physical health can influence their performance. By enabling the early detection of potential safety risks, these techniques may play a crucial role in aviation safety. Special emphasis is placed on functional Near-Infrared Spectroscopy (fNIRS), a non-intrusive optical sensing method for brain monitoring.

Monitoring physiological signals to assess pilots' conditions during flight requires advanced devices to enable a continuous real-time assessment of vital signs. Initially developed for healthcare, wearable devices have expanded into other fields, influencing lifestyles beyond medical settings. In occupational contexts, wearable sensors provide a clearer understanding of how physiological responses can impact job performance and workplace safety [6]. Despite technological advancements, current wearable health monitoring methods remain limited and often impractical. Traditional devices tend to be bulky and intrusive, thus limiting their usability for continuous everyday monitoring. Users prefer minimally visible and easy-to-operate solutions. An alternative approach for physiological monitoring lies in integrating sensors into everyday items, making them "invisible" while still effectively collecting physiological data [7].

When exploring physiological sensor technology, understanding the industry's economic dynamics is crucial, since they not only influence research funding but also ensure that innovations have practical value and a target market. The market for physiological sensors, including smartwatches and other biometric sensing devices, has grown exponentially over the last decade. In 2024, the global wearable technology market was valued at 157 billion USD and projected to reach 1.7 trillion USD by 2032 [8]. The smartwatches segment alone was projected to reach a revenue of 29 billion EUR in 2024 and grow to 37 billion EUR by 2029 [9]. Several factors drive this growth, including the rising demand for health and fitness monitoring due to an aging population, increased awareness of health and lifestyle benefits, and the expansion of the sports sector, which relies on sensors for optimizing training and recovery. Considering the economic potential of both the physiological sensor industry and the air travel sector, it becomes clear that psychophysiological monitoring in aviation presents significant opportunities for further research and innovation.

As aviation continues to expand, the economic consequences of flight accidents, and inefficiencies highlight the urgent need for advanced monitoring systems that incorporate physiological sensors to assess pilot mental and physical states in real time. Optical sensors, in particular, offer a non-invasive and effective means of detecting cognitive fatigue, stress, and alertness levels, enabling proactive interventions to mitigate aviation risks. Implementing these technologies can enhance pilot performance and reduce human error, ultimately contributing to a safer aviation industry.

2 Performance Assessment in Pilots

The aeronautics industry fosters a culture of performance monitoring, which is vital to an airline's safety management system, encompassing five key factors described in Table 1. Each of these areas plays a vital role in identifying potential risks and maintaining high flight safety standards in compliance with safety aviation performance metrics [5].

Table 1 Main aviation safety factors

Safety aspects	Description
Aircraft	Relies on systems like Flight Data Monitoring (FDM) to track critical parameters, including unacceptable aircraft states and the aircraft technical log (which records information like malfunctions and maintenance activities)
Flight voyage	Provides detailed information on fuel consumption, passenger and cargo information, flight times, routing, and documentation of safety issues
Environment	Encompasses weather conditions and the physical environment in which the aircraft operates. These are transmitted to the pilot before the flight
Crew	Acts as a coordinating interface, bridging flight operations and safety management. Crew health monitoring remains limited, focusing primarily on tracking fatigue and duty times
Pilot	In compliance with regulatory requirements pilot's health is assessed annually by official aeromedical examiners, to determine if a pilot is fit or not to fly. For a regular assessment, the three methods previously described can be used

Adapted from [5]

Although all five aspects are critical, as previously stated, pilots are the primary cause of aviation accidents and will be the focus on this chapter. Their performance serves as one of the last lines of defense in the prevention of accidents, highlighting its importance in the aviation safety framework [5].

The occupational demands of airline pilots encompass a range of challenges—fatigue, shift work, irregular schedules, sleep disruption, physical inactivity, and psychological stress—which can lead to adverse health outcomes and negatively impact pilots' performance and decision-making abilities. The nature of a pilot's career involves long periods of low workload often causing boredom, followed by sudden operations that require immediate and intense responses. This continuous shift between low and high cognitive workloads leads to alternating periods of low and high stress. During a flight, stress levels are typically heightened during high workload activities such as take-off and landing, while the autopilot phase is characterized by boredom. If unaddressed, these stress factors can contribute to significant issues, like mental workload and mental fatigue. The first emerges from an increased cognitive load due to overlapping tasks during a flight, and the second is a gradual and cumulative process resulting from a prolonged sensation of exhaustion. Both diminish pilot vigilance and impair their ability to respond to unexpected situations [10].

A systematic review by [5], which analyzed 48 studies from 20 countries, identified a significant prevalence of over 50% obesity, insufficient physical activity, elevated fatigue, and regular alcohol consumption among pilots. Recent studies focused on measuring the prevalence of depression and other mental disorders in pilots found that approximately 12.6% of airline pilots were dealing with symptoms of depression and 4.1% with suicidal thoughts. Given the demanding nature of their profession, maintaining both physical and mental health is essential for pilots. However, the high percentages of pilots experiencing health issues exposes a concerning reality; they face problems that can adversely affect their job performance and increase the risk of flight accidents [7].

There are three main methods to assess and evaluate pilots' physical and mental health: subjective evaluations; behavioral analysis/mission performance; and physiological measures [11]. Subjective evaluations, such as the NASA Task Load Index (NASA-TLX) and the Situation Awareness Global Assessment Technique (SAGAT), are post-performance questionnaires used to assess how various factors correlate with a pilot's performance. Behavioral analysis consists of comparing the pilot's actions during the flight with a predefined mission plan, focusing on identifying deviations or errors in the execution of tasks. However, both methods have significant limitations, since they are often constrained by respondent bias, becoming subjective. Physical indicators, while easily collected, are unreliable for monitoring due to factors such as personality, mood, social context, and the potential to be faked. In contrast, physiological signals such as heart rate, respiration, and pupil dilation, are harder to manipulate, providing an intrinsic measure of a pilot's cognitive and emotional state. Unlike the previous methods, physiological monitoring offers a more objective approach, enabling continuous real-time data acquisition without relying on self-reporting. This makes it more suitable for assessing the impact a flight has on a pilot, ensuring a reliable and uninterrupted evaluation of their health and performance [11].

To evaluate pilot performance, we propose to explore the methodology centered on the continuous monitoring and collection of physiological data to assess a pilot's physical and/or mental state during a flight [7]. A critical aspect for *monitoring based on physiological sensing* is the effective placement of sensors on the pilot. Nowadays, there are two main approaches: "wearables", i.e., as body-worn devices attached directly to the individual (e.g. smartwatches or smart shirts [12]); and "invisibles", i.e., embedded in objects or the surrounding environment (e.g. computer peripherals or smart chairs). Each approach has advantages and disadvantages. Wearables can be bulky, intrusive, and uncomfortable during daily activities; in addition, they typically require some form of user collaboration. In contrast, invisibles seamlessly integrate into routine tasks, requiring minimal user effort; however, their performance can be compromised by increased noise, interference, intermittent contact, and variations in user posture or positioning [11]. In this study, we will focus on wearable sensors, as they provide direct and continuous physiological monitoring. Despite their potential drawbacks, such as bulkiness and the need for user compliance, their ability to deliver precise, real-time measurements makes them a valuable tool for capturing relevant physiological and behavioral information.

3 Theoretical Background

This chapter explores physiological monitoring, which examines physiological responses linked to an individual's cognitive, emotional, and behavioral states. It also covers the corresponding physiological signals indicators that reflect these internal states, providing a foundation for understanding their role in assessing human performance and well-being.

3.1 Physiological Signals

A physiological signal is defined as a signal that originates from a human body, either as part of normal homeostasis or in response to external stimuli. These signals can be categorized based on their physical characteristics and the way they are naturally generated by stimulation [13]:

- Bioelectrical signals, arising from the electrical activity that exists in the membrane of cells.
- Bioacoustic signals, involving the measurement of sounds produced by some organs due to mechanical or fluidical movements within the body.
- Biomechanical signals, encompassing measurements of changes in position, velocity, pace, acceleration, flow rates, and pressures.
- Biochemical signals provide information on the concentration of various substances in the body, as well as their pH levels.
- Body temperature, which reflects the metabolic processes within the body.

Table 2 provides a comprehensive overview of key physiological signals, outlining their anatomical origins, measurement methods, and a brief description of their significance. These signals can be monitored using electrical and optical sensors, both capable of continuously monitoring physiological signals in real-time. Among these, some signals are of particular importance to this work and will be discussed in greater detail. These signals serve to assess various physiological functions:

- Electrocardiography (ECG): to monitor heart activity and electrical patterns of the heart.
- Electrodermal Activity (EDA): to measure variations in skin conductance.
- Functional Near-Infrared Spectroscopy (fNIRS): to assess the brain's hemodynamic response, by detecting changes in oxygenated and deoxygenated hemoglobin levels.

3.2 Electrocardiography

Electrocardiography (ECG) records the electrical activity of the heart, producing an electrocardiogram. The heart's activity is driven by the contraction and relaxation of the cardiac muscle in response to electrical stimulation, generating potential differences that propagate to the skin's surface [13]. The ECG is measured using surface electrodes that are placed on the skin, typically on the limbs or chest, capturing essential information on heart rate and overall electrical function, which are crucial

Table 2 Physiological signals

Physiological signals	Origin	Measurement location	Description
ECG (Electrocardiography)	Heart	Surface electrodes placed on the chest or clavicles	Records the electrical activity of the heart, producing an electrocardiogram
EDA (Electrodermal Activity) or GSR (Galvanic Skin Response)	Skin	Surface electrodes placed on fingers or palm of the hand	Captures changes in the amount of sweat secretion at the skin surface
EEG (Electroencephalography)	Brain	Surface electrodes placed on the head	Records the electrical activity of the brain, producing an electroencephalogram
EGG (Electrogastrography)	Gastric muscles	Cutaneous electrodes placed on the abdominal skin	Records the gastric myoelectrical activity, producing an electrogastrogram
EMG (Electromyography)	Muscle fibers	Surface electrodes placed on the face or body	Records the excitable membranes of muscle cells activity, producing an electromyogram
EOG (Electrooculography)	Eye cornea	Surface electrodes placed around the eyes	Records the corneal-retinal position potential, producing an electrooculogram
fNIRS (Functional Near-Infrared Spectroscopy)	Brain	Sensor placed on the forehead	Captures changes in the blood oxygenation and hemodynamics within the brain
PCG (Phonocardiography)	Heart sounds	Sensor placed on the chest	Records the time series of heart sounds, generated by the mechanical heart activity
PPG (Photoplethysmography) or BVP (Blood Volume Pulse)	Blood pressure	Photodetector sensor placed on the skin	Records the variations of blood perfusion in the superficial vascular tissue
RESP (Respiration)	Lungs	Chest belt worn in the thorax	Records the changes in the volume of air inhaled and exhaled

for its proper circulatory effectiveness. The ECG has an amplitude range of 10 μV–4 mV and a frequency range of 0.05–500 Hz [14]. The typical cardiac cycle produces a pattern, and its evaluation involves analyzing subtle changes in the time-series data, focusing on morphological features like: P wave, PQ segment, QRS complex, ST segment, and T wave, as illustrated in Fig. 1 and explained on Table 3.

ECG is increasingly applied in psychophysiological monitoring to evaluate the heart's responses to stress, workload, and fatigue in high-stakes environments like aviation. Higher mental workload results in increased heart rate, and lower heart

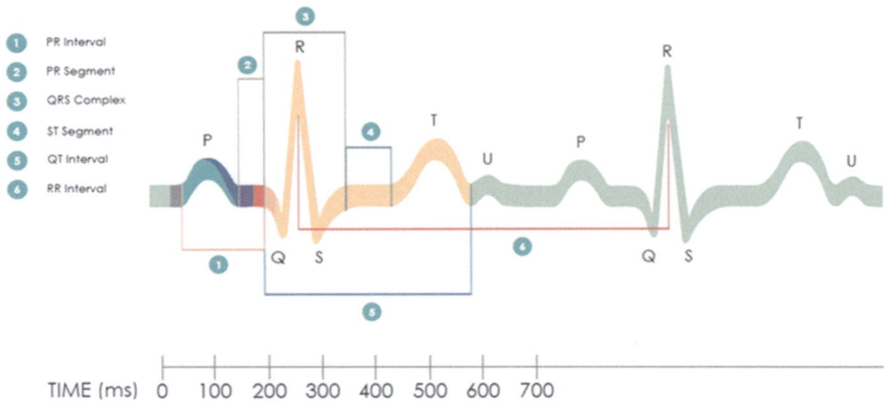

Fig. 1 Sample of an ECG signal trace in the time domain [15]

Table 3 Main ECG Parameters in the time series data [13]

ECG parameters	Description	Time duration (ms)
P wave	Result of the depolarization of heart atria	60–80
PQ interval	Isoelectric part of curve after the P wave	60–80
QRS complex	Combination of three of the graphical deflections seen on a typical ECG	80–120
ST segment	Manifestation of the action potential plateau phase of the ventricles	100–120
RR interval	Temporal gap between R peaks of two consecutive QRS (generated by heartbeat)	600–1200
T wave	Reflects cardiac ventricular repolarization	120–160

rate variability. The morphology of the signal also changes for example with the prolongation of the QT interval. Analyzing ECG data provides valuable insights to detect cardiovascular diseases, heart rhythm irregularities, stress levels and emotional states, acting as a key indicator of physiological stimulation [16].

3.3 *Electrodermal Activity*

Electrodermal Activity (EDA), also known as Galvanic Skin Response (GSR) is a biosignal that reflects changes in the skin's electrical properties by measuring variations in current or voltage between the sensors leads on the skin, essentially equivalent to a variable resistor. The human body contains approximately 1.6–4 million sweat glands, located in the reticular dermis and hypodermis, which are classified as exocrine because there is a direct secretion onto the skin surface. They can be sub-categorized into eccrine glands or apocrine glands. Sweating is a physiological process mediated by the sympathetic nervous system that occurs at the eccrine glands, scattered through the body, but with more prevalence on the hand palms and soles of the feet. The sympathetic and parasympathetic nervous system are part of the autonomous nervous system, which is responsible for regulating involuntary or automatic functions closely linked to emotional responses [17]. When the sympathetic nervous system is activated in response to stress, there is an increase in sweat production, which enhances skin conductance due to its high conductivity. EDA is divided into two key components:

- Electrodermal Level (EDL): Reflects long-term baseline levels of skin conductance.
- Electrodermal Response (EDR): Represents short-term fluctuations in conductance triggered by immediate responses to stimuli [14] (Fig. 2).

As a bioelectrical signal, EDA captures changes in the amount of sweat secretion at the skin surface, typically represented as changes in skin conductance. It is usually measured by placing two surface electrodes on areas with high density of sweat glands, such as fingers or palms of the hand. These regions are ideal because ionic sweat significantly enhances conductivity compared to dry skin.

The skin's resistance can be determined by R = V/I, and from this skin conductance is derived with G = 1/R. Despite being influenced by factors like temperature, humidity, skin hydration, and electrode placement, EDA remains a valuable biomarker. By analyzing changes in electric skin conductance, it is possible to infer a person's arousal, since EDA is a sensitive marker for evaluation of changes during the activation of the sympathetic nervous system [19].

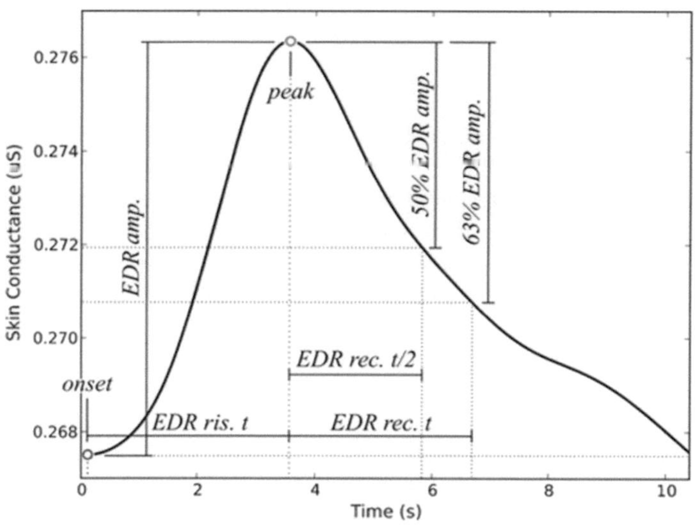

Fig. 2 Typical electrodermal response waveform [18]

Mental workload directly influences EDA, as increased cognitive demand heightens sympathetic arousal, leading to greater sweat production, and lower skin resistance.

3.4 fNIRS

Functional Near-Infrared Spectroscopy (fNIRS) is an optical neuroimaging technique that measures cortical hemodynamic responses to neural activity, by detecting changes in cerebral oxygenation and blood flow. fNIRS measures variations in the absorption of red and infrared light, which differ based on the oxygenated hemoglobin (HbO) and deoxygenated-hemoglobin (HbR). It operates within the Near-InfraRed (NIR) spectrum (650–900 nm), since light in this range is absorbed by HbO and HbR much more strongly than the surrounding and overlying tissues. The pattern of variations in their concentrations over time defines the Hemodynamic Response Function (HRF).

Oxygen is vital for living organisms, and its transport is facilitated by erythrocytes containing hemoglobin. Each hemoglobin molecule can carry up to four oxygen molecules in HbO form and, after delivering oxygen, carbon dioxide molecules are collected to be removed in HbR form. Human tissues are relatively transparent to NIR light, with hemoglobin serving as the primary absorber in the human brain. HbO and HbR exhibit distinct absorption characteristics, depending on the wavelength of the NIR light. The absorption of infrared light as a function of wavelength is different for these two chromophores, with HbR absorbing more below 790 nm and HbO more above 790 nm. By emitting NIR light at two or more wavelengths and detecting the reflected light, fNIRS can effectively measure changes in HbO and HbR levels, enabling the detection of brain activity in specific cerebral regions through fluctuations in blood oxygenation [20].

The fNIRS sensor utilizes two light-emitting diodes (LEDs), one emitting red light and the other emitting infrared light, positioned on the forehead to measure light reflectance from the brain's cortical tissue. The reflected light is absorbed by a photoreceptor/photodiode, then generating an electrical current that is used to estimate blood oxygen saturation levels within the monitored brain region. By detecting fluctuations in HbO and HbR concentrations, fNIRS effectively tracks cerebral hemodynamic activity, providing insights into cognitive workload and neural activation [20].

The brain is a high energy-demand organ and neuronal activation correlates with increases in cerebral blood flow and volume, which is a phenomenon known as neurovascular coupling—the fundament of functional neuroimaging techniques, including fNIRS. Though limited to cortical regions, fNIRS has many advantages, including portability, non-invasiveness, cost-effectiveness, and suitability for long-term real-time monitoring, making it an alternative to traditional neuroimaging methods, complementing techniques like EEG.

This technology is particularly valuable for mental workload assessments, as tasks requiring greater cognitive effort demand increase oxygen delivery to the brain. When mental workload increases, both the blood flow and HbO in the brain increase. Since HbO absorbs more infrared light, less light will be reflected and absorbed by the photodiode that is in the sensor, which leads to a decrease in the generated current. These variations are registered in the fNIRS waveforms, enabling the identification of heightened mental workload in the subject.

3.5 *Noise and Interference*

Physiological signals are often contaminated by noise and interference, both from the patient's body and external sources, complicating their analysis. Factors such as movement, physiological dysfunctions, electrode disconnection, unstable temperature, and electrostatic artifacts degrade signal quality during data acquisition [14]. These signals are challenging to interpret due to their non-stationary nature, low amplitude, and limited frequency range. To enhance signal interpretation and minimize interference, signal post-processing techniques should be applied. Table 4 outlines the main sources of interference for bioelectrical signals [12].

We used Python to analyze physiological signals, employing a range of signal processing techniques to enhance data quality and reduce noise. Specifically, we applied downsampling to reduce the sampling rate and computational load while preserving signal integrity. A bandpass filter was implemented to remove unwanted frequency components and isolate the relevant signal range. Additionally, we used

Table 4 Main types of interference and sources in physiological signals

Type of interference	Source	Causes	Frequency range
Narrowband signals	Power line interference	Electrical wiring	50 Hz
	Baseline wander	Shifts in electrode–skin interface and breathing	up to 1 Hz
	Slight movement artifacts	Movements of adjacent body parts	up to 30 Hz
Broadband signals	Myoelectrical potentials and motion artifacts	Body movements from eyes, limbs and muscles activity	20 Hz–10 kHz
	Electrosurgical noise	Other medical equipment	500 Hz–10 MHz
Impulse noise	Technical artifacts	Switching other electrical equipment	–
	Sharp changes of baseline	Loss of electrode–skin contact	–
	Stimulus artifact	Electrical stimuli in evoked activity measuring	–

smoothing techniques to reduce high-frequency noise and enhance signal clarity. Baseline correction was also performed to minimize drift and ensure accurate trend analysis. These preprocessing steps were essential to obtain reliable and interpretable signals for further analysis.

4 State of the Art

Current literature provides valuable insights into how different stimuli influence human performance, particularly in pilots. To assess these findings, the selected studies are organized into two tables, both derived from a scoping review conducted through electronic databases to ensure the inclusion of high-quality papers.

Table 5 summarizes physiological monitoring studies across diverse scenarios, with the following inclusion criteria: (i) only papers published between 2019 and 2024 were considered; (ii) studies should focus on physiological monitoring systems designed for continuous data collection; (iii) they must involve signals such as ECG, EDA, fNIRS; and (iv) the research should explore the integration of sensors either directly on the subject and/or within an interface across various areas of application. The reviewed studies are categorized based on the: physiological signals; measured parameters; study purpose; integration approach; and area applied.

Table 6 focuses specifically on aviation-related studies, with the following inclusion criteria: (i) only papers published between 2019 and 2024 were considered; (ii) primarily investigate physiological signals; (iii) propose direct practical applications for monitoring physiological signals; and (iv) aim to establish connections between physiological responses and mental states such as workload or stress in aviation. The reviewed studies are categorized based on the: measured parameters; instrumentation used; brief description of the central objectives or topics; and experimental context (Simulated Task or Real-Flight Task).

The structured synthesis presented in the tables not only showcases the diverse approaches within the field but also highlights key opportunities for advancing physiological monitoring methods in real-world applications. Collectively, the evidence gathered emphasizes the transformative potential of physiological monitoring systems and their ongoing evolution.

5 Methodology

This section outlines the methodology, organized into instruments and materials, the experimental setup used to demonstrate the operating principle of the proposed monitoring system, and the simulator environment. The primary objective of this study is to develop a physiological monitoring system for pilot data acquisition in an aviation context, by integrating wearable sensors into an Airbus A320 flight

Table 5 Summary of studies from different areas in the field of physiological monitoring

Authors	Physiological signals	Parameters	Purpose	Placed on subject	Integrated in interface	Application field
Silva et al. [21]	ECG	Heart activity—Heart Rate Variability (HRV)	Proof of concept of the new device designed: keyboard armrest and mouse ("invisible" approach)	Surface electrodes on the chest (Einthoven's Lead I positioning)	Dry electrodes with a conductive nappa on keyboard armrest and mouse	E-sports
Pereira et al. [22]	ECG	Heart activity—HRV	Proof of concept of the new device designed: smart chair system ("invisible" approach)	NA	Dry electrodes with a conductive nappa on chair armrest	Office work
Silva et al. [23]	ECG	Heart activity—HRV	Proof of concept of the new device designed: toilet seat ("invisible" approach)	NA	Polymeric electrodes of different textures on toilet seat	Sanitary facilities
Wagner et al. [24]	ECG	Heart activity—(HRV)	Comparison between BITalino toolkit and established medical-ECG BrainAmp-ExG	Surface electrodes (Einthoven's Lead II positioning)	NA	Medicine
Najafi et al. [15]	ECG, EDA, EEG	Heart, skin and brain activity	Recognize different states of attention	Electrodes placed on the chest, hands and scalp	NA	Automotive (driving)
Ferreira et al. [25]	EDA	Skin conductance	Proof of concept of the new device designed: wearable smart sock	NA	Textile conductive electrodes on wearable smart sock in the foot	Clothing
Xu et al. [20]	fNIRS	Oxygenation of hemoglobin, deoxygenated-hemoglobin	Examine workload in high-fidelity simulation training for anesthesiologists	Head worn sensor probe embedded in head	NA	Medicine

Table 6 Summary of studies of aviation in the field of physiological monitoring

Authors	Physiological signals	Parameters	Description	Context
Mohanavelu et al. [27]	ECG, EEG	Heart and brain activity—Heart Rate Variability (HRV) and EEG features	Using machine learning algorithms to identify the pilots' cognitive workload during takeoff, cruise, and landing phases	ST
Cao et al. [28]	Chest-strap for ECG	Heart activity—HRV	Examines how pilot stress (measured via HRV) impacts performance during flight simulations with varying CO_2 levels	ST
Mohanavelu et al. [29]	Smart vest with single-lead ECG	Heart activity—HRV	Analyzes fighter pilots' cognitive workload using HRV and NASA-TLX assessments, under varying visibilities and multitasking	ST
L. Kinney et al. [30]	Single-lead ECG, head-mounted tracker	Heart and eye activity—Heart Rate (HR) and eye tracking	Exploring behavioral responses to expected and unexpected flight events by analyzing heart rate and pupil dilation	ST
Alaimo et al. [31]	EcgMove 3 sensor	Heart activity—HRV	Evaluate workload through HRV by assessing physiological changes in five air maneuvers with different difficulty levels	ST

(continued)

Table 6 (continued)

Authors	Physiological signals	Parameters	Description	Context
Li et al. [32]	ErgoLAB Human Factors Physiological Recorder	Heart, respiration, and electrodermal activity. Temperature and strength	Improve human–machine interaction systems with the understanding of the pilots' status to prevent human error	ST
Masters et al. [33]	ECG, fNIRS, EDA, respiration rate, eye tracker	Heart, brain, electrodermal, respiration, and eye activity	Multimodal approach to predict the pilot's mental workload in a simulated helicopter environment	ST
Hormeño Holgado et al. [34]	Single-lead ECG, EEG, spirometer, thermometer, EMG	Heart, brain, and respiration activity. Temperature and strength	Biosignal analysis when a pilot performs offensive and defensive maneuvers	RF
Bustamante-Sánchez et al. [35]	ECG, EEG, spirometer, thermometer, EMG	Heart, brain, and respiration activity. Temperature and strength	Analyze how night flights produce stressful contexts for aircrews	RF
Hamann et al. [36]	EEG, fNIRS	Brain activity—frequency bands	Comparing the EEG and fNIRS for mental fatigue measurements	ST

simulation environment. This integration enables system testing under conditions that closely resemble real-world aviation scenarios.

To demonstrate the underlying principle of the proposed system, both electrical and optical sensors were used. This study not only explores the physiological signals accessible through these conventional methods but also highlights alternative sensing technologies, paving the way for potential implementations benefiting from the optical domain. The increasing adoption of optical sensors marks a significant advancement in physiological monitoring, valued for their non-invasive nature and ease of use. Unlike traditional electrical sensors, optical sensors cause minimal discomfort to participants, making them particularly suitable for long-term monitoring in out-of-the-lab applications.

The integration of both electrical and optical sensing technologies offers a more robust approach to physiological data acquisition, enabling a more comprehensive understanding of human health. ECG, EDA, and fNIRS were selected as physiological signals for data acquisition, given their significant role in assessing a participant's physiological state [37, 38].

5.1 Instruments and Materials

The materials used for ensuring accurate biosignal data acquisition include:

- Hardware devices: assembled BITalino (r)evolution and 4-channel biosignalsplux Hub;
- Software: OpenSignals;
- Sensors: ECG, EDA, fNIRS; for the electric sensors (i.e., ECG and EDA) 2-lead and 3-lead electrode cables were used, together with disposable pre-gelled electrodes;

The hardware devices, sensors, and software were developed by PLUX—Wireless Biosignals, S.A., Lisbon, Portugal [39]. The devices used are the BITalino (r)evolution device and the 4-channel biosignalsplux Hub device (Fig. 3). The sensors used are Electrocardiography (ECG), Electrodermal activity (EDA), and Functional Near-Infrared Spectroscopy (fNIRS) (Fig. 4).

BITalino is a modular hardware platform, available in two main configurations: the board version and the plugged version. The board version is an all-in-one, ready-to-use kit with all sensors blocks pre-connected with the BITalino core, while the plugged version separates the sensors from the main board, allowing users to connect any sensors they wish to use, even if they are from a third-party [40].

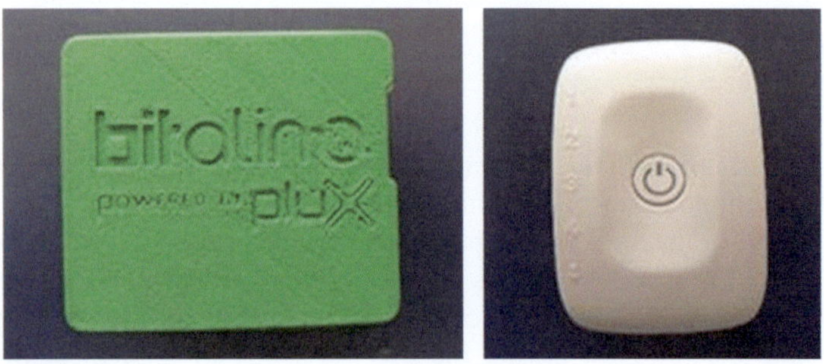

Fig. 3 Hardware devices: (left) BITalino (r)evolution; (right) 4-channel biosignalsplux Hub

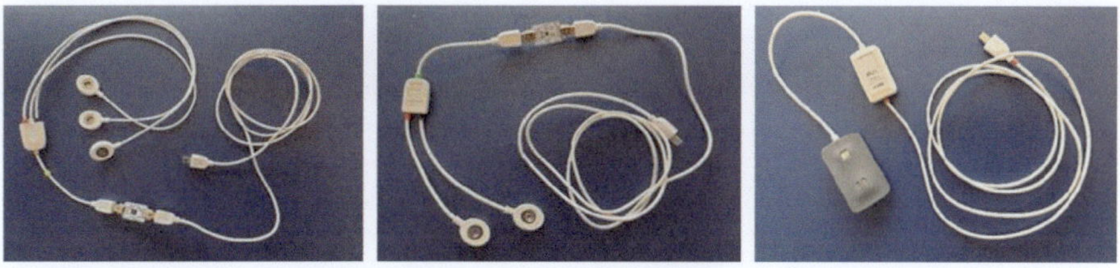

Fig. 4 Sensors: (left) ECG sensor with 3-lead electrode cable; (center) EDA sensor with 2-lead electrode cable; (right) fNIRS sensor showing the optical sensors on the monitoring probe (gray pad)

Fig. 5 Airbus A320 simulator general overview

5.2 Simulator

The flight simulator (Fig. 5) replicates the cockpit of the Airbus A320 family, and it is based on the X-Plane 11 software from Laminar Research [41] with the cockpit hardware form Skalarki [42]. It has flight panel controls and instrumentation, enabling the participants to fly the aircraft, providing a realistic and immersive environment for data acquisition.

5.3 Experimental Setup

The experimental environment, designed to simulate an Airbus A320 flight setting, used wearable sensors applied directly to the pilot's body for data collection. This setup enables continuous physiological monitoring, providing the potential for valuable real-world applications for aviation, by allowing the early detection of potential health impairments, facilitating timely interventions that can ultimately enhance passenger safety. Details regarding the sensor placement are outlined next (Fig. 6):

Fig. 6 Sensor placement on the body while using the simulator

- ECG—three pre-gelled electrodes are applied to the participant's body, following Einthoven's Lead I: IN- on the left clavicle; IN+ on the right clavicle; and the reference at the C6/C7 cervical vertebrae. Then, electrodes are connected to the 3-lead electrode cable of the ECG sensor. This cable is then connected to one of the analog inputs of the BITalino device.
- EDA—two pre-gelled electrodes are placed on the palm or the fingers of the participant's non-dominant hand and connected to the 2-lead electrode cable of the EDA sensor. This cable is then connected to one of the analog inputs of the BITalino device.
- fNIRS—the sensor is placed on the participant's forehead, targeting the prefrontal cortex, and it is fixed with an adjustable headband to remain in permanent contact with the skin. Then, the sensor cable is connected to one of the input ports of the biosignalsplux Hub device.

6 Experimental Results

This section presents the physiological data collected during the experiment. Once the sensors were correctly positioned and connected to the designated hardware devices, the participant's physiological signals were recorded using OpenSignals, a software designed for real-time data acquisition and visualization. OpenSignals connects via Bluetooth with PLUX hardware devices, including the BITalino (r)evolution and the 4-channel biosignalsplux Hub, both operating at a sampling rate of 1000 Hz. The acquired biosignals were exported as ASCII text files and stored on the computer, allowing subsequent data analysis. By recording multiple signals simultaneously and providing real-time data display, the system ensures reliable measurements throughout the experiment.

The following images present waveforms captured using OpenSignals. The signals were analyzed in Python, and the resulting graphs visually represent the data collected during the simulated flight. Each physiological signal is color-coded for clarity:

- ECG signal (red)—represents heart rate fluctuations.
- EDA signal (blue)—reflects skin conductance changes.
- fNIRS signal (orange)—indicates variations in oxygenated hemoglobin levels.

Each signal is depicted in two graphs; the first displays a small sample of the signal, while the second shows the full recording. These signals were captured in response to various stimuli experienced in the Airbus A320 simulator. The graphs highlight three key flight phases delimited by colors, namely: takeoff (in blue); autopilot (in grey); and landing (in green).

6.1 *Physiological Signals Collected*

The ECG signal (Fig. 7) exhibits quasi-periodic waveforms characteristic of cardiac cycles, aligning with the expected pattern. This signal can be used to investigate if elevated heart rates or irregular heart rhythms correlate with pilot stress, fatigue, or underlying health conditions. The complete ECG signal is not present as the obtained graph was distorted due to interference, resulting in an inaccurate representation of the ECG signal.

The EDA signal (Fig. 8) starts with a tonic baseline, followed by a phasic component characterized by oscillations in response to stimuli, reflecting sweat activity. The signal can be used to assess whether increased stress levels in pilots are linked to higher sweat activity. The complete EDA signal is in Fig. 9.

The fNIRS signals are divided into two channels: one corresponding to the electric current generated by the emitted red light (Fig. 10); and the other related to the emitted

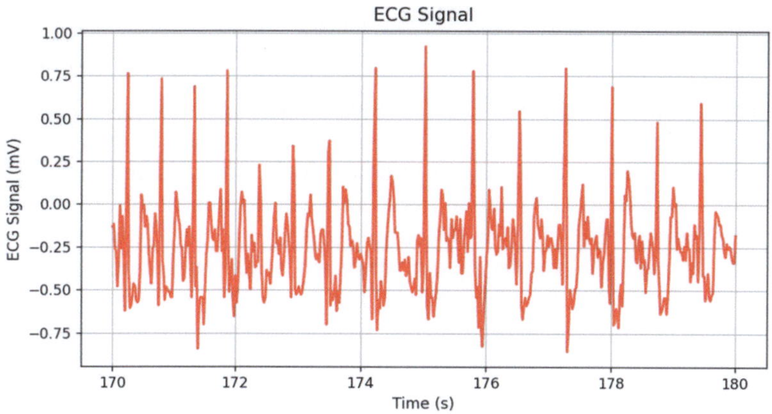

Fig. 7 ECG signal sample. The x axis refers to arbitrary time scale

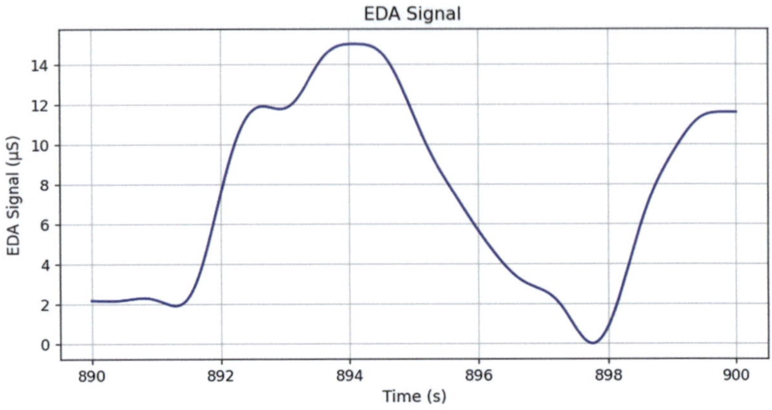

Fig. 8 EDA signal sample. The *x* axis refers to arbitrary time scale

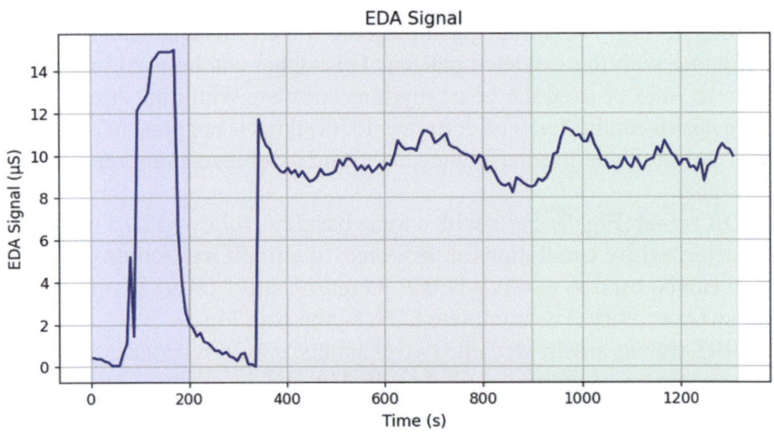

Fig. 9 EDA complete signal. The *x* axis refers to arbitrary time scale

infrared light (Fig. 12). The complete fNIRS signals are shown in Figs. 11 and 13. The fNIRS sensor produces a stable signal with minimal susceptibility to motion-related interference, due to its secure placement on the forehead, making it well-suited for the dynamic environment of aviation. As mental workload increases, blood flow and oxygenated hemoglobin (HbO) in the brain also increase. Since HbO absorbs more infrared light, less light is reflected and absorbed by the photodiode in the sensor, resulting in a decrease in signal amplitude. Therefore, fNIRS signals can be used to assess reductions in cognitive performance during high-workload phases, providing valuable insights into pilot mental states in challenging scenarios.

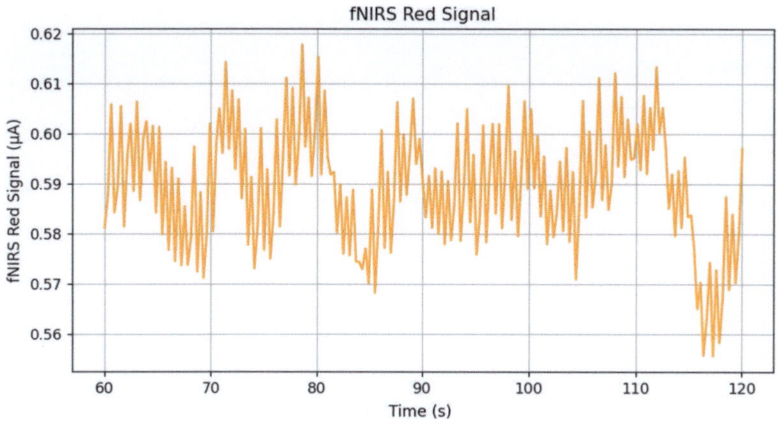

Fig. 10 fNIRS red signal sample. The x axis refers to arbitrary time scale

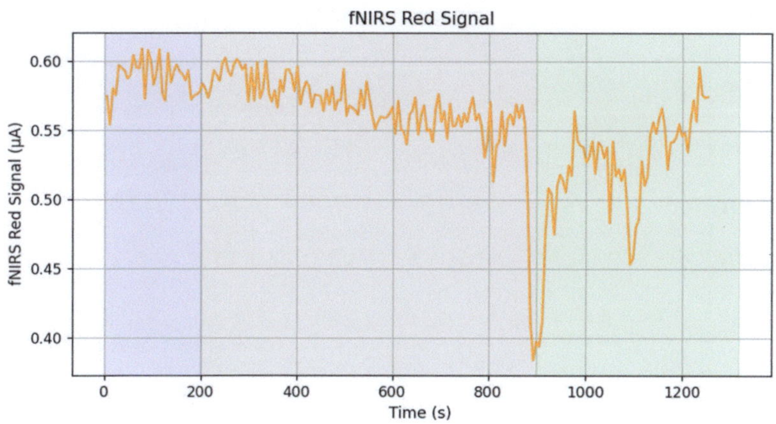

Fig. 11 fNIRS red complete signal. The x axis refers to arbitrary time scale

6.2 *Physiological Signals Analysis*

Signal preprocessing and analysis were performed with BioSPPy and MNE. BioSPPy is an open-source toolbox for BioSignal Processing in Python, and it was used to process ECG and EDA data, allowing an extraction of meaningful physiological characteristics from the data recorded. MNE is an open-source Python toolbox for neuroimaging data processing, and was used to process fNIRS data, providing insights into brain function from the recorded data [43].

The heart rate signal (Fig. 15) is expected to increase during high mental workload phases (takeoff and landing) and decrease during periods of lower mental workload (autopilot phase). The recorded heart rate graph reflects this anticipated physiological

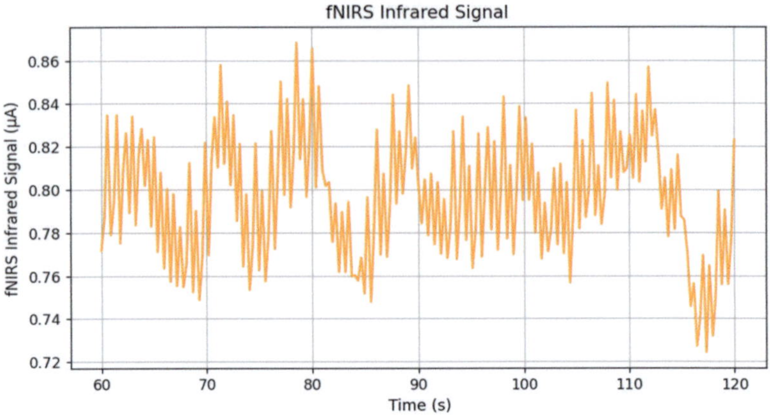

Fig. 12 fNIRS infrared signal sample. The *x* axis refers to arbitrary time scale

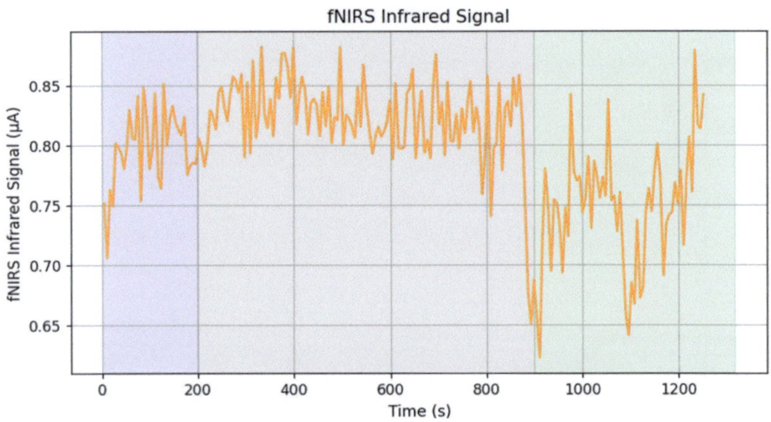

Fig. 13 fNIRS infrared complete signal. The *x* axis refers to arbitrary time scale

response, with elevated values observed during the takeoff and landing phases—generally associated with heightened cognitive and physiological stress—while lower stable values are seen during the autopilot phase.

High mental workload triggers an Electrodermal Response (EDR), characterized by increased sweat production, which enhances skin conductance and strengthens the EDA signal (Fig. 16). Since takeoff and landing demand greater cognitive effort than the autopilot phase, a more pronounced EDR is expected, which is confirmed by the recorded data in the graph.

The raw electrical current signals obtained with the fNIRS sensors cannot be directly interpreted as oxygen saturation, however, the acquired fNIRS data can be used to estimate oxygen saturation levels. Higher mental workload is linked to an increase in oxygenated hemoglobin, which absorbs more infrared light, reducing the

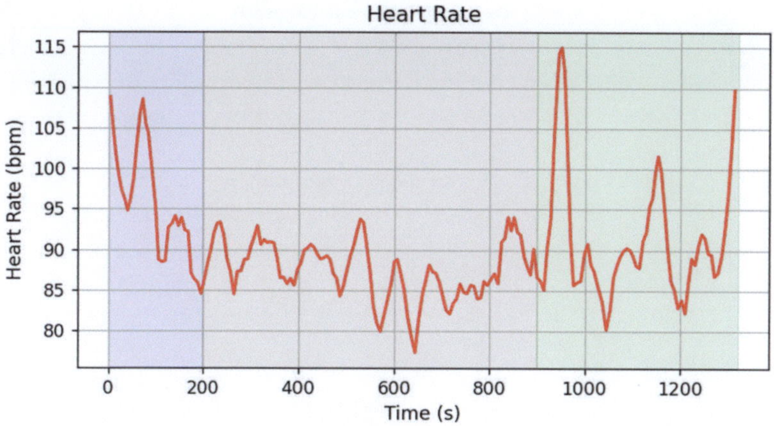

Fig. 15 Heart rate throughout the experiment. The *x* axis refers to arbitrary time scale

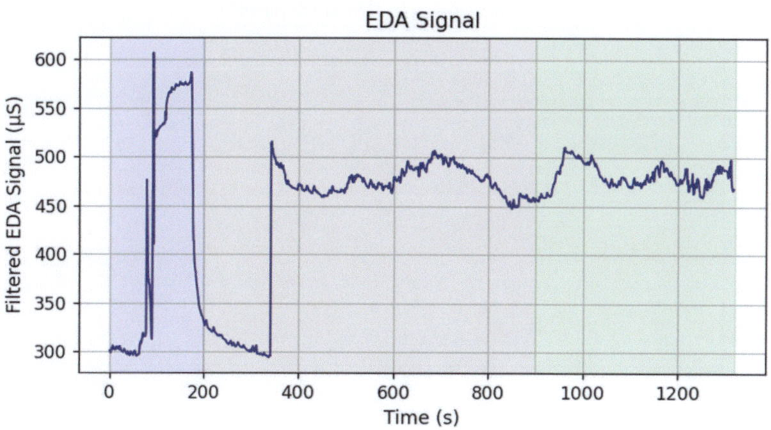

Fig. 16 Filtered EDA signal throughout the experiment. The *x* axis refers to arbitrary time scale

amount reflected and captured by the photodiode. This leads to a decrease in the generated current, with a shift from higher to lower current indicating a transition from lower to higher mental workload, as observed in the change from the autopilot phase to the landing phase. The concentration levels of oxygenated hemoglobin (Fig. 17) and deoxygenated hemoglobin (Fig. 18) were calculated from the measured currents [44].

The hemodynamic response function (Fig. 19) is calculated by taking the ratio of the red and infrared signals, followed by smoothing the result to capture the hemodynamic response over time.

To estimate blood oxygen saturation (Fig. 20) using functional near-infrared spectroscopy (fNIRS), it is necessary to process the raw signals obtained from red and infrared light absorption. Since fNIRS operates on similar principles to traditional

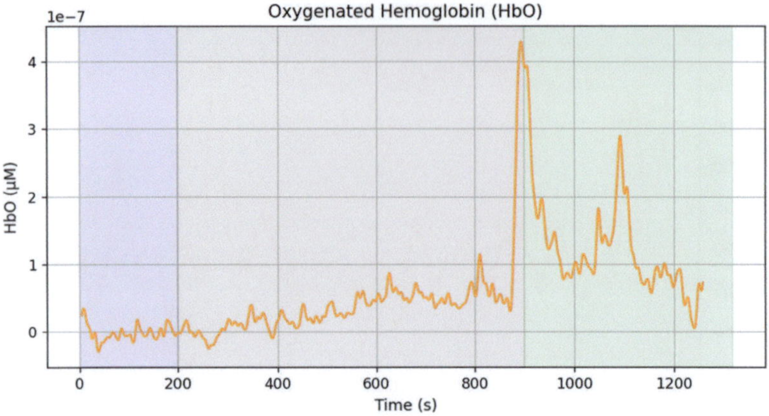

Fig. 17 Evolution of oxygenated hemoglobin during the complete simulated flight

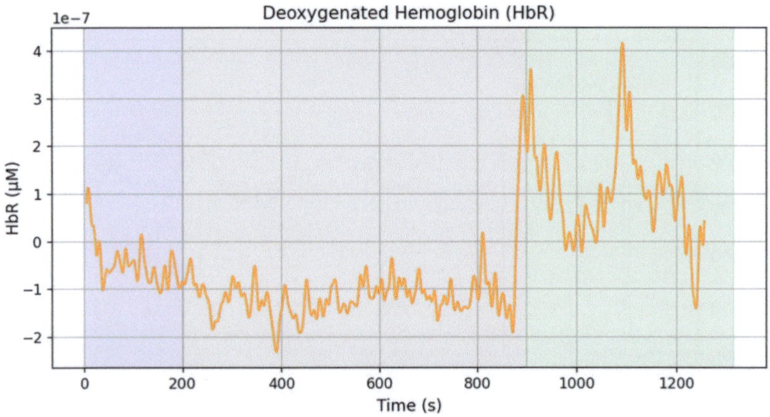

Fig. 18 Evolution of deoxygenated hemoglobin during the complete simulated flight

pulse oximetry, SpO_2 values can be derived by analyzing the ratio of modulated red and infrared signals [44].

The modulation ratio R is a key parameter in determining SpO_2 and is inversely proportional to blood oxygen saturation. It is calculated using the following equation [44]:

$$R[i] = \frac{V_{ppR}[i] \times V_{avgIR}[i]}{V_{avgR}[i] \times V_{ppIR}[i]} \tag{1}$$

- V_{ppR} e V_{ppIR} represent the peak-to-peak of the red and infrared signals, respectively.
- V_{avgR} e V_{avgIR} correspond to the average voltages of the red and infrared signals.

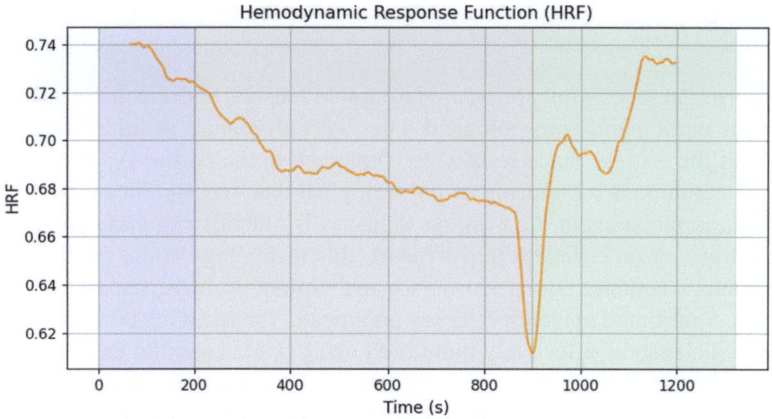

Fig. 19 Evolution of hemodynamic response function during the complete simulated flight

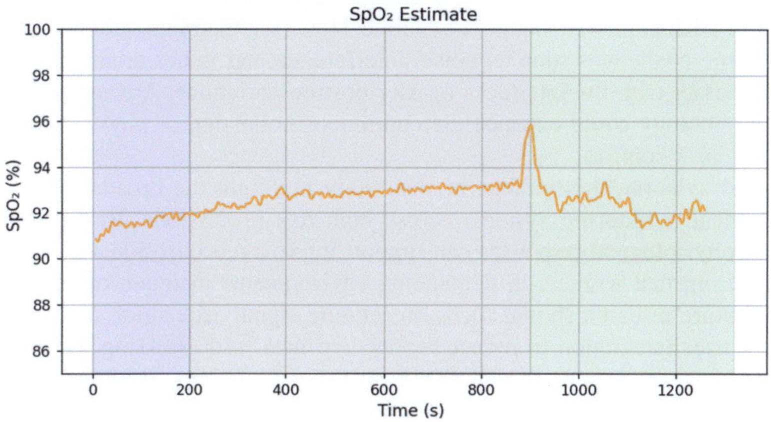

Fig. 20 Evolution of oxygen saturation during the complete simulated flight

Using a standard calibration function, SpO_2 can be estimated based on the modulation ratio R [44]:

$$SpO_2[i] = 110 - 25 \times R[i] \tag{2}$$

This equation allows for continuous monitoring of blood oxygen saturation, providing real-time insights into oxygenation levels in the brain or muscle tissue. This serves as the foundation for converting the raw fNIRS signal data into oxygen saturation values, essential for interpreting the collected data [44].

7 Conclusions

This experiment successfully integrated wearable sensors, both electrical and optical, to monitor participants' physiological data, with all signals acquired successfully throughout the testing. Since tests were conducted in the Airbus A320 flight simulator, this continuous monitoring of physiological data has significant potential for real-world applications in the aviation industry. By identifying early signs of pilot fatigue, stress, or other health impairments, this technology could facilitate timely interventions to enhance pilot performance and, ultimately, improve passenger safety. This work contributed to identify the key parameters for assessing pilot impairments, some of which can be effectively monitored using optical sensing techniques.

Although wearable sensors can be effective in continuous monitoring, devices can be bulky and intrusive, which may compromise the pilots' comfort and performance. To address this limitation, future research should explore the potential of 'invisible' sensors, i.e., sensors integrated into objects or embedded within the aviation environment. These sensors could provide seamless and unobtrusive monitoring with minimal disruption to the pilot's routine. However, their integration would need to overcome challenges such as noise, interference and issues arising from intermittent contact with the interfaces or user posture variations. Advancing invisible sensor technology could enhance the continuous collection of physiological data, improving user comfort.

Beyond aviation, the findings of this study highlight the broader potential of physiological monitoring systems across high-risk professions. Real-time monitoring of physiological responses can support improved occupational health strategies, ensuring that workers in demanding environments maintain optimal performance. Future research should focus on refining signal acquisition methods, optimizing sensor integration to reduce motion artifacts, and assessing the long-term viability of these systems in real-world operational settings. Additionally, incorporating machine learning models for automated anomaly detection could further enhance the effectiveness of physiological monitoring in aviation and other fields.

Acknowledgements This work was developed within the scope of the project Instituto de Telecomunicações (UID/50008) financed by national funds through the FCT/MCTES (PIDDAC). The support of the Department of Electrical and Computer Engineering (DEEC) and Técnico pedagogical council are gratefully acknowledged.

References

1. G. Luzzani, I. Buraioli, D. Demarchi, G. Guglieri, Preliminary study of pilot stress and mental workload monitoring through physiological signals. J. Phys. Conf. Ser. **2716**. https://doi.org/10.1088/1742-6596%2F2716%2F1%2F012075
2. Statista, Airline industry—passenger traffic worldwide 2024 (11 Oct 2024) [Online]. Available: https://www.statista.com/statistics/564717/airline-industry-passenger-traffic-globally/

3. I. Alreshidi, I. Moulitsas, K.W. Jenkins, Multimodal approach for pilot mental state detection based on EEG. Sensors **23** (2024). https://doi.org/10.3390/s23177350
4. J. Shively, *If Human Error is the Cause of Most Aviation Accidents, Then Shouldn't We Remove the Human*? (NASA, 2004)
5. J. Cahill, P. Cullen, S. Answer, K. Gaynor, S. Wilson, The requirements for new tools for use by pilots and the aviation industry to manage risks pertaining to work-related stress (WRS) and wellbeing, and the ensuing impact on performance and safety. Technologies **8** (2020). https://doi.org/10.3390/technologies8030040
6. D. Bustos, J. Guedes, J. Baptista, M. Vaz, J. Costa, R. Fernandes, Applicability of physiological monitoring systems within occupational groups: a systematic review. Sensors **21** (2021). https://doi.org/10.3390/s21217249
7. S. Junaid, A. Imam, A. Balogun, L. De Silva, Y. Surakat, G. Kumar, M. Abdulkarim, A. Shuaibu, A. Garba, Y. Sahalu, A. Mohammed, T. Mohammed, B. Abdulkadir, A. Abba, N. Kakumi, S. Mahamad, Recent advancements in emerging technologies for healthcare management systems: a survey. Healthcare **10** (2022). https://doi.org/10.3390/healthcare10101940
8. Fortune Business Insights, Wearable technology market (2024) [Online]. Available: https://www.fortunebusinessinsights.com/wearable-technology-market-106000
9. Statista, Smartwatches—Worldwide (Jun 2024) [Online]. Available: https://fr.statista.com/outlook/dmo/digital-health/digital-fitness-well-being/fitness-trackers/smartwatches/worldwide
10. G. Borghini, L. Astolfi, G. Vecchiato, D. Mattia, F. Babiloni, Measuring neurophysiological signals in aircraft pilots and car drivers for the assessment of mental workload, fatigue and drowsiness. Neurosci. Biobehav. Rev. **44** (2012). https://doi.org/10.1016/j.neubiorev.2012.10.003
11. M.S. Fernandes, *Photonic Platform for Bioelectric Signal Acquisition on Wearable Devices.* Ph.D. dissertation, Department of Biomedical Engineering, University of Minho, Braga (2012)
12. R. Martinek, M. Ladrova, M. Sidikova, R. Jaros, K. Behbehani, R. Kahankova, A. Kawala-Sterniuk, Advanced bioelectrical signal processing methods: past, present and future approach—Part I: cardiac signals. Sensors **21** (2021). https://doi.org/10.3390/s21155186
13. R. Rangayyan, Biomedical Signals and Sensors I (2015) [Online]. Accessible at: https://doi.org/10.1007/978-3-642-24843-6_1
14. P.J. Bota, C. Wang, A.L. Fred, H.P. Silva, A review, current challenges, and future possibilities on emotion recognition using physiological signals. IEEE Access **7** (2019). https://doi.org/10.1109/ACCESS.2019.2944001
15. I. Pinto, A. Fred, C. Rodrigues, H. Silva, *Electrophysiology of the Heart and the Electrocardiogram: Visual Depictions* (Institute of Telecommunications, 2020). Accessible at: https://www.researchgate.net/publication/346910888_Electrophysiology_of_the_Heart_and_the_Electrocardiogram_Visual_Depictions
16. B.K. Pradhan, B.C. Neelappu, J. Sivaraman, D. Kim, K. Pal, A review on the applications of time frequency methods in ECG analysis. J. Healthc. Eng. (2023). https://doi.org/10.1155/2023/3145483
17. A. Banganho, M. Santos, H.P. Silva, Electrodermal activity: fundamental principles, measurement and application. IEEE Potentials **41** (2022). https://doi.org/10.1109/MPOT.2020.2983381
18. H. P. da Silva, A. Fred and A. Lourenço, Electrodermal response propagation time as a potential psychophysiological marker, in *Proc. of the Annual Int'l Conf. of the IEEE Engineering in Medicine and Biology Society (EMBC)*, (2012), pp. 6756–6759. https://doi.org/10.1109/EMBC.2012.6347545
19. D.S. Bari, H. Yacoob, C. Tronstad, H. Kalvøy, Influence of relative humidity on electrodermal levels and responses. Skin Pharmacol. Physiol. **31** (2018). Accessible at: https://doi.org/10.1159/000492275
20. J. Xu, J. Slagle, A. Banerjee, B. Bracken, M. Weinger, Use of a portable functional near-infrared spectroscopy (fNIRS) system to examine team experience during crisis event management in clinical simulations. Front. Hum. Neurosci. **13** (2019). https://doi.org/10.3389/fnhum.2019.00085

21. A. Silva, M. Correia, H.P. Silva, Invisible ECG for high throughput screening in eSports. Sensors **21** (2021). https://doi.org/10.3390/s21227601
22. L. Pereira, H.P. Silva, A novel smart chair system for posture classification and invisible ECG monitoring. Sensors **23** (2023). https://doi.org/10.3390/s23020719
23. A.S. Silva, H. Almeida, H.P. Silva, A. Oliveira, Design and evaluation of a novel approach to invisible electrocardiography (ECG) in sanitary facilities using polymeric electrodes. Sci. Rep. **11** (2021). https://doi.org/10.1038/s41598-021-85697-2
24. G. Finkelstein, Mechanical neuroscience: Emil du Bois-Reymond's innovations in theory and practice. Front. Neurosci. **9** (2015). Accessible at: https://doi.org/10.3389/fnsys.2015.00133
25. A.F. Ferreira, H.P. Silva, H. Alves, N. Marques, A. Fred, Feasibility of electrodermal activity and photoplethysmography data acquisition at the foot using a sock form factor. Sensors **23** (2023). https://doi.org/10.3390/s23020620
26. R. Varandas, R. Lima, S.B. Badia, H. Silva, H. Gamboa, Automatic cognitive fatigue detection using wearable fNIRS and machine learning. Sensors **22** (2022). https://doi.org/10.3390/s22114010
27. K. Mohanavelu, P. Srinivasan, J. Arivudaiyanambi, Machine learning-based approach for identifying mental workload of pilots. Biomed. Signal Process. Control **75** (2022). Accessible at: https://doi.org/10.1016/j.bspc.2022.103623
28. X. Cao, P. MacNaughton, L. Cadet, Heart rate variability and performance of commercial airline pilots during flight simulations. Int. J. Environ. Res. Public Health **16** (2019). Accessible at: https://doi.org/10.3390/ijerph16020237
29. K. Mohanavelu, P. Srinivasan, D. Ravi, Cognitive workload analysis of fighter aircraft pilots in flight simulator environment. Defense Sci. J. **70** (2020). Accessible at: https://www.proquest.com/docview/2384828945?sourcetype=Scholarly%20Journals
30. L. Kinney, D. O'Hare, Responding to an unexpected in-flight event: physiological arousal, information processing, and performance. Hum. Fact. **62** (2019). Accessible at: https://doi.org/10.1177/0018720819854830
31. A. Alaimo, A. Esposito, *Human Heart-Related Indexes Behavior Assessment for Aerospace Human-Machine Systems* (IEEE, 2022). Accessible at: https://doi.org/10.1109/ACCESS.2022.3145043
32. Y. Li, K. Li, S. Wang, Pilot behavior recognition based on multi-modality fusion technology using physiological characteristics. Biosensors **12** (2022). Accessible at: https://doi.org/10.3390/bios12060404
33. M. Masters, A. Schulte, *Physiological Sensor Fusion for Real-Time Pilot Workload Prediction and Overload Prevention* (ARC, 2021). Accessible at: https://doi.org/10.2514/6.2022-2344
34. A. Hormeño-Holgado, V. Clemente-Suárez, Effect of different combat jet manoeuvres in the psychophysiological response of professional pilots. Physiol. Behav. **208** (2019). Accessible at: https://doi.org/10.1016/j.physbeh.2019.112559
35. A. Bustamante-Sánchez, V. Clemente-Suárez, Psychophysiological response in night and instrument helicopter flights. Ergonomics **63** (2020). Accessible at: https://doi.org/10.1080/00140139.2020.1718772
36. A. Hamann, N. Carstengerdes, Investigating mental workload-induced changes in cortical oxygenation and frontal theta activity during simulated flights. Sci. Rep. **12** (2022). Accessible at: https://doi.org/10.1038/s41598-022-10044-y
37. R.Y. Shah, Y.K. Agrawal, Introduction to fiber optics: sensors for biomedical applications. Indian J. Pharmaceut. Sci. (2011). Accessible at: https://doi.org/10.4103/0250-474X.89752
38. C.S.J. Leitão, P.F.d.C. Antunes, J.A. Bastos, J. Pinto, P.S.d.B. André, Plastic optical fiber sensor for noninvasive arterial pulse waveform monitoring. IEEE Sens. J. **15**(1), 14–18 (2015). https://doi.org/10.1109/JSEN.2014.2336594
39. PLUX, *What are Biosignals? Get Started Here!* (2022). Available: https://www.pluxbiosignals.com/blogs/informative/what-are-biosignals-get-startedhere?srsltid=AfmBOoof0kEwduVEo45EUP7WWA22ns6himngVOlmIb_6XXoJ8NfNO_5w
40. PLUX Biosignals, What is BITalino? (2022, May 6) [Online]. Available: https://www.pluxbiosignals.com/blogs/informative/what-is-bitalino

41. Xplane12 (2024). Available: https://www.x-plane.com
42. Skalarki.e (2024) Available: https://skalarki-electronics.com
43. P. Bota, R. Silva, C. Carreiras, A. Fred, H.P. Silva, BioSPPy: a python toolbox for physiological signal processing. SoftwareX **26** (2024). Available: https://doi.org/10.1016/j.softx.2024.101712
44. PLUX - Wireless Biosignals, S.A., *Functional Near-Infrared Spectroscopy (fNIRS) During Apnoea* (Technical Report TN 29082023, 2023)

Emerging Trends and Future Directions in Aerospace Optical Sensing

Than Singh Saini and Manish Kalra

Abstract Optical sensors have become indispensable in modern technology, offering rapid, accurate, and reliable data across a wide range of applications. Among these, Fiber Bragg Grating (FBG) sensors are particularly prominent for measuring temperature and strain, with extensive use in aerospace applications such as composite aircraft wings. The spectral characteristics of short and long FBGs enable the precise assessment of local stresses and crack detection on aircraft bodies. Additionally, the International Space Station (ISS) presents a unique platform for the advancement of optical materials, reducing gravity-induced defects and fostering the development of high-purity, flawless optical fibers for in-space manufacturing. These investigations pave the way for new specialty fibers and scalable commercial production in microgravity environments. Looking ahead, the continued miniaturization of optical sensors will catalyze their integration into wearable devices, Internet of Things (IoT) systems, and biomedical implants, expanding their impact across emerging technological frontiers.

1 Introduction

Sensor is the characteristic of any device or material to sense the existence of a specific physical quantity. The output of the sensor is the signal, which is converted to the human readable form. Sensor performs some function of input by sensing or feeling the physical changes in the characteristics of a system in response to stimuli. Therefore, sensors are supposed to be the artificial senses of the technological instruments just like the nose, ear, skin, toung and eyes in human body. In general, sensors

T. S. Saini (✉)
Department of Physics, National Institute of Technology Kurukshetra, Kurukshetra, Haryana, India
e-mail: tss@nitkkr.ac.in

M. Kalra
Department of Electronics and Communication Engineering, GLA University, Mathura, Uttar Pradesh, India

C. Marques et al. (eds.), *Advanced Optical Sensors for Aerospace Applications*, Progress in Optical Science and Photonics 37,
https://doi.org/10.1007/978-981-95-1626-1_13

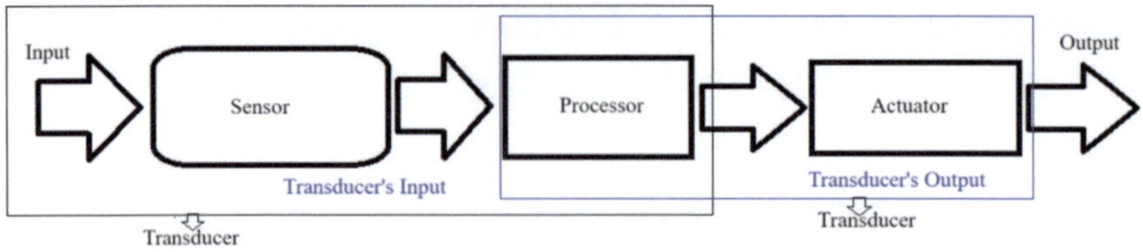

Fig. 1 A block diagram representing the common principle of sensing process

are the device that can sense and provide response to changes in the environment. Such changes could be in the form of moisture, motion, temperature, and light or any other physical parameters. The sensor converts such physical alterations into the signal that can be measured. The sensors provide the connection or bridge between the logical world and physical world. The sensors play a very important role in the technological development in most of the research area. There are various kind of sensors are available in the market. Based on the principle of operation some are mechanical, electrical, and optical sensors. Using the sensors one can estimate the temperature, pressure, revolution, motion, condition of instruments, etc. The block diagram shown in Fig. 1 represents the basic principle of operation of a sensor.

As shown in Fig. 1, initially a transducer converts the input signal from one physical form to another physical form. Transducer is also called an energy converter. For an example, a microphone converts sound to the electrical signal. It is based on the principle of conservation of energy. There are various classes of the sensors. Based on the power requirement sensors are classified into two categories: (a) active sensors and (b) passive sensors. Active sensors require an external excitation signal or power source to work. However, passive sensors do not require any external power source and it can directly generate the output response. Further, sensors can also be classified based on the means of the detection; i.e. according to the detection method they use such as optical, electrical, chemical, radioactive, or biological detection. Furthermore, sensors can be classified based on the conversion phenomenon: (a) Photo-electric: in which light is converts into electrical signal, (b) Thermo-optic: in which change in temperature converts into electrical signal, (c) Thermo-electric: in which temperature difference converts into electrical voltage, (d) Electrochemical: in which chemical reactions converts into electrical signal, and (e) Electromagnetic: in which magnetic fields converts into electrical signals.

In this chapter, some specific type of sensors that is optical sensors for aerospace applications would be discussed. An optical sensor transforms light energy into an electrical energy. The aim of an optical sensor is to measure the physical quantity of light and then converts it into a form that is readable by an integrated measuring device. Optical sensors can be employed for contact-less detection, counting or positioning of various parts. Moreover, optical sensors can be either internal or external. The external sensors collect and transfer a required quantity of the light, whereas internal sensors are used to measure the bends and other small changes in direction. Various kinds of optical sensors are illustrated in Fig. 2.

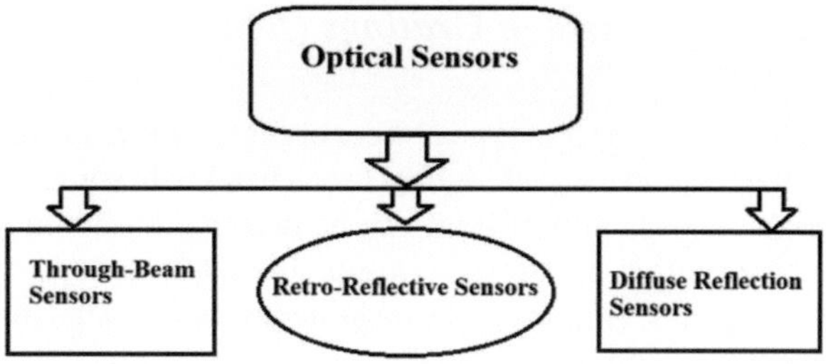

Fig. 2 Classification of optical sensors

For the application point of view of optical sensors, they are applicable ranges from computers to the motion detectors. For optical sensors to work efficiently, they must be the right type for the particular application, so that they retain their sensitivity to the parameter they need to measure. Optical sensors are integral parts of many common devices, including computers, copy machines (Xerox) and light fixtures that turn on automatically in the dark. And some of the common applications include alarm systems, synchros for photographic flashes and systems that can detect the presence of the objects.

The future engagements in the optical sensing for aerospace comprise advancements in miniaturization, light detection and ranging (LiDAR) technology, engine performance, development on fiber optic sensors for harsh environments and structural health monitoring, and AI integration. The benefits of optical sensors include electromagnetic interference immunity, lightweight and compact, versatility, high accuracy and sensitivity, integration (embedded sensor) and cost-effective. Optical sensors cannot have affected by the electromagnetic interference. Therefore, optical fiber sensors are ideal for usage in environments with high levels of electromagnetic radiations. The fiber optic sensors are significantly lighter weight and smaller in size, which make them ideal for aerospace applications where weight is a crucial factor. Optical sensors can be used to measure multiple parameters including temperature, pressure, strain, and vibration. Optical sensors can offer highly precise and profound measurements enabling precise monitoring of critical parameters. Furthermore, fiber optic sensors can be more cost effective than conventional sensors, particularly for the large-scale deployments.

2 Emerging Technologies and Research Directions

The key technologies based on optical sensing emerging for the applications in aerospace are briefly summarized as under.

2.1 Fiber Bragg Gratings (FBGs)

FBGs are a conjoint type of fiber optic sensor that is based on the principle of change in the reflective wavelength of broad band light source by stress or strain. The changes in the wavelength of light can be detected based on the any alteration in the pitch of the grating by stress or strain. To ensure safety of the aircraft it is necessary to monitor any fatigue cracks on structure of As metal aircrafts age, fatigue cracks will eventually appear in a number of components, which need to be monitored timely to ensure safe operation. There are a number of ways to observer their structural health of the aircraft by non-distractive analysis using x-rays or ultrasonic directly on the aircraft or in dedicated laboratories.

FBG can be made along the single fiber at multiple locations, sanctioning for the multi-point measurements. FBF based sensors can monitor the integrity of aircraft structures and engines, ensuring safe and reliable operation. The FBG fabrication setup is illustrated in Fig. 3.

The indispensable features of FBG based sensors as a tool for internal strain measurements have discussed in Refs. [1–3]. The methods to measure extremely localized intense stress fields, generated at the initialization point of a crack, or crack-tip, using FBG inscribed in highly photosensitive hydrogenated germanium and boron co-doped fiber have been reported earlier [4, 5].

The writings of gratings in the core region of an optical fiber require mainly interferometric methods on different photosensitive optical fibers, which typically have higher germanium concentration in the core region [6, 7]. The photosensitivity of the optical fiber can be determined by the alteration in the refractive index. The photosensitivity can be calculated by using the Bragg condition given below:

$$\Delta n = \frac{\lambda_1 - \lambda_2}{2\Lambda} \tag{1}$$

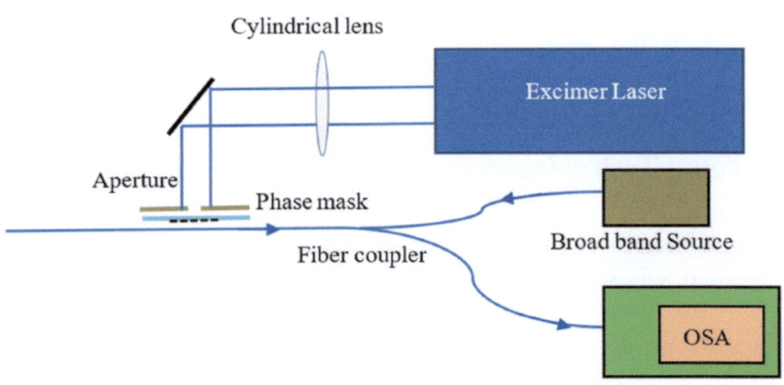

Fig. 3 Experimental setup for inscription of Fiber Bragg Gratings (adapted from [5])

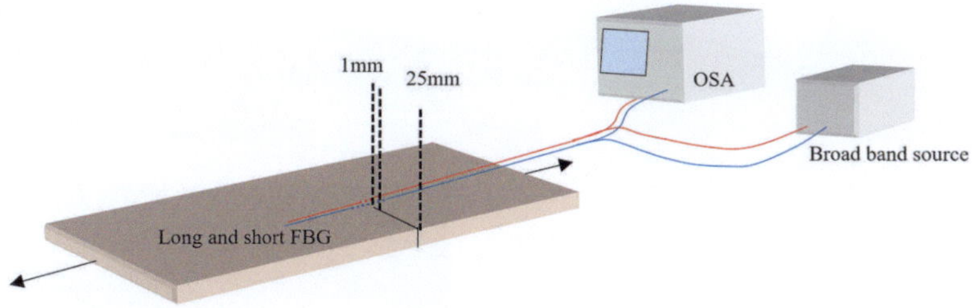

Fig. 4 Aluminum coupon with short as well as long FBGs (adapted from [5])

$$\Delta\lambda = 2\Lambda\Delta n \quad (2)$$

where, Δn is the change in the refractive index; λ_1 and λ_2 are the initial and final inscription wavelengths; $\Delta\lambda$ indicates the shift in the Bragg wavelength; and Λ is the grating period. If there is any change in the grating period due to any crack or abnormality in the structure, there is a shift in the Bragg wavelength. Based on this principle any crack or abnormality in the aircraft structure can be easily detected.

Based on the principle of FBG, a stress sensor can be developed using the grating inscripted in the core of the optical fiber. The experimental setup for the measurement of stress in Aluminum Coupon (or patch) is depicted in Fig. 4. The stress near a slot cut at the centre of an aluminum coupon, which signified a crack, can be measured by bonding the high-reflectivity FBGs close to the crack-tip. The FBGs can be attached to the coupon with two-part epoxy resin, which provides strong bonding. One short and one long FBGs are accurately positioned so the crack-tip lie in the midpoint of the gratings. A short and a long FBG, with respective wavelengths of 1529 and 1548 nm and grating period of 2 and 25 mm, have been glued perpendicular to the crack at a distance of 1 mm from the crack-tip and within 1 mm of each other. The estimated initial spectral linewidth of the short and long FBGs are 0.31 and 0.192 nm with peak relativities of 70 and 97%, respectively.

Using such arrangement, the short FBG supposed to measure only the point stresses at the crack-tip, whereas the long FBG measured both the average stresses of the coupon under tension and the stresses near the crack-tip.

In an another step, a composite patch comprised of a six-layer 0 to 90^0 boron prepreg matrix, covering approximately 65% of the total area of the aluminum coupon have attached as illustrated in Fig. 5. To examine the stress characteristics of the aluminium coupon with the bonded patch under load, FBGs were surrounded in between the third and fourth plies. Specifically, short (2 mm) and long (25 mm) FBGs were embedded centrally over the crack, 2.5 and 3 mm from the edge of the patch, respectively, indicated by *'a'* and *'b'* in the Fig. 5.

Likewise, as shown as *c* in Fig. 5, manifold fiber array grating containing an array of five short FBGs (each 1.2 mm long) with the separation of 2.5 mm from each other, can also implanted in the composite, about 3.5 mm from the edge of the patch. The protocol for the fabrication of such short multiple FBGs is provided in [8]. Furthermore, the choice of the FBGs locations can be estimated by the stress analysis

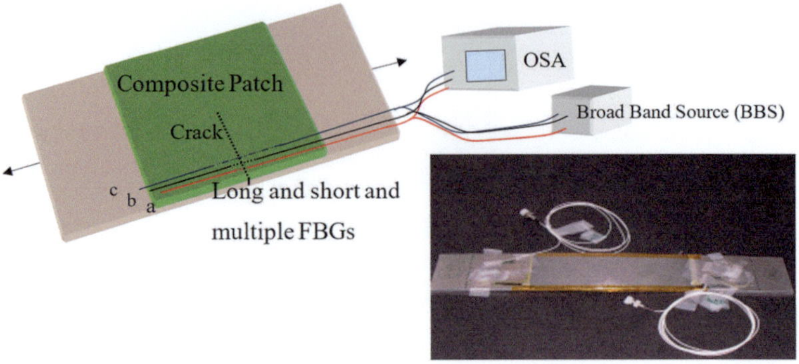

Fig. 5 Aluminum coupon with a 6-ply composite patch and short (**a**), long (**b**), and multiple (**c**) FBGs embedded in between ply 3 and 4 over the crack (adapted from [5])

of the coupon above the crack region. Therefore, gratings have been positioned over the crack with FBG (a, b, and c in Fig. 5) located above the crack as shown in Fig. 5. Such arrangement of gratings is subjected to the comparable stress distributions in the composite, measuring the variations in the peak wavelengths and spectral broadening as a function of applied stress on the coupon assembly.

To estimate the induced stresses at the crack-tip, the aluminum patch with the bonded short (grating pitch: 2 mm) and the long (grating pitch: 25 mm) FBGs has been exposed to forces and their spectra have been measured. The central wavelengths were 1530.1 and 1548.5 nm (at 20 C), respectively. The patch was placed in a computer controlled hydraulic pulling-jig, capable to applied tension between 0 and 24 kN.

The spectral variations for the short period grating (grating pitch: 2 mm) is illustrated in Fig. 6a. It indicates the reflection peak gently moving towards the longer wavelengths with increasing force. While, the spectral variations for the long period grating (grating pitch: 25 mm) has been shown in Fig. 6b. The interference fringes have been observed in between two peaks.

2.2 Optical Interconnect Solutions

For tough aerospace situations, optical interconnect solutions including optical fiber cables and connectors provide reliable, high-speed, immune to electromagnetic interference, lightweight and space saving, and longer distance data transmission. Such characteristics are very crucial for lightweight, high-performance aircraft and systems [9]. It is expected that the ruggedized optical connectors and the interconnect solutions are needed for unwavering integration of the optical sensors in harsh aerospace environments. Optical interconnects have been designed to withstand high temperatures, vibrations, and other harsh circumstances common in aerospace applications.

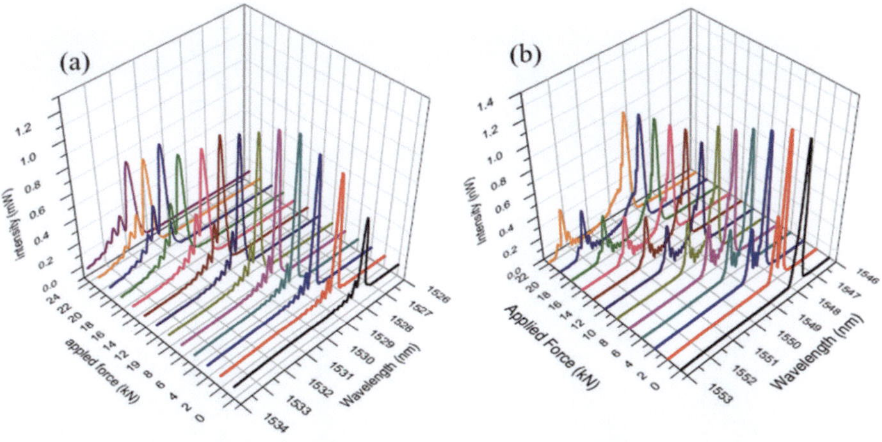

Fig. 6 **a** The spectral response using a short FBG (grating pitch: 2 mm) with applied force near the crack-tip without composite patch; **b** The spectral response using a long FBG (grating pitch: 25 mm) with applied force near the crack-tip without composite patch (adapted from [5])

2.3 Autonomous Vehicles and Unmanned Aerial Vehicles (UAVs)

Optical sensors are essential for navigation, collision avoidance, and autonomous operation of self-driving cars and unmanned aerial vehicles. Optical sensors, LiDAR and cameras are crucial for autonomous vehicles and UAVs, enabling navigation, hurdle detection, and ecological mapping through their capability to capture 3D spatial information and process visual data [10]. Weather conditions have been an obstruction to the deployment of automated driving systems and it is crucial to first acknowledge their impacts on the sensors. Figure 7. Illustrates the connections amongst weather conditions, adverse weather models, and awareness and sensing modules.

Furthermore, UAV remote sensing has steadily attracted the attention of scientific researchers and industry, due to its extensive application scenarios [11]. It has been widely used in forestry, agriculture, mining, and other activities. UAVs can be flexibly equipped with various sensors, such as optical, infrared, and LiDAR, and become a crucial remote sensing surveillance platform. Based on UAV remote sensing, researchers can obtain many high-resolution images, with each pixel being a cm or mm. Beside the high cost and lack of supply, remote sensing satellite data acquisition is also reserved by several influences that affect the observation competence and direction of research. For example, (a) the capability of detection of a remote sensing satellite is depends on its cameras. A satellite can have only one or two cameras as the sensors. Also, the cameras cannot be substituted after launching the satellite. Hence, the detection performance of the satellite cannot be enhanced in its lifetime. (b) The resolution of remote sensing imagery data is restricted by the distance between

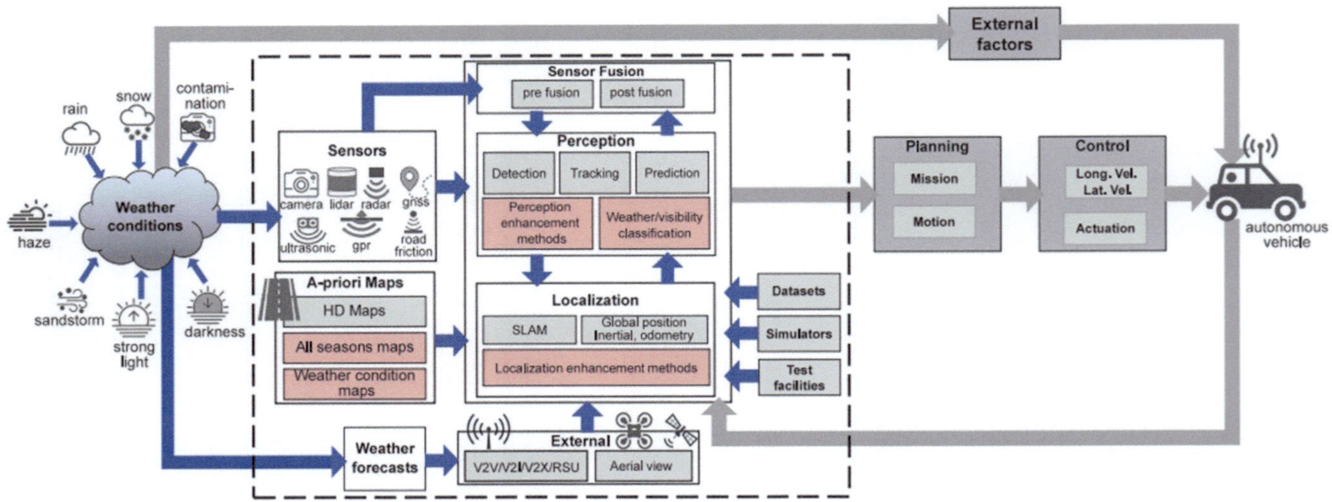

Fig. 7 An architecture for self-driving vehicles agnostic to adverse weather conditions. Red blocks denote weather-related modules. Blue arrows denote the relationships among weather and perception and sensing modules. Gray arrows denote the relationships among ADS modules including external weather factors such as wind and wet road surfaces (adapted from [10])

the target and the satellite. Since remote sensing satellites are far from the ground targets, their resolution is moderately low. (c) The remote sensing satellites can only detect objectives when winged over the face-to-face area above the target and along the satellite's orbit, which confines the capability to perceive objectives from the specific angle. (d) The meteorological conditions including cloud cover can also affect the surveillance results for optical remote sensing satellites, which confines the selection of remote sensing images for research purposes; Such constraints not only restrict the scope of the remote sensing research but also affect research directions. (e) Optical remote sensing satellites run-through visible and infrared light reflected by observation targets as a medium, such as colored, panchromatic, multi-spectral, and hyper-spectral remote sensing satellites. For such satellites, the target illumination surroundings utterly disturb the surveillance quality. The actual remote sensing imagery data can be acquired when the satellite is flying over the observation target and when the target has good lighting situations.

The hardware for UAV remote sensing platform contains two main parts including the flight platform of the drone, and secondly the sensors they are equipped with. One of the most noteworthy benefits of UAV remote sensing is the flexibility in standby of the sensors, which allows investigators to use the same drone to study the properties and characteristics of different objects by using different types of sensors. The drone's flight platforms and the different types of sensors are listed in Fig. 8. However, four distinctive UAV-carried sensors are illustrated in Fig. 9.

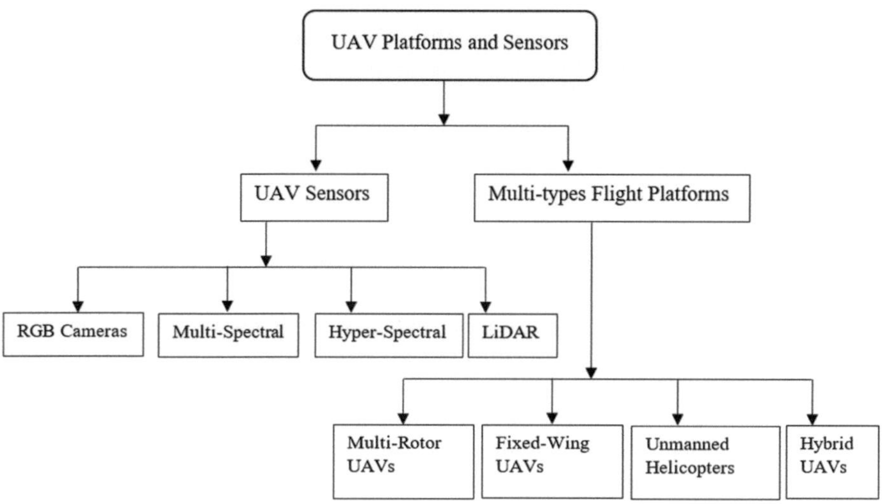

Fig. 8 The UAV platforms and sensors

(a) (b) (c) (d)

Fig. 9 Sensors carried by UAVs: **a** RGB Camera, **b** Multi-spectral Camera, **c** Hyper-spectral Camera, **d** LIDAR (adapted from [11])

2.4 *Space Missions*

Optical sensors can play a crucial role in space missions, enabling remote sensing, navigation, and communication. Optical fibers are applicable on earth and in space for uses in telecommunications, defense, medicine, and cybersecurity. Earlier reported results showed that optical fibers produced in microgravity can be higher quality than those made in normal gravity. The International Space Station (ISS) can be a potential platform for commercial production of high purity optical fibers. The production of flawless space optical fiber investigation is using the space station to demonstrate new manufacturing technology developed by flawless photonics to improve the quality and length of optical fiber produced in space [12, 13].

For several important applications high quality ZBLAN, a glass alloy made of zirconium, barium, lanthanum, sodium, and aluminum fluorides, material is needed to draw the optical fiber. The unique characteristic of ZBLAN is to allow light to travel through a fiber over a broader bandwidth, given that more than ten times the capability of conventional silica-based optical fibers and transmitting considerably more data over the same length of the optical fiber. It is desirable to made long enough and of with high quality, the increased efficiency of ZBLAN can translate into noteworthy energy savings by reducing the need to boost the signal on long-distance transmissions. However, when ZBLAN is drawn into fibers on the ground, crystals form that scatter signals and make the optical fiber brittle. Because crystals grow more slowly in microgravity, the approach is to cool drawn fibers before crystals have a chance to completely form. Microgravity also counters effects of sedimentation, convection, and buoyancy that limit the length and quality of optical fibers drawn on the earth. Optical fiber manufacturers use drop towers to manufacture ZBLAN on the earth, but in-space manufacturing provides much more time to draw longer and eventually better optical fibers [14, 15].

The key technologies and deliberations for the space mission includes (i) high-precision sensors to ensuring precise and trustworthy data collection in the harsh space environment; (ii) adaptive optics for amending for the atmospheric alterations to improve the image quality; (iii) compact and lightweight designs are required since minimizing the size and weight of sensors for easier deployment and operation in space; (iv) radiation hardening for protecting sensors from the harmful effects of space radiation; (v) fiber optics is essential for utilizing fiber optic sensors for structural health monitoring and other applications. Some of the breakdown of optical sensors in space missions is given below:

(a) Panchromatic Sensors:

A panchromatic sensor in space mission captures the broadband light, typically spanning a extensive range of the electromagnetic spectrum, such as the entire visible spectrum or a portion of the visible and near-infrared capture images, resulting in high resolution, black and white imagery, often used for detailed mapping and analysis [16].

(b) Multispectral Sensors:

Multispectral sensors in space missions capture data and images across multiple bands of electromagnetic spectrum, allowing for analysis of different wavelengths and features. Such optical sensors enable detailed analysis of the surface of the earth, atmosphere, and oceans, used for applications like vegetation monitoring, land cover mapping, and disaster response [17].

(c) Imaging Spectroscopy Sensors:

Imaging spectroscopy (or hyperspectral remote sensing) is based on the characterization of earth surface materials and processes through spectrally-resolved measurements of light interacting with matter. It captures images with a high spectral resolution, enabling detailed analysis of materials and their properties [18].

(d) Fourier Transform Spectroscopy Sensors:

Fourier transform spectrometers usage interferometry to measure the light spectrum, providing comprehensive evidence about the composition of the observed space objects. Ground-based and airborne Fourier transform interferometers explore a diversity of topics in various fields of physics, geophysics and astrophysics, including solar processes, radiative energy transport in the atmosphere, the geographic distribution of greenhouse gases and air pollutants and spectroscopic studies using the atmosphere as a natural laboratory [19].

(e) LiDAR and Active Optical Sensors:

LiDAR emit light pulses and measure the reflected light to create 3D maps and measure distances. At present, we anticipate an excellent stage for LiDAR in space technology. Current LiDAR systems have extended a high degree of maturity and superiority that progressively allows for their fruitful implementation in the space missions [20].

(f) Spectrometers and Radiometers:

Spectrometers and radiometers measure the intensity and the spectral distribution of the light, providing evidence about the composition and temperature of objects. Furthermore, spectrometers and radiometers are critical instruments in the space missions, allowing technologists to understand the temperature, composition, and various other characteristics of the celestial bodies and their atmospheres by analysing the light spectrum and other electromagnetic radiation they emit or reflect [21].

3 Role of Artificial Intelligence (AI) and Machine Learning (ML)

AI and ML can be used to simulate and analyse data from the optical sensors, providing discernments into performance of aircraft structures, engines, enabling predictive maintenance and reducing potential failures, if any. Further, AI and ML algorithms can be gradually integrated into the LiDAR systems to permit real-time data analysis, path planning, and object identification. AI is expected to enable optical sensors to analyse data in real-time, leading to the faster decision-making and enhanced situational awareness.

AI and ML technologies can ensure the most efficient bandwidth use, optimise network performance, and predict maintenance needs. Furthermore, AI-driven analytics can improve safety by offering real-time insights into the potential hazards, weather patterns, and communication system [9].

4 Development of New Sensor Technologies

Research and development are ongoing to develop new optical sensor technologies with improved performance and capabilities. Progressive developments in the nanotechnology and micro and nano fabrication techniques will lead to the design and development of smaller in size, more adaptable optical sensors which can be suitable for deployment in a broader series of aerospace applications. Optical fiber sensors can be very useful to monitor the vibration, stress, strain, and temperature of the aircraft structures and engines, enabling early detection of damage and ensuring structural integrity. Further, for the aircraft safety and security purposes, the gas and chemical sensing in the cabin of aircraft and cargo bay can be monitored by employing optical fiber based sensors. Distributed fiber optic sensor use the full length of the optical fiber as a sensor and trust on the evolution of its scattering behaviour, which can be measured or monitored through the optical time-domain reflectometry. Such fiber optic sensors are suitable for the long distance and slow changing parameters. Single-point Fabry–Perot, interferometry-based sensors follow the evolution of environmental parameters at hard to access locations with high sensitivity monitoring.

A set of mirrors separated by a cavity detect any micro displacement induced by the environment as it changes the interferometry pattern. Furthermore, multi-point Fiber Bragg Grating sensors are the most widely deployed technology. Intra-core fiber gratings can be made along the same fiber in several locations. Any change in the environment will impact the fiber, modifying the grating characteristics and the light propagation. Optical sensing technology uses multi-mode and single-mode fibers. Angled-polished interfaces may be required for more reliable measurements. Optical sensors typically measure strain, temperature and pressure. As they gain traction in the aerospace market, fiber optic sensor and the corresponding interconnect solutions must be adapted to the specific requirements of the aircraft environment.

5 Conclusion

The optical sensors play a very significant role in modern technology. It provides fast, accurate and reliable data in the various potential applications. From detecting environmental changes to ensuring the safety and efficiency of aerospace technological instruments and spacecraft. FBG sensors have been broadly useful for measurements of temperature and strain in various applications including aerospace in composite wings, etc. The local stresses near the cracks occurred on the body of aircraft can be assessed and measured from the spectral characteristics of short and long FBGs. ISS can provide reduction in gravity-induced defects into optical glass products developed on the earth and advance in-space manufacturing models. The investigation also opens the door to creating other valuable specialty fibers in space. Therefore, ISS can be a potential platform for commercial production of high purity optical fibers. The production of flawless space optical fiber investigation is using the space station to demonstrate new manufacturing technology developed by flawless photonics to increase the quality and length of optical fiber produced in space. Furthermore, the miniaturization will open up new avenues for optical sensors in wearable devices, IoT applications, and biomedical implants.

References

1. J.A. Guemes, J.M. Menendez, Response of FBG-optic sensors when embraced in composites laminates. Compos. Sci. Technol. **62**, 959–966 (2002)
2. J. Botsis, L. Humbert, F. Colpo, P. Giaccari, Embedded fiber Bragg grating sensor for internal strain measurements in polymeric materials. Opt. Lasers Eng. **43**, 491–510 (2005)
3. M. Studera, K. Petersb, J. Botsis, Method for determination of crack bridging parameters using long optical fiber Bragg grating sensors. Compos. Part B Eng. **2003**(34), 347–359 (2003)
4. Y. Okabe, R. Tsuji, N. Takeda, Application of chirped fiber Bragg grating sensors for identification of crack locations in composites. Compos. Part A **35**, 59–65 (2004)
5. A.A. Ikiades, Strain measurement on cracks using fiber bragg gratings for use in aircraft composite skin repairs. Aerospace **7**, 138 (2020)
6. E. Udd, An overview of fiber-optic sensors. Rev. Sci. Instrum. **66**, 4015–5030 (1995)

7. I. McKenzie, R. Jones, I.H. Marshall, S. Galea, Optical fiber sensors for health monitoring of bonded repair systems. Compos. Struct. **50**, 405–416 (2000)
8. M. Konstantaki, G. Tamiolakis, A. Argyris, A. Othonos, A. Ikiades, Efects of Ge concentration, B co-doping and hydrogenation on fiber Bragg grating characteristics. Microw. Opt. Technol. Lett. **44**, 148–152 (2005)
9. https://www.gemcable.co.uk/how-fiber-optics-are-changing-the-aerospace-industry
10. Y. Zhang, A. Carballo, H. Yang, K. Takeda, Perception and sensing for autonomous vehicles under adverse weather conditions: a survey'. ISPRS J. Photogramm. Remote Sens. **196**, 146–177 (2023)
11. Z. Zhang, L. Zhu, A review on unmanned aerial vehicle remote sensing: platforms, sensors, data processing methods, and applications. Drones **7**(6), 398 (2023). https://doi.org/10.3390/drones7060398
12. https://www.nasa.gov/humans-in-space/optical-fiber-production/
13. D. Starodubov, K. McCormick, M. Dellosa, E. Erdelyi, L. Volfson, Facility for orbital material processing, in *Sensors and Systems for Space Applications XI, Orlando, Florida* (2018), vol. 2, p. 10641106410T. https://doi.org/10.1117/12.2305830
14. https://issnationallab.org/upward/exotic-glass-fibers-from-space-the-race-to-manufacture-zblan/
15. M.N. Ott, Space flight applications of optical fiber; 30 years of space flight success, in *2010 Avionics, Fiber-Optics and Photonics Technology Conference, Denver, CO, USA* (2010), pp. 3–4. https://doi.org/10.1109/AVFOP.2010.5637515
16. S. Nagarathinam, A. Vasuki, K. Paramasivam, PAN and MS satellite image fusion using fast guided filter and PSO segmentation. Sādhanā **48**, 199 (2023). https://doi.org/10.1007/s12046-023-02262-5
17. https://airbornescience.nasa.gov/instrument/Multispectral_Atmospheric_Mapping_Sensor
18. L. Guanter, H. Kaufmann, K. Segl, S. Foerster, C. Rogass, S. Chabrillat, T. Kuester, A. Hollstein, G. Rossner, C. Chlebek et al., The EnMAP spaceborne imaging spectroscopy mission for earth observation. Remote Sens. **7**, 8830–8857 (2015). https://doi.org/10.3390/rs70708830
19. https://www.optica.org/events/meeting_archives/2024/fourier_transform_spectroscopy_(fts)
20. A. Fix, G. Naletto, I. Hutchinson et al., Space lidar and space optics. CEAS Space J. **11**, 359–362 (2019). https://doi.org/10.1007/s12567-019-00288-2
21. https://science.nasa.gov/mission/voyager/instruments/

Optical Metamaterials and Artificial Intelligence in Aerospace Sensing

V. B. Ankith, Farhan Khan, T. R. Harsha, Likewin Thomas, K. Prabhanjana, and Rajath R. Pai

Abstract The aerospace industry is expanding at a rapid tempo. The necessity of this investigation is underscored by the growing demand for precision and efficiency in sophisticated detecting technologies. The critical importance of optical metamaterials (OMM) is the subject of this article. The Aerospace systems are also being altered with the addition of artificial intelligence (AI) capabilities, thermal imaging, and infrared (IR) sensors. It gives as a solution to space challenges, which is vital in avionics, environmental monitoring, and structure health evaluation. Thermal imaging and IR sensors combined with metamaterials provide higher sensitivity and accuracy. This is important for night-time operations, distant detection and low visibility circumstances which widens the spectrum of aerospace applications AI boosts the capability of the sensor systems to provide real-time data processing. Predictive maintenance and adaptive control systems play a key role to increase safety and efficiency. AI methods leverage the machine learning algorithm and computer vision to interpret enormous data sets generated by thermal imaging and IR sensors. In comparison, with standard methods of anomaly detection and enables unparalleled levels of precision for AI, metamaterials, and specialized sensors. This convergence permits complicated applications like as autonomous navigation, atmospheric analysis, and with high-resolution 3D positioning, the UAV and satellites to respond to environmental changes in real-time. Apart from the above-discussed elements, this paper also addresses the factors that compound the efficiency of these technologies, including the fabrication of metamaterials and the available computational power for implementing AI as well as the performance of the sensors in very extreme aerospace conditions. It also presents solutions, among them hybrid sensor networks and adaptive AI, that can be deployed to overcome such limitations. Real-time scenarios, largely focused on sustaining lower radar cross-sections, and enhanced infrared imaging for use in the air space are evaluated to demonstrate the improvements from the methodologies under inspection. To sum up, the combination of OMM, thermal cameras, and infrared sensors with that of artificial intelligence gives a framework for putting in place smart, lightweight, and durable aircraft systems. This fusion of

V. B. Ankith (✉) · F. Khan · T. R. Harsha · L. Thomas · K. Prabhanjana · R. R. Pai
Department of AIML, PESITM Shivamogga, Virupina Koppa, Karnataka, India
e-mail: ankith.badiger51@gmail.com

C. Marques et al. (eds.), *Advanced Optical Sensors for Aerospace Applications*, Progress in Optical Science and Photonics 37,
https://doi.org/10.1007/978-981-95-1626-1_14

technologies is expected to herald a transformation in the methodology of data acquisition, processing, and utilization by the aviation community, thus announcing the arrival of aircraft systems which are fully independent, efficient, and self-contained.

1 Introduction

Artificial Intelligence (AI) and Optical Metamaterials (OMM), such as metasurfaces and metalenses, represent two emerging technologies that are driving innovation across photonics and aerospace applications [1, 2]. AI plays a vital role in the design, simulation, and optimization of metamaterials, significantly accelerating their development and improving performance [3]. Machine learning algorithms, in particular, can manage and interpret large-scale datasets to design efficient metalens structures that enhance image resolution and quality across sensing applications [4].

On the other hand, metamaterials contribute to AI advancement by enabling the physical implementation of optical neural networks. These networks promise improved energy efficiency and faster data processing, which are crucial for solving complex computational problems [1]. The cooperation between AI and metamaterials extends beyond design, impacting real-time image correction, adaptive optics, and big-data-driven design strategies [1, 4].

Together, AI and OMM are reshaping fields such as photonics, aerospace sensing, and intelligent communication systems [1]. This chapter explores the multifaceted relationship between AI and OMM, focusing on their individual contributions, synergistic potential, and transformative impact on aerospace sensing and system autonomy.

Building on this introductory context, we next explore the foundational concepts of Optical Metamaterials (OMM), which serve as the material backbone for many AI-enhanced aerospace systems (Fig. 1).

1.1 Understanding OMM: Fundamentals and Design Principles

Optical Metamaterials (OMM) are synthetic materials characterized by unique electromagnetic properties that do not naturally occur in conventional materials [5]. They enable advanced functionalities such as sub-wavelength imaging, broadband manipulation of the photonic density of states, and nanoscale light control [6]. However, one of the main challenges with existing metamaterials is spatial dispersion, which limits imaging performance by causing inconsistencies in how waves propagate through the medium [7]. To address this issue, researchers are developing optimization protocols that can engineer metamaterials to achieve specific optical behaviors, such as isotropy and diffractionless propagation [7].

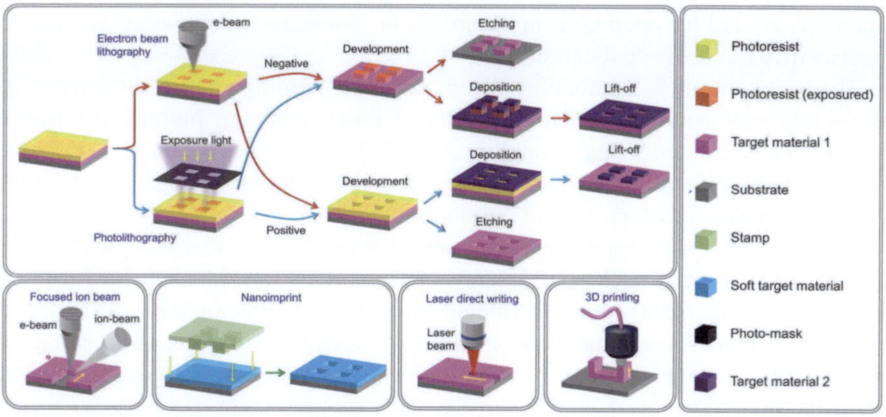

Fig. 1 Diagram of fabrication methods for metasurfaces

The application of Artificial Intelligence (AI) has significantly enhanced the design and optimization of metamaterials [3]. Machine learning algorithms can process vast datasets to identify optimal metalens configurations, thereby improving image resolution and performance across a range of applications [4]. The synergy between AI and metamaterials not only tackles issues like spatial dispersion but also boosts overall device efficiency and functionality [1].

In addition, integrating soft, reconfigurable materials into metamaterial architectures presents a promising solution to the limitations of traditional rigid structures [8]. These soft OMMs can adapt to external stimuli such as temperature or electromagnetic fields offering new, tunable functionalities [8]. When combined with AI algorithms, these reconfigurable metamaterials can be dynamically reprogrammed and optimized in real time, unlocking their full potential in adaptive optics and real-time imaging systems [1, 4].

1.2 Evolution of AI: From Concept to Revolutionizing Aerospace Technologies

Artificial Intelligence (AI) has profoundly transformed the aerospace sector over the past few decades, paving the way for autonomous navigation, intelligent operations, and more responsive space systems [9, 10]. The evolution of AI technologies now supports advancements such as fault-tolerant rocket navigation, the modernization of airport operations, and the revolutionization of space travel.

A prominent application is Trusted Autonomous Satellite Operations (TASO), where satellites make decisions independently in real time, guided by AI systems that interpret environmental data without the need for constant human input [11]. This autonomy is vital for operations in hazardous or remote areas, allowing satellites to react swiftly to unexpected conditions.

The backbone of this autonomy lies in machine learning algorithms capable of processing vast volumes of data to optimize route planning, monitor the health of space vehicles, and manage satellite constellations more efficiently [12]. These systems identify operational anomalies and predict component failures, significantly extending spacecraft lifespans and mission reliability.

Furthermore, AI's impact is amplified through its convergence with other advanced technologies such as the Internet of Things (IoT) and blockchain. Together, these technologies form secure, distributed networks of satellites and ground systems, enabling seamless real-time data exchange. Blockchain ensures data integrity and security, making it a key enabler for next-generation space communication infrastructure [12].

In recent years, AI has emerged as a catalyst for change not only in space but also in terrestrial aerospace operations. From smarter navigation systems to advanced decision support tools for mission planning, AI is helping unlock safer and more sustainable aerospace ecosystems.

However, the integration of AI into aerospace systems is not without challenges. Legal, ethical, and regulatory issues such as autonomous decision-making accountability, international space law compliance, and system transparency require collective effort and oversight from scientists, policymakers, and industry stakeholders [11, 12]. In addition, technical limitations persist, particularly in the areas of onboard real-time processing and cybersecurity.

Despite these hurdles, the potential benefits of AI in aerospace are immense. Through continued innovation and collaboration, the industry can overcome existing barriers and fully harness AI's transformative power [9]. The future of aerospace will increasingly rely on the synergy between human ingenuity and intelligent machines to enable agile, resilient, and sustainable space exploration.

As we have seen, both AI and OMM have independently made remarkable contributions to aerospace. In the next section, we delve into the synergistic potential when these two technologies converge, especially in enhancing aerospace sensing and system autonomy.

1.3 Synergistic Potential: AI and OMM in Advancing Aerospace Research

Artificial Intelligence (AI) is driving a renaissance in the aerospace industry, ushering in transformative advancements across a wide spectrum of applications. In airport environments, AI-powered systems streamline the entire passenger experience from automated check-in and biometric identification to security clearance and real-time customer support [10]. These systems reduce wait times, optimize baggage screening, and improve detection accuracy, thereby enhancing both operational efficiency and passenger satisfaction.

Beyond terrestrial boundaries, AI is playing an increasingly vital role in space exploration. It empowers scientists with advanced remote sensing capabilities to monitor phenomena such as climate change, land use, and Earth's atmospheric conditions. AI-guided robotics can carry out complex tasks in space environments that may be too dangerous or infeasible for human astronauts. Additionally, AI enhances rocket guidance systems, improving the accuracy and efficiency of satellite launches [10].

In satellite operations, AI is central to enabling autonomous navigation, system health monitoring, and dynamic management of satellite constellations [12]. By performing real-time analysis of telemetry and environmental data, AI systems can predict anomalies and initiate corrective actions before failures occur. This predictive autonomy is especially crucial for Distributed Satellite Systems (DSS), which involve groups of satellites operating in coordinated formations to achieve objectives such as global navigation, Earth observation, and broadband communication [11].

The integration of AI with blockchain technology is also reshaping aerospace supply chains. This convergence facilitates secure, transparent, and tamper-proof data exchange across the entire lifecycle of aerospace systems from manufacturing to maintenance [13]. The result is a more trustworthy and efficient supply chain, enhancing operational safety and reducing overall costs in aerospace engineering and logistics.

However, significant challenges remain before the full potential of AI in aerospace can be realized. Legal, ethical, and regulatory issues particularly those concerning AI autonomy, accountability, and compliance with international space law require coordinated efforts among policymakers, scientists, and industry stakeholders. Addressing these concerns is vital to ensure the responsible and ethical deployment of AI technologies in space operations.

The Internet of Space Things (IoST) represents another domain set to benefit from this convergence [12]. By interconnecting satellites, ground stations, and autonomous systems, IoST facilitates seamless data exchange for applications ranging from environmental monitoring to disaster response. This ecosystem enhances situational awareness and real-time decision-making across a wide range of aerospace and defense operations.

In conclusion, the integration of AI and Optical Metamaterials (OMM) holds immense promise in advancing aerospace sensing, control, and communication technologies. By fostering innovation, interdisciplinary collaboration, and regulatory foresight, these converging technologies can revolutionize aerospace systems making them more adaptive, intelligent, and resilient for the future of exploration and discovery.

2 Artificial Intelligence in Optical Metamaterial Systems

Artificial Intelligence (AI) has emerged as a transformative force in the field of Optical Metamaterials (OMM) and meta-optics, enabling unprecedented capabilities in design, optimization, and functionality [1, 3]. AI-driven tools are accelerating the development of metasurfaces and metalenses advanced materials that can manipulate light at the nanoscale with remarkable precision [2, 4].

Through AI, researchers now leverage **forward prediction** and **inverse design** to explore the interaction between light and complex metamaterial structures. Forward prediction involves modeling how specific material configurations behave under light exposure, while inverse design begins with a desired optical property and uses AI to determine the optimal structure to achieve it [4]. These capabilities are especially critical in applications requiring custom optical behavior, such as high-resolution imaging and adaptive sensing.

Machine learning algorithms analyze massive datasets and evaluate thousands of design permutations across vast parameter spaces [4]. This data-driven approach significantly accelerates innovation by identifying efficient solutions more systematically than traditional trial-and-error methods. The result: metamaterials that can surpass conventional technologies in imaging clarity, sensing accuracy, and communication performance [1].

Beyond optimization, the synergy between AI and OMM extends into the physical realm. Metamaterials are being developed to implement optical neural networks computational systems that perform calculations using light rather than electricity [1]. These photonic neural networks are expected to dramatically increase processing speed and energy efficiency, laying the foundation for a new generation of optical computing architectures [1, 3]. This interdisciplinary convergence pushes the boundaries of photonics and redefines the design of "smart" optical systems.

Moreover, the AI-OMM integration is already making an impact across industries. In telecommunications, AI-optimized metamaterials reduce signal loss and enhance data transmission rates, improving the reliability of communication networks. In healthcare, these technologies enable breakthroughs in medical imaging, allowing for earlier and more accurate diagnosis of diseases through advanced optical sensing systems.

In conclusion, the convergence of AI and OMM is catalyzing innovation across photonics and beyond. By uniting the computational power of AI with the electromagnetic manipulation capabilities of metamaterials, researchers are developing intelligent optical devices that challenge traditional limitations. This interdisciplinary synergy holds transformative potential across aerospace, telecommunications, and biomedical domains ultimately shaping a smarter and more connected future.

2.1 AI-Driven Design of OMM: Innovations in Material Engineering

Recent advances in metamaterials and meta-optics have been significantly accelerated by the integration of Artificial Intelligence (AI), which offers powerful tools for fast and efficient design and optimization. These AI-driven approaches are pushing metamaterials to reach their full functional potential and performance efficiency across various domains [1, 14].

At the core of these innovations are three transformative AI methodologies: forward prediction, inverse design, and data-driven optimization. Forward prediction allows scientists to simulate how a given metamaterial structure interacts with light, thereby extracting insights into its electromagnetic properties. Conversely, inverse design starts with the desired optical outcome and employs AI algorithms to derive the structural configuration needed to achieve it [1]. This dual capability enables the development of tailored optical components with precise control over light behavior.

These techniques are now widely applied across metamaterial research and development, contributing to rapid discovery cycles and design advancements [14]. AI's ability to process vast amounts of experimental and simulation data helps identify subtle patterns and correlations that might otherwise be overlooked. This capability is especially critical in designing advanced optical systems based on metasurfaces two-dimensional, artificially structured materials engineered to manipulate light in ways conventional optics cannot [2].

The synergy between AI and meta-optics also facilitates breakthroughs in simulation accuracy, information extraction, and device prototyping [3]. High-resolution imaging and adaptive sensing platforms are being developed through these AI-enhanced design methods, paving the way for smarter, more efficient optical systems with applications in aerospace, healthcare, and telecommunications.

Despite these advances, challenges remain. The effective application of AI to OMM design requires not only proficiency in algorithm development but also a strong understanding of the underlying physical principles of metamaterials. Additionally, there is a growing need for structured, high-quality datasets to train machine learning models and guide material discovery [2, 3]. Bridging these knowledge gaps demands interdisciplinary collaboration among experts in AI, materials science, and photonics.

Continued innovation in this field depends on fostering such collaborative ecosystems. By aligning efforts across disciplines, researchers can overcome existing barriers and unlock new frontiers in intelligent optical system design enabling technologies that have the potential to revolutionize industries and improve everyday life.

2.2 Enhancing Aerospace Sensors with AI-Enabled OMM Systems

Emerging aerospace technologies are increasingly integrating Artificial Intelligence (AI) with advanced sensing platforms to improve fault detection, health monitoring, and autonomous decision-making. The fusion of AI with Optical Metamaterials (OMM) presents a transformative approach to enhancing the performance and adaptability of aerospace sensors.

Modern aerospace systems rely on high-precision fault detection to ensure safety and mission success. Deep learning algorithms have proven particularly effective in enabling systems to detect and classify faults across complex operational conditions. These algorithms analyze vast and multidimensional sensor datasets, identifying patterns and anomalies that could indicate system failures—even in their earliest stages. This capability is critical for increasing reliability in high-stakes environments where system failure is not an option.

AI plays a key role in **Integrated System Health Management (ISHM)** by facilitating predictive maintenance and intelligent fault diagnostics [16]. By fusing data from various onboard sensors, AI-driven ISHM systems can anticipate failures before they occur, optimize maintenance scheduling, and maintain system integrity during extended missions. This proactive approach ensures the continued performance of aircraft and spacecraft systems, even under extreme operational stress.

The implementation of **Trusted Autonomous Satellite Operations (TASO)** further demonstrates the value of AI in distributed aerospace systems. TASO employs AI algorithms to support autonomous decision-making within satellite networks, granting them the flexibility to respond dynamically to changing mission parameters and unforeseen space conditions [11]. This autonomy is especially vital for operations in remote or hazardous regions, where real-time human intervention is impractical or impossible.

In addition, the integration of **Orbital Angular Momentum (OAM)**-based sensing with AI-enhanced OMM systems opens up new possibilities in environmental sensing and structural health diagnostics [17]. OAM sensors measure the angular momentum of light to capture highly detailed spatial and atmospheric data. When combined with AI, these systems can intelligently interpret complex datasets to provide precise assessments of environmental conditions and aerospace structural integrity.

Collectively, the convergence of AI and OMM technologies in aerospace sensors is driving the development of smarter, more autonomous, and highly reliable systems. These integrated solutions enhance mission adaptability and operational resilience, enabling aerospace systems to function optimally even in unpredictable environments.

In conclusion, the integration of AI-enabled OMM systems with aerospace sensing technologies represents a major advancement in aerospace engineering. By leveraging the strengths of AI in data processing and decision-making alongside the unique electromagnetic manipulation capabilities of OMMs, researchers are paving the way for the next generation of intelligent, adaptive aerospace platforms. This interdisciplinary evolution promises to revolutionize sensing systems and expand our capabilities in space exploration, atmospheric research, and beyond.

2.3 *Intelligent Optimization for Adaptive and Real-Time Metamaterial Performance*

Recent advances in the field of metamaterials have focused on developing intelligent optimization techniques that enhance adaptive performance across various domains. These smart and reconfigurable materials are poised to transform the design and operation of systems involving thermal, mechanical, optical, and electromagnetic functions.

For instance, thermal metamaterials can now be tailored in real time to meet specific thermal management requirements [18]. Deep learning models play a central role in this capability by enabling the design of metamaterials with arbitrary geometries. These models allow for rapid analysis and optimization of thermal properties based on environmental changes. This real-time adaptability is essential for applications such as electronics cooling and energy-efficient architectural systems that demand precise temperature regulation.

In the domain of vibration control, adaptive piezoelectric metamaterial beams with autonomous tuning capabilities have demonstrated significant potential [19]. These smart structures can dynamically reconfigure their attenuation zones without human intervention. This allows them to dampen vibrations effectively in complex environments, which is especially valuable in aerospace and automotive systems where vibration control is critical to both performance and safety (Fig. 2).

In optics, physics-informed, data-driven optimization methods have greatly improved the efficiency of designing large-aperture metalenses [20]. One such technique utilizes local effect-driven lattice coupling to fabricate high-performance optical devices capable of advanced light manipulation. These breakthroughs have major implications for compact imaging systems and optical communication technologies.

In the electromagnetic domain, intelligent reconfigurable metasurfaces have shown the ability to autonomously switch functions in response to variations in incident power [21]. These metasurfaces integrate sensing and feedback components, enabling them to modify their electromagnetic response in real time. Such functionality is vital for modern wireless communication and radar systems, where rapid environmental changes demand immediate and precise adjustments.

The integration of AI-based optimization with adaptive metamaterials marks a significant milestone in materials science. These intelligent systems are not only multifunctional and efficient but also capable of operating autonomously. This evolution expands the boundaries of what is possible across multiple fields and opens new avenues for application-driven innovation.

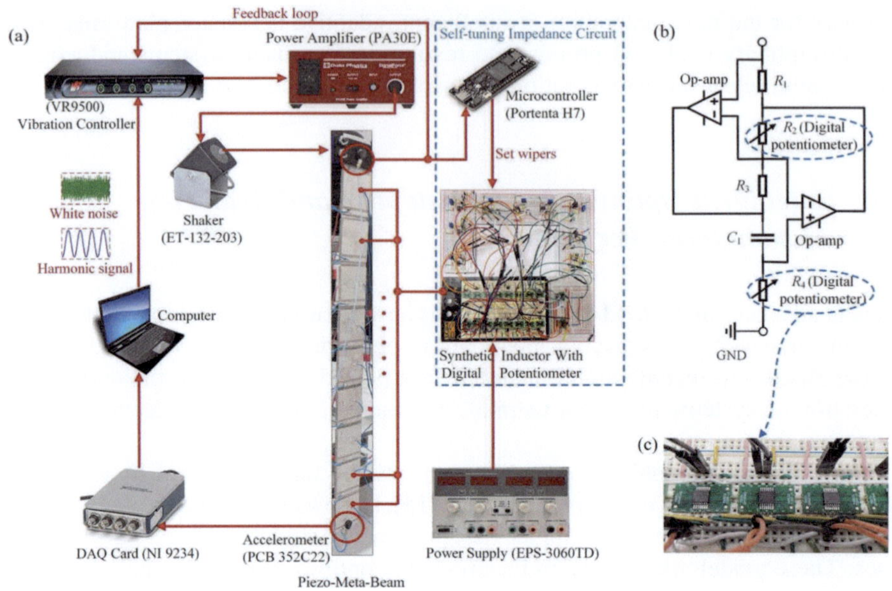

Fig. 2 **a** Experimental setup for testing the adaptive piezo-meta-beam; **b** circuit diagram of the synthetic inductor; **c** illustration of the digital potentiometers controlled by the microcontroller [19]

In summary, intelligent optimization for adaptive and real-time performance in metamaterials is reshaping the landscape of material engineering. By combining smart responsiveness with AI capabilities, researchers can tackle some of the most demanding technical challenges. This synergy supports the development of next-generation materials that are efficient, adaptable, and autonomous, with broad implications for industries ranging from aerospace and defense to energy and healthcare.

2.4 AI-Integrated Simulation Frameworks for OMM Analysis

Recent studies emphasize the transformative potential of integrating Artificial Intelligence (AI) with metamaterials and meta-optics, particularly in redefining the design and optimization processes. AI enables researchers to establish precise mappings between device structures and their optical properties, supporting both forward modeling and inverse design within the field of metaphotonics [22]. This capability is critical for developing next-generation optical devices capable of manipulating light with unprecedented precision.

Through AI-enabled simulation frameworks, researchers can accurately predict how metamaterials will behave under specific conditions before they are physically fabricated. These virtual design environments allow for in-depth exploration and iterative refinement of metamaterial configurations. For example, machine learning techniques are now being applied to optimize the thermal performance of metamaterials, enabling the development of highly efficient heat-management systems [23]. Such systems are particularly important in electronics and energy applications, where precise thermal control is essential.

One major advancement is the use of hybrid AI frameworks to design complex metamolecules with enhanced light-manipulation capabilities [24]. These frameworks allow exploration across vast design spaces, identifying structures that maximize desired optical effects while minimizing undesired interference. The result is a new class of high-performance metamaterials suited for demanding tasks in imaging, sensing, and light-based computing.

Among these innovations, metalenses have emerged as pivotal tools in imaging applications. Meta-optics, driven by AI, supports the rapid design and optimization of metalens structures [4]. Researchers can now achieve higher resolution and better image quality, improving technologies ranging from advanced microscopy to non-invasive medical diagnostics. These breakthroughs contribute directly to earlier detection of diseases and more accurate clinical assessments.

The integration of AI within OMM simulation workflows is pushing the boundaries of what can be achieved in nanoscale light and thermal manipulation. AI assists in streamlining design processes and unlocking new design paradigms across multiple fields. From enhanced energy efficiency in smart systems to breakthroughs in imaging resolution, the combination of AI and metaphotonics is shaping the future of applied research and engineering.

Additionally, AI-powered simulation frameworks offer improved scalability and cost-efficiency. These tools accelerate the development timeline and reduce the trial-and-error phase of traditional design. As a result, researchers are now able to create higher-performing optical systems with expanded functionalities and integration potential. This interdisciplinary collaboration between AI and metamaterials is expected to drive major technological shifts across industries and deepen our understanding of the physical universe.

3 Revolutionizing Aerospace with AI and OMM

Recent technological advancements have accelerated innovation across aerospace engineering, unlocking new research opportunities and expanding the scale of real-world applications. At the heart of this transformation lies the integration of Artificial Intelligence (AI) and Optical Metamaterials (OMM), which together promise to redefine performance, agility, and operational efficiency in aerospace systems.

One area witnessing significant impact is the aerospace supply chain. Abdulrahman et al. demonstrated how combining the analytical power of AI with the decentralized security of blockchain technology enhances operational transparency and efficiency [13]. This fusion enables streamlined workflows, improved cost-effectiveness, and greater integrity throughout the supply chain. Moreover, these advancements contribute to resolving legal and regulatory challenges by fostering compliance-ready infrastructure that is more robust and secure.

In parallel, breakthroughs in hyperbolic metamaterials are redefining imaging and light manipulation capabilities. According to Lee et al., these materials support phenomena such as super-resolution imaging and negative refraction, allowing optical devices to achieve resolutions beyond the diffraction limit [25]. This advancement presents new opportunities in scientific research and industrial applications, although further investigation is still required to fully realize their potential.

Krasikov and colleagues introduced the concept of intelligent metaphotonics, which leverages machine learning to enhance the site-specific design of photonic devices [26]. This emerging field merges AI with photonics to enable the development of optical systems that exhibit increased functionality, efficiency, and adaptability. Such innovation holds promise across domains ranging from telecommunications to biomedical imaging.

Deep learning is also being employed to design and optimize artificial electromagnetic materials. As reviewed by Khatib et al., these AI-enhanced methods have improved the flexibility and performance of electromagnetic material systems, making them more suitable for deployment in harsh aerospace environments [27]. This capability is crucial for the development of resilient aerospace platforms that must operate under extreme conditions while maintaining high performance.

Together, these advancements exemplify the growing synergy between AI and OMM technologies. Their integration in aerospace engineering represents a technological leap that is reshaping how systems are designed, built, and operated. This convergence lays the groundwork for a new era of adaptive, intelligent aerospace systems capable of addressing complex future challenges across science, industry, and defense.

3.1 Autonomous Navigation Systems Powered by AI and OMM

Artificial Intelligence (AI) and Optical Metamaterials (OMM) are rapidly transforming the capabilities of autonomous navigation systems and imaging technologies. The convergence of these two fields is enabling systems to perform with greater precision, adaptability, and intelligence than previously thought possible.

Consider the case of small unmanned aircraft systems (sUAS) navigating through highly complex environments with exceptional accuracy. AI techniques, including neural networks and machine learning algorithms, are enhancing navigation performance by improving real-time decision-making [28]. These systems excel in two key areas: sensor fusion, which combines data from multiple sources for better environmental awareness, and obstacle detection and avoidance, which allows for dynamic path adjustments to prevent collisions.

AI also plays a vital role in predictive monitoring, helping detect early signs of failure across system components. This proactive approach enhances the reliability and safety of autonomous navigation systems, especially in mission-critical operations such as search and rescue or environmental surveillance.

Simultaneously, AI is accelerating the design and functional optimization of metamaterials, enabling the development of advanced optical components such as metalenses [1, 4]. These components perform complex optical functions, including edge detection and shape recognition, which are critical for high-speed navigation across unpredictable terrains [29]. The ability of metamaterials to process information at the speed of light provides a significant leap forward in achieving real-time environmental responsiveness in autonomous systems.

Beyond navigation, the AI-driven design of metamaterials holds immense promise for next-generation imaging technologies. AI-optimized metamaterials enhance image resolution and clarity, with applications ranging from advanced microscopy to medical diagnostics [4]. This improvement in imaging can contribute to earlier disease detection and more accurate diagnostic procedures, ultimately improving clinical outcomes.

Furthermore, the integration of AI and OMM is not limited to aerospace or medical fields. These technologies are also poised to impact consumer electronics, leading to the development of high-performance cameras and sensors in devices such as smartphones and wearable systems. Their ability to execute complex optical functions at high speed makes them highly applicable to robotics and autonomous vehicles, where real-time data processing is essential.

In summary, the fusion of AI and OMM is revolutionizing autonomous navigation and imaging by delivering intelligent, responsive, and multifunctional systems. This convergence opens up exciting possibilities across industries, setting a new benchmark for performance and adaptability in the age of intelligent machines.

3.2 Intelligent Environmental Monitoring and Structural Integrity Assessment

Advancements in Artificial Intelligence (AI) have significantly enhanced environmental monitoring and structural health assessment. By integrating AI with advanced sensing technologies and intelligent materials, researchers are developing innovative solutions that promise to improve both safety and sustainability across infrastructure systems.

Imagine real-time monitoring of the health of bridges, buildings, and other critical infrastructure, with the ability to detect structural issues before they evolve into serious failures. AI-assisted, non-contact sensing methods such as photogrammetry and infrared thermography have proven effective in structural health monitoring (SHM) [32]. These techniques eliminate the need for physical access to infrastructure components, making inspections more efficient, safer, and accessible.

Smart electronics and intelligent packaging, when combined with AI models like convolutional neural networks, further enhance SHM capabilities [33]. Real-time monitoring using camera-based, smartphone-based, and UAV-based platforms enables early detection of structural weaknesses. This proactive approach ensures timely maintenance and extends the lifespan and reliability of critical civil infrastructure.

In addition, AI-supported design methods have accelerated the development of active metasurfaces capable of visually responding to strain by changing color [34]. These smart materials function as visual indicators of structural stress and deformation, aiding in visual inspections where traditional sensor-based evaluations may not be feasible.

AI is also transforming the optimization of thermal metamaterials by predicting their thermal properties and boosting computational efficiency during design processes [23]. These advancements support the development of energy-efficient and environmentally friendly infrastructure through better thermal management (Fig. 3).

The integration of intelligent structures is gaining momentum in aerospace and photonics applications [36]. These structures incorporate sensors and actuators that respond to mechanical, thermal, and optical changes, enabling real-time adaptation and enhanced damage detection. By combining the predictive capabilities of machine learning with metaphotonics, researchers are now able to engineer systems using nanoantennas and metasurfaces that can manipulate light with extraordinary efficiency [26].

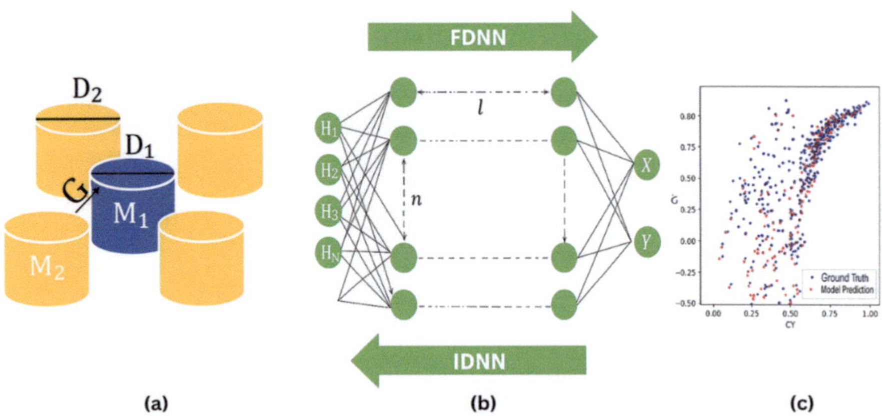

Fig. 3 **a** The structure of metasurface composed of metal and dielectric resonators where blue and yellow represent dielectric and metal NRs respectively. Materials, gap, diameter 1 and 2 are marked as M1, M2, G, D1, D2 respectively. **b** Fully connected DNN, where l and n are the number of hidden layers and neurons in each hidden layer, respectively. FDNN denotes forward design neural network and IDNN for inverse design neural network. **c** The colour coordinate from the validating dataset is shown, where red points are ground truth (input data) and blue points are model prediction [34]

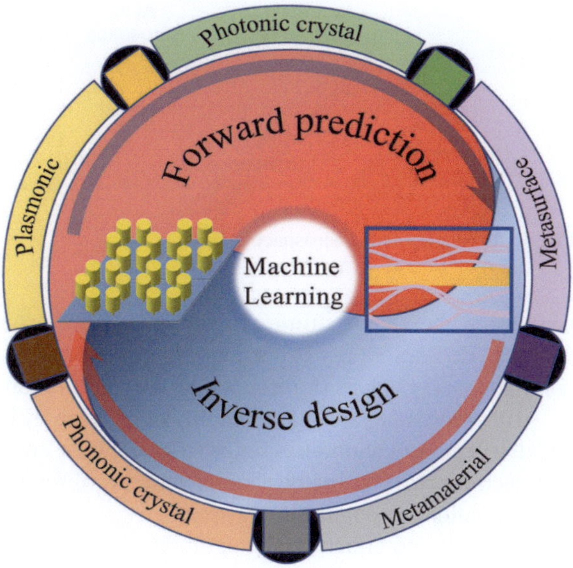

Fig. 4 Diagrammatic sketch of machine learning for properties characterization and structural design of artificial materials. The machine learning algorithms are applied to connect the structure (such as phononic crystal, metamaterial, metasurface, photonic crystal and plasmonic) information with the response information (for example, band gap, transmission etc.) [35]

Machine learning has also contributed to the design of phononic metamaterials, which are used to manipulate sound and elastic waves [35]. These materials support advanced functionalities such as noise reduction, acoustic insulation, and vibrational control, all discovered through efficient computational exploration of complex structural configurations.

Together, these AI-powered innovations are reshaping the landscape of SHM and environmental monitoring. They offer safer, more intelligent, and longer-lasting infrastructure solutions while enabling adaptive systems that can respond to environmental and operational demands. This convergence of AI, metamaterials, and smart sensing is guiding the way toward a more sustainable and resilient built environment (Fig. 4).

3.3 AI in Space-Based Optical Systems for Deep Space Exploration

Artificial Intelligence (AI) has become an integral part of modern space missions, contributing significantly to areas such as autonomous navigation, real-time decision-making, and spacecraft health monitoring. As missions venture farther into deep space, the demand for intelligent and adaptive systems capable of operating independently becomes more pressing. Through machine learning (ML) and deep learning (DL), AI is well-positioned to meet these evolving challenges.

Imagine spacecraft navigating autonomously and making mission-critical decisions based on their immediate environment. AI enables onboard information

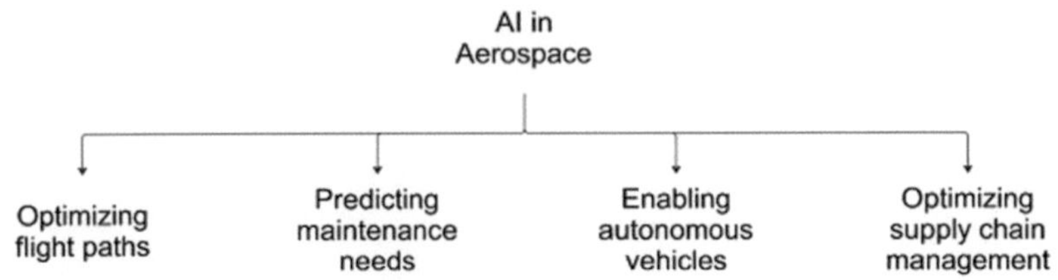

Fig. 5 AI benefits in the aerospace industry

processing in real time, eliminating the need for constant communication with ground stations. This autonomy is particularly valuable during deep space missions, where communication delays or disruptions are common due to vast distances.

AI also enhances spacecraft health monitoring by managing sensor data to predict and detect system anomalies before they escalate into critical failures. This proactive approach supports early intervention and risk mitigation, which is crucial in the harsh and unpredictable conditions of space.

Despite its potential, the integration of AI into space missions presents several challenges. These include legal implications, compliance with international space regulations, and ethical considerations surrounding autonomous decision-making [12]. Addressing these concerns requires a collaborative approach involving policy-makers, scientists, and industry leaders to establish responsible guidelines for AI use in space.

Recent developments in commercial-off-the-shelf (COTS) system-on-chip (SoC) technologies are improving onboard processing capabilities [37]. These enhancements enable real-time data analysis directly within spacecraft, reducing the need to transmit large volumes of data back to Earth. This not only conserves bandwidth but also accelerates the delivery of actionable insights for Earth observation and remote sensing tasks (Fig. 5).

AI is also proving valuable in hyperspectral imaging applications, where it can process large datasets efficiently and extract detailed environmental and resource-related information [38]. While current limitations include hardware constraints and the need for effective selection strategies, advancements in AI algorithms and onboard computing power are helping overcome these barriers to maximize mission returns.

Overall, the integration of AI into space-based optical systems offers immense opportunities for deep space exploration. These intelligent systems can improve spacecraft efficiency, autonomy, and adaptability, enabling more ambitious missions into previously unreachable territories. However, realizing this potential will require ongoing innovation, cross-sector collaboration, and a strong commitment to legal and ethical standards.

In conclusion, AI is transforming the landscape of space exploration. By supporting autonomous navigation, enhancing spacecraft monitoring, and improving data processing efficiency, AI is enabling a new generation of resilient, intelligent space missions. This technological evolution holds the promise of expanding humanity's reach into the cosmos and deepening our understanding of the universe.

4 AI-Driven Simulations and Results Analysis

Artificial Intelligence (AI) is revolutionizing the way simulations and results analysis are conducted across various high-impact industries, particularly in aerospace and photonics. Traditional simulation methods, while powerful, are often constrained by high computational costs, limited scalability, and long design iteration times. AI integration into simulation frameworks addresses these challenges by enabling rapid, adaptive, and highly efficient optimization processes.

In the domain of **metaphotonics**, AI facilitates both forward modeling and inverse design, allowing researchers to establish accurate mappings between a device's structure and its resulting optical properties [22]. Forward modeling predicts how a given metamaterial behaves when interacting with light, while inverse design begins with a desired optical response and derives the structure required to achieve it. This AI-guided approach expands the design space and accelerates the development of complex optical devices such as metalenses [4].

AI-based methods also enhance the **optimization of thermal metamaterials**, which are essential in managing heat flow in aerospace and energy systems. Machine learning algorithms can predict thermodynamic properties with high accuracy and optimize the configuration of materials to meet specific thermal performance criteria [23]. These models reduce the need for time-intensive finite element analysis and can rapidly converge on optimal designs even in highly non-linear thermal environments.

In simulation contexts, AI supports **data-driven design exploration** and **automated parameter tuning**, enabling rapid prototyping and self-adaptive functionalities. For example, researchers can use generative models or neural networks to simulate thousands of design iterations within minutes, a task that would take traditional solvers significantly longer. These advancements are especially impactful in fields such as microscopy, medical diagnostics, and consumer electronics, where rapid customization of optical components is essential [4].

Moreover, AI can process and analyze large datasets generated by simulations to uncover hidden patterns and correlations. These insights facilitate material innovation and help researchers better understand structure–property relationships in complex systems [24]. This feedback loop enhances not just design efficiency but also scientific discovery in material science and optical engineering.

AI-driven simulations are now being implemented using powerful computational frameworks, combining physics-based solvers with machine learning libraries like TensorFlow, PyTorch, and Scikit-learn. These platforms enable hybrid models that combine analytical equations with AI-predicted surrogate functions, significantly reducing computational burden.

Despite these advances, certain challenges remain. High-quality data is essential for training reliable AI models, and interpretability of AI-generated designs is still a concern. Moreover, transferring learned models across different design domains can be difficult without robust generalization techniques.

Nonetheless, the integration of AI into simulations represents a paradigm shift in design methodology. By facilitating predictive modeling, reducing development timelines, and enabling real-time adaptability, AI-driven simulations are pushing the boundaries of what is achievable in aerospace, photonics, and beyond.

In conclusion, AI-enabled simulation and analysis tools empower researchers to explore complex material behaviors, optimize high-performance devices, and develop intelligent, adaptive systems. As this interdisciplinary approach continues to mature, it holds the potential to redefine the landscape of engineering, leading to breakthroughs across aerospace, healthcare, environmental monitoring, and consumer technologies.

4.1 Case Studies: AI-Enhanced Performance in OMM Applications

Over the past decade, Artificial Intelligence (AI) has emerged as a transformative force within aerospace engineering, enabling significant advancements in design, manufacturing, assembly, testing, and operational service. The integration of AI and machine learning techniques has led to the creation of smart aerospace systems characterized by enhanced productivity, efficiency, and safety.

Imagine aerospace systems that respond to real-time data by detecting potential hazards, optimizing fuel consumption, and adapting to changing environmental conditions. This vision is becoming reality through AI's ability to analyze vast datasets and simulate complex scenarios, leading to autonomous systems and advanced data visualization platforms. These capabilities not only improve passenger safety and experience but also enhance operational efficiency, reduce environmental impact, and optimize cost management [39].

A pivotal breakthrough also comes from the intersection of AI and materials science. Through micromechanics-based modeling and AI algorithms, researchers can now analyze microstructural characteristics of materials at high resolution. This enables the design of advanced aerospace materials with essential properties such as high energy absorption, longer service life, structural versatility, and load-controlled deformation [40]. These AI-assisted technologies contribute directly to improving the durability and reliability of aerospace structures, especially in extreme environments such as space missions or high-altitude flight.

Optical Metamaterials (OMMs) present another compelling domain where AI is making a significant impact. These engineered materials can manipulate electromagnetic waves in ways that were once considered impossible, offering breakthroughs in optics, healthcare, acoustics, and energy systems. Machine learning models, particularly deep neural networks (DNNs) and convolutional neural networks (CNNs), have accelerated the design of complex optical components such as lab-on-chip meta-gratings and metalenses [41]. These AI-enabled designs allow precise control over electromagnetic, mechanical, and thermal properties, opening the door to multifunctional and application-specific metamaterial systems.

In the area of **meta-optics**, components like metalenses and metasurfaces are particularly promising. These structures interact with light at the nanoscale and require highly complex three-dimensional design and optimization processes. AI methods, including synthetic neural networks and natural language processing, are being used to analyze extensive datasets, explore a wide array of design permutations, and identify key fabrication parameters. These insights are essential for building adaptive optical systems that respond dynamically to changing imaging conditions. As a result, image resolution and quality are significantly enhanced, supporting advanced applications in microscopy, medical diagnostics, and consumer electronics.

These case studies collectively highlight how AI-enhanced Optical Metamaterials are setting the stage for next-generation aerospace systems. By unifying adaptive materials with intelligent algorithms, researchers are developing solutions that meet the growing demands of modern aerospace challenges while opening new avenues across technology, healthcare, and environmental sensing (Table 1).

Table 1 Case studies—AI-enhanced performance in OMM applications

Case study	Application	AI technique used	Outcome
Adaptive optical sensors	Real-time environmental monitoring	Deep learning, convolutional neural networks (CNNs)	Improved accuracy and responsiveness in detecting environmental changes
Stealth technology	Reducing radar visibility of aircraft	Machine learning, optimization algorithms	Enhanced stealth capabilities, reducing detectability in military operations
Health monitoring systems	Structural integrity assessment in aerospace	Predictive analytics, deep learning	Proactive maintenance and improved safety through early fault detection
Energy-efficient metamaterials	Thermal management in aerospace structures	AI-driven design optimization	Reduced energy consumption and improved thermal regulation
Advanced imaging systems	High-resolution medical and industrial imaging	Deep neural networks (DNNs)	Enhanced image quality and diagnostic capabilities
Autonomous navigation	Self-navigating unmanned aerial vehicles (UAVs)	Reinforcement learning, deep learning	Improved autonomous decision-making and navigation in complex environments
Optimized communication systems	Enhanced signal transmission in satellites	AI-based signal processing	Increased data transmission efficiency and reduced interference

4.2 Virtual Testing Environments for AI and OMM Integration

In the development of artificial intelligence and OMM, virtual testing environments are rapidly becoming powerful instruments. These operate in a digitally controlled environment where materials, designs, and configurations can be developed and tested without the need for physical prototyping. These environments make optimizing OMM designs possible with a really fast turn-around time through AI-led simulations.

Being able to investigate how a metamaterial interacts with light and how it reacts to changes in temperature in any digitally controlled environment is amazing. AI algorithms analyze such interactions, predict their outcomes, and point out those configurations that would be suitable for given applications [22]. All this will be particularly appreciated when designing complex metasurfaces and metalenses, for it would otherwise be both time-consuming and expensive for a method to rely otherwise on trial and error [4].

Virtual testing environments allow machine learning models to be integrated such that they learn from simulated data to improve their predictive capabilities and robustness over time [23]. The iterative learning process ensures the reliability of AI models and strengthens their potency in numerous real-world applications.

These environments are further necessary to study extreme conditions that are problematic or unsafe to reproduce within a physical lab space, for example, in high temperatures or outer space-like settings. Hence, researchers can demonstrate the behavior of metamaterials, thus indirectly injecting confidence into the designs that they will be solid and trustable across sizable scenarios [11].

Combining artificial intelligence with optimization and risk assessment skills in a virtual environment allows researchers to stretch the boundaries of creative thought further. By providing a safe and controlled space where experimentation can occur, these environments allow researchers to take risks, explore new ideas, and generate breakthroughs that would otherwise be perceived as too challenging or resource-intensive to pursue.

4.3 Comparative Analysis of Traditional Versus AI-Optimized OMM Systems

By comparatively analyzing through traditional and artificial intelligence (AI)-optimized models of OMM, the positive influence of AI upon materials science and Optical technologies can be clearly emphasized. Most of the traditional OMM systems use iterative trial-and-error approaches for designing OMMs. This approach is time-consuming and resource-intensive. These methods typically require manipulating some parameters and fabricating the relevant designs, followed by testing them for their optical properties. The process of optimizing these circuits may take months or even years.

On the contrary, OMM systems maximize artificial intelligence-aware optimization that makes the OMM design process more manageable and faster. With the fast processing speed by machine learning algorithms, AI can mine vast databases and simulate numerous design variations rapidly, thus optimizing configurations for meta-materials [14]. This paves extensive design possibilities from which better and efficient alternatives can be provided for researchers.

Additionally, AI-driven optimization improves the precision and performance of OMMs. Machine learning models can provide insight into how certain changes in material composition or structure will impact optical properties, thus rendering metamaterials with tailored responses to light [20]. Such levels of customization don't seem possible with traditional means, which often cannot take such material properties into consideration in relating to how they perform optically.

Furthermore, the AI-optimized systems will keep learning and thereby improving their functionalities with the passage of time. With the increasing amount of data for analysis by the AI models, they become more accurate and suitable for use, thereby tightening the performance lapses of OMMs [24]. This continuous evolution is one of the distinct advantages over the old methods, where integration of new data might be impeded, along with the incongruity to meet changing conditions.

This, in turn, allows for real-time monitoring and alteration of OMMs, which is very important for application areas in the dynamic environment such as aerospace and telecommunications [11]. Most of the earlier ones were incapable of giving this flexibility, as they operated from a more or less hard-wired design, insensitive to growing environmental variables.

It's evident from the above analysis that AI-enabled OMM systems are far better than conventional methodologies with higher efficiency, accuracy, and adaptability. By unlocking AI's potential, the researchers could actually solve historical stubborn problems in material design and optical engineering, which in turn may lead to the conception of advanced technologies that expand the threshold of realization (Table 2).

4.4 Predictive Modeling: AI Algorithms for Future Aerospace Scenarios

Recent research on aerospace predictive modeling and AI algorithms is creating pathways for major innovations in fault diagnosis, control prediction, and spectrum management. Innovations such as these are key in enhancing the safety, efficiency, and capacity of future aerospace operations.

Imagine predicting and addressing potential faults in aircraft wings before they become major challenges: particle filter-based prognostic algorithms make just this possible, by predicting flaw grow rate in the wings of the aircraft, which will help them

Table 2 Comparative analysis of traditional versus AI-optimized OMM systems

Feature	Traditional OMM systems	AI-optimized OMM systems
Design process	Iterative, trial-and-error	Data-driven, simulation-based
Time efficiency	Slow, can take months to years	Rapid, significantly reduced design time
Resource use	High, requires multiple prototypes	Lower, leverages virtual testing environments
Precision	Limited, manual adjustments	High, precise control over optical properties
Adaptability	Low, static designs	High, real-time adjustments and learning
Data incorporation	Difficult to incorporate new data	Continuously improves with new data
Application range	Limited by manual design constraints	Broad, explores vast design spaces
Example use cases	Basic optical components	Advanced metalenses, adaptive metasurfaces

to plan maintenance and keep them alert against any unpleasant surprises [43]. Proactive aircraft maintenance in turn ensures that they operate under optimal conditions enhancing safety and efficiency in operation.

Deep learning techniques combined with optimization algorithms appear highly promising in satellite power system and aircraft engine fault diagnosis [15]. Data mined from huge bases of operational data allow these algorithms to identify even the most subtle anomalies that give rise to maintenance needs before catastrophic failure occurs. This is particularly valuable for space missions where reliability and longevity become paramount.

An interpretable deep kernel learning model has been suggested in the control prediction domain for spacecraft missions [44]. The model yields better prediction accuracy and interpretability compared to standard deep neural networks and Gaussian process models, with clear implications for the success and safety of missions.

With the growing demand for aviation communications, intelligent spectrum management techniques are being studied to increase spectrum efficiency and improve airspace capacity [45]. These techniques utilize AI and machine learning to enhance the usage of the spectrum with less congested channels and with higher communication reliability-as these are vital for supporting the growing number of aircraft that need a steady approach and safe flight to operate (Fig. 6).

These advances in predictive modeling and AI algorithms, taken together, will help to take aerospace forward, enhancing safety, efficiency, and capacity. Incorporating this technology will enable aerospace operations to become proactive, adaptive, and reliable in enhancing the future with innovations.

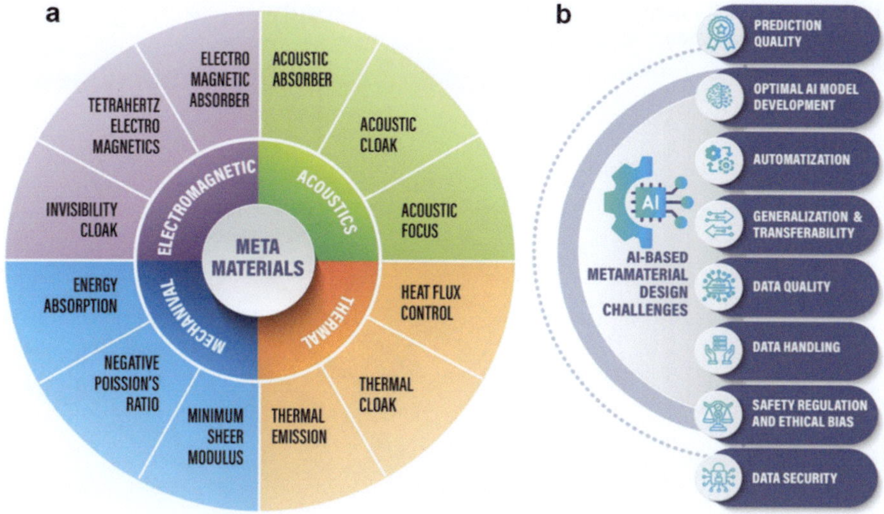

Fig. 6 **a** Overview of metamaterial design and applications. The metamaterial technologies are categorized as mechanical, thermal, acoustic, and electromagnetic based on the properties they manipulate and control. **b** Overview of some challenges in AI-based metamaterial design including the quality of AI models in predicting metamaterial function and performance, generation of optimal AI models, achieving automatization, the prospects of generalization and transferability across different metamaterials, data quality, data handling, constructing models regulated for safety, ethical conduct, and bias, and achieving data security. Some elements in this figure were designed using resources from freepik.com [42]

5 Challenges and Transformative Opportunities

See Fig. 7.

5.1 *Overcoming Fabrication and Deployment Challenges in AI-Integrated OMM*

Although some challenges must be addressed, the incorporation of AI into OMM must first tackle fabrication and deployment problems. Conventional fabrication approaches seldom achieve the requisite precision and consistency when producing designs optimized by AI methods of the OMM. Such materials are evidently elaborate, nanoscale structures that present uncertainty in their fabrication [1]. Consequently, to counter such challenges, researchers are exploring advanced manufacturing technologies such as 3D printing or nanolithography, which could promise high-fidelity production of complex metamaterial structures in the near future.

There are additional hurdles in actualizing the deployment of these AI-integrated OMM for real-world applications. OMMs need durability and versatility; otherwise,

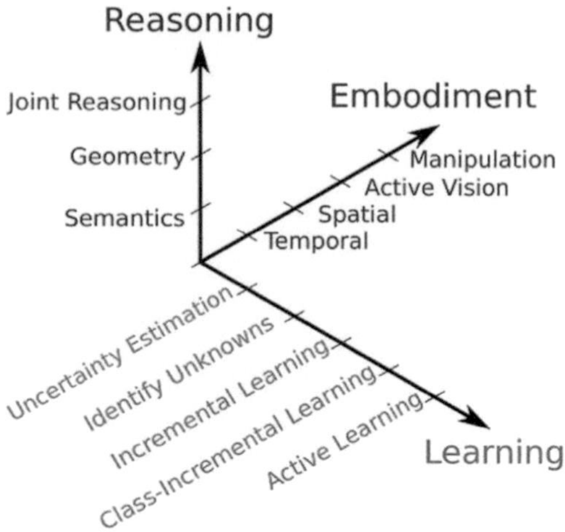

Fig. 7 Challenges for deep learning in robotic vision [46]

they face adverse effects from the changes in their environment that will compromise their optical properties [2]. Therefore, its longevity and the consistency of such materials are critical towards their viable realization into useful applications, for example telecommunications and sensing technologies.

5.2 Mitigating Computational Overhead in AI Models for Aerospace Applications

AI models, especially deep learning, are constantly pushing the limits when it comes to CPU overhead. Real-time processing is critical for any aerospace applications. Their rapid data analysis and real-time decision-making for autonomous systems and predictive maintenance require efficient algorithms and optimized hardware structures [15]. Found in the direction of reducing that overhead rests within leveraging advancements in commercial off-the-shelf (COTS) embedded system-on-chip (SoC) technologies to enable increased onboard processing to lessen reliance on ground-based systems [37].

Additionally, using plastic AI models and edge computing techniques can help ease the computational burden without compromising performance [11]. This type of approach allows aerospace systems to process data in situ: local processing is less latently responsive to the needs of the fast-moving environment.

5.3 Hybrid AI Architectures for Scalable and Efficient Aerospace Systems

Hybrid AI architectures work by synergizing those conflicts of different AI approaches such as deep learning and evolutionary algorithms in aerospace systems, thus providing scalability and efficient capability. They allow integrating diverse sources of data with adaptive learning to enhance the system's provisions to address challenging and dynamic environments. By taking in hybrid AI, aerospace systems have benefited from the improved capabilities of fault diagnosis, health monitoring, and autonomous decision-making, which lead to more predictable and efficient operations.

The requirement for new hybrid architectures should entail collaboration among multiple disciplines and continuous innovations that confront the unfurling circumstances of the aerospace industry, while hybrid architectures encourage new applications in aerospace science as AI technologies progress.

5.4 Ethical Considerations and Data Privacy in AI-Driven Aerospace Solutions

The use of AI in aerospace contexts raises several important ethical considerations and computability issues regarding privacy. As autonomous machines become more common, it is also imperative to maintain the utmost levels of transparency, accountability, and fairness in their decision-making processes [12]. Similarly, questions must be answered by stakeholders regarding the ethical implementation of AI and decisions affected by autonomous systems for human lives.

In further relation to such, it is utmost priority that sensitive data be protected and that privacy be ensured, especially on applications that are connected to public safety and national security [13]. Hence, it becomes paramount that strong ethical frameworks and data governance policies are laid out, whereby AI will be implemented into aerospace with the utmost responsibility. Collaboration efforts by the government, researchers, and industry stakeholders should set out a plan that guides seamless regulation and standards promoting ethical AI within the confines of innovation.

Table 3 Potential applications of AI-optimized metamaterials in aerospace

Application area	Benefits
Stealth technology	Reduced radar visibility for military and civilian aircraft
Satellite communications	Enhanced signal strength and reduced interference
Structural health monitoring	Real-time diagnostics and proactive maintenance of aerospace structures
Thermal management	Improved heat dissipation and temperature control in aerospace systems
Optical sensing	High-resolution imaging and detection capabilities
Energy harvesting	Efficient conversion of ambient energy into usable power
Adaptive optics	Dynamic adjustment of optical properties for varying conditions

6 Future Directions and Recommendations

6.1 Smart Metamaterials: The Next Frontier in Aerospace Applications

Smart metamaterials are an innovation in aerospace technology that have never been so flexible and functional. These can be designed to respond actively based on environmental changes, including changes in temperature, pressure, or electromagnetic fields. Enhanced performance and resilience can be achieved in aerospace structures due to more robust and versatile capabilities in various operating conditions. For example, smart metamaterials can be embedded with sensors that grant real-time monitoring and diagnosis, thus ensuring proactive maintenance and enhanced safety. Further, their unique optical properties can be exploited for stealth and cloaking applications to minimize aircraft visibility to radar and other detection systems. These aspects might be very suitable for both military and civilian aerospace, where stealth technology is crucial for safety and operational success (Table 3).

6.2 AI-Driven Autonomous and Self-Healing Aerospace Systems

The advent of AI into aerospace systems has revolutionized the autonomous and self-healing features monitored by the sensors. These AI-enabled autonomous systems make real-time decisions based on sensor data and predefined objectives, reducing human input and enhancing operation efficiency. Such systems can navigate complicated environments, optimize flight paths, and react to emergencies more promptly than human operators. Besides, AI can self-heal aerospace materials by allowing them to detect and repair harm independently. This enhances the good work life of components and helps in minimizing maintenance costs and downtime. Predictive maintenance through AI algorithms can detect failures before they occur in real time to ensure the safety or efficiency of the aerospace systems while at the same time driving their dependability with a proper level of safety. Leveraging AI thus allows the aerospace industry to move up on innovation and resilience parameters improving overall performance with higher levels of automation.

6.3 Cross-Disciplinary Collaboration for Accelerated Innovation

Cross-disciplinary collaboration is thereby urged to accomplish the acceleration of innovation in the aerospace industry, especially in the areas of OMM and AI. Setting up partnerships between material scientists, AI researchers, aerospace engineers, and other disciplines allows completely new ideas and technologies to emerge more rapidly. This allows the sharing of knowledge, resources, and expertise, leading to breakthroughs that couldn't be made within one field. No doubt industry-academia partnership plays a critical role here in pushing the latest research into commercial application. Such industrial-academic partnerships can help bridge the gap between theoretical advances and real-world implementation, thus ensuring that the innovative solutions one comes up with are efficiently deployed. Moreover, open innovation platforms are encouraged within it, enabling raw ideation and development of resources amongst the collaboration in lieu of progress, addressing aerospace technology-related challenges.

6.4 Sustainable AI and Metamaterial Solutions for a Greener Aerospace Future

The aerospace sector is now aiming for sustainability, and AI incorporated with metamaterials could find a foothold for a greener future. AI can optimize the design and operation of aerospace systems to reduce energy consumption and emissions, thus assisting in achieving a sustainable aerospace sector. By analyzing related data, algorisms can suggest a number of possibilities, besides producing further opportunities to reduce cost, improve fuel saving, or impact carbon emissions. Metamaterials, via their exclusive properties, can be designed lightweight and strong, so enabling fuel efficiency through a weight reduction of aerospace structures and design also involving sustainability features in an even more accentuated sense. Besides, the chances of developing recyclable and environmental-friendly ones will offset the pollution burden from aerospace manufacturing to disposal. The aspects of recyclable the AI processes could optimize those processes wherein other materials would be fungible, thus ensuring an efficient and sustainable platform for material use. Sustainable solutions allow aerospace to support the environmental cause even whilst innovating and at the same time growing.

7 Conclusion

The detailed exploration done in this research has unlocked the possibility of synergizing artificial intelligence (AI) with OMM in the unmanned aviation field. In fact, this synergistic field provides a great deal of developments and opportunities for multifarious changes in the aerospace industry.

This study reveals, in a most scientific manner, how AI-based smart metamaterials will improve the ability and function of aerospace structures. This class of material reacts to environmental changes, imparting a multifarious durability and versatility to ensure enhanced performance across a range of operating conditions. During this process, the sensors and smart metamaterials enable monitoring and diagnosis in real time, providing a proactive maintenance strategy for great safety and operational efficiency.

In addition, this study has endorsed the AI machine system with autonomous self-healing capabilities in aerospace technology. This integrated self-healing AI system will be able to perform complicated decision-making for routing, evasive landings, etc. with remarkable speed and accuracy. In itself, an AI-integrated material would be capable of self-healing, thus prolonging the work life of aerospace structure parts, resulting in a lowering of maintenance costs and minimizing the downtime required for such operations while ensuring reliability on critical systems.

Cross-disciplinary collaboration has been underscored throughout the research. By encouraging partnerships between the materials scientists, AI researchers, and aerospace engineers, it has also shown how faster solutions arise. The collaboration, therefore, utilizes the collective know-how of various disciplines to create breakthroughs for more complex problems in aerospace technologies.

Moreover, the aerospace industry is stressing its commitment to sustainability. There are prospects for green solutions that can enable energy- and emission-reduction in future geometries through the integration of AI with metamaterials. Lightweight, recyclable metamaterials are also capable of goodness-thus- fuel efficiency and low environmental impact-in tandem with the industry goals for sustainable development.

References

1. L. Si, R. Niu, C. Dang, X. Bao, Y. Zhuang, W. Zhu, Advances in artificial intelligence for artificial metamaterials. APL Mater. **12** (2024). https://doi.org/10.1063/5.0247369
2. Y. Fu, X. Zhou, Y. Yu, J. Chen, S. Wang, S. Zhu, Z. Wang, Unleashing the potential: AI empowered advanced metasurface research. Nanophotonics (2024). https://doi.org/10.1515/nanoph-2023-0759
3. M.K. Chen, X. Liu, Y. Sun, D.P. Tsai, Artificial intelligence in meta-optics. Chem. Rev. (2022). https://doi.org/10.1021/acs.chemrev.2c00012
4. N.L. Kazanskiy, S.N. Khonina, I.V. Oseledets, A.V. Nikonorov, M.A. Butt, Revolutionary integration of artificial intelligence with meta-optics-focus on metalenses for imaging. Technologies **12**, 143–143 (2024). https://doi.org/10.3390/technologies12090143
5. Y. Liu, X. Zhang, Metamaterials: a new frontier of science and technology. Chem. Soc. Rev. **40**, 2494 (2011). https://doi.org/10.1039/c0cs00184h
6. P. Shekhar, J. Atkinson, Z. Jacob, Hyperbolic metamaterials: fundamentals and applications. Nano Convergence **1**(2014). https://doi.org/10.1186/s40580-014-0014-6
7. T. Paul, C. Menzel, C. Rockstuhl, F. Lederer, Advanced optical metamaterials. Adv. Mater. **22**, 2354–2357 (2010). https://doi.org/10.1002/adma.200903865
8. Y. Chen, B. Ai, Z.J. Wong, Soft optical metamaterials. Nano Convergence **7**(2020). https://doi.org/10.1186/s40580-020-00226-7
9. D. Izzo, M. Märtens, B. Pan, A survey on artificial intelligence trends in spacecraft guidance dynamics and control. Astrodynamics **3**, 287–299 (2019). https://doi.org/10.1007/s42064-018-0053-6
10. N.C. Vincent, R.R. Bhakar, S.R. Nadarajan, A. Syamala, J. Varghese, *Impact of Artificial Intelligence in the Aviation and Space Sector* (CRC Press eBooks, 2021), pp. 209–229. https://doi.org/10.1201/9781003095910-15
11. K. Thangavel, R. Sabatini, A. Gardi, K. Ranasinghe, S. Hilton, P. Servidia, D. Spiller, Artificial intelligence for trusted autonomous satellite operations. Prog. Aerosp. Sci. **144**, 100960 (2024). https://doi.org/10.1016/j.paerosci.2023.100960
12. P.A. Oche, G.A. Ewa, N. Ibekwe, Applications and challenges of artificial intelligence in space missions. IEEE Access **1–1** (2021). https://doi.org/10.1109/access.2021.3132500
13. Y. Abdulrahman, E. Arnautović, V. Parezanović, D. Svetinovic, AI and blockchain synergy in aerospace engineering: an impact survey on operational efficiency and technological challenges. IEEE Access **11**, 87790–87804 (2023). https://doi.org/10.1109/ACCESS.2023.3305325
14. D. Piccinotti, K.F. MacDonald, S. Gregory, I. Youngs, N.I. Zheludev, Artificial intelligence for photonics and photonic materials. Rep. Prog. Phys. **84**, 012401–012401 (2020). https://doi.org/10.1088/1361-6633/abb4c7
15. D. Ezzat, A.E. Hassanien, A. Darwish, M. Yahia, A. Ahmed, S. Abdelghafar, Multi-objective hybrid artificial intelligence approach for fault diagnosis of aerospace systems. IEEE Access **9**, 41717–41730 (2021). https://doi.org/10.1109/ACCESS.2021.3064976
16. K. Ranasinghe, R. Sabatini, A. Gardi, S. Bijjahalli, R. Kapoor, T. Fahey, K. Thangavel, Advances in Integrated System Health Management for mission-essential and safety-critical aerospace applications. Prog. Aerosp. Sci. **128**, 100758 (2022). https://doi.org/10.1016/j.paerosci.2021.100758
17. Y. Weng, Z. Pan, Orbital angular momentum based sensing and their applications: a review. J. Lightwave Technol. **41**, 2007–2016 (2022). https://doi.org/10.1109/jlt.2022.3202184
18. Y. Wang, W. Sha, M. Xiao, C. Qiu, L. Gao, Deep-learning-enabled intelligent design of thermal metamaterials. Adv. Mater. **35** (2023). https://doi.org/10.1002/adma.202302387
19. Y. Jian, G. Hu, L. Tang, Y. Shen, Y. Zhan, K. Aw, Adaptive piezoelectric metamaterial beam: autonomous attenuation zone adjustment in complex vibration environments. Smart Mater. Struct. **32**, 105023–105023 (2023). https://doi.org/10.1088/1361-665x/acf62f

20. Y. Ha, Y. Luo, M. Pu, F. Zhang, Q. He, J. Jin, M. Xu, Y. Guo, X. Li, X. Li, X. Ma, X. Luo, Physics-data-driven intelligent optimization for large-aperture metalenses. Opto-Electron. Adv. **6**, 230133–230133 (2023). https://doi.org/10.29026/oea.2023.230133
21. Y. She, J. Chen, C. Huang, Z. Zhang, J. Liao, J. Wang, X. Luo, Intelligent reconfigurable metasurface for self-adaptively electromagnetic functionality switching **10**, 769–769 (2022). https://doi.org/10.1364/prj.450297
22. S. Lee, C. Park, J. Rho, Mapping information and light: trends of AI-enabled metaphotonics. Curr. Opin. Solid State Mater. Sci. **29**, 101144–101144 (2024). https://doi.org/10.1016/j.cossms.2024.101144
23. C. Zhu, E.A. Bamidele, X. Shen, G. Zhu, B. Li, Machine learning aided design and optimization of thermal metamaterials. Chem. Rev. **124**, 4258–4331 (2024). https://doi.org/10.1021/acs.chemrev.3c00708
24. Z. Liu, D. Zhu, K.S. Lee, A.E. Kim, L. Raju, W. Cai, Compounding meta-atoms into metamolecules with hybrid artificial intelligence techniques **32**, 1904790–1904790 (2020). https://doi.org/10.1002/adma.201904790
25. D. Lee, S. So, G. Hu, M. Kim, T. Badloe, H. Cho, J.-K. Kim, H.-Y. Kim, C.-W. Qiu, B. Lee, Hyperbolic metamaterials: fusing artificial structures to natural 2D materials. **2** (2022). https://doi.org/10.1186/s43593-021-00008-6
26. S. Krasikov, A. Tranter, A. Bogdanov, Y. Kivshar, Intelligent metaphotonics empowered by machine learning. Opto-Electron. Adv. **5**, 210147–210147 (2022). https://doi.org/10.29026/oea.2022.210147
27. O. Khatib, S. Ren, J. Malof, W.J. Padilla, Deep learning the electromagnetic properties of metamaterials a comprehensive review. Adv. Func. Mater. **31**, 2101748 (2021). https://doi.org/10.1002/adfm.202101748
28. R. Zecca, D.L. Marks, D.R. Smith, Symphotic design of an edge detector for autonomous navigation. IEEE Access **7**, 144836–144844 (2019). https://doi.org/10.1109/access.2019.2945245
29. S. Bijjahalli, R. Sabatini, A. Gardi, Advances in intelligent and autonomous navigation systems for small UAS. Prog. Aerosp. Sci. **115**, 100617 (2020). https://doi.org/10.1016/j.paerosci.2020.100617
30. P. Jiao, A.H. Alavi, Artificial intelligence-enabled smart mechanical metamaterials: advent and future trends. Int. Mater. Rev. **66**, 365–393 (2020). https://doi.org/10.1080/09506608.2020.1815394
31. L.V. Molina, R.C. Morales, J. Zhang, R. Schiek, I. Staude, A.A. Sukhorukov, D.N. Neshev, Enhanced infrared vision by nonlinear up-conversion in nonlocal metasurfaces. Adv. Mater. (2024). https://doi.org/10.1002/adma.202402777
32. A. Sabato, S. Dabetwar, N.N. Kulkarni, G. Fortino, Noncontact sensing techniques for AI-aided structural health monitoring: a systematic review. IEEE Sens. J. **23**, 4672–4684 (2023). https://doi.org/10.1109/JSEN.2023.3240092
33. V.B. Sharma, S. Tewari, S. Biswas, B. Lohani, U.D. Dwivedi, D. Dwivedi, A. Sharma, J.P. Jung, Recent advancements in AI-enabled smart electronics packaging for structural health monitoring. Metals **11**, 1537 (2021). https://doi.org/10.3390/met11101537
34. R. Fang, A. Ghasemi, D. Zeze, M.K. Hedayati, Inverse design of lateral hybrid metasurfaces structural colour: an AI approach. RSC Adv. **14**, 25678–25684 (2024). https://doi.org/10.1039/d4ra04981k
35. Y. Jin, L. He, Z. Wen, B. Mortazavi, H. Guo, D. Torrent, B. Djafari-Rouhani, T. Rabczuk, X. Zhuang, Y. Li, Intelligent on-demand design of phononic metamaterials. Nanophotonics **11**, 439–460 (2022). https://doi.org/10.1515/nanoph-2021-0639
36. E. Crawley, Intelligent structures for aerospace—a technology overview and assessment. AIAA J. (2017). https://www.semanticscholar.org/paper/Intelligent-structures-for-aerospace-A-technology-Crawley/2235ab127bc317fd8d64d22771ffc75be44c8532. Accessed 17 Feb 2025
37. A.H. Bhat, Space missions out of this world with AI. Nat. Mach. Intell. **5**, 183–183 (2023). https://doi.org/10.1038/s42256-023-00643-3

38. G. Furano, G. Meoni, A. Dunne, D. Moloney, V. Ferlet-Cavrois, A. Tavoularis, J. Byrne, L. Buckley, M. Psarakis, K.-O. Voss, L. Fanucci, Towards the use of artificial intelligence on the edge in space systems: challenges and opportunities. IEEE Aerosp. Electron. Syst. Mag. **35**, 44–56 (2020). https://doi.org/10.1109/MAES.2020.3008468
39. A.M. Wijata, M.-F. Foulon, Y. Bobichon, R. Vitulli, M. Celesti, R. Camarero, G. Di Cosimo, F. Gascon, N. Longépé, J. Nieke, M. Gumiela, J. Nalepa, Taking artificial intelligence into space through objective selection of hyperspectral earth observation applications: to bring the "brain" close to the "eyes" of satellite missions. IEEE Geosci. Rem. Sens. Mag. **11**, 10–39 (2023). https://doi.org/10.1109/mgrs.2023.3269979
40. Y. Jiang, T.H. Tran, L. Williams, Machine learning and mixed reality for smart aviation: applications and challenges. J. Air Transp. Manag. **111**, 102437 (2023). https://doi.org/10.1016/j.jairtraman.2023.102437
41. T. Duenas, Computation and AI unlock detailed studies of materials. Aerosp. Am. (2023). https://aerospaceamerica.aiaa.org/year-in-review/computation-and-ai-unlock-detailed-studies-of-materials/. Accessed 4 Mar 2025
42. E. Tezsezen, D. Yigci, A. Ahmadpour, S. Tasoglu, AI-based metamaterial design. ACS Appl. Mater. Interfaces (2024). https://doi.org/10.1021/acsami.4c04486
43. W.B. Yousuf, T. Khan, T. Ali, Prognostic algorithms for flaw growth prediction in an aircraft wing. IEEE Trans. Reliab. **66**, 478–486 (2017). https://doi.org/10.1109/tr.2017.2676722
44. H. Li, W. Yao, Y. Dong, Q. Gao, Y. Deng, Deep kernel-based optimal control prediction in aerospace missions. IEEE Trans. Aerosp. Electron. Syst. **58**, 1621–1633 (2021). https://doi.org/10.1109/taes.2021.3133225
45. E.J. Knoblock, R.D. Apaza, M.R. Gasper, H. Li, R. Han, Z. Wang, N. Schimpf, N. Rose, N. Adams, Intelligent spectrum management for future aeronautical communications. IEEE Aerosp. Electron. Syst. Mag. **38**, 46–59 (2023). https://doi.org/10.1109/maes.2022.3233817
46. N. Sünderhauf, O. Brock, W. Scheirer, R. Hadsell, D. Fox, J. Leitner, B. Upcroft, P. Abbeel, W. Burgard, M. Milford, P. Corke, The limits and potentials of deep learning for robotics. Int. J. Robot. Res. **37**. https://doi.org/10.1177/0278364918770733

MIX
Papier aus verantwortungsvollen Quellen
Paper from responsible sources
FSC® C105338

If you have any concerns about our products,
you can contact us on
ProductSafety@springernature.com

In case Publisher is established outside the EU,
the EU authorized representative is:
Springer Nature Customer Service Center GmbH
Europaplatz 3, 69115 Heidelberg, Germany

Printed by Libri Plureos GmbH
in Hamburg, Germany